Tatsachen und Probleme der Grenzen in der Vegetation

Bericht
über das Internationale Symposion
der Internationalen
Vereinigung für Vegetationskunde
in Rinteln 8. - 11. April 1968

Herausgegeben von

Reinhold Tüxen

Redigiert von
W. H. Sommer und R. Tüxen

SPRINGER-SCIENCE+BUSINESS MEDIA, B.V.
1974

by Strauß & Cramer GmbH, 6901 Leutershausen

ISBN 978-94-011-7596-8   ISBN 978-94-011-7595-1 (eBook)
DOI 10.1007/978-94-011-7595-1

Additional material to this book can be downloaded from http://extras.springer.com.

WERNER LÜDI

dem unermüdlichen Geobotaniker

und Naturschützer

in dankbarem Gedenken

Die Teilnehmer am Internationalen Symposion
über Tatsachen und Probleme
der Grenzen in der Vegetation.

INHALT

VIII

Am Abend des 9.April 1968 wurde auf dem Empfang des Kreises
und der Stadt Rinteln und der Glashütte STOEVESANDT von Herrn
Dr. STURLA FRIDRIKSSON,Reykjavik, ein Farbfilm von der Entstehung der Vulkaninsel Surtsey gezeigt.
Am 10.April erläuterte Herr Dr.SCHWABE,Plön, einen Farbfilm aus
japanischen Nationalparken, den der Vorsitzende des Vereins Na-
turschutzpark, Herr Dr.h.c.A.TOEPFER, freundlicherweise zur Verfügung gestellt hatte.

X

TEILNEHMERVERZEICHNIS

Belgien

NOIRFALISE,Prof.Dr.A., Brüssel. 10,rue Maesschlaek

Chile

WEISSER-SIEVERS,P., Santiago de Chile, Univers.de Chile

CSSR

BALÁTOVÁ-TULAČKOVÁ,Frau Dr.Emilie, Brno 16, Stara 18

BLAŽKOVA,Frl.Dr.Denisa, Průhonice u Prahy, Bot.ûstav ČSAV

KÜHN,Dr.F.,Dozent, Brno, Katedra Bot.a Mikrob.Zemědělska 1

MORAVEC,Dr.J., Průhonice u Prahy, Bot.ústav ČSAV

Deutschland

ALLGAIER,H.,Dipl.hort., 7303 Neuhausen/F. Schloßstr.24

ALTEHAGE,C., Mittelschullehrer i.R.† 45 Osnabrück

BIERHALS,S.,cand.rer.hort., 3 Hannover, Schierker Weg 12

BOEDEKER,R.,Landsch.-Arch., 402 Neandertal bei Düsseldorf

BÖTTCHER,H., Dipl.-Gärtner, 3 Hannover,Institut f. Vegetationskunde

BÖTTCHER,Frau Ingeborg,Dipl.-Gärtner,Barrigsen Nr.14

BRACKER,Dr.H.,OLRat, 225 Husum, Lehranst.f.Grünlandwirtschaft

BURRICHTER,Dr.E., Oberkustos, 44 Münster/W., Bot.Inst.Schloßgarten 3

BRAUN,Dr.W.,Reg.Rat, 8 Karlsfeld b.München,Lessingstr.24

DAMMANN,Frau Dr.Hildegard, 314 Lüneburg, Wilschenbrucher Weg 87

DIERSCHKE,Dr.H., 34 Göttingen. Geobot.Institut,Untere Karspüle 2

DIERSSEN,K.,stud.rer.nat., 3252 Bad Münder, Wermuthstr.31

DUTHWEILER,Dr.H., 3 Hannover, Institut f.Landschaftspflege.
                                        Herrenhäuserstr.2

ERNST,Prof.Dr.W., Amsterdam-Buitenveldert,NL.,Biol.Labor.Vrije
                                Universiteit,De Boelelaan 1087

FAHRENHOLTZ,Frau Dr.Käte,Apothekerin, 4964 Kleinenbremen

FEISE,Dr.J.,OLRat i.R., 29 Oldenburg, Sodenstich 115

FEISE,Frau Irmgard,   29 Oldenburg, Sodenstich 115

FOERSTER,Dr.E., 4119 Kleve-Kellen,Forsch.Stelle f.Grünland u.
                                        Futterbau,Dammstr.15

FÖRSTER,M.,Forstmeister, 351 Hann.-Münden,Inst.f.Waldbau-Grund-
                                        lagen,Schloß

FÜLLEKRUG,E.,Studienrat, 3352 Bad Gandersheim,Dr.Leonardi-Weg 8

GÖRS,Frl.Dr.Sabine, 714 Ludwigsburg, Favoriteschloß

GRIES,Frl.Dr.Brunhild 44 Münster/W.,Landesmuseum f.Naturkunde,
Himmelreichallee 50

HABER,Prof.Dr.W.   805 Freising,Inst.f.Landschaftspflege

HACKER,Dr.E., 3 Hannover, Amt f.Bodenforschung,Sven Hedinstr.

HAEUPLER,H.,cand.rer.hort., 3201 Ochtersum, Wunramstr.5

HARMS,H.,Lehrer, 3261 Krankenhagen über Rinteln, Nr.243

HARTMANN,W.,Dipl.hort., 3 Hannover,Inst.f.Landschaftspflege,
Herrenhäuserstr.2

HERMS,R.Dipl.hort., 242 Plön, Eutinerstr.73

HÖLL,Dr.K.,Reg.Rat a.D., 325 Hameln, Kreuzstr.15

Hülbusch,K.-H.,Dipl.-Gärtner,28 Bremen, Bückeburgerstr.16

HÜLBUSCH,Frau Inge, Dipl.-Gärtner, 28 Bremen,Bückeburgerstr.16

JAHN,Frau Prof.Dr.Gisela, 34 Göttingen-Weende, Büsgenweg 1

JANIESCH,Dr.P., 44 Münster/W. Bot.Inst.,Hindenburgplatz 55

JENSEN,Prof.Dr.U., 5 Köln,Bot.Inst.Gyrhofstr.15

KAULE,G.,wiss.Assistent, 805 Freising, Inst.f.Landschaftspflege

KNYPHAUSEN,Fürst zu 2981 Lütetsburg b.Norden

KOHLER,Dr.A., 805 Freising, Institut f.Landschaftspflege

KULKE,W.,Rev.Förster, 3011 Benthe, Lakefeldstr.11

LANG,Dr.G.,Priv.Doz., 75 Karlsruhe,Landessamml.f.Naturkunde,
Erbprinzenstr.13

LÖTSCHERT,Prof.Dr.W., 6 Frankfurt/Main, Siesmayerstr.70

LUCHTERHAND,J.,Dipl.-Ing.,Bds.Bahn-Abt.Präs.i.R.,56 Wuppertal-
Elberfeld, Müllerstr.107

MÜLLER,G.,OSTRat, 326 Rinteln, Eichendorffstr.14

MÜLLER,Dr.Th., 714 Ludwigsburg, Favoriteschloß

MUHLE,Dr.H., Ottawa 2, Dept.of Biology,University, Canada

OBERDORFER,Prof.Dr.E., 78 Freiburg-St.Georgen,Brunnstubenstr.31

OBERDORFER,Frau Cläre, 78 Freiburg-St.Georgen,Brunnstubenstr.31

PFADENHAUER,Dr.J., 8 München,Waldbauinstitut,Schellingstr.12 II

PHILIPPI,Dr.G., 75 Karlsruhe,Landessamml.f.Naturkunde,Erbprin-
zenstr.13

PREISING,Prof.Dr.E.,Baudirektor, 3 Hannover,Nieders.Landesverw.
Amt,Leisewitzstr.2

REHAGEN,Dr.F., 415 Krefeld, Geol.Lds.Amt, Steinstr.203

RÖDEL,Dr.H., 75 Karlsruhe, Kolbergstr.11

RUNGE,Dr.F.,Kustos, 44 Münster/W., Vinzensweg 35

RUTHSATZ,Frl.Dr.Barbara, 34 Göttingen,Geobot.Inst.,Untere Karspüle 2

SCHREITLING,Dr.K.,Dozent, 3011 Gehrden, Langsederstr.11

SCHROEDER,Dr.F.-G., 34 Göttingen,Geobot.Inst.,Untere Karspüle 2

SCHULTZ,Dr.J., 8 München,Geogr.Inst.d.Universität,Luisenstr.37

SCHWABE,Dr.G.H., 232 Plön,Max Planck-Inst.,Postfach 165

SCHWERDTFEGER,Dr.G.,Ldw.Rat, 3113 Suderburg,Staatl.Ingenieur-schule

SEIBERT,Dr.P.,Prof.ORRat, 8 München,Höslstr.9

SINDERMANN,E.OSTRat, 3261 Krankenhagen über Rinteln Nr.197

SPEIDEL,Dr.B.,Prof., 643 Bad Hersfeld, Eichhof

STOEVESANDT,W.,Fabrikant, 326 Rinteln, Hafenstr.

SUKOPP,Dr.H., 1 Berlin, Rüdesheimer Pl.9

TRENTEPOHL,Dr.M.,Doz.O-Rat, 744 Nürtingen,Staatl.Ing.Schule für Landbau

TÜXEN,Dr.J., 3 Hannover,Amt f.Bodenforschung, Hedinstr.

TÜXEN,Prof.Dr.Drs.h.c.R., 3261 Todenmann über Rinteln

ULLRICH,Prof.Dr.Chr., 58 Hagen, Pädagogische Hochschule

VOLLRATH,Dr.H., 8051 Unterzolling, Palzingerstr.16

WILDENHAIN,Dr.J., 805 Freising, Institut f.Landschaftspflege

WILMANNS,Frau Prof.Dr.Otti, 78 Freiburg/Br.,Bot.Institut,Schänz-lestr.9-11

WINTERHOFF,Dr.W., 74 Tübingen,Im Rotbad 34

England

APINIS,Prof.Dr.A., BRU Sautaines,Groesfford,Bryncrug Nr.Towyn, Merioneth, Wales U.K.

BRIDGEWATER,P., Durham,Dept.of Bot.Science Labor.

PROCTOR,Dr.M., Exeter,Hatherly Labor,Dept.of Botany Prince of Wales University

SHIMWELL,Dr.D., Geogr.Deptm.University Manchester M13 9P2.

Frankreich

CARBIENER,Prof.Dr.R., Strasbourg,Faculté de Pharmacie,2,rue St.Georges

GUILLERM,J.L., Montpellier,Institut Botanique,5,rue Auguste Broussonet

LINDER,Prof.Dr.R., Lille, Fac.des Sciences de l'Université Boîte de postale 36

METTAUER,Dr.H., Colmar, Station d'Agronomique, 28 rue de Herrilsheim

ROMANE,Dr.F., Montpellier, Institut Botanique, 5 rue Auguste Broussonet

WURCH,Dr.med., Colmar

WURCH,Mme Dr.med., Colmar

Griechenland

LAWRENTIADES,Prof.Dr.G.J., Thessaloniki,Bot.Inst.der Universi-tät

Irland

MOORE,Dr.J.J.,Pater, Dublin 4,Univers.College,Dept.of Botany
O'SULLIVAN,Dr.A., Wexford, Johnstown Castle, Agric.College

Island

FRIDRIKSSON,Dr.S. Reykjavik, Agric.Research Institute

Italien

CRISTOFOLINI,Dr.G., 34 Trieste, Via Cumano 2
HOFMANN,Prof.Dr.A., Torino, Via Duchesssa Jolanda 17
LAUSI,Dr.D., 34 Trieste, Via Cumano 2
+MARCHESE-POLI,Frau Prof.Dr.Emilia, Catania,Bot.Institut, Via a Longo 19
PIGNATTI,Prof.Dr.S., 34 Trieste,Via Cumano 2

Japan

SASAKI,Dr.Y.†, Hiroshima,Bot.Inst.d.Univers.Higashi-senda
+SUZUKI-TOKIO,Prof.Dr., Ooita,University,Dept.of Botany,Danoharu
YOSHINO,Prof.Dr., Tokyo,Dept.of Geography,Tokyo Education University, Otsuka, Bunkyo-Ku.

Jugoslavien

WRABER,Prof.Dr.M.†, Ljubljana, Slovenska Akademia, Novi trg 3

Niederlande

BARKMAN,Prof.Dr.J.J.,Dozent, Wijster (Dr.), Kampsweg 29
BEEFTINK,Dr.W.G., Yerseke, Vierstraat 28
BLIJSWIJK,M.V.van, Den Haag, Verlag Dr.Junk N.V.13,van Stolkweg
BOS,E.Drs., Sleewyk,N.Br., Vlietstraat 7
DIEMONT,Dr.H.,Landforstmeister†, Maastricht, Waldeckstr.41
DOING,Dr.H., Wageningen,Labor.v.Plantensystematiek,Gen.Foulkesweg 37
FRESCO,W.F.M.,Wiss.Ass., Groningen,Wijsterbeslaan 6
GLERUM,B.B.,Forstmeister, Maastricht, Veldspaatstraat 32
KOP,L.G.,Ir., Wageningen, J.P.Thijsselaan 63
LAAN,D.van der, Oostvoorne, Biol.Stat."Weevers Duin"
LEEUWEN,Chr.van, Zeist,"RIVON", Laan van Beek en Royen 41
LONDO,G.Drs.,   Zeist,"RIVON", Laan van Beek en Royen 41
MAAREL,Dr.E.van der, Nijmegen, Driehuizer Weg 200
PLASTERK,K.,Direktor, Den Haag, Uitgeeststraat 32
SCHOOF-van PELT,Mevrouw Dr.Margriet, Nijmegen,Bot.Lab. Driehuizer Weg 200
SEGAL,Dr.S., Wageningen,Agricult.University,Transitorium, De Dreijen 11

XIV

SISSING,Dr.G.,Oberforstmeister, Schaarsbergen,Kempersberger Weg
SLOET,Mejuffrouw Drs.Clara, Wageningen,Landbouwhogeschool,
                                        Gen.Foulkesweg 37

STOFFERS,Prof.Dr.A.H., Wageningen, Bot,Institut, Arboretumlaan 4
WESTHOFF,Prof.Dr.V., Nijmegen, Driehuizerweg 200, Bot.Institut
                                        der Kath.Universiteit

WOLTERSON,K.,Direktor, Bennekom, Alexanderweg 34
ZONNEVELD,I., Sleewyk N.Br., Vlietstr.7

Oesterreich

HOLZNER,Dr.W., Wien, Bot.Inst.d.Hochschule für Bodenkultur,
                                        Gregor Mendelstr.33

Schweiz

OCHSNER,Dr.F., Winterthur, Seuzacherstr.28

Spanien

RIVAS MARTINEZ,Prof.Dr.S., Madrid 3, Fac.de Ciencias, Catedra
                                de Botanica y Ecologia vegetal

Türkei

GENÇKAN,Dr.S.,Dozent, Borneval-Izmir, Ziraat Fakültesi Agronomi
                                                Kürsüsü

KÖSEOGLU,M.,Dipl.agr.ing., z.Zt.Hannover, Institut für Land-
                        schaftspflege, Herrenhäuserstr.2

Ungarn

JAKUCS,Dr.P., Budapest, Geograph.Forschungs-Institut,
                                Népköztársaság 62

USA

+KÜCHLER,Prof.Dr.A.W., Lawrence,Kansas, Dept.of Geography Univer-
                                                sity

+ Teilnehmer sendet Manuskript, kann aber nicht persönlich
  anwesend sein.

ERÖFFNUNG DES SYMPOSION

R e i n h o l d   T ü x e n

Meine sehr verehrten Damen und Herren!

Im Namen unseres verehrten Präsidenten, Herrn Professor LEBRUN,
darf ich Sie alle hier herzlich wilkommen heißen, die alten
Freunde und unsere jungen, die es werden. Aus 19 Ländern sind
wir zusammengekommen, in einer Rekordzahl von bedeutenden älteren und jüngeren Forschern aus der ganzen Welt, kann ich sagen;
von Chile bis Japan sind wir hier vereinigt in dieser Aula, und
wir werden 35 Vorträge hören, von Spitzbergen und Island bis
nach Australien und von Japan bis nach Südamerika. Wir werden
also einen wahrhaft umfassenden Überblick bekommen über pflanzensoziologische Probleme der verschiedensten Arten auf unserem
Erdball, von den Bakteriengesellschaften und ihren Gesetzen bis
zu tropischen Waldgesellschaften.
Ich kann Sie nicht im einzelnen begrüßen und bitte um Ihre Nach-
sicht, wenn ich das nicht tue, aber ich will Ihnen danken, meine Damen und Herren, daß Sie von so weit hergekommen sind, daß
Sie nicht Kosten, Kräfte und Zeit gescheut haben, sich hier,wie
so oft schon, mit einander und mit uns zu versammeln. Ich möchte Ihnen allen dafür danken. Ich muß ganz besonders meinen Dank
Ihnen, Herr Oberstudiendirektor Dr. ROTH sagen, daß Sie wieder
in so freundschaftlicher Weise der Retter in der Not gewesen
sind; denn wenn Sie uns nicht diesen Raum zur Verfügung gestellt
hätten, dann hätten wir irgendwo auf dem Marktplatz oder vielleicht unter den Linden auf dem Kirchplatz tagen müssen. Ein
wenig kühl! Und so sind wir Ihre Gäste, und sind Ihnen sehr
dankbar dafür. Ebenso den Herren Oberstudienräten Dr. MÜLLER
und Dr. HENSEL, die mit ihren Helfern dies alles vorbereitet
haben.
Viele unserer Freunde fehlen. Sei es, daß sie aus dienstlichen
oder Arbeitsgründen verhindert sind, sei es auch, daß sie nicht
kommen konnten, wie unsere Freunde aus der DDR, aus Polen, und
auch die Russen, die uns im letzten Jahr durch 4 Gäste dieses
weiten Landes so viel Freude gemacht haben, sind diesmal nicht

erschienen. Auch unser verehrter Herr Präsident kann nicht kommen, er hat mir am 4.April einen Brief geschrieben, den ich Ihnen vorlesen darf.

Cher Collègue,Cher Secrétaire Général,Cher Ami,
Je ne pourrai pas, cette année encore, participer au prochain Colloque de l'Association ni aux réunions qui l'accompagnent.

Tous me pardonneront,à commencer par vous-même, quand ils sauront que c'est essentiellement pour des raisons de travail sur le terrain que je ne puis m'absenter à cette époque. Mon Laboratoire, en effet, poursuit des recherches écologiques qui comportent régulièrement des périodes d'observation de 48 heures et l'une de ces périodes tombe précisément dans cette première semaine des vacences de Pâques.
Sachez donc que, pendant vos Débats, je fais du travail écologique dans la nature....
Mais si je ne puis être présent, mon coer sera avec vous! Voulez-vous le dire à tous nos Amis et leur faire part des voeux que je forme à leur intention personelle comme pour le succès de leurs travaux.
Je souhaite que le Colloque connaise la réussite habituelle, et sachant, mon Cher Ami, que la direction en est entre vos mains, je suis convaincu de l'entière réalisation de ce voeu.

Veuillez croire, Cher Collège, Cher Secrétaire Général, à mes sentiments très sympathiques
Signé: J.LEBRUN

Wir werden Herrn Präsidenten LEBRUN wie vielen anderen einen Grußbrief senden.
Ich muß aber noch einen Freund erwähnen, der nicht mehr unter uns weilt. Das ist WERNER LÜDI. Er starb am 1.März, also vor gerade vier Wochen, nach langer, schwerer Krankheit, obwohl er hoffte und noch Anfang des Winters schrieb, er würde heute hier unter uns sein. Wir haben, die Älteren unter uns, einen Lehrer verloren; alle von uns haben einen Freund verloren, einen großen Organisator - ich darf nur daran erinnern, daß er alle IPE's in den letzten Jahren organisiert hat - , einen treuen Verwalter großer wissenschaftlicher Schätze in dem von ihm so lange geleiteten Geobotanischen Institut Rübel. Er wird uns unvergessen und immer ein Vorbild wissenschaftlicher und menschlicher Treue bleiben. Wir setzen sein Bild an den Anfang dieses Berich-

tes.

Aber, meine Damen und Herren, das Leben geht weiter, und die
Jugend dominiert hier in unserem Kreise, und das ist uns eine
große Freude. Ich sehe - und bitte die anderen Länder um Nach-
sicht, wenn ich nur eins erwähne - zu meiner besonderen Freude
eine größere Zahl von jungen Vertretern der Britischen Inseln,
sowohl von Irland als auch von England. Es gab Länder, wo große alte Männer so starken Einfluß ausübten, daß neue Ideen und
neue Richtungen der Wissenschaft keinen Eingang finden konnten.
Es gibt noch solche Länder. Aber wir sehen, daß unsere Gedanken,
die wir aufgenommen und gelernt haben von JOSIAS BRAUN-BLANQUET,
weitergetragen worden sind und weitergetragen werden bis in die
entlegenen Gebiete, und wir wissen, daß es uns gelungen ist,
und weiter gelingen wird, eine Vereinheitlichung und eine Harmo-
nisierung unserer Arbeitsweise, unserer Begriffe und Methoden
auch dort zu erreichen, wo man lange gezögert hat, sie anzuneh-
men, wo man Schwierigkeiten hatte, sie zu verstehen. Darum bin
ich glücklich, daß gerade die jungen Engländer hier sind.
Unser Freundeskreis hat sich stetig vergrößert, er hat sich auch
immer von neuem verjüngt, er ist enger zusammengeschmolzen, und
umfaßt jetzt mehrere, 3-4 wissenschaftliche Generationen schon.
Wenn ich 40 Jahre zurückblicke, so sehe ich im Geiste vor mir
eine Schar begeisterter junger Männer; heute sind das Professo-
ren, Direktoren, Landforstmeister und andere hohe Persönlich-
keiten der Wissenschaft, diese Leute, die früher die gleichen
Jünglinge waren wie die, die heute neu zu uns gestoßen sind.

Wir haben auch, vor einigen Jahren, eine Krise unserer Wissenschaft erlebt. Man hat dieses Wort gebraucht, die 'Krise der
Charakterartenlehre'; man hat auch von einer Inflation gesprochen. Aber wir haben diese Gefahren erkannt und dadurch, daß
wir sie sahen, waren sie zum größten Teil schon überwunden. Und
die letzten beiden Tage intensiver Arbeit, über die Herr Professor OBERDORFER berichten wird, haben gezeigt, daß wirklich kaum
noch Reste einer solchen Krise, einer solchen Zersplitterungs-
und Inflationsgefahr übriggeblieben sind. Wir sind am Vorabend
einer neuen und ganz starken Synthese. Wir sind, und das ist
meine ganz große Genugtuung und Freude, alle jung genug, uns
nicht den modernen und fruchtbaren Entwicklungs-Aussichten unserer Wissenschaft, wie den mathematischen und technischen Möglichkeiten zu verschließen, sondern sie zu assimilieren und

ihre Vorteile auszunützen, die sie uns bieten. Wir sind dabei
ganz und gar treu geblieben den alten Begriffen, den alten Methoden, wie wir sie von unserem, uns allen gemeinsamen Lehrer
und Freund JOSIAS BRAUN-BLANQUET gelernt haben. Und so werden
wir unseren Weg weitergehen, so lange es uns vergönnt ist, als
ein enger internationaler Freundeskreis - wir alle kennen ihn
aus so vielen Ländern - in das wissenschaftliche Neuland. Wir
wissen, daß die Pflanzendecke, der Gegenstand unserer Bemühun-
gen, die Grundlage, der Ausgangspunkt allen weiteren Lebens auf
Erden ist. Und in diesem Sinne, meine Damen und Herren, darf
ich das 12. internationale Symposion eröffnen, nochmals Ihnen
allen danken, Ihnen vier Tage intensiver Arbeit,großer Freuden
und leuchtender Erkenntnisse wünschen.
Et maintenant je vous pris, Monsieur NOIRFALISE, de prendre la
parole comme président pendant cet avant-midi.

A.NOIRFALISE:
Mesdames, Messieurs,
Avant d'aborder l'ordre du jour de cette réunion, et avant d'é-
couter les premières communications, je voudrais vous informer,
qu'une réunion préparatoire a été tenue les deux jours précé-
dents par un certain nombre de membres de l'Association Internationale de Phytosociologie. Il a été discuté les problèmes
de classification, il a aussi été pris quelques résolutions.
J'ai prié le Dr. OBERDORFER de bien vouloir vous donner lecture
des principales conclusions, avant d'aborder l'ordre du jour de
la réunion.

E.OBERDORFER:
Ergebnisse des vorbereitenden synsystematischen Kolloquium, die
wir gestern in kleinem Kreise zusammengestellt haben:
1. Die Teilnehmer gaben für ihre Länder und benachbarten Gebiete einen ausführlichen Überblick über den Stand der pflanzensoziologischen Grundlagen für die Erarbeitung von Prodromi der
Pflanzengesellschaften. Die Ergebnisse wurden von PIGNATTI in
einer Skizze dargestellt. Dabei stellte sich heraus, daß die
Arbeiten in verschiedenen Ländern, z.B. Spanien, dem ostalpin-
dinarischen Raum, Deutschland, CSSR, Polen, bereits weit vor-
geschritten sind.
2. Die Teilnehmer sind der Überzeugung, daß die Zeit für die

Erarbeitung eines Prodromus der Pflanzengesellschaften im westlichen und zentralen Raum des europäischen Kontinents reif sei.
Mit eingeschlossen werden sollen geeignete Angaben aus angrenzenden Gebieten.

Von dieser Arbeit erwarten wir eine zunehmende Vereinheitlichung der pflanzensoziologischen Begriffe und Methoden und der
Nomenklatur für diesen Raum, so wie eine genauere Kenntnis der
Areale der pflanzensoziologischen Einheiten und ihrer Variabi-
lität innerhalb dieses Raumes.

3. Die Bewältigung der für den Einzelnen unübersehbaren Fülle
von Aufnahmen und Tabellen bedarf der Benutzung moderner Doku-
mentations- und Verarbeitungstechniken. Die Redaktion der verschiedenen Einheiten soll durch intensive Zusammenarbeit ihrer
besten Kenner erfolgen. Die an verschiedenen Orten in Europa bestehenden Karteien sollen für dieses Unternehmen zur Verfügung
gestellt werden. Sie umfassen z.Z. weit über 100 000 einzelne
Literaturnachweise der verschiedenen Gesellschaften.

4. Die Vorschläge von PIGNATTI, CRISTOFOLINI und LAUSI über
Verwendungsmöglichkeiten einer Komputer-Anlage für die pflanzensoziologische Dokumentation wurden eingehend diskutiert. Die
geäußerten Bedenken gegen die Kosten, die Schwierigkeiten der
Kodifizierung, und den Zeitaufwand und die Zeitdauer werden
durch die mit Sicherheit zu erwartenden Vorteile bei weitem
überwogen. Die bisher an verschiedenen Orten vorliegenden Erfahrungen lassen erwarten, daß auch die zukünftige pflanzensoziologische Forschung durch ihre Verwendung reiche, neue, bisher
unzugängliche Erkenntnisse gewinnen wird.

5. Die Grundlagen für den Prodromus bilden die Aufnahmen, die
grundlegende Einheit bildet die Assoziation. Die Redaktion des
Prodromus setzt eine einheitliche Auffassung der systematischen
Einheiten voraus, soll jedoch kritisch durchgeführt werden. Alternative Auffassungen sollen gebührend erwähnt werden.

6. Zur Nomenklatur machte Dr.MORAVEC Vorschläge,die diskutiert
wurden. Sie behandelten u.a. den starting point, die Priorität
als Leitprinzip, Typifizierung von Vegetationseinheiten, Diagno-
se der Einheiten. Alle Beteiligten waren darüber einig, daß Nomenklaturregeln für die Ausarbeitung notwendig sind, und schließen sich im wesentlichen den Vorschlägen von MORAVEC an.Es wird
gebeten,diese Vorschläge eingehend zu prüfen und dem Autor kritische Bemerkungen zuzuleiten.

7. Auf der Grundlage dieser Pläne wurde beschlossen, bis zum
nächsten Symposion den Versuch durchzuführen, eine oder mehrere
genügend umfangreiche Vegetationstypen, wie etwa A l n e t e a
g l u t i n o s a e, Q u e r c e t e a r o b o r i - p e t r a -
e a e, K o e l e r i o n a l b e s c e n t i s oder A r r -
h e n a t h e r e t a l i a, probeweise für den zukünftigen Prodromus zu bearbeiten. Die Bearbeitung soll die Dokumentation
für die Aufnahmen, die Bibliographie, die Durchführbarkeit der
Nomenklatur-Regeln, sowie die Verwendungsmöglichkeiten des Kom-
puters erproben. Die Durchführung des Unternehmens erfordert
angemessene finanzielle Unterstützung.
8. Auf Vorschlag von Herrn Professor PREISING wurde beschlossen,
daß die Internationale Vereinigung für Vegetationskunde eine
Kommission für die Erhaltung schutzwürdiger Vegetation in Europa ernennt, deren Vorsitz Professor NOIRFALISE führt. Sitz der
Geschäftsstelle ist die Landesstelle für Naturschutz und Land-
schaftspflege in Hannover. Die Aufgabe der Kommission ist:
Eine Aufstellung einer Liste schutzwürdiger Vegetation für den
Europarat, und, in akuten Notfällen durch Eingabe an die zustän-
dige Landesregierung international begründete Schutzanträge zu
stellen.

ORDINATION,CLASSIFICATION AND
VEGETATIONAL BOUNDARIES

M.C.F. P r o c t o r

In recent years there has been a considerable controversy over
the relative merits of classification and ordination in the study of vegetation.Classification implies the division of the individuals or units being classified into categories or classes,
whose members show a greater or less degree of similarity a-
mongst themselves, and differ from members of other categories.
Ordination (whose development is associated particularly with
the name of J.T.CURTIS and his school in Wisconsin) treats the
vegetation being studied as a continuum, and seeks to arrange
the stands so that their spatial relation to one or more axes
will display as much information as possible about their relationships. This controversy has become associated with, and to
a large extent heir to, an older controversy - that between those who (like CLEMENTS) held a pseudo-organismal concept of the
plant community, and those who (like GLEASON) held that a plant
community is a fortuitous aggregation of plants separately dependent on the factors of the environment. The belief that a
choice between classification and ordination must be determined
by the inherent nature of vegetation is implicit in a great deal
of recent writing on ordination and classification, and is put
forward explicitly, for instance by WHITTAKER (1967) and YARRA-
TON (1967). In this paper I shall argue that beliefs about the
nature of vegetation are irrelevant to the point at issue; that
the two approaches are complementary and appropriate to different purposes, rather than mutually antagonistic alternatives.
A similar point is made in a slightly different context by WEBB
(1954,p.362).

However, first let us examine the nature of our material. As
GOODALL(1963) and LAMBERT & DALE(1964) have pointed out,we are
dealing with a three-fold system,of sites,species and environmental factors.In some circumstances it maybe appropriate to consider as the basic vegetation units sites, in their topographical context, bearing vegetation of paticular composition. This
is the viewpoint appropriate to the mapping of specific small
areas of vegetation, or if the principal interest is centred on

vegetational variation in relation to topography and other major environmental trends within the landscape.(cf.WHITTAKER's
"gradient analysis"; see GREIG-SMITH 1964, WHITTAKER 1967). Al-
ternatively, the vegetation can be considered in the abstract;
the plant assemblages found in the sample sites may be considered in terms of their intrinsic resemblances and differences,
without reference to their spatial relationships on the ground.
This is the viewpoint appropriate to the development of phytosociology as an abstract systematisation of vegetation of general
applicability. Rather than think in terms of site-types, it may
sometimes be appropriate (as GOODALL has suggested) to define
groups of species which have similar patterns of distribution
in relation to sites (or environmental factors); groups of species faithful to a particular community (or differential for a
particular boundary) will tend to be groupings of this kind.
LAMBERT & WILLIAMS (1962, and see LAMBERT & DALE 1964) have
suggested that "plant communities" as they are generally recognised are "plant-in-habitat" units defined in both directions.
I doubt that this is generally so (see IVIMEY-COOK & PROCTOR
1966), but often the precise concept that was in a particular
author's mind is not evident.
Vegetation and vegetation boundaries may thus be considered in
either concrete or abstract terms. In vegetation mapping we
shall wish to find boundaries which we can mark on our map[1]).
It maybe that within a definite area we shall find such boundaries; but (as GOODALL (1963) has pointed out) this is not to
say that discontinuities will necessarily exist within the vegetation considered as an abstract field of variation - and it
is discontinuities of that kind which are of interest in the development of an abstract system of vegetation.
To the best of my belief it has never been demonstrated that
vegetation is - objectively - made up of discrete vegetation
types. Sharp vegetation boundaries undoubtedly exist; they may
arise from the juxtaposition of gregarious dominant species of
different life-form, or from abrupt changes in environmental
factors. Fig.1 shows clearly a sharp vegetational boundary
associated with a spring-line. Vegetations boundaries might

[1]) Often,of course,we shall hope to be able to interpret these
in terms of abstract vegetation units.

conceivably also arise at the meeting points of different mutually adapted groups or "communities" of species - of different
"competitive structures". It would be reasonable to expect mutual adaptation between species regularly growing together in
the same environment, but (apart from obvious cases of depen-
dence) it is only recently that effects of the kind that might
be looked for in natural communities have been demonstrated in
simple experimental situations (DE WIT 1960, HARPER 1964,1967).
There seems to be little field evidence for boundaries of this
kind, and a good deal which suggests that in the presence of
continuous variation in environmental factors, continuous variation in vegetational composition is the rule; the exception is
at points of abrupt change in dominant species, especially if
this also involves a change in dominant life-form.

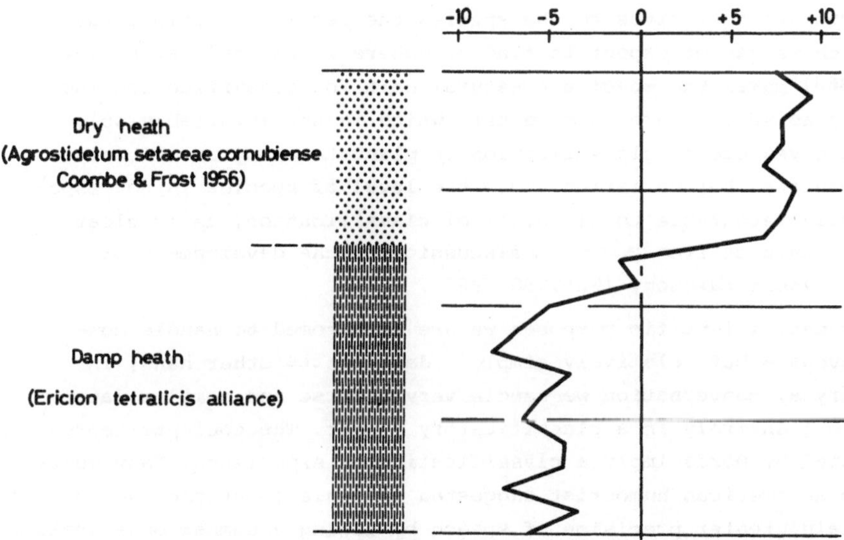

Fig.1. A well-defined vegetation boundary. The curve on the
right shows the scores of successive quadrats on the first component of a principal components analysis of data from 21 quadrat samples (each of 0.25 sq.m.) taken at approximately 3 m.
intervals along a transect on Aylesbeare Common,east Devon,
England.

More-or-less smooth transitions are common in vegetation. The
appearance of continuity is accentuated by avoiding major environmental bondaries, by the use of sampling-patterns unrelated
to minor environmental discontinuities and by taking large samp-

les; it may be still further accentuated by the form of the
subsequent treatment of the data.

If pressed, the most extreme advocate of classification would
admit the existence of "transitions" or "mixtures", and the kee-
nest exponent of ordination would admit to the existence of
boundaries across which vegetational change may be very rapid.
If the structure of vegetation is to be the criterion for choosing our technique of analysis we must agree with WEBB (1954)
that "the pattern of variation shown by the distribution of species among quadrats of the earth's surface chosen at random
hovers in a tantalising manner between the continuous and the
discontinuous."

Ecological data are often compared unfavourably with the data
of plant taxonomy as material for classification. The "genetic
structure that gives to the species the peculiar distinctness
which we cannot expect to find elsewhere in nature" (WEBB 1954,
p.368) makes the species a natural unit for classification (so
long as species are this nature, which is not invariably so),
and gives use of classification in plant taxonomy an initial
advantage. But variation above the level of species is not especially favourable to hierarchical classification, as is clear,
for instance from Walters's discussion of the development of
Angiosperm taxonomy (WALTERS 1961).

For many scientific purposes we are accustomed to handle continuous - but relatively simple - data.On the other hand, in
everyday conversation we handle very diverse and complex data
almost entirely in a classificatory manner. The concepts represented by words imply a classification of experience. Many years
ago an American humourist suggested that his countrymen would
attain greater precision of speech by adding a number on a scale
from 1 to 10 after every adjective, thus avoiding the imprecisi-
on of comparatives and superlatives. I think the fact that we
find this idea so whimsically ludicrous sufficiently makes my
point. It is worth noticing that in everyday speech we find no
difficulty in talking in classificatory terms about colour,
which can be completely and precisely specified by ordination
in relation to three axes, each showing perfectly continuous
variation. Nevertheless, for many scientific and industrial
purposes we choose to ordinate colour - to specify colour in

terms of parameters for hue, saturation and brightness. Under
whar circumstances do we prefer which approach?

Let us suppose we have a set of data of the form shown in Fig.2.
The quadrats can be considered as points in a hyperspace whose
axes represent the different species. The projection of quadrat
Qj on the axis for species S_i is the entry for species i in quadrat j in the data table. This geometrical model (see GOODALL
1963) is an ordination of the quadrats in as many dimensions as
there are species. It contains all the information in the data;
but it will almost certainly be incomprehensible.

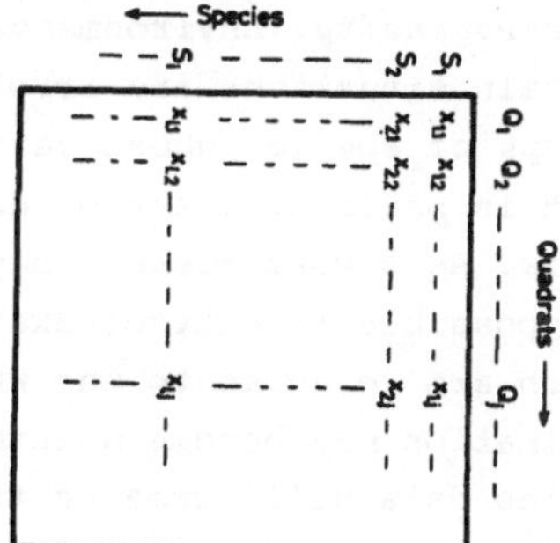

Fig.2 Scheme of a set of vegetational
data - in effect, a generalised form
of a phytosociological table.

Classification implies the simplification of this geometrical
model by dividing the hyperspace into cells. We can group the
ultimate cells into larger cells, to yield a hierarchical classification. This enables us to isolate from the whole complex of
variation groups of substantially similar stands,and to ignore
what is not relevant to us at the moment.

Ordination implies simplification of the model by projecting
the points representing the stands into a spce of fewer dimensions. This may of course be done by choosing dimensions on ex-
trinsic criteria (which is the approach of gradient analysis),
or by seeking within the data the directions of greatest variation in the vegetation - or, in other words, selecting axes
which retain the greatest proportion of the variance in the original data. For most ordinary purposes, an ordination must be
reduced (or at least reducible) to a number of dimensions which
can be readily visualised.

It is clear, then, that both classification and ordination lose

information from the original data. Classification loses information on relationship within and between groups. Ordination
loses the information that cannot be expressed in terms of a
few major directions of variation embracing the data as a whole.
Some loss of information must be accepted as the price of making
the data comprehensible. The question to be asked in a particular instance is "What information is expendable, and what should
be retained?" To this may be added a further question "What com-
promise should I choose between easy intelligibility and retention of maximum information?"

A common feature of ecological (and also taxonomic) data is its
 heterogeneity. Environmental factors which are important in
certain situations are irrelevant in others. Correspondingly,
groups of species whose variation and co-variation are important in parts of a set of data will often be absent in other
parts. As a consequence, beyond a few major trends, it is often
not possible to extract axes - or directions of variations -
which are relevant to the whole of the data; and the axes of an
ordination may become uninterpretable while much of the variance
in the data still remains to be accounted for.

This difficulty is overcome in practice by classification: by
subdividing the data into groups which show an acceptably low
level of heterogeneity (LAMBERT & DALE 1964,etc.). It is interesting to reflect that this approach is adopted in practice
even by those advocates of ordination who argue most strongly
against classification. The point I would like to emphasise
here is that classification thereby makes ordination practicable; it does not replace it.

Well established numerical techniques exist for both classification and ordination; they are reviewed by GREIG-SMITH (1964),
WILLIAMS & DALE (1964), LAMBERT & DALE (1964) and McNAUGHTON-
SMITH (1965). WILLIAMS & LAMBERT's method of "association analysis" has proved effective in a wide variety of practical situations in seeking classificatory groupings within unstructured
sets of data. Its use is discussed in various papers by WILLIAMS
and his associates (see LAMBERT & DALE 1964 for references) and
by IVIMEY-COOK & PROCTOR (1966), and programs are available for
a number of different computers. The more recently described
and simpler technique of (CRAWFORD & WISHART (1967) may prove

of equal or greater value, but has not yet been widely tested.
This are monothetic classification techniques in the sense of
SOKAL & SNEATH(1963).Polythetic methods are also available(see
WILLIAMS et al.1966; WEBB et al.1967; McNAUGHTON-SMITH 1965)
but their advantages are perhaps rather doubtfully worth the lar-
ge amounts of computer time and store capacity they require.

If electronic computing facilities are available,ordination may
be carried out by principal component analysis (see KENDALL
1957, HARMAN 1960, SEAL 1964) or, if the number of species is
large relative to the number of sites to be ordinated the equivalent "principal coordinates" technique of GOWER (1966) may be
used. The ordination method of BRAY & CURTIS (1957) and the alternative technique described by ORLOCI (1966) give solutions
of broadly the same type,and may be used as expedients when a
principal component solution is not practicable. I am doubtful
of the theoretical value of departing from a simple principal
component solution by writing communalities on the principal di-
agonal of the correlation matrix if an ordination of principal
component type(with successive orthogonal axes extracting maximum possible variance) is the end-product desired.
LAMBERT & DALE point out that classification and ordination may
be combined, but consider that "an efficient primary classification is likely to render subsequent ordination unprofitable."
From what has been said above it will be clear that I consider
this opinion to be based on a misconception; and from practical
experience it seems clear that ordination both within and between groups may be of great value (see e.g.ORLOCI 1967,PROCTOR
1967). In general, the variation within groups (or within the
data provided by a few related groups), is likely to be rather
fully determined by quite a small number of factors, and to be
fairly homogeneous, so that a large proportion of the variation
can be displayed in a few dimensions. I should like to emphasise that the divisions produced by classification techniques
like association analysis do not necessarily reflect discontinuities in the data - that is, regions of low density of stands
in the geometrical model.[1])

[1]) The definition of discontinuity given by LAMBERT & DALE
(1964,p.73) seems to be faulty,and not satisfactory separable
from the concept of heterogeneity as it is defined earlier
(p.71) in the same paper.Both concepts are important,and the
difference between them must be clearly grasped.

Such discontinuities may be sought by principal component analysis (cf.Fig.1) or, where groups can be defined beforehand, by
discrimant or canonical variate analysis (see RAO 1952, SEAL
1964). But in many instances, continuous variation will probably be found,and ordination techniques are potentially of
great value in exploring relationships at the lower classificatory levels. Their value is obvious in the case of groups
of related sites within a single area; it is perhaps less obvious, but probably more important, in the elucidation of relationships between corresponding communities in different
areas.

So far we have considered the relation of classification and
ordination in terms of using a classificatory technique to reduce heterogeneity to a point at which ordination becomes pro-
fitable. There is, however, another possibility, and that is to
design ordination techniques inherently more suited to dealing
with heterogeneous data than the principal component-type methods we have considered so far. This is to some extent possible
using factor rotation techniques. An initial principal component-type solution is computed treating the quadrats as variables.[1] This solution may then by rotated by an appropriate
method.

IVIMEY-COOK & PROCTOR (1967) have presented a rotated solution,
using KAISER's varimax criterion, for a small set of data from
an east Devon heath. The rotated solution allows a clear interpretation to be given to five axes, while only three axes could
be interpreted satisfactorily in the principal component solution. The data that we used in this analysis were of limited
extent and unusual form in that the original data matrix took
the form of a species x species multiple contingency table. I
have since obtained satisfactory results with a more extensive
and more orthodox set of a data taken from the analysis of the

---

[1] Either quadrats or species will provide homogeneous vectors,
and they are equivalent for analytical purposes.In this respect,
a species x quadrats matrix is fundamentally different from a
taxonomic species x characters matrix,or a psychometric subjects
x tests matrix;the R/Q terminology which is applicable to these
two cases in inapplicable to the species x quadrats case. (cf.
the subjects x occasions and occasions x subjects matrices in
psychology, see CATTEL 1952, HORST 1967).

"CENSUS CATALOGUE OF BRITISH HEPATICS" already referred to
(PROCTOR 1967). This analysis usefully illustrates some of points
discussed in the present paper. The initial data consist of
the records of 317 species and varieties of Hepaticae in the
152 "vice-counties"into which the British Isles were divided
by H.C.WATSON for biological recording purposes. Association
analyses yielded 25 vice-country groups and 47 species groups
(taking a 1% level of χ^2 as an arbitrary termination level).
Ordination of the vice-country groups and species groups by
principal component analysis yielded three axes reflecting
(a) abundance or richness, (b) the contrast between the moun-
tainous north and west and the south-eastern lowlands, and (c)
that between oceanic and montane distributions. Subsequent
components were not readily interpretable.
Varimax rotations of the species-groups component loading matrix
for various numbers of components showed that at least five
rotated factors could be interpreted readily, corresponding to
five main distributional trends within the British Isles. The
results will be readily appreciated by comparing maps showing
the distribution of the factor scores of the vice-country
groups with maps of individual species representative of groups
with high loadings on each of the five factors (see Fig.3).

A rotated factor analysis of this type combines the continuous
variables of a more orthodox ordination with the possibility
of selecting and isolating for study particular aspects of the
variation which is a valuable feature of classification.I think
this technique may prove to be of rather general value in vegetation study, both as a tool in the development of classifications, and as a research technique in its own right. However,
it is only one more among the various possible compromises,
and I believe that the one conclusion that must be emphasised
is that in the extremely complex field of vegetation analysis
the prime need is for flexibility; no one technique will serve
all purposes, and the investigator who condemns and rejects
without very careful thought does so at his peril.

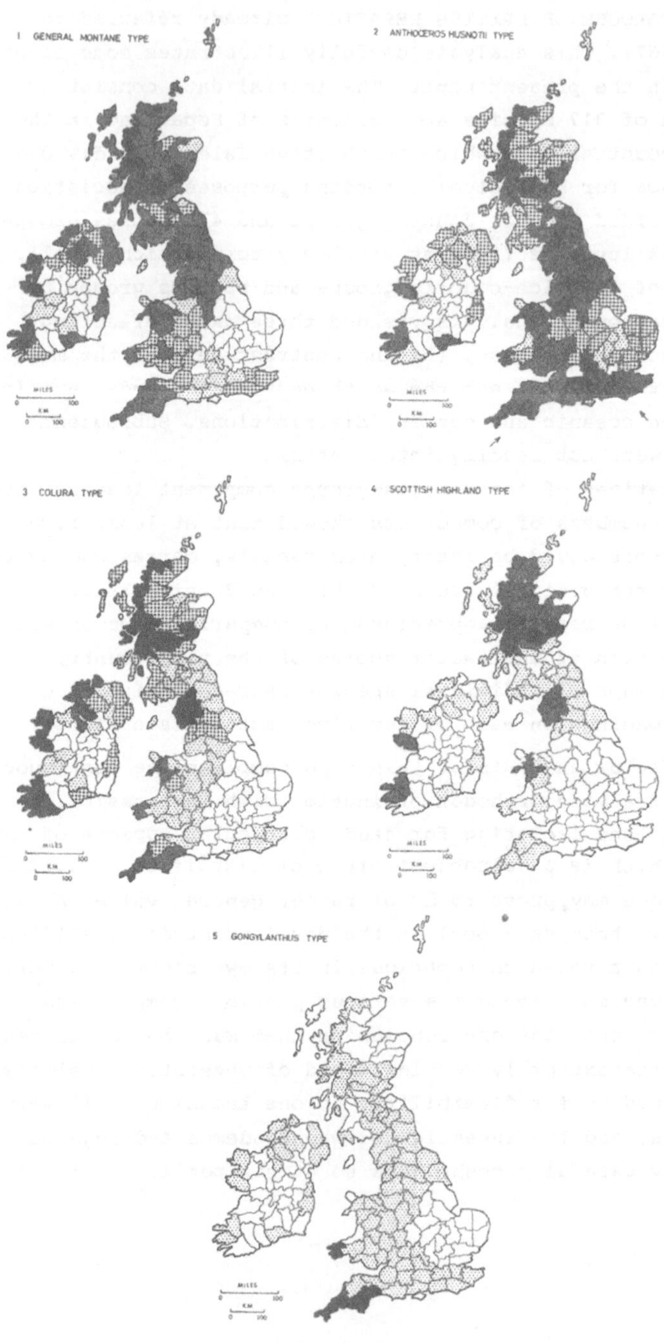
1 GENERAL MONTANE TYPE
2 ANTHOCEROS HUSNOTII TYPE
3 COLURA TYPE
4 SCOTTISH HIGHLAND TYPE
5 GONGYLANTHUS TYPE
MILES
0 100
K.M.
0 100

SUMMARY

Classification and ordination represent two different ways of
bringing about the simplification necessary to render intelli-
gible a complex set of vegetational data. Both approaches necessarily lose a proportion of the information present in the
original data, but the information lost, and the information
retained and emphasised,is different in the two cases. Certain
features of vegetation data lend themselves to treatment by ordination, while other features favour the use of classification,
and it is suggested that the most effective scheme of analysis
for any extensive set of data will make complementary use of
techniques of both kinds. The potential value of factor rotation
methods, combining certain features of established classification and ordination techniques, are briefly discussed.

ZUSAMMENFASSUNG

Klassifizierung und 'Ordination' sind zwei verschiedene Wege
zur notwendigen Vereinfachung, um komplexe Vegetationsangaben
verständlich zu machen. Beide Schritte führen zu einem Verlust
an Information der ursprünglichen Werte, aber die verlorenen
und zurückbehaltenen Angaben sind in beiden Fällen unterschiedlich. Gewisse Charakteristiken der Vegetation neigen zu einer
'Ordination-Methode', andere werden durch die Klassifizierung
besser ausgedrückt. Es wird vorgeschlagen, daß das wirksamste
Schema zur Analyse einer ausgedehnten Reihe von Angaben beide
Verfahren angewendet werden. Der potentielle Wert der Faktoren-
Rotations-Methode, die gewisse Charakteristiken von beiden be-
kannten Klassifizierungs- und 'Ordinationstechniken' einschließt,
wird kurz besprochen.

Fig.3. Distribution of scores for five components from a rotated
analysis of data on the distribution of Hepaticae in the British
Isles. For the purpose of these maps, the total range of scores
on each component has been arbitrarily divided into five equal
intervals, represented by the five densities of shading. The
maps represent five major patterns or trends in the distribution of Hepaticae. Compare Map 1 with the distribution of,e.g.,
Bazzania trilobata, Solenostoma triste, Nowellia curvifolia; 2
with Fossombronia wondraczekii, Leiocolea turbinata, Riccia
glauca, Anthoceros husnotii etc.; 3 with Colura calyptrifolia,
Lejeunea lamacerina, Adelanthus decipiens, Cephaloziella pear-
sonii, Acrobolbus wilsonii etc.; 4 with Gymnomitrion obtusum,
Scapania ornithopodioides, Mastigophora woodsii, Chandonanthus
setiformis etc.; 5 with Fossombronia husnotii, Cephaloziella
massalongoi, Porella pinnata, Gongylanthus ericetorum etc.

REFERENCES

BRAY,J.R.& CURTIS,J.T.   -1957- An ordination of the upland
  forest communities of southern Wisconsin.- Ecol.Monogr.27:
  325-349. Durham,N.C.
CATTELL,R.B.   -1952- Factor Analysis.- New York.
CRAWFORD,R.M.M. & WISHART,D.   -1967- a rapid multivariate
  method for the detection and classification of groups of
  ecologically related species.- J.Ecol.55: 505-524. Oxford.
DE WIT,C.T.   -1960- On competition.- Versl.Landbouwk.Onder-
  zoek.66 (8):1-82. 's Gravenhage.
GOODALL,D.W.   -1963- The continuum and the individualistic
  association.- Vegetatio 11:297-316. Den Haag.
GOWER,J.C.   -1966- Some distance properties of latent root
  and vector methods used in multivariate analysis.- Biometri-
  ka 53: 325. London.
GREIG-SMITH,P.   -1964- Quantitative Plant Ecology.Ed.2. London.
HARMAN,H.H.   -1960- Modern Factor Analysis.- Chicago.
HARPER,J.L.   -1964- The nature and consequences of interference
  amongst plants.- Proc.XIth int.Conf.Genet.:465-481.
HARPER,J.L.   -1967- A Darwinian approach to plant ecology.-
  J.Ecol.55:247-270. Oxford.
HORST,P.   -1967- Factor Analysis of Data Matrices.- Chicago.
IVIMEY-COOK,R.B. & PROCTOR,M.C.F.   -1966- The application of
  association analysis to phytosociology.- J.Ecol 54:179-192.
  Oxford.
IVIMEY-COOK,R.B. & PROCTOR,M.C.F.   -1967- Factor analysis of
  data from an east Devon heath: a comparison of principal com-
  ponent and rotated solutions.- J.Ecol.55: 405-413. Oxford.
KENDALL,M.G.   -1957- A Course in Multivariate Analysis.- London.
LAMBERT,J.M. & DALE,M.B.   -1964- The use of statstics in phy-
  tosociology.- Adv.Ecol.Res.2:59-99.
LAMBERT,J.M. & WILLIAMS,W.T.   -1962- Multivariate methods in
  plant ecology.IV.Nodal analysis.- J.Ecol.50:775-802. Oxford.
McNAUGHTON-SMITH,P.   -1965- Some statistical and other numerical techniques for classifying individuals.- Home Office Res.
  Rep.No.6. H.M.S.O. London.
ORLOCI,L.   -1966- Geometric models in ecology. I. The theory
  and application of some ordination methods.- J.Ecol.54:193-
  215. Oxford.
ORLOCI,L.   -1967- An agglomerative method for classification
  of plant communities.- J.Ecol.55:193-206. Oxford.
PROCTOR,M.C.F.   -1967- The distribution of British liverworts:
  a statistical analysis.- J.Ecol.55:119-135. Oxford.
RAO,C.R.   -1952- Advanced Statistical Methods in Biometric
  Research.- New York and London.
SEAL,HILARY -1964- Multivariate Statistical Analysis for Bio-
  logists.- London.
SOKAL,R.R. & SNEATH,P.H.A.   -1963- Principles of Numerical
  Taxonomy.- San Francisco and London.
WALTERS,S.M.   -1961- The shaping of Angiosperm taxonomy.-
  New Phytol.60:74-84. Oxford.
WEBB,D.A.   -1954- Is the classification of vegetation either
  possible or desirable?- Bot.Tidsskr.51:362-370. København.
WEBB,L.J., TRACEY,J.C., WILLIAMS,W.T. & LANCE,G.N.   -1967-
  Studies in the numerical analysis of complex rain forest
  communities. I. A comparison of methods applicable to site/
  species data.- J.Ecol.55:171-191. Oxford.
WHITTAKER,R.H.   -1967- Gradient analysis of vegetation.- Biol.
  Rev.42: 207-264. Cambridge.

WILLIAMS,W.T. & DALE,M.B.   -1964- Fundamental problems in numerical taxonomy.- Adv.Bot.Res.2:35-68.
WILLIAMS,W.T., LAMBERT,J.M. & LANCE,G.N. -1966- Multivariate
  methods in plant ecology. V. Similarity analysis and information analysis.- J.Ecol.54:427-445. Oxford.
YARRANTON,G.A. -1967- Organismal and individualistic concepts
  and the choice of methods of vegetational analysis.- Vegetatio
  15:113-116. The Hague.

G.LAVRENTIADES
Can we use this method of ordination-classification for every
vegetation type, or are there methods for each vegetation?

M.C.F.PROCTOR
No, I think they are of a quite general applicability. We use
methods of this type in heath vegetation, I used the same methods in other vegetation and of course, the Americans use ordination methods with a considerable variety of different vegetation types. Ecologists in the British Isles have used them
for a variety of types,grasslands,dune vegetation,and so on.

E.van der MAAREL
Vielleicht ist es gut, nochmals zu betonen, wie wichtig es ist,
ohne Dogmatik irgendeine Methodik anzuwenden, sei es Klassifikation, sei es Ordination. Und es freut uns, daß für die Anwendung dieser neuen Techniken für phytosoziologische Untersuchungen nun auch in England ein Interesse besteht. Die letzten zwei Tage haben wir oft darüber gesprochen. Und es wird
auch die Engländer freuen, daß in vielen Ländern des Kontinents
schon viel darüber gedacht und gearbeitet wird. Wir halten es
für äußerst interessant und auch notwendig, alle diese Methoden, die zum größtrn Teil aus England und den Vereinigten Staa-
ten kommen, in unseren Kreisen anzuwenden.
Zu den Techniken selbst gibt es natürlich viele Fragen und Bemerkungen, aber ich glaube, daß es vielleicht besser ist, dies
im kleinen Kreis mit Herrn PROCTOR zu diskutieren, denn es handelt sich um rein technische Fragen, wie z.B.: Sollen wir Kom-
munalitäten anwenden oder nicht? Sollen wir Cluster-Analysen
anwenden oder nicht?
Jedoch möchte ich hier eine Diskussion über die Beziehungen
zwischen Grenzen und der Methodik anregen. Sie haben ganz richtig gesagt, daß eine Klassifikation immer zu Klassen, also zu

Grenzen führt. Aber das bedeutet nicht, daß das Ausgangsmateri-
al auch diskontinuierlich sein muß. Es ist wohl möglich, eine
komplette kontinuierliche Serie von Aufnahmen diskontinuierlich
zu bearbeiten. Aber dann bleibt noch die Frage: wo gibt es objectiv festzustellende Grenzen? Shouldn't we give some information or thinking about this problem: whenever we make a classification from apparently more or less continuous data,shouldn't
we try to develop more sophisticated methods to detect and indicate the abstract boundaries between the classes we have
erected by our classification method? I think what is called
cluster analysis goes in this direction, but it is very important for the average European phytosociologist to know more
about the techniques which are available to come to these objective boundaries between classes we have erected.

M.C.F.PROCTOR:
I think this is very important. Of course, it is a very general problem and one which affects not only phytosociology but
taxonomy as well. Unfortunately, I think no single method of
analysis will give us all the answers we want. In think in general one will have to proceed by, to borrow a phrase from POO-
RE,the method of successive approximation. The techniques which
will demonstrate discontinuities most effectively are techniques like principle component analysis which will arrange our
data in relation to a direction variation. If you can reduce
the data under consideration to the vegation types immediately
on either side of a particular boundary or possible boundary
that you are interested in, you then can carry out analyses
which will give you an indication of the density of stands in
an abstract space representing that direction variation. We
then can seek the positions of the major concentrations of the
variation, which you might regard as noda in the sense of POORE,
and you can seek those regions which are relatively empty of
sites which are naturally the regions in which one would seek
to place boundaries.

J.J.MOORE
I would like to congratulate Dr.PROCTOR on the very good summary on what is going in England and the United States. What I
think maybe useful to the people here, and I'd like to get

Dr.PROCTOR's opinion on it, is to try to see what connections
has this work on classification and ordination to what BRAUN-
BLANQUET-phytosociologists do? In other words, what is the
BRAUN-BLANQUET table in terms of classification and ordination?
Suppose we have a table; what the BRAUN-BLANQUET people are trying to do is first of all to carry out a classification, a qualitative classification,in which you see use plots of differential species,4 or 5 differential species,and this then continued to a greater or less degree is essentially a qualitative
classification.That is not all. After that,one tries to carry
out an ordination,a one-dimensional ordination.You do not merely leave these blocks in the raw state,but you try to get the
most similar ones beside one another, and you try to order the
whole table along an ecological gradient. The most usual would
be dry subassociation - typicum - wet subassociation, so that
you are going across a moisture gradient. So that is actually
trying to combine classification and ordination with a more
intuitive sort of method than the strict mathematical methods,
but it is aiming at exactly the same thing and very often giving a more visual impression with the very great advantage that
you lose no information. All the information is in your table,
you stand back from the table and you look at the general picture, as you would in a picture gallery,then you can take long
views or short views and all the information is there in the
table, both classified and ordinated (Beifall).

M.C.F.PROCTOR:
The point, of course, is that you may well find yourself with
more directions of variations than that. In saying this I am
not condemning your approach, because I have done this kind of
thing myself. The point is that you are here ordinating in one
dimension. You may well have more than one dimension; but the
information is all there, in the raw data, there is very probably more information to be extracted. Maybe, one should bear
this in mind.
There was one point which particularly interested me in this
question of the one-dimensional ordination, because in discussing the treatment of data from the Burren area in the west
of Ireland we did very briefly consider this point. One tries
to ordinate phytosociological data or the higher phytosocio-

logical units along a single dimension just as one tries to ordinate the data of plant classification along a single dimension. We take perhaps phytosociological progression with pro-
gressively increasing complexity in integration. Similarly in
the case of ordinary plant systematics we take what we believe
to be an evolutionary sequence. And in both cases, of course,
one realises that this allows one to display one's results in
terms of one of the major directions of variation. You also rea-
lise very quickly that you can't represent all the important
directions of variation which you would like to in one dimension. Now, one dimension, of course, has an extremely important
property in that you can use it as an order in a book. And this
is a property which, I think, deserves more note and more sym-
pathy from its detractors.But one has precisely analogous problems to the one that arises, e.g.in representing plant relationships. In BENTHAM and HOOKER's system the Cornaceae and the
Caprifoliaceae come close together as they should;in ENGLER's
system they come wide apart. This is no condemnation of the
ENGLER system, because in the ENGLER system other clearly related groups of plants come close together. It merely reflects
the impossibility of doing everything within a single dimension.

ON ABSTRACT AND CONCRETE BOUNDARIES, ARRANGING AND CLASSIFICATION

(Über abstrakte und konkrete Grenzen,Ordnen und Klassifizieren)

I.S. Z o n n e v e l d

1. INTRODUCTION

A vegetation surveyor meets with three categories of boundaries:

(1) concrete discontinuities in the field
(2) abstract boundaries between classification units
(3) lines on a map (Cartographic boundaries)

A boundary is a line or transition zone that separates two different items. In (1) these are different vegetation stands composed of concrete plant individuals. In (2) they are brain constructed with properties abstracted from many speciments of the concrete situation. In (3) it is a concrete line on a two-dimensional piece of paper. This concrete line however is directed by a process of abstraction and reasoning, and related to the concrete, two-dimensional situation in the field.

Concrete as well as abstract boundaries may either (a) be given by nature and only need be "recognized" by man or (b) they have to be made. The first case (a) means that there are natural discontinuities or that there appear "gaps" in the abstract arranging that precedes every classification. In the second case (b), nature or abstract model of arranging (a vegetation table or any other statistical arranging of data) shows only very gradual, rather continuous transitions. At the extreme ends of the transitions there are clear differences, but between these any boundary is artificial, because differences between the parts of the two separate bodies, situated very close to the boundary are smaller than the maximum difference within each unit itself (see the example of the classification of seats in a theatre). This means that in order to "draw" a boundary we need more norms than only existing differences. From this point of view it is quite equal whether we cut a certain continuous concrete or abstract situation into two, three or a hundred subdivisions. The nature of vegetation places a limit somewhere. A vegetation unit cannot be smaller than the size of the smallest plant that helps

to compose it, and usually is much bigger. But between this limit and the non-subdivided continuum, boundaries can only be
drawn on the basis of arbitration. We need guiding principles
here, not only for the type of subdivision (see below)but also
for the place and the amounts of steps we want to create.

From this we have learned that for abstract as well as concrete
boundaries we may distinguish between boundaries coinciding
with a "natural" discontinuity (sometimes not quite correctly
called natural boundaries) and boundaries arbitrarily drawn in
a gradual transition (sometimes called artificial boundaries).[1]

Although generally speaking there may not be any differences of
opinion concerning this point there is still a large controversy regarding the question of whether classification systems can
be based on units that are separated by continuous (artifical)
boundaries, or whether "natural" discontinuous boundaries are
required.

In the discussion on this subject the principal differences
between concrete and abstract boundaries do not seem always to
be recognized. Moreover, the special character of mapping units
compared with general classification units do not seem always
to be understood either. Differences of opinion on the character of cartographic boundaries are one of the results of this.
In this paper we shall compare the three types of boundaries.

2. <u>CONCRETE BOUNDARIES</u>

The existence of both rather continuous transitions and sharp
discontinuities between clearly different vegetation stands
is not a point of discussion. It can be observed easily in the
field, both in pure natural landscapes, and in cultivated landscapes. The steep bank of a (natural) river forms a knife-sharp
boundary between the waterplant communitie(s) in the river water
and the dry land vegetation on the river levee. The boundary of
an arable land parcel forms a clear-cut vegetation boundary

[1]) Note. The terms "natural" and artifical are dangerous because
the differences within the gradual transition and the direction
of the boundary is also guided by nature; it cannot be compared
with a fence in homogeneous grassland. This is a purely artificial boundary, constructed to create a difference that did not
previously exist.

with another parcel with a differnt treatment, or with a forest,
pasture on garden.

Especially the cultivated land is rich in sharp and straight-lined boundaries. This is so because of the dominance of the
(straight-lined) cultural forces complex (cultural environment).
In less human-influenced countries sharp boundaries are especially caused by the relief (landform). The other vegetation and
land-forming factors are less spatially confined. Climate and
animals have a more random influence unless landform (water)
and/or man in combination with the vegetation itself stratify
these influences.

The most classical example of very gradual (continuous) concrete
transitions is the tropical rain-forest. Here the time dimension is very large (tertiary), the influence of the most extreme
random factor - the climate - strong (high temperature and high
moisture), and a condition of intensive chemical alteration of
rock is created. This combination forces a levelling of chemical and physical soil differences. The long duration of the same action (stability in time) has eliminated already long ago
the exceptions in a system that is trying to reach an equilibrium. It is remarkable that the existence of such continuous
transitions has been denied by earlier ecologists (compare
SCHARFETTER 1932).

The opposite is the case with young acreations in a sedimentation area. Few species may occupy large territories in a succession of a few years forming monotonous vegetations with dominance of one or a few species. The place where a certain dominant species grows is partly caused by the accidental priority
of that plant, and not always only because of a real discontinuity in the environmental factors.

The discontinuity of an environmental factor or complex of
factors does not necessarily give rise to a discontinuity in
vegetation. The vegetation itself may cause a more continuous
transition, if the plants that compose it are able to change
(level) the environmental forces complex considerably. Also
the reverse case maybe true; a very gradual transition in one
environment factor may give a sharp boundary in the vegetation.
Endogenic factors, such as masseffect, priority etc. maybe the
main cause of this. See the example of sedimentary areas, also

the forests- or wood boundary in mounteneous areas. A gradual
transition may even give rise to a special vegetation that cannot merely be considered as a transition between two adjacent
ones,but has its own characteristics(e.g.own characteristic species or structure).These are"Saumvegetationen"or borderline vege-
tations.The last situation may be also found with a more discontinuous environmental transition. If the transition zone indeed
starts to "live a life of its own" the problem of the boundary
is not difficult. In these cases there is not one but two boundaries, on both sides of the "transitional" zone, which appears
to be only a transitional zone in physico-chemical environment,
but not necessarily in the vegetation cover itself.

The above-mentioned concrete boundary situations in the field
are described by VAN LEEUWEN and WESTHOFF (see WESTHOFF,this
Symposion) as limes convergens (sharp transitions) and limes
divergers (diffuse transitions).

The characteristics of these boundaries will be treated in detail by WESTHOFF at this symposium and do not therefore need
to be discussed in this paper. It should be emphasized that
each discontinuity in the field is caused by an unique combination of all local environmental factors.

Too often comparisons are made between the systems of biological
individuals (species) and vegetation types. This can only be
done without danger to a certain extent. However, as soon as
the individuality of the concrete vegetation stands is exagge-
rated and considered to be of the same order a the individuality of an oak tree, an elephant or an ant, then the way to
misunderstanding is open. Then too much value is given to discontinuous boundaries in the field (does an elephant not also
have a "discontinuous boundary" seems to be the reasoning!).

Summarizing, it can be stated that concrete (discontinuous)
boundaries between vegetation stands often occur (in earlier
days it was even supposed that it was a general rule). They are
very common in cultivated areas and in areas with strong relief.
There the vegetation forming factors man and relief are the ma-
ster factors determinating in the place and the character of the
boundaries. In flat areas and those with an extreme dominating
climate, more gradual transitions may occur between different
vegetations.

3. ABSTRACTION

So far we have discussed transitions between two concrete vegetation stands. Comparison of these stands and the transition
between them, does not require consideration of other stands.
Problems can only be solved fully by considering these two
stands themselves. It is irrelevant whether, a few hundred
metres further on, similar vegetation stands also form a boundary. These stands may be similar, but not equal. Some plants
may be missing, others may occur that did not appear in the
first. It may be even that the second group of adjacent stands
together is almost identical with the first considered pair,
but that the separating discontinuouty lies on a different level. For example, the first group of stands may the following
composition:A-B-C-D/F-G-H-I. There is a discontinuouty between
D and F because E is missing; That is a boundary. The second
group may have the following "formula" A-B-C/E-F-G-H-I. Here
the discontinuity lies between C and E because D is missing.
In another place the same plants may occur, but now in the formula A-B/D-E-F-G-H-I. This phenomeon is not a problem as long
as we are only interested in boundaries. However, as soon as we
direct our attentions to the stands that are separated by the
discontinuum, certain questions arise. These questions have
their source in our scientific mind, which always tries to compare items according to their similarities and differences in
properties. We have learned to see the phenomena on the earth's
surface not only in their spatial relation but also in their
other intrinsic properties.

No single object is fully equal to another object; in many cases different objects do not differ in all their properties. We
compare, that means we look for differences and for similarities. Systematic comparison means selecting certain properties
that are interesting in this respect. These properties start to
have a "life of their own" in our considerations; we set these
properties "apart", we separate them from the others in our
mind, we list them on a piece of paper, we "abstract" them from
the total. This leteral "abstraction" is used for a first systematic putting into order. The objects with one ore more similar
properties are put together. Others with different properties
from the former group are grouped elsewhere. We have made

"classes". The process in which we have insolved ourselves is "classification".

The type of arrangement in this case is that we considered the combination A-B occurring in one of the situattions in the field on one side of a concrete continuum as a unit that could easily be distinguished from the stand on the other side of the "natural" boundary by the occurrence of the combination (association" A-B and the absence of D-E-F-G-H-I.etc. But we also found that there is a concrete composed of A-B-C and one of A-B-C-D. So the problem we met with in comparing the stands on both sides of the concrete discontinuum in the various places described above, is that the stands A-B, A-B-C-D etc. can be arranged after abstraction in such a way that they form a gradual series, an abstract continuum, in spite of the fact that in the field discontinuous boundaries can often be observed.

Before we continue the discussion on boundaries, it is necessary first of all to discuss some of the basic principles of arranging and subsequent classification.

4. CLASSIFICATION

4.1 General

In any science it is necessary to create order amidst the multitude of items and phenomea. This usually results in an arrangement of the material in a number of groups, each consisting of related items. Such groups may or may not be arranged into hierarchical systems.

Every group receives a certain place which maybe regarded as a pidgeon hole in a cupboard, or a hook of a hat-stand. The only purpose is to have arranged the material in such a way that it can easily be handled and compared. Such a "cupboard" or "hat-stand" is called a classification system.

In literature (GOODALL 1954, CURTIS 1959, v.d.MAAREL 1966) a principal distinction is sometimes made between "ordination" and "classification". The latter would require a high degree of discontinuity of the material. The first would be the only possibility in a continuum. We do not agree with this conception. Classification is indeed more easy, and maybe more natural in a discontinuum, but also a continuum can and should be classified.

There are, however, a few main points or requirements that are
important for any classification, and especially for vegetation
(in soil the situation is similar). They follow directly from
the procedur of classification shetched in the former paragraph (3).

a. Classification unit needs "characteristics" which can be
   used for determination and recognition of the units.

   These characteristics should be properties of the item to
   be classified, and moreover they should be measurable, that
   is they should be "morphometric".

   (Any measurement has an element of estimation, even the use
   of a ruler. So the possibility of estimation of properties
   which are difficult to measure may also fulfil the morphometric requirement).

b. Any clasification is an abstraction. We cannot use all the
   properties of an item for comparison. Therefore we must select, that means we "abstract" (= separate from the total of
   properties) some properties that are thought to be representative.

   (See chapter 3. A simple example derived from planimetry:
   two triangles are congruent if each of the three angles and
   thé three sides of a triangle is equal to the corresponding
   angles and sides of the other triangle. In total these are
   six items. It appears, however, (in this case it can be proved) that if only three sides or two sides and one angle, or
   two angles and one side are equal, the triangles are congruent. So it is sufficent in this case to compare only three
   items instead of all six).

c. A classification needs one or more guiding principles.
   These guiding principles guide:
   (1) the selection (abstraction) of characteristics from many
   properties of the basic material;
   (2) the arrangement of the classification units. That means
   the making of the shape, the size of the "cupboard" or "hat-
   rack"and its "pigeon holes" and hooks.
   (3) the "mapping" practice. This will be treated in Chapter 7.

d. Items with a spatial character with distinct horizontal extension occuring at the earth's surface, e.g.soil,vegetation,
   can be classified principally in two ways.

                     1.) The pure or general classification
                     2.) The "chorological classification.

The latter is characterized by the dominance of the spatial   -

(=chorological and, as a consequence, the horizontal maco-pattern and size as a guiding principle. Therefore the concept scale also plays an important role. Map legends are (should,be) directly derived of, and are usually almost identical with these chorological   classifications. The principal difference between "chorological " (= legend-) and the "general or pure" classifications is not always well understood. It should be taken in mind that for the first type of classification concrete boundaries in the field are irrelevant.

A sharp distinction between both types of classification would increase clearness both in map legends and general (worldwide) pure classification systems. Only if a sharp distinction is made can they be combined or even integrated in an appropriate way (see paragraph 7).

4.2 <u>Guiding principles</u>

Guding principles can be derived from many disciplines.
They can be purely morphometrical. For instance we divide according to a one, two or three-dimensional form (shape) into planimetric or stereometric terms. Or we consider in particular the composition. In this case the central concepts and boundaries of the units are selected only in a morphometrical way.

If there are certain aggregations and subsequent discontinuities, this can easily be done. If the material to be classified has a continuum-character it is more difficult to use morphology alone as a guiding principle. Even if there are clear "aggregations" with discontinuities (="natural" boundaries) in between, it may still be usefull to have guiding principles of another type. For instance the translation (interpretation) of the indication value of a certain vegetation type for one environment factor or another can be done more easily if that factor is used as a guiding principle. Therefore in general it is more useful to take forming factors and processes as guiding principles. This means, in vegetation, vegetation-forming factors (mesological ecological factors).

So climate may be an excellent guiding principle for world-wide classification systems. It is indeed widely used as such although it surely is not strictly necessary.

For local classification systems another factor may be more

appropriate as guiding principle,e.g. hydrology, eutrophy(nutri-
ent status) of the soil, landform, or human influence etc.

Usually more than one guiding principle is used.In this case
the system may have a hierarchical structure. For instance at
a high level climate may be used, at the next lower level hydrology; lower still eutrophy may occur. But the reserve maybe also more suitable for the special purpose of the classification
in that special area.

So the choise of a guiding principle depends on:

(a)   the special purpose
       (e.g.if we want mainly to study hydrology with the help
       of indication by the vegetation, we should take hydrology
       as guiding principle).

(b)   The items to be classified.
       If we cannot find any correlation of vegetation with the
       factor that we originally intended to have a guiding prin-
       ciple, then it is not possible to use that one and we
       should try another.

So the selection of a guiding principle may be a matter of trial
and error. Consequently, the same applies to the selection (abstraction) of characteristics. Usually, however, the nature of
an area shows without too much trouble the directions the direction that leads to success, right from the beginning.

Some discussions on the dillema whether a vegetation classification should not be based on ecological considerations suffer
from the fact that the difference between characteristics and
guiding principles seems to be unclear to the opponents in such
discussions.

We whish to state that a vegetation classification should only
use morphological properties of vegetation itself as characteristics and not for instance the soil or climate. We should try
as much as possible to adapt the nomenclature of the classes to
this principle. Names such as salt marsh, woodland, scarecely
vegetated dunes, tropical rain forest etc. invite misunderstanding and misuse; even if the units are properly defined morpho-
metrically.

A classification system should be such that a concrete vegeta-

vegetation stand can be recognized, determined in that classification system, not with the help of land(scape) soil or climate or any other non-vegetational property (factor) of the landscape.

5. <u>ARRANGING AND "ORDINATION"</u>

Before we can start to the classify we should look over our basic data and arrange them. If we have accepted that morphometric properties have to be used as characterizing properties, this means that in each classification units, items which are morphologically related should be combined. So the arranging should be done on the basis of morphological relationships.Each of the items to be classified (relevés, lists of plants) should be compared with all the others. The table method used by the BRAUN-BLANQUET school is up to the present one of the most comprehensive semi-statistical method for this comparison.Other systems may be more objectiv but are more time-consuming and obscure extreme and other border cases[xx])that often are the most important for scientific research.

For long-term research under normalised working conditions computer statistics are of great value. For practical survey purposes these still present considerable organisational difficulties.

In fig.1 (ovals and triangles) we have a simplifdied example of the principle of arranging. Items with variation in colour (black and white), shape (ovals and triangles), size (large, medium and small) and "behaviour" (standing straight, with the broadest end below) or at the top (with the point pointing downwards) are arranged (ordered) in such a way that we could "cut" the total into pieces containing similar things. In this case there are discontinuities that "guide the cutting knife". The most natural divisions is line (a). This one divides not only according to one property (colour) but also according to "behaviour". On the left-hand side all the figures, with only one. occasional exception, stand straight with the narrowest end pointing upwards; on the right-hand side the reverse is the case. This may be the reason for taking dividing line (a) as the first (the highest categories)in the hierarchy.

[xx]) The author applied the method of CHEKANOFSKI,of de VRIES and similar methods using modern computers. See ZONNEVELD,BANNING and LEYS (in prep.)

The figur of eggs and triangles is a kind of statistical model
of a special structure.

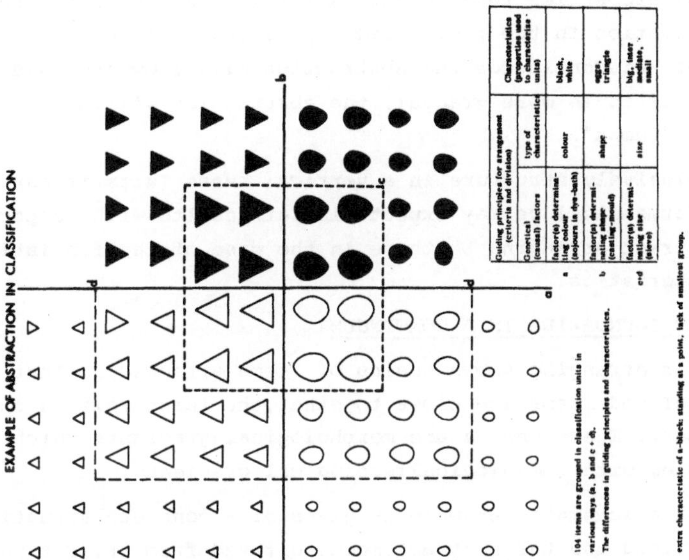

Fig.1 Example of abstraction in classification

The result of each statistical treatment is a model. This is
usually a more dimensional one. The vegetation table of BRAUN-
BLANQUET as well as the more sophisticated established (so-cal-
led "ordination") model diagrams of GOODALL, CURTIS and many
others are models of such type.

Before we discuss the situation of boundaries in such models,
compared with concrete ones in the fiels, we shall go back for
a while to our example in paragraph 4.

In this example we have abstracted the species combination.With
the mentioned combinations A,B,C etc. as characterizing properties it appears that each of these combinations may occur: A
or A-B or A-B-C or A-B-C-D or A-B-C-D-E etc., but also I-H and
I-H-G and I-H-G-F etc. Possibly also others like C-D-E or E-F-
G.

We did not select other properties as characteristics. We could
have taken the size of stands, or even the type of boundary
with adjacent stands or any other spatial criterion in the ho-

horizontal plane. There is a fair chance that we should have obtained different results in the relation to the concrete situation in the field. Especially the use (abstraction) of horizontal spatial properties tend to give a result that reflects the
spatial situation in the field (see map legend). If this situation is very discontinuous the abstraction will show the same
character. If it is more gradual, the abstraction will also
show fewer "jumps".

Taking exclusively structure in a vertical sense (stratificati-
on, life-forms etc.) we may expect similar results with respect
to the abstract arranging to those in the case of the floristical characteristics.

6. <u>ABSTRACT BOUNDARIES IN A CONTINUUM</u>

Any model of arranging shows chains of items arranged so that
the most related items are close together,the least related ones
are far apart. These chains are morphological gradients which
maybe related with (mesological)ecological gradients.

In figure 2 a schematic example is given of a concrete situation in the field and the abstract model dirived from it. For purposes of simplicity the example is constructed as a two-dimensional continuum. The left part is the concrete situation in the
field. The right part is the rearrangement in an abstract model.
The morphometrical differences are expressed in two gradients:
horizontally there is a gradual change from triangle to circle
(9 triangles on the extreme left, 9 circles on the extreme
right side. Vertically there is a gradual change from black below to white at the top.)

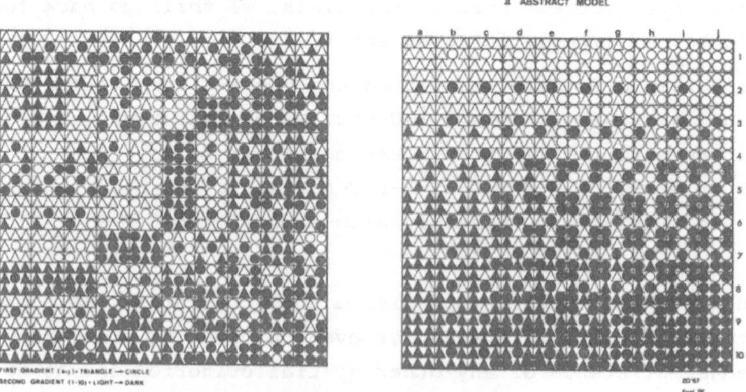

Fig.2 Scheme of a continuum with two gradients

Classification in this continuum means the cutting of into pieces. The place of the boundaries appears however completely arbitrary. If for example we cut this continuum into four blocks,
not one of the four units is fully homogeneous because within
each units the gradients still occur. Each of the thus formed
4 classification units has parts that are more related to a part
of an other's unit (close to the boundary) than to the most remote parts.

This is clear from the arrangemant of seats in a theatre or
cinema. All the seats together may form a continuum (in this
case even a concrete one). Still a subdivision is made. Say
first class consist of 7 rows, and the second class also out
7 rows. There is a difference in price between the classes.
The difference in quality of position to the screen is larger
between the first and the last of the seven rows of first class
rows!than between the adjacent rows of both classes (one row
only!). Everybody who books too late feels this relative in-
justice, but accepts it at the same time.

Between the ground floor and the balcony there is usually a
discontinuity. There the difference of quality is less arbitrary and the difference in price accepted with less reluctan-
ce.

With respect to the vegetation of the model, the situation is
similar. So far our model and also the example of paragraph 3
is hypothetical. It assumes that the units of any abstract division of vegetation or soils of a not too small area (classification) form a continuum.

Another hypothesis might suppose that there is always aggregation and by consequence there are "natural" delimitations (discontinuities) between the units.

Continuity versus discontinuity is a typically mathematical
subject. If the morphometric requirement is fulfilled, the properties of the units can be expressed in figures. Consequently
the units as such can be expressed in combinations (or frequencies) of properties in figure form.

So theoretically, the question of which of both is right
could be answered with the help of mathematics. However the
first requirement of a mathematical treatment is a "random"

sampling of the <u>total</u> vegetation types.

Furhermore the item is so complex, there are so many variables, that we need an immense number of data. Automatically some of the combinations and frequencies of properties will not be represented in the material if we take too little data. Moreover we cannot take fully random samples because the size of the concrete stands would start to play a part. If we select floristic composition, life form or vertical (and micro-horizontal) structure as characteristics, this is not allowed. The consequence is the appearance of discontinuities, unless we would not give attention to the density of clusters in the model. The relation between concrete and abstract boundaries is illustrated in fig.2. The right part is constructed out of the same units as the left part. They are, however, arranged in such a way that the most related units are as close as possible to each other in respect to both gradients (black-white, and circle-triangle respectively).

There seems to be a perfect continuum in this example. In the field (left figure) there are many discontinuities.

If we take only half of the left figure, then it is clear that a similar arrangement of the patches of that figure must necessarily show the discontinuities, because many transitional stages in either "black-white" or "circle-triangle" are lacking.

In small areas a mathematical treatment usually shows discontinuities[x]) but even there a continuum situation has been proved to exist (CURTIS 1959).

For the total vegetation of a continent, let alone the whole earth, sufficiently extensive random sampling will never be possible.

So the hypothesis can never be proved strictly. There is much evidence that the continuum may be the right concept. The contirious quarrel and fight on the boundaries of abstract units *is* evidence of this.[xx])

[x]) Viz.de VRIES 1953. Table method of French-Swiss school, see ELLENBERG 1956, GOUNOT 1955, ZONNEVELD,BANNINK and LEYS,i.p.

[xx]) An excellent new example of this theme has been given at this symposion by JAKUCS. (Das Wald-Mantel-Saum-Problem in SO- und Mittel-Europa.'

On the other hand, in "coarse-grained"vegetations not all the
possible combinations of properties may be realized. So "natural"
discontinuity may exist there. The same holds for rare vegetations with extreme environment demands or which grow in isolated places where divergent development (also floristic development and divergence) takes place.

The floristics between the continents which are isolated by
oceans or high mountain chains differ in a discontinuous way,
and consequently so are units of classification based on them.
Classification based on structure only are less influenced by
isolation. (there are vegetations known from similar climates
or hydrological conditions in different continents, which have
an identical structure but no species in common).

A well-known and generally uncontested observation is that vegetation differences caused only by climate show gradual transitions. WESTHOFF (this symposion) calls this series of vegetation showing such continuous variation "clines". It should however be understood that exactly the same type of clinal variation can be caused by hydrological, pedological or other factors. They only do not show so easily on world-wide (plantgeo-
graphical!!) maps because the distribution of hydrological and
soil phenomea is much more complex than macro-climate.

From a practical point of view we wish to state that it is appropriate always to start from the concept of a continuum. If later
on it appeared that indeed a "natural" discontinuity existed,
then good arranging of the material would automatically reveal
this.

In the reverse case an assumption of natural discontinuity will
lead to much confusion, and indeed has already done so in many
cases (fruitles fight on boundaries).We   applied this principle in creating an applied classification system for arable
land and forest plantations.Here communities are arranged in
gliding scales according to general eutrophy and moisture conditions. (ZONNEVELD,BANNINK and LEYS i.p.).

7. CARTOGRAPHIC BOUNDARIES

So far we have discussed purely concrete boundaries which are
discontinuities between stands, and abstract boundaries which
may also be discontinuities but often appear to be lines cut

arbitrarily through a continuum; but both types of abstract
boundaries are related to pure brain constructions of properties abstracted from the reality.

Lines on a map are both, concrete and abstract. They are concrete because they consist of ink and are drawn or printed on
paper. However they are also abstract because they are nothing
other than a spatial, concrete expression of the map legend.The
map legend is a kind of classification. Everything that has
been written on classification (paragraph 4) is also true of
the map legend.
The difference between a general classification (typology,taxo-
nomy) of vegetation and a map legend is mainly caused by a different guiding principle. In general classification, the spatial aspect is left out or is only of low importance. In the legend, however, the spatial relation should be important guiding
principle. The more detailed the map, the less it may count.

In fig.3 this can be easily understood from the series of small
maps seen from below to the top. Map I shows individual plants.
The system of classification used is the well-known system of
plants. There is no need of spatial relation, because we read
the situation directly on the map. It is even an aim of mapping
to study spatial relation. In Map II we map communities,single
vegetation stands. We see boundaries in the field,locate those
in space and draw them after reduction on the map.

The units inside the boundaries are single stands, that is to
say the basic material on which the classification will be built.
there is no need for spatial guiding principles here either.
The map can be described with the aid of the units of the general systems. In general these maps have a scale between 1:100
and 1:10 000. Usually however, the situation changes even with
mapping scales smaller (less detailed) than 1:5000.

In map III only mosaics and higher complexes can be mapped.The
purely homogeneous stands representing plant communities characterized by floristic or life-form-composition and vertical
or micro-horizontal structure, are too small to be mapped on
this scale.Within one mosaic or complex plant communities may
occur that are morphometrical and/or ecologically not related.
(Compare a strong relief where acid leached dry sand vegetations
may alternate with wet, saline or boggy depressions). Neverthe-

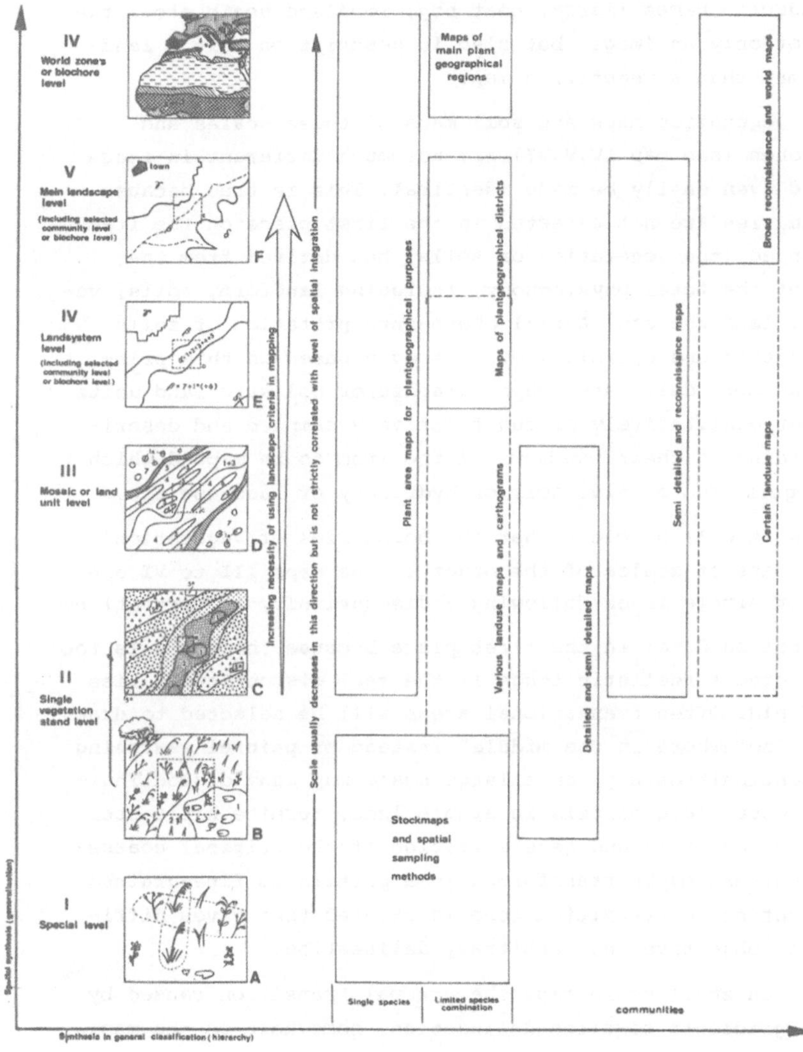

gether they form a specific landscape. In different mosaics the
same vegetation, associations of the general system may take
part. The guiding principle of spatial relationship is at the

same time a "Landscape" or chorological   guiding principle.

The legend unit can and should be described with the help of the
general classification. If it is done (as is often the case)
with landscape names (marsh, peat bog, woodland heath etc.) the
map is not only an image, but also in description more a land
(scape) map than a vegetation map.

In fact, vegetation maps and soil maps of these scales and
smaller ones (see map IV,V,VI) are not much different in image
and could even easily be made identical. This is true because
the boundaries are not selected in the first place on the item
to be mapped (the vegetation or soils) but derived from the
pattern of the total physiognomy, including landform, soils, vegetation, land use etc. Aerial photo-interpretation of soils
and vegetation and hydrology etc.,is even based on this principle. The thus delineated physiographic or holistic land units
are either qualitatively or quantitatively sampled and described according to their "content" of the item to be mapped,which
may be vegetation, or also soil or hydrology or land-use etc.

From this it will be clear that the boundaries of vegetation
units on maps of scales of the order of the maps III to VI are
surely not simple lines following a discontinuity in vegetation.

This cannot be done; in the first place because the scale is too
small to give a realistic image if the real discontinuity lies
in the field. Often transitional areas will be selected to draw
the line "somewhere in the middle" instead of painful searching
for discontinuities.E.g. on a large scale map many discontinuities may occur (e.g.parcels in arable land, termite hills etc.)
that after reduction and generalization of the original coarse-
grained pattern maybe transformed in a pattern so fine-grained
that a continuous transition zone is created that gives difficulties in objective, non-arbitrary delineation.

Moreover, on small scale maps the gradual transition caused by
climatic gradients requires delineations somewhere on the map.
Such a delineation is usually fully arbitrary as far as the general classification is concerned. As soon however as this general system has been established, the line in the field is fixed. This may often not be a discontinuous line. It may divide
a smaller or larger rather homogeneous forest or prairie area

into two, not according to discontinuity, but according to the description in the classification systems that prescribes the boundary in abundance or frequency of species, forms and/or structures. Vegetation scientists with an interest in vegetation survey (mapping) have for a long time been trying to design a classification system that could simultaneously be used as general system and mapping legend.

This wish is still widely found. One often hears that a general system must be constructed in such a way that the orders of high hierarchical rank may coincide with legend units of high hierarchical rank. The logical results of such trials is that climate influencing properties of vegetation are used as characteristics at a high level. This is not always very useful for other purposes.[x]

One may even suppose that the succes of the monoclimax theory (that has proved to be not true) was partly caused by a kind of wishful thinking on a communal main hierarchy in general classification, legend classification and vegetation dynamics. Indeed if the monoclimax theory were true, the ultimate stage of succession would also be the main unit in the small-scale map legend, and also the most highly developed classification unit in the general system.

One avoids many problems if one recognizes that a general system using morphometric characteristics is different in principle from a legend that must use spatial, chorological guiding principles. For each map one should design a separate legend. A general classification system can be used to describe the units and will also be helpful in comparison with other maps.

A special problem is formed by the shifting boundaries (see KÜCHLER 1967). These have the same character in relation to abstraction and concreteness as the not-so-quickly shifting ones. Only the cartographic representation forms a problem - for this see KÜCHLER 1967.

---

[x])Soil classification systems have the same trouble. The damage done by the uncritical application of the Davisian peneplain theory can be felt here, the zonal soil concept and monoclimax theory.

<u>ZUSAMMENFASSUNG</u>

Der Vegetationskartierer hat es mit drei verschiedenen Arten von
Grenzen zu tun:1. Konkrete Grenzen (Diskontinuitäten), 2. Abstrakte Grenzen in der Klassifikation und 3. Kartografische Grenzen.

In der Vegetation bestehen k o n k r e t e   D i s k o n t i -
n u i t ä t e n, die man im Gelände klar beobachten kann. Doch
existieren auch Stellen, wo die Vegetation sich allmählich in
horizontaler Richtung verändert.

Die Ursache einer scharfen konkreten, oder einer diffusen Grenze
ist immer das Resultat der Wirkung des Umweltfaktoren-Komplexes
im Zusammenhang mit endogenen Faktoren. Dieses Resultat ist für
jeden Punkt der Erdoberfläche einmalig. Dies gilt ebenfalls für
jeden Vegetations bestand. Vegetationsbestände, die zu ein und
derselben Klassifikationseinheit gerechnet werden, haben keine
"Blutsverwandschaft" wie es z.B. bei Pflanzen oder Tier-Indivi-
duen der Fall ist, die einer und derselben Art angehören. Sie
sind einmalige Dinge und durch die aus der Flora resultierenden
Umwelt-Bedingungen determiniert.Allein die Flora einer Gegend,
also der Baustein einer Vegetation, hat einen "geerbten Charakter. Die Vegetation an sich jedoch entbehrt ihn.

Aus diesem Grunde haben konkrete Diskontinuitäten zwischen Vegetationsbeständen einen vollkommen anderen Charakter als die
konkrete Diskontinuität individueller Pflanzen. Für diese ist
die Diskontinuität mit dem Boden, dem Wasser oder der sie umge-
benden Luft ein Beweis ihrer Individualität! Für die Vegetation
ist die Diskontinuität nichts anderes als eine Angabe, daß mindestens einer der determinierenden Umweltfaktoren einen Sprung
macht. Im Übergangsgebiet zwischen zwei Vegetationseinheiten
kann es an verschiedenen Stellen Sprünge geben, die Diskontinuitäten zur Folge haben. Es wäre aber nur ein Zufall, wenn die Diskontinuität immer an der gleichen Stelle in jeder konkreten Übergangszone vorkäme. Unter der Voraussetzung, daß man nach einer
statistischen Bearbeitung (z.B. die Tabellen-Methode BRAUN-BLAN-
QUET's oder Verwandschaftsanalysen verschiedener Art) von groß-
und kleinflächigen Vegetationsbeständen gleiche Anzahlen von
Mustern verarbeiten würde, könnte man sich vorstellen, daß es
in den Grenzen zwischen den Einheiten so viele Unterschiede gibt,
so daß man ein Kontinuum bekäme. In Abb.2 ist eine schematische

Darstellung eines konkreten Zustandes und eines hiervon abgeleiteten abstrakten Kontinuums gegeben. Der konkrete Zustand
weist viele Diskontinuitäten auf. In beiden Gradienten (schwarz-
weiß und Kreis-Dreieck) sind aber alle Übergänge anwesend. Eine
Einordnung in ein System, laut welchem die meisten verwandten
Muster beieinander und die mindest verwandten so weit als möglich voneinander entfernt sind, hat ein absolutes zweidimensi-
onales Kontinuum zur Folge; das heißt, daß alle Grenzen in Abstraktion diffus oder kontinuierlich geworden sind.

Es ist schwer zu beweisen, ob die ganze Vegetation ein solches
Kontinuum bildet. Man müßte alle vorkommenden Strukturen und
sämtliche Arten-Kombinationen auf der Welt auf die gleiche Weise bemustern und dann mathematisch-statistische verarbeiten.
Auf der Welt besteht kein menschliches Wesen und keine Maschi-
ne, die im Stande wären, dies zu tun. Falls man nicht alle Ve-
getationsunterschiede bemustert, erhält man auf jeden Fall ein
Diskontinuum (beim Versuch nur einen Teil der Figur 2 einzuordnen, fehlen dann viele Übergänge).Außerdem sollte dann das
Problem der mathematischen Methode völlig gelöst sein, was bis
jetzt noch nicht gelungen ist. Man kann aber voraussetzen, daß
es im allgemeinen in der Abstraktion Kontinua und weniger klare Diskontinuitäten gibt (siehe GOODALL,CURTIS,van der MAAREL
usw.). Es ist gefährlich, zwischen "ORDINIEREN" und "KLASSIFI-
ZIEREN" einen Unterschied zu machen. Das sogenannte "ORDINIE-
REN" (to ordinate) (GOODALL 1954) ist prinzipiell nichts neues,
sondern nur die möglichst objektive Ordnung, die jeder Klassifikation vorausgehen sollte. To ordinate ist ein Begriff, der
außerhalb der Vegetationskunde eine ganz spezifische Bedeutung
hat[x]); vielmehr sollte der herkömmliche Ausdruck: "ordnen"
beibehalten bleiben. Abstrakte Klassifikationseinheiten, die
durch das "In Stücke schneiden" eines Kontinuums entstehen, verhalten sich zueinander wie die Sitzordnungen in einem Kino oder
Theater. Kategorien haben einen bestimmten Wert und in einem
Kontinuum mit graduell verlaufendem Abstand zur Bühne sind die
Grenzen willkürlich gezogen, da es unmöglich ist, für jeden
Sitz einen eigenen Tarif festzulegen. Der Endeffekt ist dann,
daß zwischen den schlechtesten Sitzen einer teueren und den
besten Sitzen einer billigeren Kategorie nur ein kleiner Un-

[x]) z.B.,wenn sich jemand zu einer religiösen Ordnung,einem
Orden bekennt.

Unterschied besteht. Innerhalb einer und derselben Kategorie
finden wir allerdings einen größeren Unterschied.

Es ist sehr wahrscheinlich, daß in einem Weltsystem mit floristischen Merkmalen diskontinuieliche Grenzen vorkommen (z.B.
zwischen Kontinenten). Allem Anschein nach, ist jedoch der kontinuierliche Übergang allgemeiner, doch darf dies kein Anlaß
dafür sein, nicht klassifizieren zu können. Klassifikation hat
mit kontinuierlichen oder diskontinuierlichen Grenzen nichts zu
tun. Man benötigt Eigenschaften, um sie als Merkmale zu nutzen,
man braucht Leitfäden, um die Merkmale aus der großen Gruppe
der Eigenschaften auszuwählen, und die Hierarchien bestimmen
zu können. Wenn es in der Abstraktion keine Diskontinuitäten
gibt, hat man noch Leitfäden nötig (Normen, die man arbiträr
feststellen kann), um den Ort und die Menge der anzunehmenden
Grenzen bestimmen zu können. Je lokaler man arbeitet, desto
einfacher ist das Problem, da eine lokale Ordnung von Daten beinahe ohne Ausnahme eine genügende Anzahl von Diskontinuitäten
liefert.

<u>KARTOGRAFISCHE GRENZEN</u> sind konkrete Linien auf einem Stück Pa-
pier, das eine Widerspiegelung von zwei Vorgängen gibt: 1. das
Nachzeichnen einer konkreten Diskontinuität im Gelände und 2.
das Nachzeichnen einer im Gelände eingeschätzten künstlichen
Linie, die durch eine abstrakte Grenze in einer Klassifikation
determiniert ist (allgemeine Typologie oder eventuell eine Legende).
Beide Arten kartografischer Grenzen finden wir auf Karten mit
großem und kleinem Maßstab. Je kleiner dieser Maßstab ist, desto weniger kann das Nachzeichnen der topographischen Grenzen
auch Grenze zwischen den einzelnen Vegetationsbeständen sein.
Bei einem Maßstab 1: 20 000 lassen sich nur noch Vegetations-
komplexe kartieren. Grenzen zwischen den einzelnen Kartenein-
heiten sind Grenzen zwischen Großgebieten und folgen meistens
(speziell bei Luftaufnahmen) Landschaftsgrenzen, die nicht durch
Vegetationseinheiten, sondern durch andere, die Landschaft prä-
gende Elemente, bedingt sind. Bei sehr flachem Land, Steppen
oder tropischen Wäldern, kann auch bei dicsem Maßstabe eine
durch klimatisch bedingte Unterschiede zusammenhängende Grenz-
linie völlig arbiträr sein. Man wird aber versuchen, das Kartenbild so einfach wie möglich zu halten, damit solche Grenzen mög-

möglichst mit realen (z.B. edaphisch bedingten) Diskontinuitäten
zusammenfallen.

SUMMARY AND CONCLUSIONS

C o n c r e t e   b o u n d a r i e s   are discontinuities in the
field. In less discontinued transitions between vegetation
stands, even if a fully gliding gradual gradient exists, boundaries can be delineated on the basis of a general (abstact)
classification system.

Discontinuities in the vegetation in the field are usually caused by
a discontinuity in an environmental factor in combination with
the endogenic factors of the vegetation. Relief (landform) often causes abrupt differences in hydrology or climate, and related with both, in soils, especially in more detailed study
one will also discover discontinuities caused purely by (endogenic) vegetational forces. Especially in coarse-grained structures (often occuring in young vegetations which are not well
balanced) the type of growth and expansion of certain species
growing in typical herds or aggregates, causes local discontinuities. In no place of the world is the interaction of all environmental and endogenic factors completely indentical. In each
place the environment is being "made" by local unique powers.
There is not such a thing as a complete set of hereditary properties is the case with living indivuduals which can be classified in species. These similar individuals are connected with genetic lines; They have a true "blood relation". This is not so
with vegetation stands. Any individual stand has been formed by
the local environmental conditions which have selected a certain number of taxa from the available flora. Only these flora
have some hereditary aspects.

Therefore discontinuities may occur on different positions within the species combination and not necessarily always in an indentical situation.

A b s t r a c t   b o u n d a r i e s   are the delineations in a
classification system. They are in principle not related to concrete boundaries in the field. Only few worldwide environmental
boundary conditions with an extreme character maybe found back
in abstraction (differences between continents). Usually, however, abstract boundaries are most probably continuous.

This means that parts of units close to the boundary may be more
related to parts of another unit than with a remote part of their
own unit.

It also means that there is no "natural" boundary. For each region or area the boundary may be different if one tries to get
as close a correlation as possible with the landscape units
(discontinuities in the landscape).

If one starts from the point of view that there is no discontinuity in the abstraction, whereas one does appear to exist, no
harm is done. In the reverse, however, difficulties may arise.
An absolute proof of the existence or non-existence of a continuum is very hard to give. The only certainty is that there are
rather discontinuous transitions between certain units and the
rest of vegetation, but that many parts show continuous transitions.

It is probable that the latter case dominates. Mathematical research in this respect should take samples in such a way that
the average of the individual stands does not disturb the relationships. This means that an equal number of samples should be
taken out of each group of similar vegetation stands. If this is
not done, the vegetation type that is correlated with dominating
environmental conditions would be treated differently in relation to vegetation covers in places with rare combinations of environmental factors.

Cartographic boundaries usually will coincide with the place
of discontinuities in the field. However, in the legend they can
be described (and therefore prescribed) by the general (abstract)
classification. Therefore they may also fall somewhere halfway
the gradient in a concrete continuum. Legends are also abstract
classifications that differ only in one respect of the general
typological classification, in that they have to take the spatial relationship as main guiding principle.

In the higher cathegories of a legend, units are often combined
which have no morphometric (nor ecological) relationship. They
are connected because of the mere fact that they occur together
in such an intricate or (in relation to the mapping scale) such
a fine-grained pattern that they cannot be separated cartogra-
phically. Guiding principles for the way of combination of these
units have a spatial character. Landscape criteria are used.

On small scale maps the delineations between patterns which are
always discontinuities are used as cartographic boundaries.

Any line on a map may represent either a rather   continuous
transition, or a discontinuity in the field. Map users should
be made aware of this fact, which is not always understood.

C i t e d   l i t e r a t u r e

CURTIS,J.T.   -1959-   The vegetation of Wisconsin,an ordination
    of plant communities. University of Wisconsin Press. Madison.

ELLENBERG,H.   -1956-   Aufgaben und Methoden der Vegetations-
    kunde. In: Walter,H.: Einführung in die Phytologie 4(1).
    Stuttgart.

GOODALL,D.W.   -1954- Objective methods for the classification
    of vegetation.- Australian J.2 (2):304-324.   Melbourne.

GUINOCHET,M   -1955- Logique et dynamique du peuplement vége-
    tal.- Paris.

KÜCHLER,A.W.   -1967- The unstable climax and some aspects of
    mapping it.- Internationales Symposion über Fragen der Ge-
    sellschafts-Entwicklung (Syndynamik). Rinteln. (im Druck).

MAAREL,E.v.d.   -1966- Over vegetatiestructuren, -relaties en
    -systemen in het bijzonder in de duingraslanden van Voorne.
    (On vegetational structures, relations and systems, with
    special reference to the dune grasslands of Voorne (Nether-
    lands).Diss. Groningen.

SCHARFETTER,R.   -1928- Die kartographische Darstellung der
    Pflanzengesellschaften.- Abderhalden's Handb.Biol.Arbmethod.
    11 (4):77-164.   Berlin.

de VRIES,D.M.   -1953- Constellation of frequent herbage plants.
    Based on their correlation in occurrence.   Berlin,Wien.

WESTHOFF,V.   -1968- Stufen und Formen von Vegetationsgrenzen
    und ihre methodische Annäherung.- Dieser Band,p.44.   Rinteln.

ZONNEVELD,I.S.,BANNINK,J.F.& LEYS,H.N.   -1973- Vegetatie, gro-
    eiplaats en boniteit in Nederlandse naaldhoutbossen.
    (Vegetation habitat and site class in Dutch conifer fo-
    rests.)- Bodenkundige studies 9.
-- -- --   -i.p.- Akkeronkruidgezellschappen als indicators of
    the environment. (Arable weed communities as indicators
    of the environment).- Netherlands Soil Survey Institute,
    Wageningen (preparation).

E.van der MAAREL:

Ich glaube, Sie haben keine Bedenken gegen das Wort Klassifikation. Aber es gibt auch das Wort Klassierung, und es besteht
derselbe Unterschied zwischen "klassieren" und "klassifizieren"
wie zwischen "ordnen" und "ordinieren". Also Ordination und
auch Klassifikation sind Techniken, und Klassieren und Ordnen
sind allgemeine Methoden. Wenn wir es so sehen, dann gibt es
gar keine Schwierigkeiten mit dem Begriff Ordination.

I.ZONNEVELD:

Das ist eine sprachliche Sache, auf die wir hier nicht eingehen
wollen, denn im Englischen gibt es kein Äquivalent für Klassieren. Sind Sie mit mir einig, daß kein fundamentaler Unterschied
besteht?

E.van der MAAREL:

Nein, es ist Methodik gegenüber Technik. Aber was sagen die
Engländer?

J.J.MOORE:

Ich möchte nur sagen, daß bei uns Ordination ganz eingebürgert
ist. Wir verstehen alle, was mit Ordination gemeint ist.

STUFEN UND FORMEN VON VEGETATIONSGRENZEN
UND IHRE METHODISCHE ANNÄHERUNG

V. W e s t h o f f

Der Vortrag enthält eine Übersicht der verschiedenen Stufen und
Formen von Vegetationsgrenzen, also eine Darstellung, die als
Einführung zum Gesamtthema zu betrachten wäre. Es werden kaum
neue Tatsachen und keine neue Methoden vorgeführt; es ist nur
versucht worden, schon bekanntes in einem Gesamtrahmen zu fassen.

Der Ausdruck "Vegetationsgrenzen" kann sich auf zwei wesentlich
verschiedene Begriffe beziehen, und zwar einmal auf konkrete,
andererseits aber auf abstrakte Grenzen. Konkrete Vegetations-
grenzen sind im Gelände gegebene, topographisch nachweisbare
Grenzen zwischen Beständen oder Bestandeskomplexen beliebigen
Ranges.Mit abstrakten Grenzen dagegen sind syntaxonomische Grenzen zwischen Vegetationseinheiten gemeint. Man kann sich z.B.
fragen, wo die Grenze zwischen dem Phragmition-Verband und dem
Magnocaricion-Verband (bzw.den Ordnungen Phragmitetalia und
Magnocaricetalia) zu setzen sei. Ein Beispiel einer Assoziation
die beiden Verbänden zugeordnet werden könnte, ist das Marisce-
tum serrati Zobrist 1935, also die Assoziation von Cladium
mariscus, die fast nur aus dieser stark dominierenden Art besteht. Diese Grenze ist also in Schwankung begriffen. Das Problem hat jedoch mit der konkreten Grenze eines C l a d i u m-
Bestandes nichts zu tun: diese tritt im Gelände meistens eben
besonders scharf und klar hervor.

Obwohl abstrakte und konkrete Vegetationsgrenzen somit wesentlich verschieden sind, gibt es dennoch einen Zwischentypus,
also einen Grenzfall im doppelten Sinne, der zu beiden Kategorien gestellt werden kann. Das ist der Fall, wo zwei vikariierende Assoziationen im Gelände aneinander grenzen, d.h.ineinander
übergehen und besonders, wo es Assoziationen betrifft, die sich
durch ein zusammenhängendes, langes, streifenförmiges Areal
kennzeichnen, wie die Küstendünen-Assoziationen des Ammophilion
borealis. Die Grenze zwischen zwei einander geographisch erset-
zenden Gesellschaften ist hier sowohl eine abstrakte syntaxonomische zwischen Typen wie eine synchorologische Grenze zwischen

langgestreckten Beständen. Wir werden auf diesen Fall noch zurückkommen.

Zuerst sei nun darauf hingewiesen, daß es Forscher gibt, hauptsächlich aus der Wisconsin-Schule, welche die Existenz von Vegetationsgrenzen und besonders von abstrakten Grenzen schlechthin leugnen (CURTIS 1959, CURTIS & McINTOSH 1951, McINTOSH 1967).
In Zusammenhang damit stellen sie auch die Möglichkeit einer
Klassifikation der Vegetation in Abrede. Sie betrachten die Ge-
samtvegetation als ein "continuum", und meinen, man könne da nur
zu einer gewissen Ordnung kommen, indem man die allmähliche
räumliche Variation der Vegetation in Zusammenhang mit dem Um-
welt-Gradienten in einem räumlichen oder mathematischen Modell
wiedergibt. Diesen Vorgang nennt man Ordination. Es ist keinesfalls zu leugnen, daß es derartige allmähliche Übergänge vieler-
orts gibt, und daß diese auch in der BRAUN-BLANQUET-Schule nicht
unbeachtet geblieben sind. Schon vor einem Vierteljahrhundert
(WESTHOFF 1947) wurde für diese Übergangsbereiche der Terminus
"cline" vorgeschlagen, nach Analogie des von HUXLEY (1938) ein-
geführten Begriffs "cline" für eine Sippenreihe, deren Merkmale
allmählich von einem Typus zum anderen überleiten. MEYER DREES
(1951) hat diesen Begriff übernommen und verschiedene Fälle unterschieden, wie "topocline" oder "climacline" für einen vom
Großklima bedingten allmählichen Übergang, z.B.den zwischen zwei
vikariierenden Gebietsassoziationen,weiterhin "ecocline" für örtliche, edaphisch bedingte allmähliche Vegetationsübergänge. Will
man den pflanzensoziologischen Kline-Begriff von dem taxonomi-
schen trennen, so kann man den ersten als "Coenokline" bezeichnen.

Die Existenz derartiger klinaler Variation ist jedoch nicht eine
so universelle Erscheinung, daß die von CURTIS c.s. vertretene
Auffassung, es seien in der Vegetation nur Kontinua da,darum berechtigt wäre. Ordination und Klassifikation sind beide sinnvoll,
sei es, daß in einem Falle diese, im anderen Fall jene methodische Annäherung am meisten angebracht sei (WHITTAKER 1962,1972;
VAN DER MAAREL 1969; VAN DER MAAREL,WESTHOFF & VAN LEEUWEN 1964).

Soweit ich unterrichtet bin, wird Dr.ZONNEVELD sich hier besonders über abstrakte Vegetationsgrenzen äußern. Wir werden uns
daher nun weiterhin mit konkreten Vegetationsgrenzen beschäftigen.

Man kann vor allem räumliche und zeitliche Grenzen unterscheiden, oder besser gesagt, Grenzen im Raum und in der Zeit.Gren-
zen in der Zeit begegnen wir z.B. in der Periodizität, der Fluk-
tuation und der Sukzession. Diese Grenzen in der Zeit gehörten
dem Thema des letztjährigen Symposion an. Wir werden uns weiterhin mit Grenzen im Raum befassen, sei es, daß man sich zu ver-
gegenwärtigen hat, daß es zwischen Raum- und Zeit-Qualitäten
Wechselbeziehungen gibt, sich äußernd im Unterschied zwischen
stabilen und unstabilen Grenzbereichen. Wir kommen darauf noch
zurück.

Eine Grenze ist kaum je eine scharfe Linie. Meistens findet sich
ein schmalerer oder breiterer Grenzstreifen, also ein Grenzbereich vor. Wie man diesen erfaßt, sei in erster Linie an einem
pflanzengeographischen Beispiel, und zwar dem Grenzbereich zwischen Florenregionen oder Florenprovinzen, dargestellt.Wenn man
z.B. die klassische Reise vom kühlen Norden nach dem Mittelmeer
unternimmt, und mit konstanter Geschwindigkeit fährt, passiert
man innerhalb einer gewissen Zeit Arealgrenzen einer Anzahl von
Arten. Diese Grenzen können pro Zeiteinheit mehr oder weniger
zahlreich sein. Anders ausgedrückt: innerhalb eines gewissen Ab-
standes, z.B.100 km., gibt es mehr oder weniger Arealgrenzen.
Die Zahl der Arealgrenzen der Arten nennt man Florengefälle. Der
Bereich innerhalb dessen dieses Gefälle deutlich ansteigt, nennen wir den Grenzbereich zwischen eurosibirischer und mediterraner Region.

Wenn wir von diesem pflanzengeographischen, also floristischen
Beispiel auf Vegetationsgrenzen übergehen,sei zuerst auf einen
grundsätzlichen Unterschied hingewiesen. Die arealgeographischen
Grenzen sind nicht konkret in dem Sinne, in dem Vegetationsgrenzen konkret sind. Sie beziehen sich nicht auf Pflanzen-Massen,
sondern auf Arten ohne Rücksicht auf die räumliche Verteilung
der dazu gehörigen Individuen. Arten sind jedoch typologische
Abstrakta. Es gibt keine konkreten Grenzen zwischen Arten, nur
zwischen Individuen oder Beständen. Nichtsdestoweniger sind die
floristischen Grenzen keinesfalls abstrakt: sie sind chorologischen Charakters und in der naturräumlichen Wirklichkeit gegeben. Man kann diesen Sachverhalt erkenntnistheoretisch so aus-
drücken, daß jene floristischen Grenzen zwar _real_ sind, jedoch
nicht _konkret_. Dies läßt sich auch folgendermaßen formulieren:

Florengrenzen sind eindimensional, Vegetationsgrenzen dagegen
dreidimensional. Diese entsprechen dem Übergangsbereich zwischen
Raum-Qualitäten.

Nach Analogie mit dem Fall der floristischen Grenzen kann man
daher sagen, daß eine Vegetationsgrenze vorliegt, wo sich gewisse qualitative oder quantitative Änderungen räumlich häufen,
also dort, wo das Gefälle gewisser Änderungen stärker ist als
beiderseits dieses Bereiches. Das Gefälle kann sich dabei jedoch
auf verschiedene Kriterien beziehen, und es ist bei jeder Forschung angebracht und notwendig, diese Kriterien zu unterscheiden und sie nicht zu vermischen. Es kann sich z.B. handeln um:
1. ein Gefälle der Abundanz der Individuen einer Population
einer gewissen Art; 2. ein Gefälle im Lebensformen-Spektrum,
also ein Ansteigen der Abundanz der zu bestimmten Lebensformen
gehörigen Individuen zusammen mit einer Verringerung der Abundanz der zu anderen Lebensformen gehörigen Individuen; 3. ein
Gefälle der örtlichen Arealgrenzen, in dem Sinne, daß in kurzer
Strecke eine relativ große Anzahl Arten verschwindet, indem relativ viele neue Arten hinzutreten. Den letzten Grenztypus meinen wir meistens, wenn wir von Vegetationsgrenzen im allgemeinen
reden. In allen Fällen ist das Gemeinsame eine Diskontinuität.

Es ergibt sich nun zuerst der Unterschied zwischen senkrechter
und waagerechter Blickrichtung. In senkrechter Richtung ist die
Vegetationsstruktur von der Schichtung bedingt. Die einzelnen
Schichten können mit Synusien zusammentreffen, jedoch ist dies
nicht unbedingt der Fall. Die Grenzen, oder besser Grenzflächen,
zwischen den einzelnen Schichten können ziemlich scharf und klar
sein, wie in europäischen Betriebsforsten oder in den borealen
und subarktischen Nadel- und Birkenwäldern; sie können aber auch
allmählich, verwickelt und unscharf sein, wie im tropischen Regenwald (RICHARDS 1952, CAIN & CASTRO 1959). Man braucht sich
daher nicht zu wundern, daß die Grenzen zwischen den Schichten
besonders im kühlen Norden zu jener methodischen Annäherung geführt haben, wobei die einzelnen Schichten als die Grundeinhei-
ten der Vegetation betrachtet worden sind. Die Schichtungsein-
heiten sind von LIPPMAA (1935) Unionen genannt worden, von
DU RIETZ und Mitarbeitern Sozietäten und später Konsozionen
(DU RIETZ 1930). Im Prinzip wird dabei so vorgegangen: eine
Sozietät ist eine als Gesellschaft betrachtete Schicht, in

welcher eine einzige Art dominiert. Gibt es mehrere Schichten,
deren jede von einer Art beherrscht wird, dann werden diese Sozietäten in senkrechter Richtung zu einer Soziation zusammengestellt. Ein einfaches Beispiel ist die Pinus sylvestris-Vacci-
nium myrtillus-Pleurozium schreberi-Soziation, die Kiefernforst-
gesellschaft auf trockenen nährstoffarmen Sanden. Ist nur eine
Schicht da, dann fallen Sozietät und Soziation zusammen. In der
nächsthöheren Einheit, der Konsoziation, braucht nur in einer
der Schichten eine einzige Art zu dominieren; die anderen Schichten bestehen dann, jede für sich, aus einem Mosaik-Komplex verschiedener Sozietäten. Das klassische westeuropäische Beispiel
einer Konsoziation ist das von SCHEYGROND (1931) genau studier-
te Sphagnum-reiche Schilfgelände der holländischen Brackwasser-
Sümpfe: Die Phragmites-Schicht ist hier die verbindende Schicht,
indem in der Krautschicht sowohl wie in der Moosschicht verschiedene Dominanten mosaikartig abwechseln, in der Moosschicht z.B.
Sphagnum palustre, S.plumulosum und Dicranum bonjeanii. Dieser
Gesellschaftstyp ist von senkrechten, sowohl wie wagerechten,
Grenzflächen klar strukturell gegliedert, und die methodische
Annäherung der Vegetationsgrenzen nach der skandinavischen Methodik liegt in diesem Fall auf der Hand. Allerdings kann man,
wie NORDHAGEN(1937) vorgeschlagen hat, solche Konsoziationen
auch in das BRAUN-BLANQUET-System einordnen, wie es für die
obenerwähnte Phragmites-Sphagnum-Konsoziation auch gemacht worden ist (WESTHOFF 1949).

Öfter werden die Schichtungsgrenzen mit Synusien-Grenzen gleich-
gesetzt oder mit diesen verwirrt. Wie schon gesagt, brauchen
diese und jene aber nicht zusammenzutreffen; sie sind grundsätzlich verschieden. Unter Synusie versteht man einen aus
ähnlichen Lebensformen oder einer einzigen Lebensform-Gruppe
zusammengesetzten Verein. Als MIYAWAKI & J.TÜXEN (1960) die
Lemna-Gesellschaften des Süßwassers als Lemnetea von den Pota-
metea trennten, lag nicht so sehr eine Schichtungseinheit, als
vielmehr eine Synusie vor. In derselben Schicht wie die Lemni-
den finden sich ja auch die Schwimmblätter der Nymphaeiden (wie
Polygonum amphibium fo.natans und gewisser Laichkräuter wie
Potamogeton natans), die man in den Lemnetea-Aufnahmen nicht
unbedingt zu berücksichtigen braucht. Wenn DEN HARTOG & SEGAL
(1964) die von MIYAWAKI & J.TÜXEN angefangene Aufspaltung der
Wasserpflanzen-Gesellschaften methodisch konsequent weiterführ-

weiterführten und zu Klassen wie Ceratophylletea,Utricularietea
und Stratiotetea kommen, sollte man dieses System nicht als ein
Schichtungssystem im Sinne von LIPPMAA und DU RIETZ betrachten,
sondern vielmehr als ein synusiales System, wie DEN HARTOG und
SEGAL auch selbst betonen. Es ist hier nicht beabsichtigt, den
Wert oder die Brauchbarkeit des letzterwähnten Systems zur Diskussion zu stellen, da dieses nicht zu unserem Thema gehört,son-
dern nur zu zeigen, welcher Typ von Vegetationsgrenzen in diesem Fall als diagnostisches Merkmal angewandt worden ist.

Auch bei gewissen terrestrischen Phanerogamen-Gesellschaften
ist das synusiale Kriterium in unserer Methodik angewandt worden, und zwar bei der Aufstellung der Schleier-Gesellschaften
des Calystegion sepium von TÜXEN (1950). Analytisch sind diese
Gesellschaften allerdings schwer zu fassen, und dieses ist hauptsächlich ein Problem der Vegetationsgrenze. Es ist nämlich methodisch besonders schwer, die Lianen von den anderen Arten,die
mit ihnen im gleichen Substrat wurzeln, zu trennen. Eine objektive Vegetationsgrenze liegt da kaum vor. Wenn man, wie man eigentlich vorgehen soll, den ganzen Artenbestand in die Aufnahme
einbezieht, werden die Kennarten anderer Gesellschaften, wie z.B.
die der Molinietalia oder des Filipendulion, normalerweise über
die Schleierpflanzen überwiegen. Wenn man jene Arten aber außer
Betracht lassen will, ist der unscharfen räumlichen Abgrenzung
wegen eine unerwünschte Subjektivität wohl kaum zu vermeiden.

Wir werden uns jetzt der waagerechten Blickrichtung zuwenden,
also die Vegetation aus der Vogelschau betrachten: wir fragen
uns also, welche Abstufungen konkreter räumlicher Grenzen in
jenem horizontalen Gefüge vorliegen, das auf englisch "pattern",
auf niederländisch "patroon" heißt, von SCHMITHÜSEN (1961) als
Anordnungsmuster bezeichnet worden ist. Wenn wir von der kleinsten Einheit zu immer größeren fortschreiten, sind dabei zu unterscheiden: die Aggregationsgrenze, die Bestandesgrenze, die
Assoziationsgrenze, die Physiochorengrenze oder Fliesengrenze,
die Grenze des Wuchsdistriktes und die damit zusammenhängende
Grenze des Gesellschaftspulks, die Formationsgrenze und die Bi-
omgrenze. Die letzte wird von DANSEREAU (1957) Biochoren-Grenze
genannt, wobei er unter Biochoren die vier Haupteinheiten Wüste,
Steppe, Savanne und Wald versteht. Da aber SCHMITHÜSEN(1961) das
Wort Biochore in anderem Sinne verwendet, werden wir diesen Terminus weiterhin vermeiden.

Die Vegetationsgrenzen im kleinsten Raum sind die Grenzen der
einartigen Bestände, die man mit BRAUN-BLANQUET (1928) Populationen nennen kann. Wegen des Mißverständnisses mit Bezug auf
den genetischen Populationsbegriff ist aber der Terminus Aggregation, von WEAVER & CLEMENTS (1938) geprägt, vorzuziehen. Ag-
gregationsgrenzen, also Grenzen einartiger Bestände, treten besonders scharf hervor in Pionier-Gesellschaften und in gestör-
ten, unstabilen Gesellschaften. Einige Beispiele: Bestände von
Calla palustris neben solchen von Carex paniculata im Verlan-
dungsgürtel eutropher Seen; Bestände von Sedum acre neben solchen von Echium vulgare in einem von Kaninchen überweideten
kalkreichem Dünengelände; Bestand von Asteriscus maritimus an
mediterranen Felsenküsten.

Weiterhin begegnet uns starke und auffällige Aggregation im
Hochmoor. In älteren, stabileren Gesellschaften sind die Grenzen der Aggregationen jedoch meistens unscharf und allmählich;
Aggregation kann da sogar ganz fehlen, wie z.B. in alten subalpinen Mähwiesen.

Es ist öfter betont worden, daß Instabilität der Umweltverhält-
nisse Aggregation fördert, indem diese in stabiler Lage zurück-
gedrängt wird, u.A. von RAMENSKY (1930), MARGALEF (1968), besonders in der Relationstheorie von VAN LEEUWEN (1966,1970);s.a.
SHIMWELL (1971), und unabhängig vom Letzteren von DI CASTRI &
COVARRUBIAS (1966). Wir werden darauf noch zurückkommen. Es
sei nun darauf hingewiesen, daß die allmähliche Verfeinerung
unserer Arbeitsweise dahin geführt hat, daß die Aggregationsgrenze in der syntaxonomischen Diagnostik immer größeres Gewicht
erlangt hat (WESTHOFF 1967,1968). Wesentlich ist dabei der Maßstab der Beobachtung und, im Zusammenhang damit, die Wahl der
Probefläche. Als Beispiel seien die Pionier-Gesellschaften der
nw-europäischen Salzwiesen erwähnt, wo die sommerannuelle Salicornia stricta und der Geophyt Spartina townsendii in Herden zu-
sammenwachsen. Ursprünglich wurde dieser Komplex zweier Aggregationen als eine einzige Assoziation beschrieben, und zwar unter den Namen S a l i c o r n i o - S p a r t i n e t u m (BRAUN-
BLANQUET & DE LEEUW 1936), hauptsächlich aus dem Grunde, weil
die beiden Arten auf demselben Standort leben. TÜXEN & OBERDOR-
FER (1958) haben aber mit Recht die Grundverschiedenheit der
beiden Teile dieses Komplexes betont. Wenn man jedoch die beiden
aus einer einzigen Art zusammengesetzten Herden als verschiedene

Gesellschaften betrachtet, stellt sich heraus, daß sie meistens
keine einzige Art gemeinsam haben; dieses hat zu der logischen
Schlußfolgerung geführt, daß sie in zwei Klassen aufgeteilt werden sollen, und zwar die T h e r o - S a l i c o r n i e t e a
R.Tx.1954 und die S p a r t i n e t e a   R.Tx.1961.

Wenn man mit demselben Kriterium z.B. die Mosaikstruktur des Ver-
landungsgürtels der eutrophen stehenden Gewässer, also des
S c i r p o - P h r a g m i t e t u m   W.Koch 1926 betrachtet,
stellt sich heraus, daß diese Assoziation meistens aus Aggregationen drei verschiedener Arten zusammengesetzt ist, und zwar
aus Herden von Scirpus lacustris bzw.Phragmites australis oder
Typha angustifolia. Es soll dazu allerdings bemerkt werden, daß
zwar die Scirpus- und Typha-Herden öfter aus einer Art bestehen,
der Phragmites-Bestand dagegen meist schon mit eingestreuten
Arten anderer Epharmonie bereichert ist (Alisma plantago-aqua-
tica, Sium latifolium, Sium erectum, Sparganium erectum usw.)
und daß diese Lage also ein Grenzfall zur nächsten Kategorie,
die der Bestandesgrenzen, darstellt. Die Grenzen zwischen den
drei oben erwähnten Beständen sind im allgemeinen scharf. Es
liegt nahe, diese drei Fazies, wie man sie gewöhnlich nennt,als
selbständige Gesellschaften und zwar Assoziationen zu betrachten,
umsomehr, da sie nicht nur in der kennzeichnenden Dominante,son-
dern auch syndynamisch und synökologisch verschieden sind: der
Scirpus-Bestand kennzeichnet die tiefste und am meisten bewegte,
bis 3 m tiefe Wasserzone und ist als Pioniergesellschaft zu bewerten; der Schilfgürtel und der Typha-Gürtel finden sich an seich-
terem und ruhigerem Standort, und zwar der Phragmites-Bestand
vorzugsweise auf hartem und sandigem, der Typha-Bestand dagegen
auf weichem und schlammigem Substrat. Diese Verhältnisse sind
schon längst bekannt und sind z.B. von BOER (1940) klar beschrieben worden. Es wäre also im Rahmen der heutigen analytischen
Verfeinerung und der heutigen Bewertung der Frage der Vegetationsgrenzen nicht unangebracht, das S c i r p o - P h r a g m i -
t e t u m   in drei Assoziationen aufzugliedern: das S c i r p e -
t u m   l a c u s t r i s ; (Allorge 1922), Chouard 1924, das
T y p h e t u m   a n g u s t i f o l i a e (Allorge 1922) Soó
1927, und eine P h r a g m i t e s - A s s o z i a t i o n, die
nach den Nomenklaturregeln als S c i r p o - P h r a g m i t e -
t u m   in engerem Sinne bezeichnet werden sollte (vergl.dazu
WESTHOFF & DEN HELD 1969). Als viertes Teilglied käme vielleicht

das E q u i s e t e t u m f l u v i a t i l i s hinzu (E-
q u i s e t e t u m l i m o s a e Steffen1931 em.Wilczek 1935).

Ein anderes Beispiel sind die Honckenya-Bestände innerhalb des
E l y m o - A m m o p h i l e t u m Br.-Bl. et de L.1936, welche jetzt analytisch abgetrennt und synthetisch zum Verband
S a l s o l o - H o n c k e n y o n R.Tx.1950 gestellt worden
sind.

Wir kommen jetzt zu den Grenzen zwischen Beständen, die aus mehreren Arten zusammengesetzt sind, und die ich hier vorläufig Bestandesgrenzen nennen möchte. Dieser Name trifft zwar nicht ganz
zu, indem einartige Herden, hier Aggregationen genannt, auch Bestände darstellen.Die Andeutungen "Assoziationsindividuum" oder
"Assoziationsexemplar" sind aus verschiedenen Gründen uner-
wünscht: erstens weil sie eine Analogie mit Organismen vortäu-
schen, zweitens weil es sich nicht unbedingt um Assoziationen
handelt, sondern auch um andere Vegetationseinheiten handeln
kann, und drittens wegen der unbequemen Länge dieser Worte.
Namen wie "phytozönotische Grenze" oder auch"Gesellschaftsgren-
ze" sind irreführend aus einem anderen Grund, nämlich, indem
sie sich auf zwei verschiedene Sachen beziehen können, und zwar
sowohl auf Bestandesgrenzen, wie auf konkrete Assoziationsgrenzen. Der Unterschied zwischen den zwei letzten Begriffen sei
wie folgt erläutert: Bei Bestandesgrenzen handelt es sich um
den Fall, der meistens gemeint wird, wenn man schlechthin von
Vegetationsgrenzen redet: ein zu einer Assoziation gehöriger
Bestand grenzt räumlich an einen Bestand einer anderen Assoziation, wobei diese nicht mit der ersten vikariiert und überhaupt
nicht mit ihr verwandt zu sein braucht.Beispiele sind die Grenze zwischen einem Wald und einer Wiese, zwischen einer Bulte
und einer Schlenke im Hochmoor, zwischen einem S a l i c o r-
n i e t u m- und einem P u c c i n e l l i e t u m -Gürtel. Bei
Assoziationsgrenzen im konkreten Sinne dagegen handelt es sich
um die Frage, wo zwei oder auch mehrere geographisch vikariierende Assoziationen sich räumlich gegen einander abgrenzen lassen. Es liegt da eine konkrete synchorologische Grenze vor, die
in diesem Fall zugleich eine syntaxonomische, also abstrakte
Grenze darstellt. Als Beispiel seien erwähnt die Grenzen in der
äußeren Meeresküsten-Dünenlandschaft zwischen dem südlichen
E u p h o r b i o - A m m o p h i l e t u m, dem nordwestlichen
S o n c h o - A m m o p h i l e t u m und dem nordöstlichen

P e t a s i t o - A m m o p h i l e t u m (s.TÜXEN 1967). Die
Grenze zwischen E u p h o r b i o - A m m o p h i l e t u m
und S o n c h o - A m m o p h i l e t u m verläuft an der bel-
gischen Küste. Im Gelände liegen in beiden Fällen konkrete Bestandesgrenzen vor, wobei z.B. das S o n c h o - A m m o p h i -
l e t u m einerseits vom E l y m o - A g r o p y r e t u m
R.Tx.1967, andererseits vom T o r t u l o - P h l e e t u m
a r e n a r i i Br.-Bl.et de L.1936 begrenzt sein kann.

Nach manchen Auffassungen stellen jene Grenzen zwischen vikariierenden Assoziationen einen Fall des sogenannten geographischen
Kontinuum dar. Es wird damit gemeint, daß derartige Grenzen subjektiv sind und willkürlich gezogen werden. Das trifft methodisch aber nicht zu. Für eine entscheidende Beurteilung ist es
notwendig, das Gesamtareal der betreffenden Arten und Gesellschaften zu betrachten. Es handelt sich um die Frage, ob alle
in Betracht kommenden Arten in ihrer Verbreitung völlig unabhängig variieren oder ob doch eine gewisse Häufung, ein Zusammentreffen mehrerer Arten über größere Teile des Areals statt-
findet. Wenn das Gesamtauftreten der betreffenden Arten einen
größeren Raum beansprucht als der Bereich, in dem die Teilarea-
le sich nicht überlagern, ist es möglich, vikariierende Assoziationen zu unterscheiden, obwohl diese sich nicht durch scharfe
Grenzlinien trennen lassen. Die Grenze stellt dann einen Übergangsbereich dar.

Kehren wir noch einmal zu der Bestandesgrenze zurück. Die Frage,
wo die Grenze zwischen zwei Beständen gezogen werden kann, hängt
unmittelbar mit der Frage der Homogenität dieser beiden Bestände zusammen. Wenn diese Homogenität absolut wäre, wäre die Grenze leicht zu bestimmen. Unter absoluter Homogenität verstehen
wir den(imaginären Fall,wo sich auf willkürlich gewählten Stichproben eines Bestandes immer genau dieselbe Verteilung der Arten und Individuen findet;die Stichproben dürften sich also in
keiner Weise unterscheiden.Eine derartige Homogenität gibt es
aber überhaupt nicht,jedenfalls nicht in mehrartigen Beständen.
Homogenität ist immer relativ und wird somit zu einer Frage des
Maßstabes.Die Arten sind meist fleckenartig verteilt,in dem Sinne,daß sie zu einer bestimmten Art gehörigen Individuen oder
Sprosse sich stellenweise zu Aggregationen häufen.Diese Sachlage wird schon von der Wuchsform,also von der spezifischen Soziabilität der Arten bedingt;so tritt C o r y n e p h o r u s

c a n e s c e n s normalerweise mit Soziabilität 2, S c h o e-
n u s n i g r i c a n s mit Soziabilität 3 auf. Die Frage,
wann man derartige Flecken als eigene Bestände einer Sonderge-
sellschaft zu betrachten hat, ist jedoch nicht nur eine Frage
des Maßstabes. Sie wird in der Methodik unserer Schule meistens
so gelöst, daß man erst zur gesonderten Analyse übergeht, wenn
nicht nur eine einzige Art örtlich ihre Individuen oder Sprosse
häuft, sondern zugleich eine oder mehrere andere Arten an die
Kleinbestände der ersterwähnten Art gebunden erscheinen, also
mit dieser Art positiv korreliert sind. Es sei ein Beispiel erwähnt aus dem westeuropäischen Dünen-Trockenrasen des G a l i o-
K o e l e r i o n, bisher als K o e l e r i o n a l b e s c e n-
t i s beschrieben (dieser Name kann nicht aufrecht erhalten
werden, weil die "Koeleria albescens" der Küstendünen sich taxonomisch nicht von der innenländischen Koeleria cristata unterscheiden läßt). Wenn sich in solch einem Dünen-Trockenrasen örtlich regelmäßig einige etwa quadratmetergroße Flecken von Salix
repens ssp.arenaria vorfinden, liegt es noch nicht ohne weiteres
nahe, diese Salix-Kleinbestände als eigene Gesellschaft zu werten. Was tun wir aber, wenn man zu seiner Überraschung fest-
stellt, daß an der niederländischen Küste eine Art wie Monotro-
pa glabra speziell in diesen Salix-Herden auftritt, oder, noch
überraschender, daß im Süden der französischen Atlantikküste,
und zwar auf der ehemaligen Insel Oléron, Cephalanthera rubra
stets in diesen Salix-Aggregationen vorkommt - Cephalanthera
rubra, die wir in Mitteleuropa ja fast nur im C e p h a l a n-
t h e r o - F a g i o n finden? In solchen Fällen wird man
kaum zögern, diese Salix-Herden als eine eigene Gesellschaft,
z.B. der Ordnung P r u n e t a l i a s p i n o s a e, zu werten, und daher als gesonderte Bestände zu analysieren. Wir gehen deshalb so ausführlich auf diese wesentliche analytische
Fragestellung ein, weil merkwürdigerweise in den üblichen Hand-
und Lehrbüchern darüber kaum gesprochen wird.

Vor allem in Trockengebieten finden sich jedoch Mosaik-Land-
schaften, wo die Häufung gewisser Arten in dem einen, die Häufung anderer Arten in dem anderen Glied des Komplexes doch nicht
ohne weiteres zur Sonderfassung der verschiedenen Mosaik-Kompo-
nenten als eigene Gesellschaften zu führen braucht. In seiner
Betrachtung der Vegetation der kanarischen Inseln unterscheidet
OBERDORFER (1970) labile und stabile Mosaiken. In jenen können

Rasen-Bestand, Ginster-Gestrüpp und Baumheide-Gebüsch jeweils
großflächig eigene Gesellschaften bilden, ohne die Bestandteile
der anderen Komponenten zu enthalten. In diesem Fall ist es
nach OBERDORFER angebracht, die verschiedenen Komponenten als
getrennte soziologische Einheiten zu behandeln. Im zweiten Fall
dagegen, dem Sukkulenten-Gebüsch, gelingt es den Zwergsträuchern
und höheren Sukkulenten-Büschen auf Grund des Wassermangels nie,
zu einem dichteren Vegetationsschluß zu gelangen. Diese Gebüsche und die dazwischen eingestreuten offenen Stellen bestehen also überhaupt nur in diesem Zusammenhang; nirgendwo gibt
es ausgedehntere eigene Bestände des betreffenden Gebüsches. In
solchen Fällen, die für subtropische Trockengebiete charakteristisch sind, ist es nach OBERDORFER (l.c.) unumgänglich, das
stabile Wuchsformenmosaik als soziologische Einheit zu behandeln. Zur selben Schlußfolgerung gelangt WERGER (1973) in seiner Arbeit über die Trockengebüsche entlang dem Oranje-Fluß in
Südafrika.

Die Grenze zwischen verschiedenen Beständen kann mehr oder weniger scharf sein; anders ausgedrückt: die Diskontinuität kann
stark oder schwach sein. Die Diskontinuität wird besonders hervorgehoben, wenn Dominanz als einziges oder doch als eins der
wichtigsten Kriterien der Gesellschaftsunterscheidung beachtet
wird, wie in der schwedischen Methodik (z.b.DU RIETZ 1922).

Diskontinuität ist, wie wir alle wissen, z.T.menschlich bedingt.
In der Kulturlandschaft findet sich mehr Diskontinuität als in
der Naturlandschaft, und zwar umsomehr, wenn der menschliche
Einfluß stärker ist; es liegen dann nur wenige und recht scharfe Grenzen vor, wie zwischen Forst und Weide, Acker und Graben.
In der Naturlandschaft findet sich starke Diskontinuität hauptsächlich dort, wo jähe Umweltgrenzen vorliegen, wie am Meeres-
strand, an Flußufern und am Fuß jäh ansteigender Abhänge. Es
wird von denjenigen Forschern, die sich Anhänger der "individualistic concept" von GLEASON (1926) nennen, behauptet, daß scharfe Vegetationsgrenzen nur vorkommen, wenn die abiotischen Standortsgrenzen scharf sind. In Fällen, wo sich ein allmählicher
Standortsübergang vollzieht, wäre auch der Vegetationsübergang
ganz allmählich und würde somit keine Vegetationsgrenze, sondern
ein Kontinuum bestehen.Diese Voraussetzung ist jedoch unrichtig.
Der hier erwähnte Standpunkt ist nicht induktiv, sondern deduktiv gewonnen: er entspricht nämlich der aprioristischen Annahme,

daß es zwischen Pflanzen in einem Bestand keine anderen Beziehungen gäbe als einerseits die direkten Abhängigkeitsbeziehungen, wie Parasitismus und Epiphytismus, andererseits gemeinsame ökologische Amplituden, die jene Pflanzen auf einem gemeinsamen Standort zusammentreffen ließen. Bei dieser Auffassung werden aber die vielseitigen Wechselbeziehungen zwischen den Pflanzen, vor allem die indirekten Abhängigkeitsbeziehungen außer acht gelassen (für eine inhaltsreiche Arbeit über diese Beziehungen sei vor allem auf KNAPP 1967 verwiesen). Als klassisches Beispiel einer ziemlich scharfen Vegetatonsgrenze ohne offensichtlich scharfe abiotische Standortsgrenze sei nur die alpine Waldgrenze erwähnt. Nach ELLENBERG (1963) ist eine derartig scharfe Grenze wohl immer durch Wechselwirkung verschiedener Faktoren bedingt. Ein anderes Beispiel einer derartigen scharfen Vegetationsgrenze trotz allmähligen Standortsverlaufs bietet die Verlandungsgrenze der eutrophen Flachmoor-Gewässer.

Es sei jetzt auf einen anderen Aspekt der Erforschung der Bestandesgrenzen hingewiesen, und zwar den, daß manche Gesellschaften sich speziell in Grenzbereichen vorfinden. Diese Sachlage bedeutet, daß die Kennarten dieser Gesellschaften von gewissen Standortsgradienten bedingt werden, daß also die Arten an sich räumlich mehr oder weniger kurzstreckig ändernde oder andererseits an instabile Standorte gebunden sind. Die Autökologen gehen vielfach von der Voraussetzung aus, daß eine Art einen uniformen, quantitativ faßbaren Standort beansprucht; in dieser Betrachtung spricht man von stenöken und euryöken, oder auch von basiphilen und azidophilen Arten; charakteristische Gradient-Toleranz leuchtet dem Autökologen weniger ein. Die Pflanzensoziologen sind aber schon während längerer Zeit mit diesem Toleranz-Typ vertraut. In den letzten Jahrzehnten haben eben die Grenzbereich-Gesellschaften immer mehr die Aufmerksamkeit auf sich gezogen. Erwähnt seien die von TÜXEN (1952) zuerst erkannten Mantel-Gesellschaften der P r u n e t a l i a s p i - n o s a e, die von Th.MÜLLER (1962) beschriebenen trockenen Saum-Gesellschaften der T r i f o l i o - G e r a n i e t e a s a n g u i n e i, die frischen Saumgesellschaften des G a l i - o - A l l i a r i o n und des A e g o p o d i o n p o d a - g r a r i a e (OBERDORFER 1967, OBERDORFER et.al 1967, TÜXEN 1967 b), die S a g i n e t e a m a r i t i m a e des wechselhalinen Grenzbereiches zwischen halophiler und xerophiler Vege-

Vegetation an der Meeresküste (TÜXEN & WESTHOFF 1963), und die
Gesellschaften des A g r o p y r o - R u m i c i o n ç r i s-
p i auf dem instabilen Pendel-Grenzbereich zwischen naß und
trocken, salzig und frisch, nährstoffreich und nährstoffarm
(TÜXEN 1950, VAN LEEUWEN 1958, WESTHOFF & VAN LEEUWEN 1962,
WESTHOFF & DEN HELD 1969).

Wir haben die zwei Haupttypen von Grenzbereichen in der Vegetation als "limes convergens" bezw. "limes divergens" unterschieden (VAN LEEUWEN 1966,1970, WESTHOFF 1969,1971 a,b). Obwohl
diese Typen in unsren bisherigen Symposien in anderem Zusammenhang schon erläutert worden sind, dürfen sie in diesem Thema
erneut kurz erwähnt werden.

Unter "limes convergens" verstehen wir Grenzbereiche mit einer
instabilen "dann-und-wann-Lage", einem Pendel-Vorgang. Sie bilden für die Organismen eine Umwelt, die durch wechselnde unge-
wisse Lebensverhältnisse mit einem hohen Grade von Verschiedenheit im Laufe der Zeit, also durch Unstabilität, gekennzeichnet
ist. Dieser Umwelttyp wird in der angelsächsischen Literatur
als "ecotone", "stress zone" oder "tension belt" bezeichnet.
Standorte dieser Art werden von relativ artenarmen Vegetationseinheiten besiedelt, deren Anordnungsmuster sich durch grobe
und scharf umgrenzte Teilstrukturen auszeichnet. Zu den Pflan-
zemgesellschaften dieses Typus zählt z.B. das A g r o p y r o-
R u m i c i o n c r i s p i.

Demgegenüber steht der "limes divergens": Grenzbereiche mit
einem stabilen "mehr-oder weniger"-Gradienten, die durch einen
hohen Grad räumlicher Variation gekennzeichnet sind, von Umwelt-
faktoren bedingt, die von einem Punkt zum anderen verschieden,
aber jeweils stabil sind. Grenzbereiche dieser Art, seien es
Zonationen oder Mosaik-Komplexe, tragen relativ artenreiche und
zugleich relativ individuenarme Gesellschaften, deren Anordnungsmuster sich durch feine und unscharf umgrenzte Teilstrukturen
auszeichnet. Charakteristische Beispiele dieses Typus sind die
Saumgesellschaften der T r i f o l i o - G e r a n i e t a
s a n g u i n e i, weiterhin auch allmähliche Übergänge zwischen
M e s o b r o m e t e n und M o l i n i e t e n, wie z.B. am
Ufer des Bodensees, oder zwischen Hochmoor- und Kalk-Flachmoor-
Gesellschaften, wie in Skandinavien oder den Alpen.

Es zeigt sich der hier erwähnte Gegensatz auch beim Vergleich

des mitteleuropäischen ombrotrophen Hochmoores mit dem borealen
Aapa-Moor. Jenes kennzeichnet sich durch weitgehende Isolation
gegen die Außenwelt: es wird in der Naturlandschaft nicht von
Störungen, die aus der Umgebung kommen, beeinflußt. Das ombrotrophe Moor ist daher stabil. Es kann da eventuell eine (kürzlich allerdings in Zweifel gezogene) Rhytmik der zyklischen
Sukzession des Bult- und Schlenken-Komplexes geben. Jedoch ist
diese Rhytmik in der Stabilität inbegriffen, wie der Wechsel der
Jahreszeiten in einem stabilen Klima(vgl. den von ODUM 1969 ge-
prägten Begriff "pulse stability"). Das soligene boreale Hochmoor dagegen, das Aapa-Moor, ist nicht nach außen isoliert; es
wird jährlich von großen Mengen fließenden Schmelzwasser des
Schnees überflutet und daher mit Nährstoffen bereichert. Das
stabile ombrotrophe Moor nun zeigt die unscharfen inneren Grenzen des divergenten Grenzbereichs; das instabile Aapa-Moor dagegen zeigt konvergente, scharfe Grenzen zwischen den Rücken, dort
Stränge genannt, und den relativ ausgedehnten Schlenken (VAN
LEEUWEN 1962, WESTHOFF & VAN LEEUWEN 1964).

Wir verlassen nun die Bestandes- und Assoziationsgrenzen. Es sei
jetzt darauf hingewiesen, daß auch Grenzen zwischen Vegetationskomplexen als Vegetationsgrenzen gewertet werden sollen. Unter
einem Vegetationskomplex verstehen wir ein räumliches Gefüge
von Beständen verschiedener und systematisch meist nicht ver-
wandter Gesellschaften, z.B. Assoziationen, die sich in gesetz-
mäßiger Wiederholung immer zusammentreffen und in ihrer Gesamtheit einen gewissen Standortsraum besiedeln, wie z.B. eine Düne
mit ihrem Nord- und Südhang und das dazugehörige feuchte Dünen-
tal. Die Bedeutung des Vegetationskomplexes für die pflanzensoziologische Forschung ist vor allem von KRAUSE (1952) und PFEI-
FFER (1958) betont und weiterhin von SCHWICKERATH (1954 seq.)
angewandt und von SCHMITHÜSEN (1961) methodisch ausgearbeitet
worden (s.auch SEIBERT (1968). In der Vegetationskartierung
sind Vegetationskomplexe als Legenda-Einheiten von weitgehender
Bedeutung.

Die erste von SCHMITHÜSEN(1961) unterschiedene Kategorie von
Vegetationskomplexen, die hier zu erwähnen wäre, ist die Fliese
oder Physiochore. Sie bezieht sich auf Standortsräume mit einer
bestimmten potentiell natürlichen Vegetation, wo die augenblick-
liche Pflanzendecke durch menschlichen Einfluß bedingt ist. Die

dynamisch zusammenhängenden Assoziationen der Physiochore bilden nach SCHWICKERATH (1954) einen Assoziationsring (vergl.
SEIBERT 1968). Jede Physiochore hat somit nicht nur ihren eigenen Waldtyp, sondern auch ihren eigenen Wiesen-, Weiden- und Ak-
kertyp usw. Die Vegetationsgrenzen zwischen den Physiochoren
sind also von abiotischen Faktoren bedingten Grenzen zwischen
Vegetationskomplexen.

Darüber hinaus unterscheidet SCHMITHÜSEN (1961) als Vegetations-
komplex zweiter Ordnung den Gesellschaftspulk, einen in der
Landschaft räumlich assoziierten Komplex von Gesellschaftsrin-
gen.

Wir kommen dann schließlich zu der Formation und der Formationsgrenze. Die Formation wird nach GRISEBACH (1872) meist definiert
als ein physiognomisch charakterisierter Vegetationstyp, wie
der tropische Regenwald, der Hartlaubwald, der Nadelwald, die
Savanne oder die Steppe. Die Grenzen zwischen den Formationen
sind also konkrete räumliche Vegetationsgrenzen. Es sei dazu
jedoch bemerkt, daß in der Kulturlandschaft auf kurze Entfernung Wald, Grünland und Acker abwechseln können, obwohl man hier
nicht von verschiedenen Formationen spricht. Die Grenze zwischen
Wald und Grünland innerhalb des Waldklimax-Bereiches wird keine
Formationsgrenze genannt, im Gegensatz zu der alpinen Baumgrenze. Im üblichen Formationsbegriff ist also der Begriff der potentiell natürlichen Vegetation implizite aufgenommen worden.
Wenn wir daher den Formationsbegriff richtig kennzeichnen wollen nach dem Sinne, in welchem er angewandt zu werden pflegt,
wäre eine zutreffende Definition: die Formation ist die physiognomisch charakterisierte, klimatisch bedingte potentiell natürliche Vegetation eines klimatisch einheitlichen Erdteils.

ZUSAMMENFASSUNG

Zuerst werden konkrete und abstrakte Vegetationsgrenzen verglichen. Die Allgemeingültigkeit des Kontinuum-Begriffs wird abge-
lehnt. Klinale Variation kann mit Hilfe des Kline-Begriffes
auch klassifikatorisch angewandt werden. Es werden weiterhin
Grenzen im Raum oder in der Zeit unterschieden. Die auf einem
Florengefälle beruhenden chorologischen Grenzen sind weder konkret noch abstrakt, sondern real. Räumliche Vegetationsgrenzen
beziehen sich auf ein Gefälle: 1. der Abundanz und (oder) der Deckung

2. des Lebensform-Spektrum, 3. der örtlichen Arealgrenzen. Es
gibt horizontale und vertikale Vegetationsgrenzen.Jene sind von
der Schichtung bedingt. Sie werden methodisch besprochen (Sozietät, Konsozion, Soziation, Konsoziation) und mit synusialen
Grenzen verglichen. Vertikale Grenzen beziehen sich auf das horizontale Anordnungsmuster. Besprochen werden: Aggregationsgrenze, Bestandesgrenze, Assoziationsgrenze, Physiochorengrenze und
Formationsgrenze. Die methodischen Vorgänge und Schwierigkeiten
bei der Erfassung von Bestandesgrenzen werden eingehend diskutiert. Der Typ der für Grenzbereiche charakteristische Pflanzengesellschaften (Gradient-Gesellschaften) wird erörtert. Der Zusammenhang zwischen Raum- und Zeit-Qualitäten in Grenzbereichen
wird erläutert: ein instabiler Standort wird von grobkörnigen,
artenarmen, individuenreichen Anordnungsmustern mit scharfen
inneren Grenzen besiedelt (limes convergens), ein stabiler
Standort dagegen von feinkörnigen, artenreichen, relativ indivi-
duenarmen Anordnungsmustern mit unscharfen inneren Grenzen (limes divergens),

SUMMARY

Concrete and abstract vegetation boundaries are compared. The
continuum concept is discussed. Clinal variation is often obvious; with the aid of the cline concept clinal variation can
be dealt with in vegetation classification. Vegetation boundaries in time and in space are discerned. Boundaries in space
are maked by a gradient (a) in abundance or cover, (b) in life
form spectrum, (c) in the number of local species area boundaries. Horizontal and vertical vegetation boundaries are distinguished. The former relate to layer structure; their methodical
approach is discussed and they are compared with synusial boun-
daries. Vertical boundaries result in horizontal pattern. Several levels are discussed: aggregation boundaries, stand bounda-
ries, association boundaries, boundaries of vegetation complexes and formation boundaries. Community types characteristic of
ecotones and ecoclines are dealt with, as well as the relation
between pattern and process.

Literatur.

BOER,A.C.   -1942- Plantensociologische beschrijving van de or-
     de der Phragmitetalia.- Ned.kruidk.Arch.52: 237-302.
     Amsterdam.
BRAUN-BLANQUET,J.   -1928- Pflanzensoziologie. Grundzüge der
     Vegetationskunde.- Berlin.
BRAUN-BLANQUET,J.& DE LEEUW,W.C.   -1936- Vegetationsskizze von
     Ameland.- Ned.kruidk.Arch.46: 359-393. Amsterdam.
CAIN,S.A.& DE OLIVEIRA CASTRO,G.M.   -1959- Manual of vegetation
     analysis.- New York.
CASTRI,D.DI & COVARRUBIAS,R.   -1966- Symposium on ecology of
     sub-artic regions.- Helsinki.
CURTIS,J.T.   -1959- The vegetation of Wisconsin. An ordination
     of plant communities.- Madison.
CURTIS,J.T.& McINTOSH,R.P.   -1951- An upland forest continuum
     in the prairie-forest border region of Wisconsin.- Ecology
     32: 476-496. Lancaster,Pa.
DANSEREAU,P.   -1957- Biogeography, an ecological perspective.-
     New York.
DU RIETZ,G.E.   -1922- Die Grenzen der Assoziationen.- Bot.No-
     tiser. Lund.
DU RIETZ,G.E.   -1930- Vegetationsforschung auf soziations-ana-
     lytischer Grundlage.- Abderhalden's Handb.Biol.Arb.Meth.11
     (5) 293-460. Berlin.
ELLENBERG,H.   -1963- Vegetation Mitteleuropas mit den Alpen.-
     Stuttgart.
GLEASON,H.   -1926- The individualistic concept of the plant
     association.- Bull.Torrey bot.Club 53: 7-26. New York.
GRISEBACH,A.   -1872- Die Vegetation der Erde nach ihrer kli-
     matischen Anordnung.- Leipzig.
HARTOG,C.DEN & SEGAL,S.   -1964- A new classification of the
     water plant communities.- Acta bot.neerl.13: 367-393.
     Amsterdam.
HUXLEY,J.   -1938- Clines: an auxiliary taxonomic principle.-
     Nature 142: 219-220. London.
KNAPP,R.   -1967- Experimentelle Soziologie und gegenseitige
     Beeinflussung der Pflanzen.2.Aufl.- Stuttgart.
KRAUSE,W.   -1952- Das Mosaik der Pflanzengesellschaften und
     seine Bedeutung für die Vegetationskunde.- Planta 41: 240-
     289. Berlin.
LEEUWEN,C.G.VAN   -1958- Enige opmerkingen over het Agropyro-
     Rumicion crispi Nordh.1940 in Nederland.- Correspondentie-
     blad Rijksherbarium 11: 117-123. Leiden.
LEEUWEN,C.G.VAN   -1962- De hoogvenen van Twente.- Wetensch.
     meded.KNNV 43: 19-36. Amsterdam.
LEEUWEN,C.G.VAN   -1966- A relation theoretical approach to
     pattern and process in vegetation.- Wentia 15: 25-46.
     Amsterdam.
LEEUWEN,C.G.VAN   -1970- Raum-zeitliche Beziehungen in der Ve-
     getation. In: Tüxen,R.(Édit.): Gesellschaftsmorphologie.
     Ber.int.Symp.Rinteln 1966: 63-68. Den Haag.
LIPPMA,T.   -1935- La méthode des associations unistrates et le
     système ècologique des associations.- Acta Inst.Horti bot.
     Univ.Tartuensis 4: 1-7. Tartu.
MAAREL,E.VAN DER   -1969- On the use of ordination models in
     phytosociology.- Vegetatio 19: 21-46. The Hague.
MAAREL,E.VAN DER, WESTHOFF,V.& LEEUWEN,C.G.VAN   -1964- Europe-
     an approaches to the variation in vegetation.- Paper 10th
     int.bot.Congress Edinburgh. Edinburgh.

MARGALEF,R.   -1968-   Perspectives in ecological theory.-
    Chicago.
MEYER-DREES,E.   -1951-   Capita selecta from modern plant socio-
    logy.- Rapp.Bosb.proefst.Bogor 52: 1-68. Bogor.
MIYAWAKI,A.& TÜXEN,J.   -1960-   Über Lemnetea-Gesellschaften in
    Europa und Japan.- Mitt.flor.-soz.Arbeitsgem.N.F.8: 127-136.
    Stolzenau/Weser.
MÜLLER,Th.   -1962-   Die Saumgesellschaften der Klasse Trifolio-
    Geranietea sanguinei.- Mitt.flor.-soz.Arbeitsgem.N.F.9: 95-
    140. Stolzenau/Weser.
NORDHAGEN,R.   -1937-   Versuch einer neuen Einteilung der sub-
    alpinen-alpinen Vegetation Norwegens.- Bergens Mus.Arbok
    Naturv.Rekke 1936: 1-88. Bergen.
OBERDORFER,E.   -1957-   Süddeutsche Pflanzengesellschaften.-
    Jena.
OBERDORFER,E.   -1970-   Pflanzensoziologische Strukturprobleme
    am Beispiel Kanarischer Pflanzengesellschaften.- In: Tüxen,
    R.(Edit.): Gesellschaftsmorphologie. Ber.int.Symp.Rinteln
    1966: 273-278. Den Haag.
OBERDORFER,E.et al.   -1967-   Systematische Übersicht der west-
    deutschen Phanerogamen- und Gefäßkryptogamen-Gesellschaften.-
    Schr.Reihe Vegetationskunde 2: 7-62. Bonn-Bad Godesberg.
ODUM,E.P.   -1969-   The strategy of ecosystem development.-
    Science 164: 262-270. Lancaster,Pa.
PFEIFFER,H.   -1958-   Über das Zusammentreffen von Pflanzengesell-
    schaften in Komplexen.- Phyton 7: 288-295. Buenos-Aires.
RAMENSKY,L.G.   -1930-   Zur Methodik der vergleichenden Bearbei-
    tung und Ordnung von Pflanzenlisten und anderen Objekten
    die durch mehrere, verschiedenartig wirkende Faktoren be-
    stimmt werden.- Beitr.Biol.Pfl.18: 269-304. Breslau.
RICHARDS,P.W.   -1952-   The tropical rain forest: an ecological
    study.- Cambridge.
SCHEYGROND,A.   -1931-   Het plantendek van de Krimpenerwaard.
    IV.Sociographie van het hoofdassociatie-complex Arundinetum-
    Sphagnetum.- Utrecht. Ned.kruidk.Arch.1932. Amsterdam.
SCHMITHÜSEN,J.   -1961-   Allgemeine Vegetationsgeographie.-
    Berlin.
SCHWICKERATH,M.   -1954-   Die Landschaft und ihre Wandlung.-
    Aachen.
SEIBERT,P.   -1968-   Gesellschaftsring und Gesellschaftskomplex
    in der Landschaftsgliederung.- In: Tüxen,R.(Edit.): Pflan-
    zensoziologie und Landschaftsökologie. Ber.int.Symp.Stolze-
    nau/Weser 1963: 48-60. Den Haag.
SHIMWELL,D.W.   -1972-   Description and classification of vege-
    tation.- London.
TÜXEN,R.   -1950-   Grundriß einer Systematik der nitrophilen Un-
    krautgesellschaften.- Mitt.flor.-soz.Arbeitsgem.N.F.2: 94-
    175. Stolzenau/Weser.
TÜXEN,R.   -1967a-   Pflanzensoziologische Beobachtungen an süd-
    westnorwegischen Küstendünengebieten.- Aquilo Ser.Bot.6:
    241-272. Oulu.
TÜXEN,R.   -1967b-   Ausdauernde nitrophile Saumgesellschaften
    Mitteleuropas.- Contrib.Bot.431-453. Cluj.
TÜXEN,R.& OBERDORFER,E.   -1958-   Eurosiberische Phanerogamen-
    Gesellschaften Spaniens.- Veröff.geobot.Inst.Rübel 32: 1-
    328. Zürich.
TÜXEN,R.& WESTHOFF,V.   -1963-   Saginetea maritimae, eien Gesell-
    schaftsgruppe im wechselhalinen Grenzbereich der europäi-
    schen Meeresküsten.- Mitt.flor.-soz.Arbeitsgem.N.F.10: 116-
    129. Stolzenau/Weser.

WEAVER,J.E.& CLEMENTS,F.E.   -1938-   Plant ecology.- 2nd ed.
   New York-London.
WERGER,M.J.A.   -1973-   Phytocociology of the Upper Orange River
   Valley,South Africa.- Diss.Nijmegen.
WESTHOFF,V.   -1947-   The vegetation of dunes and salt marshes
   on the Dutch islands of Terschelling, Vlieland and Texel.-
   Diss.Utrecht. Den Haag.
WESTHOFF,V.   -1949-   De plantengezelschappen van Botshol.- In:
   Landschap, flora en vegetatie van de Botshol nabij Abcoude:
   43-102. Baambrugge.
WESTHOFF,V.   -1967-   Problems and use of structure in the clas-
   sification of vegetation.- Acta bot.neerl.15: 495-511.
   Amsterdam.
WESTHOFF,V.   -1968-   Einige Bemerkungen zur syntaxonomischen
   Terminologie und Methodik, insbesondere zu der Struktur als
   diagnostischem Merkmale.- In: Tüxen,R.(Edit.): Pflanzenso-
   ziologische Systematik. Ber.int.Sympos.Stolzenau/Weser
   1964: 54-68. Den Haag.
WESTHOFF,V.   -1971a-   The dynamic structure of plant communities
   in relation to the objectives of conservation.- In: Duffey,
   E. and Watt,A.S.(Edit.): The scietific management of animal
   and plant communities for conservation: 3-14. Oxford.
WESTHOFF,V.   -1971b-   Choice and management of nature reserves
   in the Netherlands.- Bull.Jard.Bot.Nat.Belg.41: 231-245.
   Bruxelles.
WESTHOFF,V.& HELD,A.J.DEN   -1969-   Plantengemeenschappen in Ne-
   derland.-   Zutphen.
WESTHOFF,V.& LEEUWEN,C.G.VAN   -1962-   Ökologische und systema-
   tische Beziehungen zwischen natürlicher und anthropogener
   Vegetation.- In: Tüxen,R.(Edit.): Anthropogene Vegetation.
   Ber.int.Sympos.Stolzenau/Weser 1961: 156-172. Den Haag.
WESTHOFF,V.& LEEUWEN,C.G.VAN   -1964-   Geografische differentia-
   tie in de hoogvenen van Europa.- Jaarb.Kon.Ned.Botan.Ver.
   32-33.
WHITTAKER,R.H.   -1962-   Classification of natural communities.-
   Bot.Rev.28: 1-239. Lancaster,Pa.
WHITTAKER,R.H.   -1972- Convergence of ordination and classifi-
   cation.- In: Tüxen,R.(Edit.): Basic problems and methods
   in phytosociology. Ber.int.Symp.Rinteln 1970: 39-57.
   Den Haag.

R.TÜXEN:

Ich wollte meinem Freund Westhoff herzlich gratulieren, sehr
herzlich danken und ihm widersprechen.Er hat nämlich gesagt,
er würde gar nichts Neues sagen. Das ist für die Älteren unter
uns vielleicht bis zu einem gewissen Grade richtig. Das ist
aber nicht richtig für die Jüngeren. Ich höre seit Monaten fast
jeden Tag von meinen jungen Mitarbeitern gerade die Fragen, die
Du in Deinem inhaltsreichen Vortrag beantwortet hast, und ich
bin ganz sicher, daß Du allen von uns, den Jüngeren, sowie den
Älteren, viel Neues gesagt hast, und darum noch einmal herzlichen Dank.

E.VAN DER MAAREL:

Ich möchte auf die Begriffe "ecocline und "ecotone" zurückkommen. Der Anlaß dazu ist ein Brief, den ich von Professor A.W.
KÜCHLER aus Kansas bekommen habe. Erstens schreibt er, daß er
bis zum letzten Augenblick versucht hat, hierherzukommen. Offenbar ist das nicht gelungen. Er schreibt: "best regards to our
mutual friends in Rinteln".

Zu den Begriffen schreibt er, nachdem er bezweifelt hat, ob
WESTHOFF und VAN DER MAAREL recht hatten, den ecocline-Begriff
zu ändern: " Clements introduced the term in 1936 and defined it
as a transition (cline) between ecosystems. It has been used
ever since then this sense. So why change now and possibly create confusion?" Herr KÜCHLER meint hier die Änderung in manchen
unserer Arbeiten, wenn wir dazu neigen, den Begriff ecocline
zu beschränken auf den limes divergens-Grenztypus, während CLE-
MENTS ihn ursprünglich mehr allgemein gemeint hat. Herr KÜCH-
LER schreibt weiter: "Ecotone, too, is so well established
among ecologists and phytocenologists that a change in meaning
may not be accepted very readily. I think this is the sort of
thing that might be usefully discussed at Rinteln."

Ich bin der Meinung, wir sollten Professor KÜCHLER nicht ent-
täuschen, und schlage zur Eröffnung einer Diskussion vor, die
Begriffe ecocline und ecotone zu beschränken auf die limes di-
vergens- bezw.limes convergens-Situation. Ich glaube, daß es
nicht schwierig oder gefährlich ist, dies zu machen, weil es
nicht im Widerspruch zu den Originaldiagnosen ist. Der cline-
Begriff war ja gemeint für allmähliche Übergänge und obwohl

CLEMENTS ihn wohl nicht gerade für diesen Typus von Übergangs-
situationen gemeint hat, ist es hier nicht gefährlich, glauben
wir, ecocline zu reservieren für die limes divergens-Situation.
Der ecotone-Begriff, 1905 von dem Amerikaner LIVINGSTONE entwickelt, ist gerade für "Spannungszonen" gemeint, für Wechsel-
bedingungen, und das ist gerade der instabile Charakter, den
wir an den limes convergens-Grenzen erkennen.Ich schlage vor,
zu versuchen, diese beiden ursprünglich Amerikanischen Begriffe
auch für unsere neue Typologie der Grenz-Situationen anzuwenden.
Aber das ist vielleicht an erster Stelle eine Frage an die Groß-
brittanier, sie beschäftigen sich am meisten, auch linguistisch,
mit diesen Begriffen.

R.TÜXEN:

Ich finde, und es wird nicht allein mir so gehen, daß in der
deutschen Literatur diese Ausdrücke gar nicht oder ganz selten
gebraucht werden, und ich habe jedesmal Schwierigkeiten, wenn
ich sie lese, glaube aber, daß sie nützlich sind. Dies wäre
eine Gelegenheit, sie besser einzuführen, und deshalb bin ich
Herrn KÜCHLER und Herrn VAN DER MAAREL dankbar für diesen Hinweis.

S.SEGAL:

Über die neuen Begriffe limes divergens und limes convergens
habe ich mich immer etwas gewundert, weil sie meiner Meinung
nach ungefähr zusammenfallen mir ecocline und ecotone. Es wäre
schade, wenn es notwendig wäre, neue Termini einzuführen für
Begriffe, die schon lange Zeit existieren. Warum genügen nicht
ecocline oder Ökokline und ecotone oder Ökoton?

J.J.MOORE:

Ecocline is a term used in Experimental Taxonomy for a Taxon
which varies morphologically in a continuous fashion over an
ecological gradient. We do not use the word for communities,
and it is not an ecological term. We would probably understand
it in the literature, but we would be surprised, we would not
expect to find that term in ecology. We would always use ecotone in a quite general sense for the two types of limes. Ecotone is an ecological term used in a general way for both types

of transitional zone envisaged in **VAN LEEUWEN**s system. This is
the normal use of the terms in English, but I believe that Ecocline would be understood in a phytosociological content.

E.VAN DER MAAREL:

Es gibt ja mehrere Begriffe, die zugleich in der Idio- und in
der Synbiologie angewandt werden. Das kann man hier auch leicht
machen. Man kann sprechen von syncline - Synkline und, das hat
WHITTAKER einmal vorgeschlagen, coenocline, also Zönokline. Der
Nachteil ist, daß das Wort wieder länger geworden ist. Aber es
ist prinzipiell natürlich immer möglich, diese beiden idio- und
synsystematischen Begriffe auf diese Weise auseinander zu halten. Ich glaube, daß es nicht viel Schwierigkeiten gibt; aus
dem Context ist es doch immer klar, ob es sich um Übergänge zwischen Sippen oder synsystematischen oder allgemein vegetationskundlichen Einheiten handelt. Eine ganz andere Sache ist natürlich die, daß die Engländer und im allgemeinen die anglo-ameri-
nischen Vertreter immer noch die Absicht haben, ecotone ganz
allgemein anzuwenden, und wir schlagen gerade vor, zu versuchen,
dies nur auf einen der beiden Typen zu beschränken.

Natürlich hat Herr SEGAL recht, daß diese beiden amerikanischen
Begriffe etwas leichter eingeführt werden können. Man kann aber
dagegen sagen, daß es gerade die Schwierigkeit ist, daß sie ursprünglich eine andere Bedeutung hatten, und es ist nun nicht
ohne weiteres möglich, das zu ändern. Das soll dann mindestens
diskutiert werden. Ich bin damit einverstanden, daß die ziemlich
schwierig auszusprechenden Begriffe limes divergens und limes
convergens vielleicht ersetzt werden können durch Termini, die
etwas leichter auszusprechen sind, aber das ist natürlich kein
Bedenken gegen den Grundsatz.

R.TÜXEN:

I must confess, when I hear a lecture of an Englishman or an
American, it is very difficult for me to follow you, Dr.PROC-
TOR, because I have too many difficulties with the language
and also with your kind of problems. I do not understand enough
of the mathematical problems you spoke about. And these two
problems are very difficult for me.

Aber wenn ich jetzt die pädagogisch wunderbaren Beispiele mit

der Telefonnummer von dem Kino gehört habe, dann habe ich vie-
les verstanden. Ich fand, daß diese 3 Vorträge von Dr.PROCTOR,
von Professor WESTHOFF und von Dr.ZONNEVELD heute morgen eine
Einheit bildeten, die uns viel gegeben haben. Ich erinnere mich
an einen Ausspruch von BRAUN vor vielen Jahrzehnten. Er sagte:
"Systematisieren heißt Ordnen". Und jetzt wird hier von ordination und classification gesprochen, und wir haben eben gehört,
das eine ist eine methodische und das andere vielleicht mehr
technische Weise, dieses Problem zu lösen. Ich wäre glücklich,
wenn wir mit der Zeit -und wir sind auf dem besten Wege dazu,
glaube ich- zu einem gegenseitigen Verständnis der Betrachtungsweise in den englisch sprechenden Ländern und in den anderen,
hier auf dem Kontinent, kommen würden. Ich glaube, dieser Vor-
mittag hat viel dazu beigetragen.

ZUR MATHEMATISCHEN BESTÄTIGUNG DER TABELLARISCHEN
ABGRENZUNG VON PFLANZENGESELLSCHAFTEN
(Kurzfassung)

J.J.MOORE & A.O'SULLIVAN

Es gibt verschieden Grenzen:
    A. Konkrete, im Gelände kartierbare Grenzen:

        1. Physiognomisch sichtbar;
        2. Durch Gruppen von nicht auffallenden Trennarten
           abbegrenzt.
    B. Zwischen topographisch zerstreuten Aufnahmen:
        Abstrakte Grenzen, die am 'grünen Tisch' festge-
        stellt werden durch

        1. Synsystematische Tabellenarbeit;
        2. Multivariate Analysis.
Die beiden letzteren werden verglichen.

<u>Test-Material</u>

<u>I.Salzwiesen</u>: North Bull Island; 4 km x 0.2 km.
    28 wahllos festgestellte Aufnahmepunkte (1 x 1 m) sind durch
    die BRAUN-BLANQUET Punkt-Quadrat- und Frequenz-Quadrat-Me-
    thoden beschrieben worden. Quantitative Daten wurden mit
    einer 'Principal Components Analysis' (P.C.A.) analysiert.

Die Ergebnisse gleichen 1. Dem Tabellenbild;
                     2. Einem P.C.A.auf transformierten
                       BRAUN-BLANQUET-Zahlen.

<u>II.Grünland</u>: National Grassland Survey - 900 Aufnahmen (nach
    Braun-Blanquet mit Bodenbeschreibung und chemischen Analy-
    sen.
Wegen der großen Zahl der Aufnahmen ist eine Auslese, Co.Carlow-
als Beispiel, notwendig.
Tabellenbild = Scharfe Abgrenzung der Klassen, Ordnungen,Verbände
(115 Aufnahme) Mäßige Abgrenzung der Assoziationen,Sub-Ass.
               Unklare Abgrenzung der Varianten.
28 Aufnahmen sind wahllos aus der Tabelle ausgelesen und durch
P.C.A. analysiert worden.

Es ergibt sich 1. Klare Abgrenzung der M o l i n i e t a l i a
             2. Unscharfe Abgrenzung des L o l i o - C y n o-
               s u r e t u m.

<u>Schlußfolgerungen:</u>
1. Der immer wieder vorgebrachte Einwand, die BRAUN-BLANQUET-
   Methode sei auf einem Zirkelschluß aufgebaut, ist nicht immer

gut begründet. Die 'schönen' Tabellenbilder, mit scharfen Abgrenzungen, erscheinen nicht immer wegen der willkürlichen Auslese der Bestände.

2. Erfahrung mit den mathematischen Methoden bestätigt die klaren Ergebnisse der Tabellenarbeit.

3. Gültige mathematische Verfahren lassen sich nur auf kleineren Zahlen von Aufnahmen durchführen wegen:

   1. Komputer Zeit
   2. Größe der Matrizen

z.B. Die 28 Grünland-Aufnahmen brauchten fast 3 Stunden auf dem IBM 1620, um die Koeffizienten zu rechnen und um die P.C.A. Analyse durchzuführen (nur 3 Eigenwerte sind gerechnet worden).

J.Moravec:

Ich danke Pater MOORE für sein inhaltsreiches Referat, das uns gezeigt hat, daß man zu den gleichen Assoziationen kommen kann, auch wenn die Aufnahmen at random verteilt sind und nicht an bestimmten Stellen nach der vorherigen Besichtigung des Geländes gemacht werden. Ich glaube, dieses Ergebnis ist äußerst wichtig für die Beziehungen zwischen angelsächsischen und kontinentalen Richtungen in der Vegetationsforschung.

I.Zonneveld:

Ihre mathematische Behandlung bezieht sich auf ein kleines Gebiet. Sie finden klare Diskontinuitäten. Würde das aber auch so sein, wenn Sie alle Grünländer Europas auf dieselbe Weise untersuchen würden? Ich glaube, daß das nicht der Fall sein würde.

Eine andere Frage ist: Würde es möglich sein, viele zehntausende Aufnahmen, (die man bei einer solchen Untersuchung unbedingt braucht), mathematisch zu behandeln? Nur dann kann man "beweisen", daß es ein Kontinuum oder ein Diskontinuum in der Abstraktion gibt, wie ich es in meinem Vortrag dargestellt habe.

J.J.Moore:

Nein, auch für Irland geht es nicht. Ich wollte eigentlich für dieses Symposion nicht nur eine kleine Grafschaft, die etwa 50 km lang ist,sondern für ganz Irland eine principal components analysis machen. Aber es kam ein Streik der Elektrizität,

der Komputer funktioniert dann nicht, und ich habe nicht alles
in der letzten Woche fertig bekommen. Für Europa geht es gar
nicht mit einer so großen Zahl von Aufnahmen zu machen. Aber
man könnte vielleicht eine radom selection machen in, sagen wir,
diesem Übergangsgebiet zwischen C y n o s u r e t u m und M o-
l i n i e t a l i a, oder wo man das will, und ich glaube - ja,
ich soll ja nicht glauben! - ich ahne, daß ähnliche Resultate
herauskommen.

M.C.F.Proctor:

May I raise just one or two points which struck me in Father
MOORE's talk. First of all, I think in connection with all the
lectures we have today, the importance of correlation between
the characters we are using, the species, whatever they are,
should be emphasized. I spoke this morning of the specification
of colour in terms of three parameters. Now this is a case
where we get three uncorrelated factors and obviously we have
dark green and light green, we can have any possible combination
of hue, saturation and brightness. Our choice of classification or
ordination methods is determined purely on whether we want information on these. But when you are dealing with more complex
data, when in fact you are driven to either classify or produce
a simpler ordination, to comprehend data which is in too many
dimensions, or which have too many variables to be comprehensible
without, then correlation between the variables become extremely important. If I could draw a two-dimensional diagram, colour
gives us the situation which we saw this morning in one of Dr.
ZONNEVELD's diagrams, with points evenly distributed and we can
arbitrarily decide to divide this field up into categories, the
categories of the classification, or we can say we will specify
the position of any one of these points by giving it cordinates
on each of two axes. Now I think one of the things that became
very clear out of Father MOORE's presentation was that in biological data obviously we do not have points evenly distributed
and if you have correlation then it means for classification
purposes certain of the classification are largely empty and for
practical purposes can probably be ignored. In other words, we
can now classify with virtually the same efficiency using two
units instead of four. If we have a discontinuity there as in the
example of the P u c c i n e l l i e t u m and the J u n c e -

J u n c e t u m   g e r a r d i i, so much the better, that would
mean that our classes are now very much better divided one from
another.

Personally I think it is perhaps worth pointing out to those
who haven't used these methods that principal component analysis and the classificatory techniques we have discussed are designed to attain grouping from data in which there are no groupings in the beginning, from unstructured data.

And I think there is another point about component analysis. In
principle it is a geometrical transformation and it will work
satisfactorily with data that is not multivariate normal, but it
is rather more important that the variables should be linearly
related and the data as a whole should be reasonably homogeneous,
in the sense I used in speaking earlier this morning.

G.Lavrentiades:

Is there any correlation between the number of relevés and the
size of the angle you mentioned?

J.J.Moore:

No, I do not think so. There will be a slight difference if
particular methods of principal components are used. We have
found that strict component analysis using the variance-covari-
ance matrix of the species as the starting point will give you
almost the same picture. The axes will have different parameters along them, but the picture is almost exactly the same as
when one uses Orloci's method or when one uses the method of
Gower transformation of the distance coefficient or when one
uses the correlation coefficient between the species. But as
far as I can see, taking subsets of the data gave the same general appearance in the clusters without any very exact mathematical analysis, but the appearance of the projected clusters
looks the same no matter whether they are the full 28 relevés
or just say 19 relevés. But it is a very limited small sample
and much too small to make generalizations about.

R.Tüxen:

Pater MOORE sagte, die verschiedene Betrachtungsweise der Vegetation läge an diesen schwierigen Erscheinungen: Kontinuum-

oder nicht Kontinuum. Mag sein. Aber die verschiedene Betrachtungsweise liegt nach meiner Ansicht viel mehr in der verschiedenen Veranlagung der Menschen, die sich mit diesem Gegenstand
beschäftigen. Ich finde, wenn wir etwas Einfaches untersuchen
wollen, sollten wir es auch so einfach machen, wie möglich.

J.J.Moore:

Ich habe am Anfang gesagt, daß diese Stimme aus England, aus
den USA, so laut war, daß ich mich selbst überzeugen mußte, daß
es eigentlich mit dieser sehr einfachen Methode geht. Natürlich
wollen wir sie für die große Vegetationsforschung Irlands benutzen, aber man muß erst sich selbst und den Studenten zeigen,
daß man doch etwas mit diesem BRAUN-BLANQUET-System erreichen
kann.

SMALL-SCALE VEGETATIONAL BOUNDARIES;
ON THEIR ANALYSIS AND TYPOLOGY
Shortened version of a lecture originally delivered in German.

E. van der M a a r e l

In the meantime main results and further considerations have
been published by THALEN (Acta bot.neerl. 20: 317-349,1971)
FRESCO (in Grundfragen und Aufgaben der Pflanzensoziologie, ed.
E.VAN DER MAAREL & R.TÜXEN,1972 p.99-112) and VAN DER MAAREL
(in: The scientific management of animal and plant communities,
ed. E.DUFFEY & A.S.WATT,1971,p.45-63). See also VAN DER MAAREL
On vegetational structures, relations and systems. Thesis 1966,
Utrecht; VAN DER MAAREL & LEERTOUWER, Acta bot.neerl.16: 211-
221,1967.

1. INTRODUCTION

This paper deals with a simple method of boundary analysis and
with some attempts to interprete the results, obtained in dune
and salt marsh vegetations, with the help of the environmental
boundary typology by VAN LEEUWEN There is some need for such
an approach since very little has been published so far on this
aspect.

2. SOME GENERAL REMARKS ON BOUNDARIES.
A boundary may be seen as a separator of adjacent systems showing some sort of difference. Difference between systems con-
sists of differences between the variable that constitute the
systems.The detection of difference and consequently the detection of boundaries depends on the scale of observation. When,
on a certain scale of observation, the differences between two
adjacent systems are great as compared with the internal differences within each system, the boundary between these systems
is called sharp.When external differences only slightly exceed
internal differences we speak of a vague boundary.
The difference between sharp and vague is not absolute but gradual, whilst the distinction between these two extremes is
again dependent on the scale of observation.
Differences may be measured either qualitatively or quantitatively. The distinction between these two possibilities is again
dependent on the scale of observation,i.g.on very small scales

almost any difference is qualitative. The use of quantitative
measurements is more realistic as the scale of observation in-
creases.
Vegetational boundaries are determined by differences in vegetational variables. These are commonly divided into floristical
and structural variables. Consequently we may distinguish between floristical and structural boundaries - which may, of course, coincide in nature -. Another distinction to be made is
that between real vegetational boundaries in the field and abstract boundaries between vegetation types. This paper concen-
trates on real floristic boundaries.
In most studies of vegetation, irrespective of the School to
which an investigator maybe reckoned, the choise of stands to
be analysed, implies delineation of vegetation in the field
and hence detection of boundaries. Such detections, are mainly
based on observed differences of dominant or otherwise conspi-
cuous species.It maybe questioned whether these procedures are
always in agreement with the natural situation.

3.   A SIMPLE METHOD OF ANALYSING BOUNDARY SITUATIONS.
The method is based on a transect or grid of adjacent squared
plots of varying size.
Theoretically the smallest quadrat size to start with maybe chosen so that each quadrat contains not more than one plant unit.
In practice larger sizes will be chosen to start with, in order
to analyse a sufficiently large area.
Then the difference between adjacent quadrats is measured as
follows: presence - absence data are collected for all species
in the smallest quadrats. Data for larger quadrats are obtained
by grouping together quadrats of smaller size; then quantitative, viz.frequency data are available.
According to the Information Theory the difference between two
quadrats maybe measured as the heterogeneity in the set of two
quadrats. Each species occurring in only one of the two quadrats
contributes 1 bit of selective information or, preferably, 1/2
bit/quadrat. When quantitative data are available the heterogeneity contribution of a species maybe approximated by dividing
the difference of performance in the two quadrats by the maximum difference. This leads to:

$$H = 1/2 . \left| \sum_{i=1}^{G} p(g_{i,a}) - p(g_{i,b}) \right| / p_{max}$$

with $p(g_{i,a})$= performance of the i-th species in quadrat a, etc.
p can be measured as frequency, coverage, abundance or with a
combined scale, e.g. the BRAUN-BLANQUET scale.
For qualitative data this formula reduces to

$$H = \frac{G_a - G_c + G_b - G_c}{2}$$

with G_a and G_b = numbers of species in quadrats a and b resp.
and G_c = number of common species.
In addition the intrinsic or relative heterogeneity, H_r should
be measured. H_r follows from H_r = H/Hmax. It can be easily
shown that H_r is complementary to similarity as measured with
the Jaccard formula.
H-values for transects or grids maybe graphically presented as
one or two dimensional H-profiles or vegetation "differential
profiles".

4. THE ENVIRONMENTAL BOUNDARY TYPES OF VAN LEEUWEN AND THEIR
   DIFFERENTIAL PROFILES.
The main types limes convergens = ecotone and limes divergens
= ecocline boundary situations and some of their derivates maybe easily detected by differential profiles.

SOME APPLICATIONS IN DUNE AND SALT MARSH VEGETATIONS.
1. DUNE GRASSLAND COMPLEX NEAR OOSTVOORNE. (transect A)
   This is a permanent transect of 40 x 2 m. Main vegetational zones are determined by Corynephorus canescens +
   Cladonia foliacea, Galium verum + Thymus pulegioides,
   Anthoxanthum odoratum + Festuca tenuifolia, Calamagrostis
   epigejos + Lysimachia vulgaris, Festuca ovina + Polytri-
   chium juniperinum, Briza media + Sieglingia decumbens,
   Holcus lanatus + Lolium perenne.(VAN DER MAAREL,thesis
                                                      1966.)
2. LOW SALT MARSH ON SCHIERMONNIKOOG.(quadrat III)
   This is a transect of 16 x 4 m in a relatively homogene-
   nous vegetation of Suaeda maritima, Puccinellia maritima
   and Salicornia europaea with zones and patches of Hali-
   mione portulacoides, Triglochin maritimum, Aster tripo-
   lium and Armeria maritima. (THALEN 1971)

3. TRANSITION BETWEEN HIGHT SALT MARSH AND LOW DUNE ON
   SCHIERMONNIKOOG. (quadrat IV)
   This is a transect of 8 x 4 m across a pronounced zona-

zonation with a Juncus gerardii-Glaux maritima zone at
the lower end, bordered by zones characterized by Schoenus nigricans, Lotus corniculatus and Elytrigia pungens
respectively.(THALEN,1971).

4. DUNE SLACK COMPLEX ON SCHIERMONNIKOOG. (quadrat V)
This is a transect of 16 x 4 m in a gradient from a vegetation of Schoenus nigricans, Juncus maritimus and
Parnassia palustris and a vegetation of Drosera rotundi-
folia and Radiola linoides with an intermediate zone
characterized by Linum catharticum and Carex serotina
ssp. pulchella.(VAN DER MAAREL & LEERTOUWER 1971).

Of these transects data on differential profiles, and species
diversity on various scales from 1/16 sq.m to 4 sq.m are presented together with data on height and pH variation.

6. SOME GENERAL CONCLUSIONS.

In all transects a considerable variation in H-values was found.
Generally the differential profiles showed a distinct pattern.
Profiles based on quantitatively determined H-values did coincide more satisfactory with circumstantial evidence in the
field, esp. on quadrat sizes of 1 sq.m and larger. The quadrat
size of 1/4 sq.m appeared to be most convenient in describing
the general situation in these kinds of relatively low and dense herb vegetations.

In quadrat V the boundary pattern on the scale 1/16 sq.m was
rather irregular. The H-level was very high as compared with
those on other scales suggesting a high micro-heterogeneity.
This maybe explained by the occurrence of micro-mosaics in pH
and soil moisture which factors also built up the main gradient
within this transect.

Generally zones with high H-values, thus containing relatively
sharp boundaries, are found in relatively steep parts of apparent environmental gradients, which are mostly topographical.

In these zones also highest species diversity values are found.
Thus there is a general coincidence of boundary intensity and
species diversity, which is already discernable in very subtile
gradients.

Differences in the pattern of H and H_r are explained by differences in species diversity.

In the transects investigated so far boundaries of the limes

divergence type are predominant. Even in the relatively homogeneous salt marsh communities few sharp boundaries could be detected, although species boundaries were obvious. Evidence of
this kind favours the "continuum concept" of vegetation.

J.J.Barkmann:

Von Herrn ZONNEVELD wurde dokumentiert, daß Assoziationsgrenzen
auf kleinem Raum meist scharf sind, die abstrakten dggegen oftmals nicht. Auch auf kleinstem Raum sind die konkreten Abgrenzungen sehr schwer faßbar. Zu diesem Problem möchte ich auf das
Buch Ecology von ODUM hinweisen, der von einem positiven und
negativen Rand-Effekt spricht.
Professor WESTHOFF hat die Fixierung von Assoziationsgrenzen
verdeutlicht. Das waren konkrete Grenzen, aber auf großem Raum,
etwa der Bereich einer Assoziation in ihrem Areal. Doch auch
diese hat nicht selten fließende Grenzen. Wenn also in einem
solchen Fall eine Art nach der anderen ausfällt und neue Spe-
zies hinzukommen, wäre das ein Kontinuum. Es könnte aber auch
sein, daß die Grenze scharf ist. Nun ist aber bei solchen Fällen noch zweierlei zu bedenken: es ist nämlich möglich, daß für
jede Art, die ausfällt, eine andere hinzukommt und dies in
kleinstem Gesellschaftskomplex. Dann ist der Artenreichtum der
Gesellschaft also sowohl in diesem, als auch in jenem und im
Übergangskomplex ungefähr gleich groß, d.h. die "species diversity" ist nahezu gleich stark. Daneben gibt es natürlich auch
Arten, die "durchgehen", sonst wäre es keine homogene Assoziation. Es kommt aber auch vor, daß erst eine Art ausfällt, dann
eine zweite, eine dritte, und erst dann kommen neue hinzu. Dies
kennen wir vom E l y m o - A m m o p h i l e t u m, das an der
holländischen Küste von Süd nach Nord immer mehr Arten "ver-
liert"; Calystegia soldanella, Glaucium flavum, Eryngium maritimum, Euphorbis paralias gehen zurück, um dann auf den deutschen Nordseeinseln gänzlich zu fehlen. Die Assoziation ist artenarm geworden. In dem erwähnten Übergangsgebiet ist sie jedenfalls ärmer als im Süden Hollands, in Belgien oder Nordfrankreich. Sie wird dann wieder artenreicher in Norddeutschland und
in Jütland, wenn beispielsweise Lathyrus maritimus sich einfin-
det. Dies wäre der negative Randeffekt, wenn im Übergangsbereich
der Artenreichtum geringer als in den beiden benachbarten Gebieten ist. Gleichartiges habe ich am P h y s c i e t u m

e l a e i n a e, einer Epiphytengesellschaft auf Olea europaea
und Quercus ilex im Mittelmeergebiet studieren können. Hier ist
die Assoziation sehr artenreich, was ich über Toulouse hinaus
bis in die Gironde-Ebene beobachten konnte. In der Dordogne-Ebe-
ne verarmt diese Gesellschaft, um dann sowohl in Nordfrankreich,
als auch in Holland wieder reicher zu werden, aber das Arten-
Inventar ist ein anderes. Es handelt sich um zwei geographische
miteinander vikariierende "Rassen".
Das Umgekehrte, der positive Randeffekt, läßt sich gleichfalls
beobachten: dann finden sich im Übergangsgebiet die geographischen Differentialarten beider Ausbildungen, die "species diversity" und ist hier im Übergangsgebiet am größten. Dies kann soweit gehen, daß zwei "gute" Assoziationen miteinander verschmelzen. In Holland sind beispielsweise das P a r m e l i e t u m
a c e t a b u l a e und das R a m a l i n i e t u m f a s t i·
g i a t a e zwei scharf von einander getrennte Epiphyten-Gesellschaften. In der Auvergne verschmelzen beide zu einem nicht
trennbaren Gemisch, einer neuen Assoziation. Da aus Zeitmangel
noch nicht von der unterschiedlichen Festlegung einer Grenzbil-
dung in Gesellschaftskomplexen gesprochen und diskutiert werden
konnte, habe ich nur kurz auf diese Probleme hingewiesen.
VAN DER MAAREL sagt: dort wo das Artengefälle am größten ist,
also an diesem Beispiel auf kleinstem Raum, ist auch die "species diversity" am größten, somit gleich einem positiven Rand-
Effekt. Es wäre sehr ratsam, diese Idee einmal an anderen Vegetationstypen zu testen, ob es hier und dort einen lokal bedingten negativen Rand-Effekt gibt oder nicht. Dies ist aus der
Kurve nicht zu ersehen. Die Differenzial-Kurve zeigt uns nur
die Änderung der Artenzahl. Um aber erkennen zu können, ob der
Rand-Effekt positiv oder negativ ist, benötigt man eine zweite
Kurve, die die "species diversity" angibt, wie sie VAN DER MAA-
REL mitgeteilt hat. Meine Bemerkungen zielen nur darauf hinaus,
Anregung zu geben für Untersuchungen, um zu erfahren, welches
die Ursachen für das Phänomen positiver und negativer Rand-Ef-
fekte sein mögen.

E.van der Maarel:

Ich habe zu wenig Erfahrung mit anderen Situationen, um schon
jetzt sagen zu können, ob und wo positive und negative Rand-
Effekte gefunden werden können.

EINE METHODE RECHNERISCHER AUFSTELLUNG
UND ABGRENZUNG VON VEGETATIONSTYPEN

Fr. KÜHN

Bei dem Studium der Ackerunkraut-Vegetation in Mähren fand ich
oft Arten-Verbindungen,die man nur schwer in die Unkraut-Asso-
ziationen aus anderen Ländern einreihen kann. Ich versuchte darum,in meinen mährischen Aufnahmen die typischen Arten-Verbindungen festzustellen. Ich hatte Interesse daran, typische Arten-Verbindungen ohne subjektiven Einfluß, mit einer objektiven,
reproduzierbaren, möglichst einfachen Methode aus den Aufnahmen
herauszuarbeiten.
In der vorliegenden Arbeit wertete ich Aufnahmen, 200 Ackerun-
kraut-Aufnahmen, von Lednice (Eisgrub) in Südmähren aus, in
welchen die Unkraut-Vegetation des Katasters aufgenommen wurde.
Die Aufnahmen stammten zumeist aus der Zeit von Ende Juni und
Anfang Juli 1963 und 1964. Sie wurden mit Hilfe einer Schablone
in Hefte aus durchsichtigen Papier notiert, wodurch eine schnel-
le Verarbeitung der Aufnahmen ermöglicht wurde. Die Schablone
enthält Abkürzungen von lateinischen Namen von 205 Unkraut-Ar-
ten, drei Schablonen stehen auf einer Seite nebeneinander. Außerdem wurden Mitte April 1966 26 Aufnahmen aufgenommen, um den
Frühjahrsaspekt der Unkraut-Vegetation festzuhalten; diese wurden jedoch gesondert verarbeitet.
Die mittlere Jahrestemperatur der Lokalität ist 9,0°C, das Nie-
derschlagsmittel 524 mm, die Meereshöhe 164-213 m, Bodentyp
Schwarzerden, welche teilweise zu Braunerden inklinieren, und
humose Alluvial-Böden, die Halophyten-Standorte enthalten.
Die durchschnittliche Artenzahl in einer Aufnahme war 18,595
die niedrigste 3, die höchste 46.
Als Vergleichsbasis der Aufnahmen wurde die charakteristische
Artenkombination nach RAABE 1954 gewählt. Die Präsenz der 19
häufigsten Arten aus den Sommer-Aufnahmen (Charakteristische
Artenkombination) war: 1. Chenopodium album 159 Aufnahmen, 2.
Echinochloa crus-galli 132, 3. Galinsoga parviflora 117, 4.
Amaranthus retroflexus 104, 5. Polygonum aviculare 100, 6. Con-
volvulus arvensis 98, 7. Sinapis arvensis 96, 8. Anagallis arvensis ssp.phoenicea 86, 9. Setaria glauca 83, 10. S.viridis 82,
11. Polygonum convolvulus 81, 12. Cirsium arvense 79, 13.   .

Capsella bursa-pastoris 76, 14. Thlaspi arvense 72, 15. Polygonum lapathifolium 70, 16. Raphanus raphanistrum 68, 17. Matricaria maritima ssp.indora 68,18. Agropyrum repens 65, 19. Sonchus asper 62. Mit der charakteristischen Artenkombination wurden alle Aufnahmen verglichen, und die Anzahl der gemeinsamen
Arten ermittelt (durch Unterlegen einer Schablone im Aufnahme-
heft.) Die Ähnlichkeit mit der charakteristischen Artenkombination konnte nicht nach JACCARDS Gemeinschaftskoeffizient ausgedrückt werden, da Aufnahmen mit geringer Artenzahl bei Anwendung
von JACCARDS Gemeinschaftskoeffizient $Gp = \dfrac{Pc}{Pa + Pb + Pc}$ automatisch niedrige Koeffizienten erhalten, auch wenn sie ausnahms-
los aus Arten zusammengesetzt sind, welche in der Arten-Kombination der Vergleichskollektion vorkommen. H.ELLENBERG 1956
schlägt vor, Aufnahmen mit JACCARDS Gemeinschaftskoeffizient
über 25% zur gleichen phytozoenologischen Einheit zusammenzu-
schließen, bei niedrigeren Gemeinschaftskoeffizienten aber zu
trennen. Bei der durchschnittlichen Artenzahl in den Aufnahmen
im Material 18,595 ist JACCARDS Gemeinschafts-Koeffizient 25%
7,4464. Aufnahmen, welche bei gleicher Artenzahl (19) weniger
gemeinsame Arten mit der charakteristischen Arten-Kombination
haben,wurden als unähnlich mit der charakteristischen Arten-Kombination bezeichnet. Um bei ungleicher Artenzahl ein vergleich-
bares Kriterium der nötigen Anzahl von gemeinsamen Arten zu bekommen, wurde eine Kurve nach folgender Überlegung konstruiert:
Die Summe (Interpolation zwischen 18 znd 19 Arten) der Präsenzen der 18,6 häufigsten Arten war 1672,6. Die Summe der Präsenzen der 7,4 häufigsten Arten (Gp 25%), welche für die Einreihung in die ähnlichen Aufnahmen mit der charakteristischen Arten-Kombination in den meisten Fällen verantwortlich sind, war
844,4 (Interpolation zwischen 7 und 8). Die Wahrscheinlichkeit,
daß in einer Aufnahme mit 18,6 Arten im Material 7,4 Arten aus
der charakteristischen Arten-Kombination vorkommen, ist daher
1672,6:844,4=51%. Nun wurden für alle vorkommenden Artenzahlen
in den Aufnahmen (3-46) Zahlen errechnet, welche 51% der Summe
der Präsenzen der gleichen Anzahl der häufigsten Arten entsprachen. Zum Beispiel für die Artenzahl in den Aufnahmen 3 ist die
Summe der Präsenzen der häufigsten Arten 159+132+117 = 408; 51%
davon ist 208; 208 entspricht der Summe der Präsenzen von 1,4
der häufigsten Art im Material; wenn daher in einer Aufnahme
mit 3 Arten mehr als 1,4 Arten der charakteristischen Arten-

Kombination vorkommen, wird die Aufnahme als ähnlich mit der
charakteristischen Arten-Kombination bezeichnet. Die errechne-
ten Zahlen waren: 4:1,8, 5:2,2, 6:2,6, 7:3,03, 8:3,5, 9:3,9,
10:4,3,11:4,7, 12:5,1 usw.
Mit dieser Methode wurden etwa 50 unähnliche Aufnahmen ausgeschieden. Diese Aufnahmen waren zumeist auf ökologisch extremen
Standorten, welche meist zugleich durch das besondere Vorkommen
von Arten und Artengruppen charakterisiert waren. Die ausgeschiedenen Aufnahmen konnten daher leicht zu Gruppen zusammengeschlos-
sen werden. Von diesen Aufnahmengruppen wurde eine Aufnahmenta-
belle zusammengestellt, und in den einzelnen ausgeschiedenen
Gruppen von Aufnahmen die charakteristischen Artengruppen nach
RAABE ausgearbeitet (Einreihung der letzten Arten in Zweifelsfällen mit Zuhilfenahme der Abundanz). Die einzelnen Charakter-
artengruppen wurden danach mit der charakteristischen Artenkombination aus dem ganzen Material (200 Aufnahmen) und die einzelnen charakteristischen Artengruppen untereinander verglichen und
auf ihre Ähnlichkeit geprüft und nur die Vegetationstypen belassen, welche nach JACCARDS Gemeinschaftskoeffizient mit allen anderen Typen einen Koeffizient unter 25% hatten (mit Respektierung der Artenzahlen in den Kombinationen). Die charakteristische Artenkombination aus den 200 Aufnahmen wurde dann durch
Abrechnung der unähnlichen Aufnahmen revidiert. Da nur eine Art
nicht bestätigt wurde, (statt Sonchus asper, Melandrium nocti-
florum) wurde die Kombination nicht geändert. In allen 200 Aufnahmen wurde nun die Anzahl der gemeinsamen Arten mit den erhaltenen Vegegationstypen ermittelt (mit Hilfe von Schablonen). Für
die einzelnen Vegegationstypen wurden JACCARDS Gemeinschaftskoeffizienten für 25% für die mittlere Artenzahl ausgerechnet.
Bei fast allen Aufnahmen konnte dann leicht die Zugehörigkeit zu
den ermittelten Vegetationstypen festgestellt werden (mit Respektierung der mittleren Artenzahlen in den Kombinationen).
Einzelne Aufnahmen blieben allen Typen unähnlich und konnten
nicht eingereiht werden (zu ihnen werden in anderen Materialen
Analogen gesucht werden). Einzelne Aufnahmen wurden als intermediär zwischen zwei Typen bezeichnet. Die Resultate wurden in
eine Karte eingetragen.
Die Vegetationstypen sind im studierten Material durch edaphische Faktoren bedingt; der Einfluß von Kultur-Maßnahmen und von
Kultur-Pflanzen zeigte sich nur in einem Falle edaphischen Fak-

toren adäquat, und zwar im Unkraut-Vegetationstyp der gärtnerisch bearbeiteten Flächen, der Rosenbeete und Rabatten im Schloßpark, in der Gärtnerei und in manchen gut gedüngten Weinbergen. Der Typ der Halophyten-Standorte, der Typ von Gartenbeeten und der Typ vom "Mlýnský rybník"haben eine Anzahl von Differentialarten, welche zwar nur in einem kleineren Anteil der Aufnahmen vorkamen und nicht in die charakteristischen Arten-Kombinationen nach RAABE gehören, aber welche im Material ausschließl.ch auf den Halophyten-Standorten oder auf Gartenbeeten vorkommen.

Charakteristische Artenkombinationen der erhaltenen Unkraut-Vegetationstypen:

Artnamen	durchschnittlicher Typ im Kataster	Halophyten-Standorte im Aluvium	Typ gärtnerisch bearbeiteter Flächen	Typ vom "Mlýnský rybník"	Typ auf Feldern bei Nejdek
Chenopodium album	+	+	+	+	+
Polygonum aviculare	+	+	+	+	+
Galinsoga parviflora	+	+	+	+	
Anagallis arvensis ssp.phoenicea	+	+		+	
Matricaria maritima ssp.inodora	+	+		+	
Sonchus asper	+		+	+	
Amaranthus retroflexus	+		+		+
Convolvulus arvensis	+		+		+
Sonchus oleraceus			+	+	+
Polygonum lapathifolium	+	+			
Sinapis arvensis	+		+		
Capsella bursa-pastoris	+		+		
Agropyrum repens	+		+		
Polygonum convolvulus	+			+	
Cirsium arvense	+				+

Art					
Echinochloa crus-gallii	+				+
Equisetum arvense		+		+	
Plantago major ssp.intermedia		+		+	
Symphytum officinale		+		+	
Sonchus arvensis		+		+	
Stellaria media			+	+	
Erigeron canadensis			+	+	
Poa annua			+	+	
Raphanus raphanistrum	+				
Thlaspi arvense	+				
Setaria viridis	+				
Setaria glauca	+				
Ranunculus repens		+			
Ranunculus sardous		+			
Chenopodium polyspermum		+			
Atriplex hastata f.incana		+			
Potentilla anserina		+			
Potentilla supina		+			
Epilobium tetragonum ssp.adnatum		+			
Lythrum virgatum		+			
Rorippa silvestris		+			
Odontites rubra		+			
Mentha arvensis		+			
Gnaphalium uliginosum		+			
Juncus bufonius		+			
Amaranthus lividus			+		
Portulacca oleracea			+		
Urtica urens			+		
Oxalis stricta v.europaea			+		
Euphorbia peplus			+		
Senecio vulgaris			+		
Galinsoga ciliata			+		
Papaver rhoeas				+	
Melandrium noctiflorum				+	
Polygonum persicaria				+	
Vicia sativa ssp. angustifolia				+	
Viola arvensis				+	
Galium aparine				+	

Taraxacum officinale   +
Cichorium intybus   +
Medicago lupulina   +
Vicia villosa   +
Lathyrus tuberosus   +
Erodium cicutarium   +
Veronica persica   +
Veronica polita   +
Lamium amplexicaule   +
Stachys annua   +
Avena fatua   +

"Differentialarten" der Unkraut-Vegetationstypen im Material außer der charakteristischen Arten-Kombination.

Halophyten-Standorte im Alluvium: Juncus ranarius, J.articulatus, Cyperus fuscus, Bolboschoenus maritimus, Rumex maritimus, Polygonum amphibium, Potentilla reptans, Trifolium fragiferum, T.hybridum, Lotus corniculatus ssp.tenuifolius, Atriplex tatarica, Rorippa islandica, Lythrum hyssopifolium, Veronica anagallis, Verbascum blattaria, Limosella aquatica, Teucrium scordium, Inula brittanica, Xanthium strumarium, Bidens tripartitus, B.melanocarpus.

Typ gärtnerisch bearbeiteter Flächen: Matricaria suaveolens, Impatiens parviflora, Chelidonium majus, Convolvulus sepium, Cynodon dactylon, Setaria verticillata.

Beispiel der Auswertung:

Aufnahmenummer:	∅	60	61	62	63	64	65	66	67	68	69
Artenzahl:		20	9	12	8	16	23	17	15	27	22
Arten des Durchschnitt-Typus	18,6	9	5	7	5	4	14	12	9	11	10
Arten d.Halophyten-Standorte	25,1	1	1	1	3	4	5	3	1	2	3
Arten des Garten-Typus	18,6	8	4	6	3	4	9	4	7	10	11
Arten des Typus vom "Mlýnský rybník"	22,7	4	2	4	7	6	9	7	2	7	8
Arten des Typus der Felder bei Nejdek	16,5	6	3	6	1	5	9	6	4	4	5

Schablone für die Aufnahmen (Reihenfolge der Symbole nach dem
System von A.L.TACHTADZJAN 1954):

Nr.	Fr.	Wuchs	Bearb	Ver- unkr.
Ort				
Boden				
Wasser				
PotAmaTraViaAnaVagEri				
EqaAmhTrcResAncVarFil				
DcoAmrTrdSsoCemVheGna				
MmiPcaTprErhLyvVpeGal				
RarChaTreEreShaVpoAar				
RreChhLotBarGapVseAau				
AdoChpCovRorGspVtrAch				
ParAniVanArmGtrRhiMin				
PduApaVhiAthVd OdoMsu				
PrhSalVpaBerVolPlmArt				
FuoRcrVteAlyKnaPlcTus				
FurRllVviErvConAjuSen				
FuvRobLtuDipLapGheCir				
ArePa UrtSinLitGbiCen				
HolPmpLytRpeMyaGteJbu				
StgPc OxaRphMymGanAve				
StmPhyGdiCorEvuLamHmo				
CglPlpGpuLdrLycLpuAst				
CscPpeEroTa NonStaAte				
SagRubEryTpeSymStpApe				
SpaPanCauCmiHscMenPhl				
SruPrpBifNesSniCamPoa				
SclPsuAegCapDatLpsPtr				
AgiAlaFalMerAntCreAgr				
MlaMalAecEesLivTofPhr				
MlnMofAneEexLelLacEra				
SdiMluPasEfaLspSarEch				
Gm MsaDauEheLm SasSvi				
Vac HhuEpe SolSgl				
Sap HpeEpl Dsg				
Zuschriften				

Karte der Unkraut-Vegetationstypen:

1 : Durchschnittlicher Typ im Kataster
2 : Halophyten-Standorte im Alluvium
3 : Typ auf gärtnerisch bearbeiteten Flächen
4 : Typ vom "Mlýnský rybník"
5 : Typ auf Feldern bei Nejdek
6 : Lokalitäten abweichender Aufnahmen

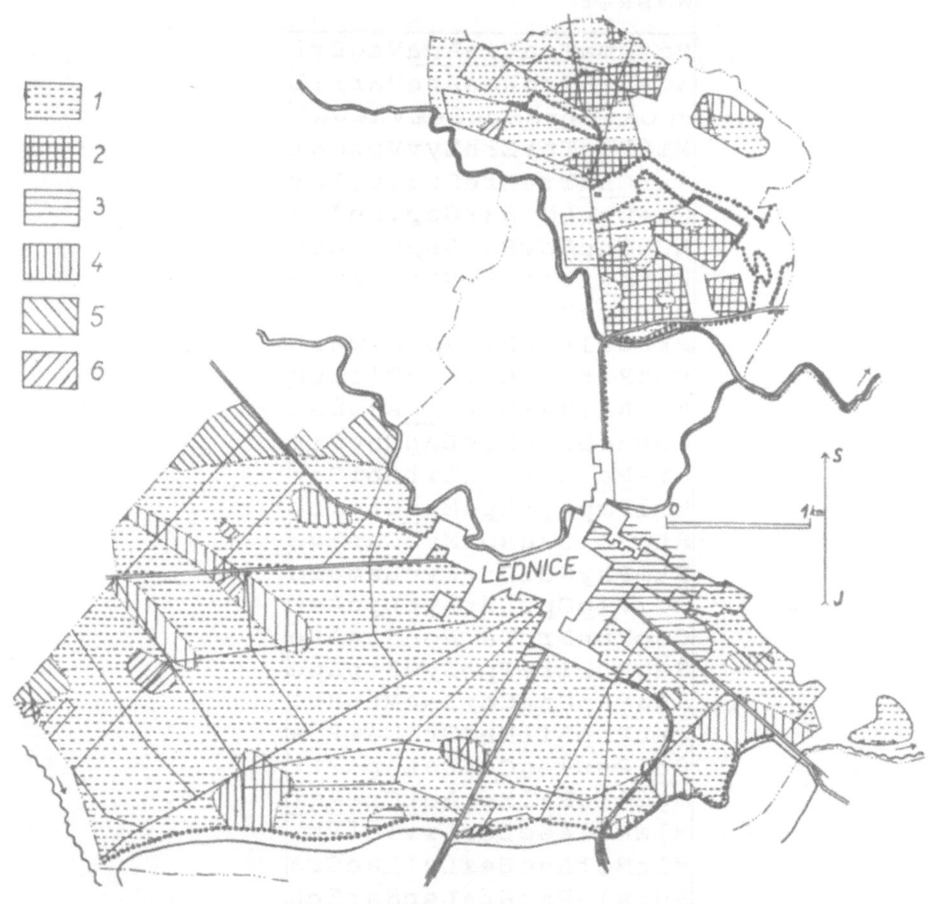

ZUSAMMENFASSUNG

Die durchschnittliche Artenzahl in den 200 verarbeiteten phyto-
zönologischen Aufnahmen war 18,6. Es wurde die charakterische
Arten-Kombination nach RAABE 1954 zusammengestellt. In jeder
Aufnahme wurde die Anzahl der gemeinsamen Arten mit dieser Kombination festgestellt. H.ELLENBERG 1956 schlägt vor, Aufnahmen
mit dem Gemeinschaftskoeffizient nach JACCARD 1928 über 25% in
gleiche phytozönologische Einheiten einzureihen.Der Gemeinschaftskoeffizient 25% in Aufnahmen von einer Artenzahl 18,6 im Material ist 7,446. Der Koeffizient kann nicht zum Vergleich von Aufnahmen mit ungleicher Artenzahl verwendet werden, da Aufnahmen
mit niedriger Artenzahl automatisch niedrige Koeffizienten aufweisen, auch wenn alle Arten mit der verglichenen Arten-Kombination gemeinsam sind.
Die Summe der Präsenzen der 18,6 häufigsten Arten war 1672,6,
die Summe der Präsenzen der 7,446 häufigsten Arten war 844,4 =
51%. Für alle vorkommenden Artenzahlen in den Aufnahmen (3-46)
errechnete ich die Artenzahl, welche 51% der Präsenzsumme der
gleichen Anzahl der häufigsten Arten entsprachen; z.B. für die
Artenzahl in den Aufnahmen 3 ist die Summe der Präsenzen der
häufigsten Arten im Material 159 + 132 + 117 = 408; 51% davon
ist 208; 208 entspricht der Summe der Präsenzen von 1,4 der
häufigsten Arten im Material. Mit dieser Methode wurden aus dem
Material 50 Aufnahmen ausgesondert, welche dem Durchschnittstyp
unähnlich waren. Diese Aufnahmen waren zum größten Teil von extremen Standorten, und waren zumeist durch spezielle Artengruppen gekennzeichnet. Es war leicht, von diesen Aufnahmen Gruppen
zusammenzustellen. Es wurde eine Tabelle der Aufnahmen zusammengestellt, und die charakteristischen Artenkombinationen der Aufnahmengruppen ermittelt. Diese Artengruppen wurden mit der charakteristischen Artenkombination des ganzen Materials,und alle
Artengruppen miteinander verglichen, und die Gemeinschaftskoeffizienten ermittelt. Nur Kombinationen mit dem Koeffizienten unter
25% wurden als selbstständige Gruppen belassen. So wurden 5 Arten-Kombinationstypen aufgestellt. Die charakteristische Arten-
Kombination für das ganze Material wurde nach Abziehung der unähnlichen Aufnahmen neu revidiert. In allen Aufnahmen des Materials wurden dann die Anzahlen der gemeinsamen Arten mit allen
aufgestellten Arten-Kombinationstypen gezählt, und die Zugehörigkeit zu den Typen bestimmt (mit Respektierung der Artenzahlen.)

Nur einige Aufnahmen waren intermediär zwischen zwei Typen, und
einige allen Typen unähnlich. Zu solchen Aufnahmen werden in anderen Materialen Analogien gesucht werden.

SUMMARY

In 200 phytocoenological samples of field weeds, the average
number of species was 18,6. The characteristic combination of
species according to RAABE 1954 has been established. In all
samples the number of species common with that combination has
been determined. H.ELLENBERG 1956 propose to put samples with
the coefficient of identic species according to P.JACCARD 1928
25% into the same phytocoenological units. This coefficient for
25% in samples of 18,6 species in one sample is 7,446. The coefficient could not be used to compare samples with different
numbers of species, because samples with a little number of species have a low coefficient even if all species are common with
the compared species complex.
The sum of presences of the 18,6 most frequent species was
1672,6; the sum of 7,446 most frequent species 844,4 = 51%. For
all numbers of species in the samples (3-46), I calculated the
number of species responding 51% of the sum of presences of that
number of most common species; for example, for 3 species in a
sample, the sum of presences of the most frequent species is 159
+ 132 + 117 = 408; 51% from it is 208; 208 responds to the presence of 1,4 most frequent species. By this method, about 50
samples dissimilar with the average type of weed vegetation have
been sorted out. These samples were in the most cases from extrem stands, and were often characterized by special groups of
weed species. It was easy to arrange the samples in groups. A
table of samples has been compiled and the characteristic species combinations of the single groups of sample have been established. These groups have been compared with the characteristic
combination of species of the whole material and all combinations one with another, and only the weed vegetation types which
were different from each other in the range of JACCARDS coefficient 25% (respecting the species number in the combinations)
remained. So 5 weed vegetation types have been established. The
average species combination of the whole material has been revi-
ded after separating the dissimilar samples. In all samples the
numbers of common species with all established types have been

counted and the vegetation type has been determined (with respecting the species number). In a few cases, the samples were intermediate between two types, and a few samples have been dissimile to all types. Those samples will be worked up in comparing with other materials.

LITERATUR

ELLENBERG,H.   -1956-   Aufgaben und Methoden der Vegetationskunde.- In: Walter,H.: Einführung in die Phytologie 4 (1). Stuttgart.

JACCARD,P.   -1928-   Die statistisch-floristische Methode als Grundlage in der Pflanzensoziologie.- In: Abderhalden,E. (Edit.): Handb.biol.Arbeitsmeth.11 (5). Berlin-Wien.

RAABE,E.   -1952-   Über den "Affinitätswert" in der Pflanzensoziologie.- Vegetatio 4 (1).   Den Haag.

APPLICATION D'UN PROCESSUS DE
CLASSIFICATION NUMERIQUE A DES
DONNEES PHYTOSOCIOLOGIQUES

J.L.GUILLERM et F.ROMANE

INTRODUCTION

ESCOUFIER (1966) propose une méthode objective pour constituer
des groupes de variables à partir de leur matrice de corrélation.

Cette méthode a été employée en utilisant comme variables l'abondance numérique de 90 espèces observées   dans 628 relevés de
terres de culture du Bas-Languedoc. Il a pu être vérifié que
les groupes d'espèces ainsi obtenus coincident bien avec les
groupes "écologiques" établis directement à partir de l'analyse
des profils écologiques des espèces pour chaque facteur du milieu.

Nous commencerons par décrire la méthode de classification puis
nous présenterons les groupes floristiques obtenus, enfin, à
titre d'exemple, nous donnerons la signification écologique de
quelques uns d'entre-eux.

I- La méthode de classification

Cette méthode a été envisagée à l'origine comme un préalable à
l'Analyse des Composantes Principales pour la recherche de groupes de variables fortement corrélées entre-elles (ESCOUFIER,
1966).

Il s'agit donc bien de construire des groupes, c'est-à-dire,
que le terme de classification est pris ici dans le sens pro-
posé par DAGNELIE (1966) et TOMASSONE (1967): définir ou "faire"
des classes.

La classification se fait à partir de la matrice des coefficients
de corrélation totale entre les variables prises deux à deux
selon le processus suivant:

> 1) Les deux variables qui présentent le plus fort
>   coefficient de corrélation sont extraites de la
>   matrice et donnent naissance au premier groupe;
>
> 2) On isole la variable dont la somme des coeffici-
>   ents de corrélation avec les variables du groupe
>   ébauché est maximale;
>
> 3) Il faut alors décider si cette variable apparti-
>   ent au groupe ébauché ou non, ce qui est le pro-

blème difficile. ESCOUFIER (1966) propose le moy-
en suivant : Soit un groupe constitué par p variables (p $\geqslant$ 2) ; une nouvelle variable j est alors
introduite suivant le critère 2). Entre les (p+1)
variables, il existe p $\frac{(p+1)}{2}$ coefficients de corrélation auxquels on applique la transformation

$$z = \frac{1}{2} \ \text{Log}\Big[(1+r) - \text{Log} \ (1-r)\Big]$$

(Log = logarithme népérien)

On calcule pour ces $\frac{p\ (p-1)}{2}$ coefficients transformés la moyenne $\underline{z}$; il est alors possible de prendre pour estimation de la
variance de $\underline{z}$, la quantité $\sqrt{\frac{1}{n-3}}$ où $\underline{n}$ est l'effectif de l'échan-
tillon à partir duquel ont été établis les coefficients de cor-
rélation. La valeur $z + t \sqrt{\frac{1}{n-3}}$, $\underline{t}$ étant le paramètre de
Student, peut alors être considérée comme une valeur maximum
d'estimation de la moyenne, valeur qui varie suivant le seuil
de probabilité choisi pour $\underline{t}$.
Le critère pour décider de l'appartenance ou non de la variable
$\underline{j}$ au groupe déjà ébauché s'établit ainsi:

- Si tous les coefficients de corrélation de la
variable $\underline{j}$ avec les variables du groupe ébauché
sont inférieurs à $z + t \sqrt{\frac{1}{n-3}}$, la variable $\underline{j}$ est
intégrée au groupe. On reprend alors le processus
en 2) pour chercher de nouvelles variables pouvant
appartenir au groupe.
- Si un au moins des coefficients est supérieur
à $z + t \sqrt{\frac{1}{n-3}}$ la variable j est rejetée. Le
groupe ébauché est donc terminé et la variable $\underline{j}$
est dite "variable essayée" à la fin de ce groupe.

L'algorithme repart alors en 1) avec les variables non encore intégrées dans un groupe, (y compris
la variable qui vient d'être rejetée).

Remarque:
Le programme de calcul actuellement existant considère les va-
leurs absolues des coefficients. Ceci entraîne que certains
groupes floristiques puissent contenir des espèces "opposées".

II- Application à 628 relevés des terres de culture du
    Bas-Languedoc

Ainsi qu'il a été dit, la matrice des coefficients de corrélation a été établie en prenant comme variable, l'abondance numérique de 90 espèces, ce chiffre de 90 étant fixé par les possibi-
lités de l'ordinateur (IBM 360-40) mis à notre disposition. Cette abondance numérique, égale à zéro si l'espèce est absente, a
été notée suivant une codification (voir Annexe)proposée par le
Centre d'Etudes Phytosociologiques et Ecologiques (C.E.P.E.,
1967).

Quant au choix des 90 espèces, parmi les 404 présentes dans l'ensemble des 628 relevés, il a été dirigé de manière à conserver
les principaux types d'espèces rencontrées, c'est-à-dire : espèces abondantes, moyennement abondantes ou rares, et espèces
à écologie stricte ou non.

A partir de ces données, un essai a été effectué, en prenant
comme valeur pour le paramètre $\underline{t}$ de Student celle correspondant
au seuil de probabilité de 5%. Les résultats de cet essai sont
donnés ci-dessous sous la forme de la liste des espèces consti-
tuant les différents groupes floristiques.

A la fin de chaque groupe le nom de l'espèce essayée puis rejetée (cf § II-3) est mentionnée, accolé au signe +.

GROUPE 1

Juncus bufonius L.	Veronica anagallis L.
Lythrum hyssopifolia L.	+Ranunculus sardous Crantz

GROUPE 2

Artemisia verlotorum Lamotte	Equisetum arvense L.
Sorgum halepense (L.)Brot.	Convolvulus sepium L.
Polygonum lapathifolium L.	+Amaranthus retroflexus L.

GROUPE 3

Anthemis maritima L.	Convolvulus soldanella L.
+ Cakile maritima Scop.	

GROUPE 4

Cakile maritima Scop.	Arundo phragmites L.
Rumex tingitanus L.	+Plantago ramosa (Gilib.) Asch.

GROUPE 5

Linaria spuria (L.) Miller	Polygonum convolvulus L.
Polycnemum arvense L.	Papaver rhoeas L.
Euphorbia falcata L.	Atriplex patulus L.

Rapistrum rugosum L.   Reseda phyteuma L.
Anagallis arvensis ssp.   +Plantago lanceolata L.
phoenicea Scop.

GROUPE 6
Lamium amplexicaule L.   Stellaria media (L.) Vill.
          + Lagoseris sancta (L.) Maly

GROUPE 7
Geranium dissectum L.   Ornithogallum ombellatum L.
            + Picris echioides L.

GROUPE 8
Bromus tectorum L.   Chondrilla juncea L.
Senecio gallicus Chaix   Silene conica L.
              + Medicago minima L.

GROUPE 9
Lagoseris sancta (L.) Maly   Sonchus oleraceus L.
            + Muscari neglectum Gussone

GROUPE 10
Lotus corniculatus L. ssp.   Alopecurus agrestis L.
decumbens (Poir.)Briq.
Ranunculus sardous Crantz   +Rumex crispus L.

GROUPE 11
Picris echioides L.   Beta maritima L.
Rumex crispus L.   +Galium aparine L.

GROUPE 12
Erodium cicutarium (L.)   Rumex pulcher L.
L'Héritier
Muscari neglectum Gussone   Setaria viridis (L.) P.B.
Muscari comosum (L.)Miller   +Senecio vulgaris L.

GROUPE 13
Carduus tenuiflorus Curtiss   Podospermum laciniatum (L).D.C.
            + Hordeum murinum L.

GROUPE 14
Anacyclus tomentosus (Gouan)   Erodium ciconium (L.) Ait.
                      D.C.
Malva silvestris L.   Euphorbia serrata L.

Hordeum murinum L   +Medicago minima L.

GROUPE 15
Diplotaxis erucoides(L.)D.C.   Sonchus asper (L.) Hill
Senecio vulgaris L.   +Amaranthus retroflexus L.

GROUPE 16
Galium aparine L.   Scandix pecten-veneris L.
             + Sherardia arvensis L.

GROUPE 17
Amaranthus retroflexus L.   Antirrhinum orontium L.
Portulacca oleracea L.   +Convolvulus arvensis L.

GROUPE 18
Plantago lanceolata L.   Aristolochia rotunda L.
Plantago major L.   +Potentilla reptans L.

GROUPE 19
Aristolochia clematitis L.   Cirsium arvense (L.) Scop.
Convolvulus arvensis L.   +Veronica hederifolia L.

GROUPE 20
Potentilla reptans L.   Sherardia arvensis L.
             + Cerastium glomeratum Thuill.

GROUPE 21
Diplotaxis tenuifolia (L.) D.C. Lactuca saligna L.
Stachys annuus L.   Lactuca scariola L.
Lithospermum arvense L.   +Equisetum ramosissimum Desf.

GROUPE 22
Fumaria officinalis L.   Veronica polita Fries
Veronica hederifolia L.   Lepidium draba L.
Fumaria parviflora Lmk.   +Lactuca virosa L.

GROUPE 23
Carduus picnocephalus L.   Cerastium glomeratum Thuill.
             + Galactites tomentosa Moench

GROUPE 24
Galactites tomentosa Moench   Medicago minima L.
             + Equisetum ramosissimum Desf.

GROUPE 25

Equisetum ramosissimum Desf.   Lactuca virosa L.
              Lamium purpureum L.

III-Signification écologique des groupes floristiques

Les relevés utilisés étant à la fois floristiques et écologiques, des "groupes écologiques" (GOUNOT,1958) ont été établis
à partir de l'étude des profils écologiques des espèces pour
les différents facteurs du milieu.

La concordance entre les "groupes écologiques et les "groupes
coenologiques; (établis sur le critère de l'abondance numérique) est très satisfaisante.

L'écologie de trois de ces groupes coenologiques ainsi que leur
correspondance avec les groupements végétaux de BRAUN-BLANQUET
est donnée ci-dessous, à titre d'exemple:

              Groupe 1:

              Juncus bufonius L.   Lythrum hyssopifolia L.
                    Veronica anagallis L.

Ce groupe se rencontre sur des sols hydromorphes, inondés en
hiver. Les espèces qui le constituent ont besoin d'un sol saturé
en eau jusqu'au printemps. Elles se retrouvent dans les groupements végétaux non cultivés des ordres I s o e t e t a l i a
et P a s p a l o - H e l e o c h l o e t a l i a (BRAUN-BLAN-
QUET et al.1951)

              Groupe 10:

              Lotus corniculatus L.   Ranunculus sardous
              ssp.decumbens (Poir.) Briq.   Crantz
                    Alopecurus agrestis L.

Ces espèces se développent sur des sols de texture fine, très
humides mais non inondés en hiver, et qui ont des conditions
hydriques moyennes en été, par suite de l'abaissement du niveau
de la nappe phréatique. Ces sols présentent même parfois un horizon désséché en surface (fentes de retrait) du à une très
forte évaporation superficielle estivale. Le groupe 10 est voi-
sin du groupe 1, comme le montre "l'espèce essayée" Ranunculus
sardous Crantz. Ils font partie, tous les deux par rapport à
l'ensemble des relevés, de la série des groupes hygrophiles. Les
espèces du groupe 10 se retrouvent dans les groupements végétaux
de l'ordre J u n c e t a l i a m a r i t i m i, (BRAUN-BLANQUET,

J. et de RAMM,C.1958)

   Groupe 8:
   Bromus tectorum L.   Chondrilla juncea L.
   Senecio gallicus Chaix Silene conica L.

Ce groupe de xérophytes se développe sur des sols de texture
sableuse et limono sableuse, bien drainés, souvent dolomitiques.
Ces espèces se retrouvent dans les groupements végétaux de
l'alliance A r m e r i o n   j u n c e a e (BRAUN-BLANQUET et
al.1951)
Silene conica L., qui est présente aussi sur les sables calcai-
res, dunaires, fixés et mis en culture, se retrouve dans l'ordre A m m o p h i l e t a l i a. (BRAUN-BLANQUET,J.1933).

Remarques:
Sur ces mêmes données du Bas-Languedoc, d'autres essais ont été
effectués d'une part, en prenant en considération les espèces
situées en extension, leur abondance numérique étant fixée ar-
bitrairement à 0,3, et d'autre part, en faisant varier le seuil
de probabilité pour le paramètre t de Student. Les résultats
obtenus avec ces différentes modifications sont pratiquement
identiques à ceux que nous avons présentés.

CONCLUSION
Aucune des méthodes de classification numérique proposées jusqu'
à maintenant n'est réellement satisfaisante pour des phytoso-
ciologues. Celle-ci ne prétend pas reposer sur un critère par-
faitement rigoureux; si nous la présentons, c'est parce qu'elle
est rapide et que les groupes d'espèces obtenus concordent bien
avec les "groupes écologiques".

ANNEXE - ABONDANCE NUMERIQUE DES ESPECES (extrait du code écolo-
                 gique, C.E.P.E.,1967)

Code	Limite des classes		
0			1 individus
1	2	à	4 individus
2	5	à	9 individus
3	10	à	19 individus
4	20	à	49 individus
5	50	à	99 individus
6	100	à	199 individus

7	200 à 499	individus
8	500 à 999	individus
9	1000 individus ou plus	

SUMMARY

The process proposed by ESCOUFIER (1966) for the working out
of groups of variables from their correlation matrix was em-
ployed using as a variable the abundance of 90 species observed in 628 releves of the Bas-Languedoc.
It came true that the groups of species obtained in this manner
agree satisfactorily with the ecological groups drawn up from
the analysis of the ecological profiles of the species for each
environment factor.

ZUSAMMENFASSUNG

Die Methode, die von ESCOUFIER vorgeschlagen war, um variable
Gruppen auf Grund ihrer Korrelationsmatrize zu entwerfen, wur-
de Anwendung der numerischen Abundanz von 90 in 628 Aufnahmen
der im Bas-Languedoc untersuchten Arten benutzt.
Es erwies sich, daß die mit Hilfe dieses Verfahrens bestimmten
Artengruppen genügende Übereinstimmung mit den ökologischen
Gruppen zeigen, die auf Grund der Untersuchung der ökologischen
Profile der Arten für verschiedene Umweltfaktoren erarbeitet
worden sind.

BIBLIOGRAPHIE

BRAUN-BLANQUET,J.   -1936- Un joyau floristique et phytosocio-
     logique "l'Isoetion" mediterranéen.- Comm.SIGMA 42,Montpel-
     lier. 23 p.
BRAUN-BLANQUET,J. & RAMM,Cl.de -1938- Les près salés du Lan-
     guedoc méditerranénne.- Comm.SIGMA 139, Montpellier. 43 p.
BRAUN-BLANQUET,J., ROUSSINE,N., NEGRE,R. & EMBERGER,L. -1952-
     Les groupements végétaux de la France méditerranénne.-
     C.N.R.S. Paris. 297 p.
C.E.P.E.   -1967- Code pour le relevé méthodique de la végéta-
     tion et du milieu.- Principes et transcription sur cartes
     perforées.- C.N.R.S. Paris. 270 p.
DAGNELIE,P. -1966- A propos des différentes méthodes de clas-
     sification numérique.- Revue statistique appliquée 16 (3).
     p.55-75. Paris.
ESCOUFIER,Y.   -1966- Analyse des composantes principales, uti-
     lisation des groupes de variables dans le recherche de la
     solution.- Thèse, Montpellier. 53 p.
GOUNOT,M.   -1958- Contribution à l'étude des groupements vé-
     gétaux messicoles et rudéraux de la Tunisie.- Ann.Serv.Bot.
     et Agron de Tunisie 3. Tunis. 275 p.

MOOR,M.   -1936- Zur Soziologie der Isoetetalia.- Bcitr.Gcobot.
    Landesaufn.Schweiz 20. Bern. 148 p.
TOMASSONE,R.   -1967- Discrimination et classement.- Séminaire
    Probabilités et Statistiques. Nancy, 3 Mai 1967.- C.N.R.F.
    14, rue Girardet, Nancy. 27 p.

DIE ROLLE DES MASSTABS BEI DER ABGRENZUNG
VON VEGETATIONSEINHEITEN

P. S e i b e r t

Aus dem Thema meines Vortrags geht schon hervor, daß es sich
bei den Grenzen, von denen hier die Rede sein soll, nur um
räumliche Grenzen handeln kann. Doch wird zu zeigen sein, daß
sich die Maßstäbe fast immer auch auf die inhaltliche Fassung
der Vegetationseinheiten auswirken.

Kriterien für die Wahl des Maßstabes
Fast jeder, der im Gelände oder nach Bildern (nach gewöhnlichen
Photos, Luftbildern) Vegetationseinheiten voneinander abgrenzt,
hat die Absicht, diese Grenzen in irgendeiner Weise auf dem Papier wiederzugeben, sei es als Aufriß oder, was die Regel ist,
in kartographischer Darstellung. Hierbei taucht sogleich die
Frage nach dem zu wählenden Maßstab auf.
Ein großes Objekt, z.B. die Vegetation eines Landes oder gar
Erdteils verlangt einen kleinen Maßstab, bei welchem Einzelheiten nicht dargestellt werden können. Kleine Objekte, wie die
Vegetation eines Landschaftsauschnittes, die im Detail dargestellt werden soll, oder gar Bestandteile von Pflanzengesellschaften wie Einzelpflanzen, Pflanzengruppen, Synusien, können
nur in großen oder sehr großen Maßstäben wiedergegeben werden.
Sollen bei großen Objekten auch gewisse Einzelheiten noch er-
kennbar bleiben, muß die Darstellung auf mehrere Kartenblätter
verteilt werden. Als extreme Beispiele seien die Karte von
BROCKMANN-JEROSCH "Klimatisch bedingte Formationsklassen der
Erde" im Maßstab 1 : 90 000 000 und das bei BRAUN-BLANQUET
(1962) wiedergegebene Kärtchen der Artenzusammensetzung der
S t i p a   b a i c a l e n s i s - K o e l e r i a   g r a c i -
l i s -Steppe im Maßstab 1 : 10 genannt.
Der Maßstab richtet sich in seiner Größenordnung also nach der
Größe des darzustellenden Objektes und den Einzelheiten, die
dargestellt werden sollen ; in der genauen Festlegung richtet
er sich nach dem verfügbaren Kartenmaterial.
Die Größe des darzustellenden Objektes richtet sich nach den
sachlichen Fragestellungen. Rein vegetationskundliche Betrachtungsweisen können alle denkbaren Maßstäbe erforderlich machen.
Sehr häufig geben aber Fragestellungen anderer Fachgebiete, für

welche die Vegetationskarten Anwendung finden sollen, den Ausschlag über den Maßstab. Vegetationskarten für die forstliche
Standortskartierung haben meist den Maßstab 1 : 10 000, solche
für Grünlandkarten, die als Grundlage für Meliorationen dienen
sollen,einen von 1 : 5 000. Bei ihnen genügt es, wenn die in
Nutzung stehenden Vegetationseinheiten unterschiedlichen Ertrags-
potentials und unterschiedlicher Standortsbedingungen voneinander abgegrenzt werden. Alle anderen, oft viel kleinflächiger
verbreiteten Pflanzengesellschaften wie die Trittvegetation der
Wege, Magerrasen auf Grenzstreifen, Röhrichte an Bächen und Gräben, Waldmäntel und -säume können in solchen Karten vernachlässigt werden. Eine Vegetationskarte, die der Landesplanung und
Raumordnung dienen soll, hat ein größeres Gebiet darzustellen
und kann vieler Einzelheiten entbehren. Ihr Maßstab liegt demnach etwa zwischen 1 : 25 000 und 1 : 100 000 oder gar 1 :
200 000. Bei ihrer Anfertigung im Gelände wendet man eine noch
komplexere Betrachtungsweise an, indem man nicht die einzelnen
Pflanzengesellschaften, Bodenmerkmale, Nutzungsformen usw. für
sich betrachtet, sondern den Landschaftsraum eines Vegetationsgebietes als Einheit nimmt und gegen andere abzugrenzen versucht.

Aus dem Gesagten geht hervor, daß in den meisten Fällen schon
die Fragestellung und das verfügbare Kartenmaterial den anzuwen-
denden Maßstab bestimmen. Damit üben sie von vornherein auch
einen sehr starken Einfluß auf die Geländearbeit aus. Denn jeder Bearbeiter legt ja Wert darauf, seine Aufgabe in einer an-
gemessenen Zeit zu erfüllen. Er wird die Gesellschaften, die er
nicht darstellen kann, gar nicht erst aufnehmen, oft nicht einmal bemerken. So habe ich z.B. vor 12 Jahren bei der pflanzensoziologischen Kartierung der Pupplinger Au an der Isar bei
Wolfratshausen (SEIBERT 1958) 2 seltene und daher beachtenswer-
te Gesellschaften des Alpenvorlandes, nämlich das J u n c e -
t u m, a l p i n i und das B e l l i d i a s t r o - S a x i -
f r a g e t u m m u t a t a e übersehen, weil sie für eine
Wiedergabe im Maßstab 1 : 5 000 zu klein waren. Und oft sind
Gesellschaften komplex gefaßt worden, weil der Bearbeiter bei
der Aufnahme und Gliederung im wirklichen Sinne einen falschen
Maßstab angelegt hat, einen kleineren nämlich, als der betreffenden Einheit angemessen war.
Wenn man oft sagt - und ELLENBERG (1956) schreibt es auch in
seinem Buch "Aufgaben und Methoden der Vegetationskunde" -,daß

die Bedeutung der Vegetationskartierung zunächst einmal darin
liegt, daß eine kartenmäßige Darstellung den Bearbeiter zwingt,
die Pflanzendecke lückenlos zu betrachten und jeden Pflanzenbe-
stand in das System einzuordnen, so ist das nur bis zu einem gewissen Grade richtig. Jeder Bearbeiter kann bei einer Kartierung diejenigen Gesellschaften vernachlässigen, die er in seinem Maßstab nicht mehr darstellen kann, und er wird manche mosaikartig verzahnten Gesellschaften als komplexe Einheit darstellen.

Es ist deshalb zweckmäßig, sich darüber Rechenschaft zu geben,
was man eigentlich tut, wenn man Zusammenfassungen irgendwelcher Art trifft oder etwas wegläßt, um die Vegetation für einen
bestimmten Maßstab darstellbar zu machen. Immer handelt es sich
um die Bildung von Gesellschaftskomplexen, die ich hier einmal
unter dem Gesichtspunkt des Maßstabes betrachten möchte. Damit
soll auch gleichzeitig ein Beitrag zur Systematik der Gesellschaftskomplexe geleistet werden.

Zuvor ist es aber notwendig, zu sagen, was unter Gesellschaftskomplex zu verstehen ist.

DIE ZUSAMMENFASSUNG VON VEGETATIONSEINHEITEN ZU GESELLSCHAFTS-
KOMPLEXEN

Der Gesellschaftskomplex

Unter Gesellschaftskomplex versteht man den Zusammenschluß von
räumlich benachbarten selbständigen Vegetationseinheiten zu
Einheiten höherer Ordnung (PASSARGE 1965).

Wesentlich ist die Selbständigkeit der Vegetationseinheiten.
Von einem Gesellschaftskomplex kann man beispielsweise nicht
sprechen, wenn im Übergangsbereich von 2 Gesellschaften bereits
einige Arten des einen Typs im anderen enthalten sind. Hier handelt es sich um Übergänge, die man systematisch als Subassoziation, Variante oder Subvariante fassen kann, indem man sie durch
Aufnahmen belegt, in denen die aus der benachbarten Einheit über-
greifenden Arten die Differentialarten gegen den Typ darstellen.
Eine solche Übergangseinheit ist, um ein einfaches Beispiel zu
nennen, das A r r h e n a t h e r e t u m   c i r s i e t o -
s u m, das zwischen dem typischen A r r h e n a t h e r e t u m
und dem A n g e l i c o - C i r s i e t u m vermittelt. Bei
der kartographischen Darstellung in einem genügend großen Maßstab würde diese Subassoziation als eigene Einheit wiedergegeben oder, wenn man sie nicht durch Aufnahmen belegt hat, als

Durchdringung von A r r h e n a t h e r e t u m und A n g e -
l i c o - C i r s i e t u m dargestellt.
Wenn dagegen, sagen wir auf einem Talboden mit kleinflächig
stark wechselndem Relief, ein vollständig ausgebildetes A r -
r h e n a t h e r e t u m mit einem vollständig entwickelten
A n g e l i c o - C i r s i e t u m kleinflächig wechselt, dann
haben wir einen Gesellschaftskomplex vor uns, weil es sich um
selbständige, voneinander unabhängige Vegetationseinheiten handelt. Hier kann ich von einem bestimmten Maßstab an nicht mehr
die einzelnen Bestandteile, sondern nur noch den Gesellschaftskomplex als solchen kartographisch darstellen.
So einfach wie bei dem genannten Beispiel liegen die Verhältnisse aber nicht immer. Man hat in manchen Fällen Gesellschaftskomplexe als Vegetationseinheiten aufgefaßt, weil man die Selbst-
ständigkeit ihrer Komponenten nicht erkannte. Als Beispiele will
ich den Röhricht-Komplex und den Steppenheidewald-Komplex nennen,
von denen der letztgenannte laut Rundschreiben von TÜXEN ja so
etwas im Mittelpunkt dieses Symposions stehen soll. Warum hat
man die Selbständigkeit so spät erkannt oder will sie z.T. auch
heute noch nicht anerkennen? M.E. weil 1. die Gesellschaften
dieser Gesellschaftskomplexe kleinflächig miteinander verzahnt
sind und man bei ihrem Studium einfach den falschen Maßstab angelegt hat. Bei der Untersuchung und Auflösung des Steppenheide-
wald-Komplexes muß eine detailliertere Betrachtungsweise, also
ein größerer Maßstab, angewendet werden als etwa bei einer Wald-
gliederung. Und für die Auflösung eines Röhricht-Komplexes wäre
ein Maßstab, der für die Auflösung des Steppenheidewald-Komple-
xes ausreicht, auch noch zu klein. Der 2.Grund für das späte
Erkennen der Selbständigkeit ist die Tatsache, daß die Komponenten der genannten Komplexe ihr Hauptvorkommen in diesen Gesellschaftskomplexen haben oder zu haben scheinen, so daß sie
ihre Selbständigkeit nicht gleich unter Beweis stellen konnten.
Erst als durch TÜXEN (1952), WENDELBERGER (1954) und Th.MÜLLER
(1962) die Selbständigkeit der Gesellschaften des Steppenheide-
wald-Komplexes nachgewiesen war, konnte man diesen Komplex überzeugend in seine Bestandteile auflösen. Der 3. Grund für die
späte Auflösung dieser Komplexe sind schließlich die wenig au-
genfälligen Standortsunterschiede ihrer Komponenten. So ist der
Hochmoor-Komplex, bei dem die Standortsunterschiede seiner Komponenten leichter erkennbar sind, schon viel früher als Komplex

verschiedener selbständiger Einheiten erkannt worden (GRISEBACH
1880, OSVALD 1923).

Gesichtspunkte für die Einteilung der Gesellschaftskomplexe

Die Gesellschaftskomplexe lassen sich nach verschiedenen Gesichtspunkten einteilen, von denen in unserem Zusammenhang
Struktur, Art und Größe wichtig sind.

Strukturen der Gesellschaftskomplexe

Über verschiedene Strukturen von Gesellschaftskomplexen haben
sich zahlreiche Autoren (DU RIETZ 1930, BRAUN-BLANQUET 1951/
1962, ELLENBERG 1956, PFEIFFER 1957, PASSARGE 1965) geäußert.
Ich möchte hier die wichtigsten herausstellen und erläutern:

1) Mosaikkomplex. Er umfaßt nach diesen Autoren einen kleinräumigen Komplex von bestimmten Gesellschaften beim Wechsel verschiedener Kleinstandorte auf engem Raum, z.B. Hochmoor-Komplex. Mir scheint es aber weniger wichtig zu sein, das Kleinräumige zu betonen, wie es die oben genannten Autoren getan
   haben. Bei kleinmaßstäblicher Darstellung von Landschafts-
   räumen, in welchen zwei oder mehr Vegetationseinheiten über
   die ganze Fläche hin miteinander abwechseln, kann man ebenfalls von einem Mosaik sprechen und die Gesellschaften zu
   einem Mosaikkomplex zusammenfassen. Wesentlich scheint mir
   das Abwechseln von zwei oder mehr Vegetationseinheiten, von
   denen jede einen beträchtlichen Anteil an der Gesamtfläche
   des Gesellschaftskomplexes haben muß, über die ganze Fläche
   zu sein.
   Zum Mosaikkomplex möchte ich auch den Durchdringungskomplex
   von PFEIFFER (1957) rechnen, der, wie der Autor selbst angibt, aus miteinander nicht verwandten Gesellschaften besteht,
   "welche mit einer gewissen Regelmäßigkeit in entsprechend
   m o s a i k -artig gegliederten Milieubedingungen verfloch-
   ten sind und einander wechselseitig durchdringen." [1]
   Auch die Zwillingsgesellschaft von TÜXEN (1962) muß hierher
   gerechnet werden.

[1] Handelt es sich beim Durchdringungskomplex um selbständige
Pflanzengesellschaften, liegt ein Mosaik-Komplex vor, besteht
die Durchdringung aber aus einer selbständigen Einheit und Teilen einer anderen, so handelt es sich um eine Durchdringung,die
als Untereinheit der selbständigen Gesellschaft gefaßt werden
kann, wie oben bereits erläutert wurde.

2) Zonationskomplex oder Gürtelkomplex. Gemeint ist die gürtel-
   förmige Anordnung von Pflanzengesellschaften, die durch den
   zunehmenden Einfluß eines Standortfaktors (z.B. Grundwasser)
   oder eines dynamischen Faktors (Sukzession) zustande kommt.
   Ein Zonationskomplex umfaßt immer eine ökologische Reihe
   (Catena) oder eine Sukzessionsserie, die beide länger sind
   als bei einem Mosaikkomplex. Auch hier ist die Kleinräumig-
   keit nicht ausschlaggebend. In kleinmaßstäblichen Karten
   könnte man die durch die Höhenstufen gegebene Zonation ebenfalls als Zonations- oder Gürtelungskomplex zusammenfassen.
3) Dominanzkomplex.Wenn ein Kartierer kleinflächige Gesellschaften, die er in dem verwendeten Maßstab nicht darstellen kann,
   vernachlässigt, läßt er sie in Wirklichkeit nicht weg, sondern in der dominierenden Kontaktgesellschaft aufgehen. Weg-
   lassen würde heißen, die Fläche abgrenzen und weiß lassen.
   Das Aufgehenlassen in der dominierenden Gesellschaft dagegen
   bedeutet die Bildung eines Gesellschaftskomplexes, der durch
   die dominierende Gesellschaft repräsentiert wird. Wenn ein
   Vegetationsgeograph in Karten kleinerer Maßstäbe großflächi-
   ge Wuchsdistrikte darstellt und dabei alle nicht darstellba-
   ren Pflanzengesellschaften in der dominierenden Leitgesellschaft aufgehen läßt, tut er auch nichts anderes.[1]) Für diese Form möchte ich den Begriff Dominanzkomplex vorschlagen.

Arten von Gesellschaftskomplexen

Von der Notwendigkeit her gesehen, die Vegetation in bestimmten
Maßstäben darzustellen, ist nur eine Komplexbildung interessant,
die zu einer räumlichen Zusammenfassung benachbarter Vegetationseinheiten und damit zu einer Vereinfachung des Kartenbildes
führt.Solche kann nach rein räumlichen Gesichtspunkten erfolgen,
sie kann damit gleichzeitig aber auch andere Gesichtspunkte berücksichtigen oder gar in den Vordergrund stellen (vergl. vor
allem SCHMITHÜSEN 1963). Dementsprechend haben wir verschiedene

---

[1]) Man wird übrigens pflanzensoziologisch verwandte und ökologisch wenig gegensätzliche Pflanzengesellschaften eher in einem Gesellschaftskomplex aufgehen lassen als Gesellschaften
entfernter Verwandschaft und ganz anders gearteter Standorte.
Bei solchen wird man bei Maßstabsverkleinerungen zunächst noch
versuchen, sie durch eine etwas größere als dem Maßstab entsprechende Darstellung im Kartenbild zu erhalten. Allzu groß
ist der Spielraum aber nicht, der durch diese Möglichkeit gegeben ist.

Arten von Gesellschaftskomplexen zu unterscheiden.

1. Synsystematischer Gesellschaftskompex.

   Wenn ein Grünland- oder Waldkartierer, der seine Subassoziationen und Varianten noch leicht im Maßstab 1 : 5 000 oder
   1 : 10 000 darstellen kann, eine Kartierung im Maßstab 1 :
   25 000 durchzuführen hat, wird er zunächst dazu neigen, auf
   seine detaillierten Untergliederungen zu verzichten und nur
   noch Subassoziationen oder Assoziationen wiederzugeben. Er
   könnte genau so gut Komplexe von Varianten bilden, Mosaik-
   komplexe,Dominanzkomplexe, sogar Zonationskomplexe, je nach
   den örtlichen Verhältnissen. Das wäre eine rein räumliche
   Zusammenfassung, die in manchen Fällen den Tatbestand sogar
   genauer wiedergeben würde; denn beim Dominanzkomplex zweier
   Varianten würde noch zum Ausdruck gebracht, welche Variante
   vorherrscht, bei der Darstellung der nächst höheren Einheit,
   der Subassoziation, ist das nicht mehr der Fall. So kann man
   durch die Wiedergabe immer höherer systematischer Einheiten
   zu einer fortschreitenden Vereinfachung des Kartenbildes und
   der Darstellungsmöglichkeit in kleineren Maßstäben kommen.
   Eine Zusammenfassung nach synsystematischen Gesichtspunkten
   zu höheren Vegetationseinheiten kann aber ohne Berücksichtigung andersartiger Komplexbildungen nur so lange stattfinden,
   wie ausschließlich zu der jeweiligen höheren Vegetationseinheit gehörende Pflanzengesellschaften benachbart sind. Bei
   manchen Gesellschaftsverzahnungen ist man damit schon sehr
   bald,d.h. bei großen Maßstäben, am Ende. Ein Röhrichtkomplex
   aus M a g n o c a r i c i o n und P h r a g m i t i o n -
   Gesellschaften läßt sich noch als Klasse P h r a g m i t e-
   t e a darstellen. Bei einem Mosaik mit B i d e n t e t e a
   oder L e m n e t e a - Gesellschaften versagt dagegen eine
   synsystematische Zusammenfassung schon beim Maßstab 1 : 100.

   Es ist in der Pflanzensoziologie nicht üblich, die Zusammenfassung von bestimmten Gesellschaftseinheiten zu solchen höherer Ordnung als Bildung von Gesellschaftskomplexen aufzufassen; von der räumlichen Betrachtung her ist es aber durchaus
   zulässig, von einem synsystematischen Gesellschaftskomplex
   zu sprechen.

2. Genetischer Gesellschaftskomplex.

In einem genetischen Gesellschaftskomplex werden Pflanzengesellschaften miteinander vereinigt, die genetisch oder dynamisch miteinander in Beziehung stehen. Die Stadien einer natürlichen Sukzession lassen sich ebenso zu einem genetischen
Gesellschaftskomplex zusammenfassen wie die anthropogenen
Ersatzgesellschaften einer Kulturlandschaft mit der natürlichen Ausgangsgesellschaft, aus der sie hervorgegangen sind
und zu der sie bei Aufhören des menschlichen Einflusses wieder hinstreben würden.
Solche genetischen Gesellschaftskomplexe sind unter dem Begriff "potentiell natürliche Vegetation" von TÜXEN (1956)
und unter dem Begriff "Gesellschaftsring" von SCHWICKERATH
(1954) herausgestellt und präzise erläutert worden. Beide
Begriffe beziehen alle in einem genetischen Beziehungsgefüge stehenden Gesellschaften auf eine Schlußgesellschaft (Klimax) oder Dauergesellschaft.Im Gegensatz zum Gesellschaftsring läßt der Begriff der potentiell natürlichen Vegetation
aber auch den Bezug auf Pionier- oder Folgegesellschaften
der natürlichen Sukzession zu, nämlich dann, wenn diese eine
längere Lebensdauer haben und nicht so rasch von der darauf-
folgenden Gesellschaft abgelöst werden. In der Definition
der potentiell natürlichen Vegetation heißt es nämlich: "Dabei muß das neue Gleichgewicht zwischen natürlicher Vegetation und Standort schlagartig eingeschaltet gedacht werden
und nicht erst, bis es sich - über Jahrhunderte hinweg -
eingespielt hat" (TRAUTMANN 1962).
Die Vereinigung von Pflanzengesellschaften zu genetischen
Gesellschaftskomplexen bringt nur dann auch eine räumliche
Zusammenfassung und Vereinfachung des Kartenbildes mit sich,
wenn die genetisch verwandten Pflanzengesellschaften räumlich benachbart sind. Das ist aber meistens der Fall, sowohl
unter den Bedingungen der natürlichen Sukzession, etwa in
Flußauen oder Hochmooren, als auch unter den Bedingungen der
Kulturlandschaft, wo viele Gesellschaftsgrenzen nur auf den
Unterschieden der menschlichen Bewirtschaftung beruhen. Inso-
fern ist der genetische Gesellschaftskomplex auch eine ganz
reale Angelegenheit. Er ist zugleich aber auch eine Abstraktion. Das wird in den Fällen ganz deutlich, in denen die Be-
zugsgesellschaft (Schluß-, Dauergesellschaft) im Gebiet fehlt,
der Gesellschaftskomplex also durch etwas repräsentiert wird,

das gar nicht vorhanden ist.

Obschon der genetische Gesellschaftskomplex in den meisten
Fällen eine Vereinfachung des Kartenbildes bringt und danßt
kleinere Maßstäbe ermöglicht, reicht auch diese Art der Zusammenfassung allein nicht aus, um zu kleinen Kartenmaßstä-
ben zu kommen. Selbst wenn man beispielsweise aus allen Gesellschaften eines Bachufers einen genetischen Gesellschaftskomplex macht, bleibt immer noch ein bestenfalls 1 - 5 m
breiter Streifen übrig, den man in Maßstäben kleiner als
1 : 5 000 nicht mehr wiedergeben kann.

3. Ökologischer Gesellschaftskomplex.

Der ökologische Gesellschaftskomplex schließt Pflanzengesellschaften zusammen,die in bestimmten ökologischen Eigenschaften übereinstimmen. Beispiele sind Wasserstufen (TÜXEN 1954,
MEISEL und WATTENDORF 1962, SEIBERT 1963), Nährstoffstufen
(u.a. JENSEN 1961) u.a., aber auch Begriffe wie Kalkbuchen-
wald-Komplex, Geröllhalden-Komplex usw. (vergl. HOFMANN 1965).
Die Zusammenfassung erfolgt unter dem Gesichtspunkt eines
bestimmten ökologischen Faktors, wobei alle anderen Faktoren
vernachlässigt werden. Es ist leicht einzusehen, daß auch
beim ökologischen Gesellschaftskomplex eine Vereinfachung
des Kartenbildes und damit die Möglichkeit einer stärkeren
Verkleinerung nur soweit gegeben ist, wie Pflanzengesellschaften, die hinsichtlich des ausschlaggebenden Standortsfaktors
gleich sind, unmittelbar benachbart liegen.

Eine Zusammenfassung als ökologischer Gesellschaftskomplex,
in welchem die Gesamtheit aller natürlichen Standortsfaktoren zum Ausdruck kommt, ist wiederum die potentiell natürliche Vegetation. Ihre Karte gibt "eine Gesamtdarstellung des
Milieus, von dem die Vegetation abhängig ist" (SCHMITHÜSEN
1963). Wegen dieser Eigenschaft vor allem ist eine Karte der
potentiellen natürlichen Vegetation für angewandte Wissen-
schaften und für die Geographie interessant.

4. Formationskomplex.

Von den Möglichkeiten der Komplexbildung, die von strukturel-
len und physiognomischen Eigenschaften der Pflanzengesellschaften gegeben sind, soll hier nur der Formationskomplex
genannt werden. Bei ihm werden Pflanzengesellschaften vereinigt, die in ihrer Physiognomie übereinstimmen. Diese Art

der Zusammenfassung geht häufig mit anderen parallel. So ist
z.B. der Formationskomplex des Buchenwaldes eine Zusammenfassung aller buchenreichen Wälder, die im wesentlichen der synsystematischen Zusammenfassung zum F a g i o n entspricht,
und die gleichzeitig auch einem ökologischen Gesellschaftskomplex parallel laufen kann, nämlich der Zusammenfassung aller montanen Wälder eines Gebietes.

5. Topographischer Gesellschaftskomplex.
Bei allen bisher genannten Komplexbildungen hört schon bei
sehr großen Maßstäben die Möglichkeit einer kartenmäßigen
Darstellung auf. Das liegt daran, daß die Größen der jeweiligen Gesellschaftskomplexe sehr unterschiedlich sind. So ist
es fast immer notwendig, zusätzlich rein räumliche Zusammenfassungen vorzunehmen, bei denen die oben genannten Gesichtspunkte nicht zum Zuge kommen. Diese Art der Zusammenfassung
bezeichne ich als topographischen Gesellschaftskomplex. Wenn
ich bei einer Grünlandkartierung 1 : 5 000 die Röhrichte eines Bachufers nicht darstelle, oder wenn ich bei der Bildung
von Vegetationsgebieten für Karten des Maßstabs 1 : 500 000
oder kleiner eine ganze Flußaue in dem Dominanzkomplex der
vorherrschenden potentiellen Waldgesellschaft aufgehen las-
se, in welchen sie weder nach synsystematischen, noch genetischen, ökologischen oder physiognomischen Gesichtspunkten
aufgenommen werden kann, bilde ich solche topographischen
Gesellschaftskomplexe. Sie können der Form nach Mosaik-, Zo-
nations- oder Dominanzkomplex sein.
Freilich bestehen auch bei ihnen zwischen ihren Kontaktge-
sellschaften noch andere als rein räumliche Beziehungen;
denn im Gelände sind die verschiedenen Pflanzengesellschaften nicht wahllos nebeneinander angeordnet. Doch liefern die
verbindenden Faktoren in diesen Fällen keinen Gesichtspunkt
für eine sinnvolle Komplexbildung.
Da der topographische Gesellschaftskomplex nur den Gesichtspunkt einer räumlichen Zusammenfassung in Kontakt liegender
Pflanzengesellschaften betont, könnte man ihn auch Kontakt-
gesellschaftskomplex nennen. Auch die Benennung als syncho-
rologischer Gesellschaftskomplex ist möglich.
Der topographische Gesellschaftskomplex ist unter allen auf-
gezählten der am wenigsten elegante. Trotzdem bleibt er in

allen Maßstabbereichen die einzige Möglichkeit, kleinflächig
verbreitete Gesellschaften aufzunehmen. Gerade wegen der rein
mechanischen Bildung des topographischen Gesellschaftskomplexes wird man aber bemüht sein, seine Aussagekraft durch
Angabe seines Inhaltes, also seines Gesellschaftsinventars,
zu erhöhen. Das kann durch einfache Aufzählung geschehen.

Einen besseren Vergleich gestattet jedoch eine tabellarische
Zusammenstellung, in welcher die Gesellschaften der einzelnen Komplexe in Spalten nebeneinander stehen. So kommt man
zu einer Gesellschaftsinventar-Tabelle, welche die Vergesellschaftung von Gesellschaften (TÜXEN 1950, SCHMITHÜSEN 1963)
erkennen läßt und die einer pflanzensoziologischen Tabelle
durchaus äquivalent ist. Wir finden in ihr die in einem bestimmten Gebiet dominierende Gesellschaft, die Leitgesellschaft (nach SCHRETZENMAYR 1961), welche der dominierenden,
also faziesbildenden Art einer Gesellschaftstabelle entspricht. Den Kenn- und Charakterarten ist die Charakterge-
sellschaft eines Vegetationsgebietes vergleichbar.Das sind
die Gesellschaften, die auf ein bestimmtes Gebiet beschränkt
sind, ohne daß sie dort flächenmäßig eine Rolle spielen
müssen. Als Beispiel sei das C o r y n e p h o r e t u m
im Vegetationsgebiet des Q u e r c o - B e t u l e t u m
genannt. Den Differentialartengruppen sind Gesellschaftsgrup-
pen, z.B. Gesellschaftskomplexe obengenannter Gesichtspunkte,
insbesondere genetische, äquivalent.
Freilich setzt die Angabe des Gesellschaftsinventars voraus,
daß seine Gesellschaften bekannt sind und mit erfaßt wurden,
die Gesellschaftskomplexe also induktiv gebildet worden sind.

Eine sehr vorbildliche Zusammenstellung des Gesellschaftsinventars von Vegetationsgebieten gibt TÜXEN (1956). Nur wer
seine Listen äußerst gründlich studiert hat, kann ermessen,
welch hoher Informationswert in ihnen steckt.

Größe der Gesellschaftskomplexe

Die Gesellschaftskomplexe sind in meinem Vortrag bisher nur
nach dem Gesichtspunkt ihrer Struktur und ihres Inhaltes behandelt worden, nicht nach dem Gesichtspunkt ihrer Größe. Wir
konnten zeigen, daß die vorgebrachten Probleme und Lösungsmög-
lichkeiten für alle Maßstabbereiche Gültigkeit haben, wenn sie

auch umso häufiger auftreten, je kleiner der Maßstab ist. Denn
da die Pflanzengesellschaften sehr unterschiedliche Bestandes-
größen haben, können Notwendigkeiten einer Zusammenfassung in
allen Maßstäben auftreten. Bei einem gegebenen Maßstab ist es
fast immer nötig, bei differenzierter Darstellung des einen
Teils von Pflanzengesellschaften einen anderen Teil zu Gesellschaftskomplexen zu vereinigen.
Es ist deshalb nicht erforderlich, auf die Größe der Gesellschaftskomplexe, ihre Einteilung, und die Benennung der verschiedenen Größenordnungen einzugehen. Darauf kann umso leichter verzichtet werden, als sich SCHMITHÜSEN (1959) in seiner
"Allgemeinen Vegetationsgeographie" ausführlich darüber geäußert hat.

ZUSAMMENFASSUNG

Zusammenfassend darf ich folgendes festhalten:

1. Der für die kartographische Darstellung gewählte Maßstab beeinflußt den Bearbeiter schon bei der Fassung und Abgrenzung
   seiner Einheiten im Gelände.

2. Die Vereinfachung des Kartenbildes und damit der Übergang zu
   kleineren Maßstäben ist nur durch Bildung von Gesellschaftskomplexen erreichbar. Schon sehr früh, d.h.bei großen Maßstäben, führt nur die Bildung von topographischen Gesellschaftskomplexen, d.h. eine rein räumliche Komplexbildung
   ohne andere Gesichtspunkte, weiter.

3. Der Informationswert bleibt auch bei kleinmaßstäblichen Karten groß, wenn das Inventar der Gesellschaftskomplexe genannt werden kann.

SUMMARY

Criterions for the selection of scale.

The order of scale is depending on the size of the object to be
represented and on the details to be reproduced, and its exact
fixation depends on the scales of the available maps.
The size of the object is depending on the real questions, that
maybe originating from science of vegetation or practical special branches. By fixing the scale, they have much influence on
the work in field and enforce to sum up plant communities to
complexes of communities.

The summing up of units of vegetation to comlexes of communities.

Complexes of communities are composed from several independent units of vegetation. The form those is necessary in all orders of scale. The complexes maybe divided according to the aspects of their structure, kind and size.

Structures of complexes of communities.

1. Mosaic complex including complex of penetration, twin community,
2. complex of zonation or complex of zones,
3. complex of dominance.

Kinds of complexes of communities.

1. A synsystematic complex of communities means the summing up of units of plant communities to units of superior rank.
2. The genetic complex of communities is enclosing genetically related plant communities, that are links of a natural succession or links of man-made replacement communities. Examples of this kind of communities are the potential natural vegetation (potentiell natürliche Vegetation) and the circle of communities (Gesellschaftsring).
3. The ecological complex of communities is enclosing plant communities, that do agree in particular or all of their ecological properties: for instance stage of water or other factors, potential natural vegetation.
4. The complex of formation joins plant communities, that do agree in their physiognomy.

However, too in larger scales forming of complexes seldom will satisfy for mapping out all small units of vegetation. For this reason in all orders of scale it's necessary to sum up relating to the space. So we shall get the

5. topographical complex of communities. The more complete the stock of units is indicated, which is done best by a table of the stock of communities, the better is the statement of this complex.

Size of complexes of communities.

The indicated points of view are of value for all sizes and orders of complexes of communities. The nomenclature of the terms is referred to SCHMITHÜSEN (1959).

Literatur

BROCKMANN-JEROSCH,H. -1930- Klimatisch bedingte Formations-
    klassen der Erde.- In: Rübel,E.: Pflanzengesellschaften der
    Erde. Bern-Berlin.
BRAUN-BLANQUET,J. -1964- Pflanzensoziologie 3.Aufl.- Wien.
    (2.Aufl.1951).
DU RIETZ,G.E. -1930- Vegetationsforschung auf soziationsana-
    lytischer Grundlage.- In: Abderhalden,E.: Handb.biol.Ar-
    beitsmeth. 11 (5). Berlin und Wien.
ELLENBERG,H. -1956- Aufgaben und Methoden der Vegetationskun-
    de.- In: Walter,H.: Einführung in die Phytologie 4 (1).
    Stuttgart.
GRISEBACH,A. -1880- Gesammelte Abhandlungen und kleinere
    Schriften zur Pflanzengeographie.- Leipzig.
HOFMANN,G. -1965- Über Vegetationskomplexe unter besonderer
    Berücksichtigung der Trockenwaldkomplexe.- Feddes Rep.
    Beitr.z.Vegkd.7. Berlin.
JENSEN,U. -1961- Die Vegetation des Sonnenberger Moores im
    Oberharz und ihre ökologischen Bedingungen.- Natursch. u.
    Landschaftspfl. in Niedersachsen 1. Hannover.
MEISEL,K.& WATTENDORF,J. -1962- Über eine von der Wirtschafts-
    art unabhängige Wasserstufenkarte.- Mitt.flor.-soz.Arbeits-
    gem.N.F.9. Stolzenau/Weser.
MÜLLER,Th. -1962- Die Saumgesellschaften der Klasse Trifolio-
    Geranietea sanguinei.- Mitt.flor.-soz.Arbeitsgem.N.F.9.
    Stolzenau/Weser.
OSVALD,H. -1923- Die Vegetation des Hochmoores Komosse.-
    Svenska Växtsoziolog.Sällsk.Handl.1. Uppsala.
PASSARGE,H. -1965- Zur Frage der Probeflächenwahl bei Gesell-
    schaftskomplexen im Bereich der Wasser- und Verlandungsve-
    getation.- Feddes Rep. Beitr.z.Vegkd.7. Berlin.
PFEIFFER,H. -1957/1958- Über das Zusammentreten von Pflanzen-
    gesellschaften in Komplexen.- Phyton 7. Buenos-Aires.
SCHMITHÜSEN,J. -1959- Allgemeine Vegetationsgeographie.-
    Berlin.
-- --   -1963- Der wissenschaftliche Inhalt von Vegetations-
    karten verschiedener Maßstäbe.- In: Tüxen,R.(Edit.): Vege-
    tationskartierung. Ber.intern.Symp.1959 Stolzenau/Weser.
    Weinheim.
SCHRETZENMAYR,M. -1961- Die Leitgesellschaft. Eine vegetati-
    onskundliche Arbeitshypothese im Rahmen der forstlichen
    Standortskartierung.- Arch.Forstw.10. Berlin.
SCHWICKERATH,M. -1954- Die Landschaft und ihre Wandlung auf
    geobotanischer und geographischer Grundlage entwickelt und
    erläutert im Bereich des Meßtischblattes Stolberg.- Aachen.
SEIBERT,P. -1958- Die Pflanzengesellschaften im Naturschutz-
    gebiet "Pupplinger Au".- Landschaftspfl.u.Vegetationskinde
    1. München.
-- --   -1963- Über eine Grundwasserstufenkarte mit Darstel-
    lung verschiedener Wassereigenschaften.- Mitt.flor.-soz.
    Arbeitsgem.N.F.10. Stolzenau/Weser.
TRAUTMANN,W. -1962- Erläuterungen zur Karte der potentiellen
    natürlichen Vegetation der Bundesrepublik Deutschland. Blatt
    85 Minden.·Maskr.- Stolzenau/Weser.
TÜXEN,R. -1950- Pflanzensoziologie als unentbehrliche Grund-
    lage der Landeswirtschaft.- Studium generale 3. Berlin.
-- --   -1952- Hecken und Gebüsche.- Mitt.geogr.Ges.Hamburg
    50. Hamburg.

TÜXEN,R.   -1954- Die Wasserstufenkarte und ihre Bedeutung für
    die nachträgliche Feststellung von Entwässerungsschäden.-
    Angew.Pflanzensoz.8. Stolzenau/Weser.
--   --   -1956- Die heutige potentielle natürliche Vegetation
    als Gegenstand der Vegetationskartierung.-Angew.Pflanzensoz.
    13.Stolzenau/Weser.
--   --   -1957- Dsgl.Bericht dtsch.Landeskunde 19 (2).Remagen.
--   --   & LOHMEYER,W. -1962- Über Untereinheiten und Verflech-
    tung von Pflanzengesellschaften.- Mitt.flor.-soz.Arbeitsgem.
    N.F.9. Stolzenau/Weser.
WENDELBERGER,G. -1954- Steppen, Trockenrasen und Wälder des
    pannonischen Raumes.- Angewandte Pflanzensoziologie.
    Aichinger-Festschr.1. Wien.

W.LÖTSCHERT:

Die Bezeichnung Formationskomplex muß man mit einer gewissen
Kritik betrachten, denn der Begriff Formation ist ursprünglich
physiognomisch gefaßt, d.h. er ist im Anschluß an ALEXANDER VON
HUMBOLDT und die nachfolgenden Pflanzengeographen allmählich
entwickelt worden. Nun möchte ich aber nur zu bedenken geben,
daß man diesen Begriff nicht ohne weiteres in diesem Zusammenhang benutzen kann. Ich möchte zunächst die Zwergstrauch-Heiden
als Beispiel nehmen. Zwergstrauch-Heiden kommen ja einerseits
im nordwestlichen Mitteleuropa, andererseits im submediterran-
mediterranen Grenzbereich und drittens auch etwa in Australien
vor. In diesen verschiedenen Gebieten gehören die Zwergstrauch-
Heiden ganz verschiedenen systematischen Einheiten an, und es
ist nicht möglich, die etwa überwiegend aus Proteaceen bestehenden australischen Zwergstrauch-Heiden mit den subatlantischen
des nordwestlichen Mitteleuropa ohne weiteres in Einklang zu
bringen. Ein weiteres Beispiel wäre etwa der tropische Regenwald. Es ist bekannt, daß der tropische Regenwald eine außer-
ordentliche Artenfülle aufweist, und es ist in diesem Falle
schwierig, etwa wie das am Beispiel von Fagus silvatica möglich wäre, eine durchgehende Einheitlichkeit zu finden. Wir
müssen also überlegen, wie wir in dieser Hinsicht eine gemeinsame Plattform finden, um den Begriff des physiognomischen Asso-
ziationskomplexes einheitlich zu fassen.

R.TÜXEN:

Ich gratuliere Herrn SEIBERT zu dem gedankenreichen Vortrag,
der auf langjähriger praktischer Erfahrung beruht.Wenn ich recht
verstanden habe, sprachen Sie von genetischen Komplexen. Ich
würde bitten oder vorschlagen, das Wort "genetisch" lieber zu

vermeiden, und eher von syndynamischen Komplexen zu sprechen.
" Genetisch " ist ja doch etwas anderes als hier das Dynamische.

Ich möchte noch zwei Dinge hinzufügen, von denen das erste eine
Jugenderinnerung ist: Als wir vor 40 Jahren mit der Kartierung
beginnen wollten, da hat man vorgeschlagen: wenn der Maßstab zu
klein ist, um Assoziationen darzustellen, dann kartiert man am
besten nach Verbänden oder Ordnungen. Das läßt sich aber gar
nicht durchführen, wie sich sehr schnell gezeigt hat. Man kann
nicht das B r o m i o n kartieren,oder das F a g i o n, sondern muß so kartieren, wie wir es eben gehört haben.
Dann wollte ich noch sagen: Um zu Tabellen von Vergesellschaftungen von Gesellschaften, von denen Herr SEIBERT gesprochen
hat, induktiv zu kommen, haben wir versucht - und ich möchte
das zur weiteren Anwendung sehr empfehlen - innerhalb eines einheitlichen Vegetationsgebietes der gleichen potentiell natürlichen Vegetation - sagen wir also im Gebiet des Q u e r c o -
C a r p i n e t u m oder eines bestimmten F a g e t u m oder
in einer A l n e t u m- oder in einer F r a x i n o - U l m e-
t u m - Landschaft - auf 1 oder 2 km Länge auf einem Linienpro-
fil alle hier angetroffenen Gesellschaften so aufzuschreiben,
daß eine richtige Aufnahme der hier vorhandenen Gesellschaften
entsteht, so wie wir auf einer Probefläche eine Aufnahme aller
Arten machen. Man bekommt dann mit der Zeit Aufnahmen der Ge-
sellschafts-Kombinationen von bestimmten Vegetationsgebieten,
die man zu Tabellen zusammenstellen kann. Man kann die Deckung
der angetroffenen Gesellschaften sehr leicht mit einfachen Zahlen, etwa 1 - 5, ähnlich wie wir die Menge der Arten schätzen,
oder auch mit bestimmten Signaturen angeben. Man ordnet die Gesellschaften in der Tabelle nach ihrer Stetigkeit. Man kann
dann diese Tabellen mit anderen aus den verschiedenen Vegetationsgebieten vergleichen und bekommt dann exakte Unterlagen dafür, was ich in meiner von Herrn SEIBERT freundlich zitierten
Arbeit (1956/1957) über die potentiell natürliche Vegetation
nur aus der allgemeinen Erinnerung habe angeben können, und was
dort gewiß mit vielen Fehlern behaftet ist. Ich möchte empfehlen, in dieser Richtung - das ist eine Anregung, die von Professor SCHMITHÜSEN stammt - weiterzuarbeiten.

EIN MECHANISCH-ELEKTROMAGNETISCHES GERÄT ZUR SCHNELLBEARBEITUNG PFLANZENSOZIOLOGISCHER TABELLEN

M.W. T r e n t e p o h l

Eine Tabelle ist ein unentbehrliches Hilfsmittel, um aus einer
Vielzahl konkreter Einzeluntersuchungen durch induktiven Vergleich zu abstrakten Aussagen zu gelangen. Dabei können sich
die Befunde der Einzelanalysen sowohl qualitativ als auch quantitativ voneinander unterscheiden. Gut bearbeitete Tabellen lassen
durch die räumliche Nachbarschaft einander ähnlich strukturier-
ter Größen die diesen gemeinsamen Kategorien deutlich hervortreten. So ist es kein Wunder, daß sie sich als unbestritten
wichtigste Werkzeuge biozönotischer und besonders pflanzensoziologischer Forschung allgemein durchgesetzt haben.
Pflanzensoziologische Tabellen schriftlich zu bearbeiten, ist
nun aber mühsam und zeitraubend, wenn man dem Einschleichen von
Fehlern mit Sicherheit vorbeugen will. Jede Spalte und jede
Zeile muß nach dem Umschreiben durch Nachzählen auf Richtigkeit
geprüft werden. Haben sich dabei Unstimmigkeiten ergeben, so
müssen die Fehler gesucht werden, was oftmals einige Zeit erfordert. V i e l schneller als durch das schriftliche Umstel-
len kommt man auch durch Auseinanderschneiden und erneutes Zu-
sammenkleben der Tabellen n i c h t zum Ziel.Vor allen Dingen kann man dies Verfahren meist nur in e i n e r Richtung,
- also entweder zeilen- oder spaltenweise - anwenden. Immerhin
gab dieser Gedanke im Zuge der sich heutzutage allgemein ein-
bürgernden Rationalisierungtendenzen den Anstoß, nach technisch
besseren Möglichkeiten der Tabellen-Bearbeitung zu suchen. Die
konsequente Weiterverfolgung dieses Weges führte schließlich -
analog dem Buchdruck vor 520 Jahren - zur Auflösung der Einheiten in ihre Elemente, also i.W. in Artnahmen und Mengensymbole,
die auf kleine Materialstückchen geschrieben oder gedruckt sind.
Nun wird die Tabelle nicht mehr geschrieben, sondern durch
A u s l e g e n ihrer Elementarbestandteile zusammengesetzt.
Ist dies geschehen, so kann man unter Zuhilfenahme entsprechen-
der Vorrichtungen ganze Zeilen mühelos und fehlerfrei (ebenso
wie ganze Spalten) herausheben und umsetzen.
Eine der zahllosen im Prinzip möglichen Formen eines derartigen
Gerätes soll hier vorgeführt werden. Zwei Vorteile verdienen,
besonders beachtet zu werden.

1) Die Bearbeitung der Tabelle geht verhältnismäßig rasch von-
   statten.
2) Die unmittelbare Anschaulichkeit und ursprüngliche Übersicht-
   lichkeit der früheren, geschriebenen Tabelle bleibt in vol-
   lem Umfange erhalten.

Auf einer Lochplatte aus Kunststoff werden die in der Pflanzensoziologie allgemein als Vegetationsaufnahmen bezeichneten Ergebnisse der einzelnen Artenbestandsanalysen ausgelegt. Die
räumliche Anordnung bleibt die gleiche wie bei geschriebenen
Tabellen. Es kommt also in jede senkrechte Spalte eine Vegetationsaufnahme, und jede horizontale Zeile ist einer Pflanzen-
art vorbehalten. Die entsprechend dem BRAUN-BLANQUET'schen
Schlüssel (r = rar; + = spärlich vorhanden; 1 bis 5 = reichlich
vorhanden mit verschiedenen Deckungsgraden) geschätzten Mengen
der in einer bestimmten Aufnahme gefundenen Arten werden nun
Spalte für Spalte ausgelegt. Alle Tabellenelemente, d.h. sowohl
die Artnahmen- als auch die Mengensymbol-Schildchen sind auf
der Rückseite mit paramagnetisch permeablen Weißblechstück-
chen beklebt. An beiden Enden trägt jedes dieser Blechstücke
einen zinkenartigen, kurzen Fortsatz, der rechtwinklig nach unten umgebogen ist. Auf diese Weise finden die Tabellenelemente
in den Vertiefungen der Lochplatte soweit Halt, daß sie nicht
verrutschen können.

Mechanisch-elektronisches Tabellen-Bearbeitungsgerät.

Zum Gerät gehört ferner ein U-förmig profilierter Elektromag-
netstab, der zum Umsetzen einzelner Zeilen oder Spalten mit

seinen Führungspaßstiften in besondere Führungslöcher an den
Plattenrändern gesteckt wird. Nach dem Einschalten des von einem 6 V-Akkumulators mit zugehörigem Ladegerät gelieferten Stromes "hängt" alles, was in der betreffenden Zeile oder Spalte
vorhanden war, an dem Stab, der nun auf eine andere Zeile oder
Spalte umgesetzt werden kann. Nach Ausschalten des Stromes bleiben alle Elemente in genau richtiger Anordnung liegen, so daß
die Tabelle in verhältnismäßig kurzer Zeit völlig fehlerfrei bearbeitet ist.

Das Endergebnis wird abfotografiert (bei Bedarf evtl.auch interessante Zwischenergebnisse), und für Veröffentlichungen
braucht man nur noch Tabellenklischees nach den Lichtbildern
ätzen zu lassen. Sie sind auf jeden Fall frei von Abschreib-
und Setzfehlern. Eine Lochplatte bietet Platz für maximal 55
Aufnahmen und 85 Arten, was natürlich nicht für alle Fälle ausreicht. Bei großen Tabellen wird man daher zweckmäßigerweise
2 Platten aneinanderlegen, so daß man dann Platz für rund 170
Arten hat.

Neben dieser Arbeitsvereinfachung bietet das Gerät noch eine
andere Verwendungsmöglichkeit. An unserer Staatlichen Ingenieur-
schule für Landbau in Nürtingen wurde neuerdings eine Studien-
Differenzierung eingeführt. Durch diese haben unsere Studenten
die Möglichkeit, in den beiden letzten Semestern in einem
Schwerpunkt-Fachgebiet nach eigener Wahl - in den meisten Fällen ist es das Gebiet, in dem sie auch ihre Ingenieur-Arbeit
schreiben -, ihre Kenntnisse eingehender zu vertiefen und abzu-
runden. Zu einem der möglichen Differenzierungs-Fachgebiete,
der Acker- und Pflanzenbaulehre, gehört auch die landwirtschaft-
liche Pflanzensoziologie. Da aber für diese in den vorgesehenen
beiden Semestern nur je eine Wochenstunde zur Verfügung steht,
wäre es rein didaktisch schlechterdings unmöglich, in dieser
kurzen Zeit den gesamten Stoff von der ersten Einführung an
bis zum Abschluß zu behandeln; also die Vegetationsaufnahme
und ihre Problematik; ökologische, geobotanische und klimatolo-
gische Fragen und Grundbegriffe; pflanzensoziologische Tabellen
und ihre Bearbeitung, Bedeutung und Aussage; das Problem der
Gesellschaften und ihrer Abgrenzungen; schließlich die Kartierung und ihre praktische Auswertung. Das alles in insgeamt etwa
25 Arbeits- bezw. Vorlesungsstunden an den Hörer zu vermitteln,
die zum weitaus größten Teil als Studienvoraussetzung die Fach-

schulreife mitbrachten und außer Bodenkunde noch nie etwas von
diesen Fragen gehört haben, setzt eine äußerst konzentrierte
Straffung des Stoffes voraus. Da die Tabellenbearbeitung aber
in diesem Lehrstoff eines der am meisten zeitraubenden Kapitel
darstellt, besteht ohne den Einsatz des mechanisch-elektromag-
netischen Tabellen-Bearbeitungsgerätes kaum eine Aussicht, dieses Pensum in der vorgesehenen Zeit befriedigend zu bewältigen.
Es ist nicht ausgeschlossen, daß sich dieses Gerät auch an anderen Fachschulen, sowie an Hochschulen und Universitäten, als
Demonstrationsmittel in Kursen und Übungen einbürgern wird.

ZUSAMMENFASSUNG

 Zur Erleichterung und Beschleunigung der mühsamen schriftlicher
Bearbeitung pflanzensoziologischer Tabellen wurde ein Gerät entwickelt. Im Gegensatz zur gewohnten Form werden die Tabellen
nicht mehr geschrieben, sondern ihre Elemente (Artnamen sowie
Deckungsgrad - und Soziabilitätssymbole) werden auf einer Lochplatte ausgelegt.
Jedes dieser Tabellenelemente ist auf ein kleines rechteckiges
Stück aus Papier, Pappe, Kunststoff oder ähnlichem Material geschrieben oder gedruckt, das auf der Rückseite mit einem kleiner
Stück aus paramagnetisch permeablem Weißblech beklebt ist. An
beiden Enden trägt jedes dieser Blechstückchen einen zinkenartigen Fortsatz, der rechtwinklig nach unten umgebogen ist und
in den Vertiefungen der Lochplatte soweit Halt findet, daß die
Tabellenelemente nicht verrutschen können.
Mit Hilfe eines U-förmig profilierten Elektromagnetstabes, der
von einer 6 V-Batterie mit Strom versorgt wird, können alle Elemente einer Spalte als Ganzes durch Einschalten des Stromes an-
gezogen und auf eine andere Spalte umgesetzt werden, wo sie nach
dem Ausschalten in genau richtiger Lage verbleiben.
Auf die gleiche Art werden auch die Zeilen bearbeitet.
Die Vorteile dieser Tabellenbearbeitungstechnik sind 1) Leichte
Bearbeitung in kurzer Zeit; 2) Anschaulichkeit der geschriebener
Tabelle bleibt erhalten; 3) fehlerfreie photomechanische Wiedergabe der Ergebnisse.
Abgesehen von der Arbeitsvereinfachung bietet sich das Gerät
auch als lehrmethodisch in Kursen vielseitig verwendbares De-
monstrationsmittel an.

A Mechanical Electro-magnetic Device
For Rapid Editing Of Phyto-sociological Tables

SUMMARY

A device has been developed to accelerate the tiresome editing
of phyto-sociological tables by repeated re-writing. In contrast
to previous methods it is not necessary for the tables to be
written out. Instead, elements (i.e. the names of species and
the symbols representing their quantities) of a table are arranged on a perforated sheet of plastic. Each element is written
or printed on a small rectangular piece of paper, cardboard,
plastic or similar material, at the back of which adheres a
small easily magnetizable piece of metal. At either end of the
plate is a small tooth. These teeth are at right angles to the
plate and fit securely into the holes of the perforated plastic
board thus preventing the plates from 'riding up'.
The apparatus it fitted with an electro-magnetic stick, U-sha-
ped in profile, which is supplied with current from a 6 volt
battery.
When the current is switched on, all elements in a column can
be attracted and transposed to another column. When the circuit
is broken the elements remain in their correct order. Rows of
the table may also be treated in a similar way.

I.ZONNEVELD:

Wir haben mit SEGAL auch eine Tabellen-Maschine entwickelt. Innerhalb einiger Monate wird sie fertig sein. Sie ist einfacher
und wirkt rein mechanisch. Sie hat ein größeres Gewicht, aber
auch Vorteile: z.B. anschließende Klötze, schwer genug, um nicht
leicht zerstört zu werden. Man kann einfach in alle Richtungen
schieben. Es läßt sich damit frei schreiben und rechnen.
Wieviel koste Ihr Apparat?

M.W.TRENTEPOHL:

Nach dem, was ich bisher für das Anfertigen ausgegeben habe,
würde ich sagen, daß das alles, also etwa ein Gerät ohne diese
Namen, nur mit den Schildern, den gestanzten Metallteilen, etwa
zusammen für 2 Platten und einen Magnetstab und das Ladegerät,

also mit Akku zusammen auf etwa 2100 Mark kommen würde.

S.SEGAL:

Die von ZONNEVELD und mir entworfene Ausführung hat einige Vorteile: 1. Viele Reihen können gleichzeitig verschoben werden.
2. Die Blöckchen sind größer und können nicht einzeln verschoben werden (keine Fehler). 3. Das Bild ist ruhig (keine Löcher)
und besser geeignet zur Photographie. 4. Elektrzität ist nicht
nötig. 5. Das Ganze ist billiger.

P.SEIBERT:

So ein mechanisches Gerät aus Kunststoff-Würfeln, wie es Herr
ZONNEVELD nennt, und auf denen auch die Soziabilität steht, hat
man in Wien, in Mariabrunn, entwickelt; man nennt dieses Gerät
Mariabrunner Tafel. Diese kostet aber 5000 Mark, da ist also
Herr TRENTEPOHL mit 2100 Mark unschlagbar.

R.TÜXEN:

Ich möchte doch zu diesem interessanten Zauberkünstler-Kongreß
auch noch etwas hinzufügen. Als wir vor vielen Jahren mit der
Herstellung unserer Tabellen begannen, standen wir vor den gleichen Problemen, wie sie Herr TRENTEPOHL geschildert hat, Fehler-
quellen, Zeitaufwand und Umwechselbarkeit usw. Da haben wir versucht - nicht wahr, DIEMONT ? - auf solchen Stäben (Zwischenruf
von Dr.DIEMONT: "Die taugten auch nichts") die Aufnahmen aufzu-
tragen: vorne war eine Leiste mit den Namen, und wir konnten
die senkrechten Spalten umwechseln, aber nur in einer Richtung.
Ich habe einmal so ein Bündel zu BRAUN nach Montpellier gebracht
ich glaube, sie liegen noch da.Sie sind nie benutzt worden. Dann
hat vor etwa 20 Jahren PINTO DA SILVA in Portugal etwas Ähnliches konstruiert, das ist auch in Vegetatio publiziert worden.
Auch Frau Professor WILMANNS hat so ein Ding konstruiert, und
als vor etwa 8 Monaten mein Freund SASAKI aus Japan kam, da hatte er ein großes Paket und hat versucht, mir deutlich zu machen,
er habe ganz originell einen Apparat konstruiert, dessen Elemente in zwei Richtungen verschiebbar seien, um Tabellen zu machen. Ich kann ihn morgen vorführen. Aber auch der ist schwer,
schwerfällig. Ich weiß nicht, wie teuer er ist, und er kommt
nicht an diese geniale Erfindung heran. Ich habe aber trotz des
billigen Preises gewisse Bedenken,etwa für uns,denn unsere Tabel

len waren manchmal so groß, die von Herrn SASAKI z.B., daß zwei
ziemlich große Zimmer gerade ausreichten, um sie auszulegen.
Wir haben sie in wenigen Tagen, nachdem sie geschrieben waren,
dann doch verarbeiten können. Ich darf sagen, daß Herr HÜLBUSCH
in zwei Monaten 1700 Aufnahmen des L o l i o - C y n o s u r e -
t u m verarbeitete, mit etwa 400 Arten. Das sind also die Größenordnungen, mit denen wir jetzt zu tun haben. Aus diesem Grunde sind die Apparate für uns gar nicht verwendbar. Eine kleine
Tabelle aber machen wir mit Hilfe unserer Teiltabellen-Methode,
mit der wir die diagnostisch wichtigsten Arten finden, die wir
herausziehen, die wir in Teiltabellen ordnen und dann nur einmal die ganze Tabelle neu schreiben, ohne Kosten für eine Appa-
ratur, und wahrscheinlich fast ebenso schnell wie mit dieser.
So fürchte ich, daß sich für den Pflanzensoziologen, der ein
großes Material zu bearbeiten hat, die Mechanisierung der Tabellenarbeit gar nicht durchführen läßt. Für Demonstrations-
zwecke ist das natürlich ganz ausgezeichnet, aber für die praktische Arbeit mit einem größeren Material muß man andere Methoden anwenden. Wir benutzen jetzt die Methode der Teiltabelle.
Wir werden in Kürze wissen, wie weit man Komputer einschalten
kann.

J.J.MOORE:
Ich mache keine Reklame für IBM, aber wir gebrauchen, z.B. für
die große Tabelle von Grünland des County Carlow, Karten. Es
hat viele Vorteile, nämlich, daß es in den meisten Universitäten Lochkarten-Maschinen gibt und auch diese printing machines.
Ich glaube, daß man am besten jetzt mit Lochkarten vorgeht. In
unseren Tabellen ist z.B. jede Art mit 1 oder 2 oder 3 Karten
vertreten, man kann die Arten sehr schnell in eine andere Richtung bringen, mit der Hand, oder man muß zum Komputer gehen, um
die anderen Spalten anzuordnen. Aber es geht sehr schnell und
es lohnt sich, glaube ich.

DIE ERSTBESIEDLUNG NATÜRLICHER KÖRPERHÖHLEN DES MENSCHEN DURCH MICROFLORA

R.Linder, R.Humbel, Germaine Wurch u.Th. Wurch.

Die Besiedlung eines vegetationslosen Neulands zeigt Grenzprobleme besonderer Art auf. Solches Neuland finden wir in den natürlichen Körperhöhlen des Neugeborenen, wie Nase, Mund, Rectum, und der Vagina beim Mädchen.

Zum ersten sehen wir das Übertreten einer Grenze in der Besiedlung als solcher. Sodann finden wir uns an einer Grenze dadurch, daß hier ein dem Soziologen ungewohnter Vegetationsboden zu betrachten ist. Schließlich steht dabei eine Microflora in Betracht, die an der Grenze des Pflanzenreiches klassifiziert wird.

Die Untersuchungen wurden an neugeborenen Mädchen innerhalb der ersten 12 Lebenstage durchgeführt.

Während der ersten 24 Stunden werden alle vier Höhlen besiedelt. In der Nase tritt regelmäßig Diplococcus auf, der hier im allgemeinen vereinzelt bleibt, während der eigentliche Nasenbewohner, Neisseria catarrhalis, beim Neugeborenen nicht zu finden ist. Bemerkenswert ist jedoch, daß grampositive Granulärformen häufig und dicht auf den exfoliierten Zellen haften.

Diplococcus kommt auch im Mund vor, hier aber meist größer und dichter gruppiert. Vielfach ist auch Staphylococcus vorhanden, der ebenfalls spärlich in der Nase gefunden wird. Vereinzelt zeigen sich Lactobacillus acidophilus, Neisseria catarrhalis und einmal sogar Candida am zweiten Lebenstage.

Die Abstriche des Rectums ergaben schon am ersten Lebenstage die üblichen Darmfloravertreter des Säuglings: Lactobacillus acidophilus, Bifidobacterium bifidum, Escherichia coli, Streptococcus foecalis, Staphylococcus. Die exfoliierten Zellen sind meist reichlich mit grampositiven Granulärformen punktiert.

Die Besiedlung der Vagina erfolgt so, daß vom zweiten Lebenstage an sich die Vaginalassoziation in der gleichen Form ausprägt, wie sie bei der reifen Frau gefunden wird, und zwar sowohl als L a c t o b a c i l l e t u m wie auch als H a e - m o p h i l e t u m, mit den üblichen Begleitarten der Haut- und Darmflora, jedoch nie mit Trichomonas und Candida. Streptococcus foecalis manifestiert sich deutlich als Erstbesiedler.

Die Dynamik des Besiedlungsprozesses kann folgendermaßen zusammengefaßt werden. Das Kind wird keimfrei geboren. In den ersten Stunden nach der Geburt werden Keime sporadisch festgestellt. Sodann entwickeln sich inselartige Kolonien in günstigen Nischen. Unter geeigneten Verhältnissen kann sich schließlich explosionsartig eine ganze Vegetationsdecke entfalten. Die häufige Beobachtung der Granulärformen, die wir als Resistenzformen der Bakterien deuten, spricht allerdi gs dafür, daß das Substrat nicht stets ohne weiteres sich als bereitwilliger Besiedlungsboden darbietet.

In der Nase des Neugeborenen bleibt es meist bei einer dispersen Verteilung. Im Mund entstehen Nester, häufig um exfoliier-

Tag nach der Geburt	A		T		J			O		E		M	
	1	3	1	7	2	4	11	2	3	5	7	6	8
NASE													
Diplococcus pneumoniae	.	.	r	+	1	+	1	3	+	r	r	r	1
Staphylococcus sp.	.	.	.	.	r	.	r	.	.	.	.	.	.
Granulärformen Gram +	+	+	+	4	.	1	1	.	.	r	.	.	4
MUND													
Diplococcus pneumoniae	r	1	r	1	1	+	+	r	+	1	1	r	1
Staphylococcus sp.	.	3	.	.	.	.	1	.	r	r	r	.	.
Lactobacillus acidophilus	.	.	.	.	+	r	r	.	.	.	.	.	.
Neisseria catarrhalis	.	.	.	.	.	r	.	.	r	.	.	.	.
Granulärformen Gram +	.	1	r	r	.	r	.	r	r	.	r	.	2
RECTUM													
Lactobacillus acidophilus	.	2	.	1	.	1	2	+	+	2	2	1	1
Bifidobacterium bifidum	.	1	.	2	.	2	+	+	+	.	2	.	.
Enterobacteriaceae	.	.	r	1	.	1	1	1	+	2	1	3	r
Streptococcus foecalis	+	1	1	+	1	+	1	r	1	1	+	1	r
Staphylococcus sp.	.	r	.	r	+	.	.	.	.	r	.	+	.
Granulärformen Gram +	.	r	1	3	+	+	+	r	4	r	r	.	1
VAGINA													
Lactobacillus vaginalis	.	1	r	4	+	1	2	.	1	2	3	r	4
Haemophilus vaginalis	.	.	.	.	.	.	.	.	.	.	.	5	.
Streptococcus foecalis	+	+	+	.	1	1	1	.	1	1	.	1	1
Staphylococcus sp.	.	.	.	.	.	+	1	.	r	.	.	+	+
Enterobacteriaceae	.	.	.	.	r	.	.	.	.	.	.	.	.
Granulärformen Gram +	.	+	+	+	+	+	r	r	2	r	.	.	+

Tabelle der Microflora-Aufnahmen in den natürlichen Körperhöhlen von 6 neugeborenen Mädchen (A,T,J,O,E,M.).

te Zellen gruppierte,fragmentarische Horste. Im Rectum und in
der Vagina verläuft die Entwicklung bis zur vollen Vegetationsdecke.

Hervorzuheben ist auch, daß wir keine Sukzession haben. Vielmehr werden gleich die - zwar wenigen - Arten vorgefunden, die
an der späteren, definitiven Gesellschaft beteiligt sind. Obwohl die ersten Aufnahmen Pionieraspekt zeigen, weil jeder Ein-
dringling sich mit unterschiedlichem Erfolg durchzusetzen versucht, stellt sich doch bald ein Gleichgewicht ein, bei dem die
typischen Gesellschaftsarten dominieren.

Speziell interessant sind die Ergebnisse im Fall M, wo die Vegetationsdecke der Vagina vom H a e m o p h i l e t u m (6.Tag)
in kurzer Frist zum L a c t o b a c i l l e t u m (8.Tag) ü-
bergeht. Dies deutet auf eine Umwälzung der ökologischen Verhältnisse hin, die durch die erste Genitalkrise beim ♀ Säugling
um den 7. - 8.Tag nach der Geburt auftritt. Ähnliche Assoziati-
onsänderung entsteht oft bei der erwachsenen Frau in der Mitte
des Zyklus, nach der Ovulation. Die unter dem Mikroskop schwer
identifizierbaren gramnegativen Bakterien sind unter der Global-
bezeichnung "Enterobacteriaceae" zusammengefaßt (Escherichia
coli, Shigella, Aerobacter, Proteus, Eggerthella).

DAS VAGINALE ECOSYSTEM UND SEINE GRENZPROBLEME

Th. Wurch und R. Linder

Im Rahmen des diesjährigen Themas zeigen wir, daß ein Micro-Ecosystem auch Grenzprobleme aufweist. Bei normalen konstanten klimatischen Verhältnissen in der Vagina besteht ein Gleichgewicht zwischen Substrat und Lactobacilletum vaginalis.

Topographische Grenzprobleme spielen sich am Übergang vom Vaginal-Epithelium auf das Haut-Epithelium einerseits und auf die Cervix-Schleimhaut andererseits ab.

Das Lactobacilletum vaginalis tritt nur im Vaginalraum als endemische Gesellschaft auf; es ist streng an das Vaginal-Epithelium gebunden und den zyklischen Veränderungen dieses Epitheliums unterworfen. Der Climax-Zustand wird rasch erreicht, er kann aber manchmal überschritten werden. Dann tritt Cytolyse ein, das heißt, die Lactobazillen greifen das Substrat an; schließlich kommt es zur Degradation des Lactobacilletums, die Bazillen gehen in perlschnurartige Granulärform über.

Ändert sich das Milieu, steigt zum Beispiel der pH-Wert auf 5 bis 6, so siedeln sich andere - auch endemische - Gesellschaften an, wie das Haemophiletum oder das Trichomonadetum. Installiert sich das Trichomonadetum, so kommt es zur Epithelial-Zerstörung und zu Entzündungsvorgängen im Substrat. Im Abstrich können wir dann krebsähnliche Zellen finden. Wird aber das Trichomonadetum durch Behandlung zerstört, so stellt sich sofort, mit dem gleichzeitigen Ausheilen des Substrats, wieder ein Lactobacilletum ein.

In der Vagina ist gerade Krebs ein Grenzproblem, denn Krebsherde kommen immer nur an der Epithelial-Grenze am Muttermund vor, wenn zum Beispiel das eine Epithel in das andere System hineinwuchert, somit das Gleichgewicht gestört ist.

Diese Befunde bestätigen den biologischen Wert der Anwendung der phytosoziologischen Methode in der Humanpathologie: maßgebend ist nicht das Vorkommen einer Microorganismen-Art, sondern die Assoziation.

ECOSYSTEME VAGINAL

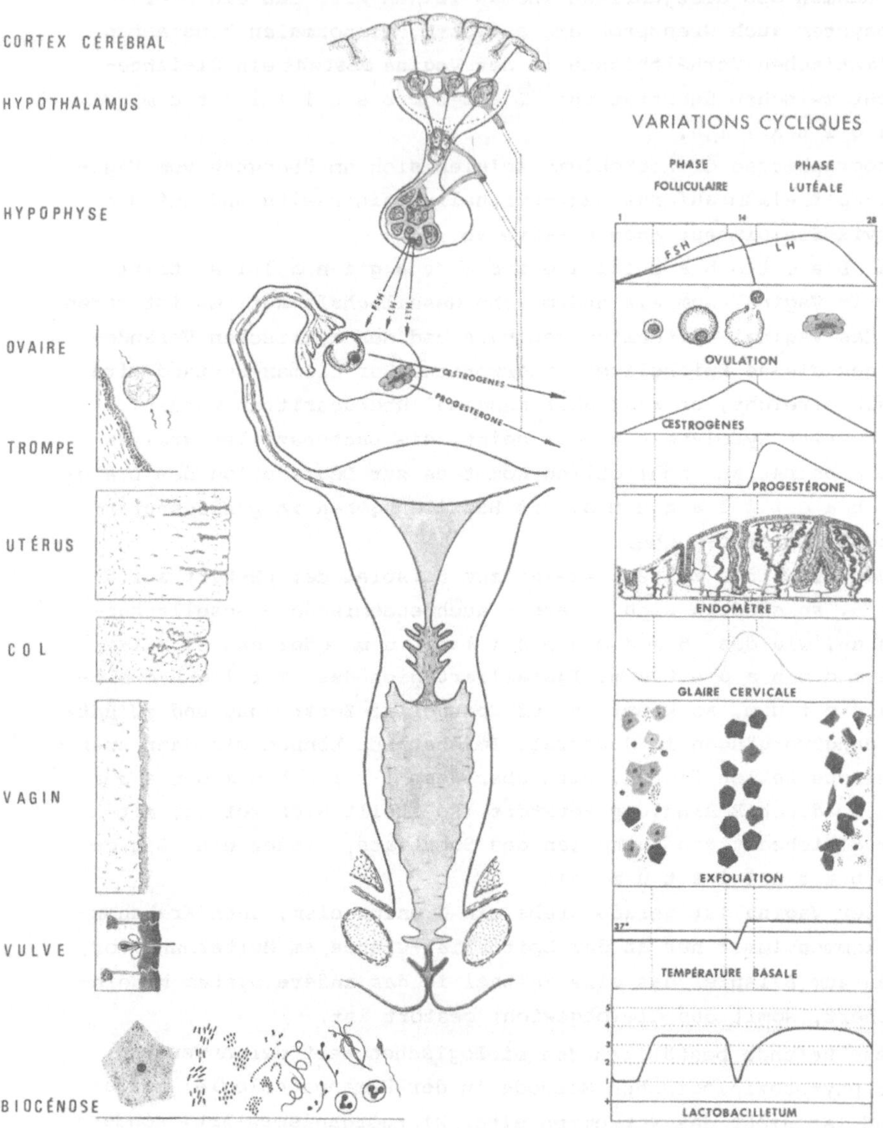

R.TÜXEN:

Ich bin überrascht - man braucht nur den Maßstab und die Namen
ändern - durch die frappante Ähnlichkeit dieser Mikrogesellschaften, sagen wir etwa mit Therophyten-Gesellschaften unserer
Küste.Wenn man statt Lactobacillus oder einem anderen Microor-
ganismus Salicornia oder Suaeda sagt, (und die Bilder nicht ganz
scharf einstellt), dann hat man die Küsten-Gesellschaften. Wenn
man den Maßstab etwas verkleinert, auch den zeitlichen, dann
hat man die gleichen Erscheinungen bei den S a l i c o r n i -
e t e n, bei den S u a e d e t e n und ähnlichen halophilen
oder nitrophilen Gesellschaften, die wir als therophytische
Spezialisten-Gesellschaften an unseren Küsten ja seit langem
kennen. Das ist einer der stärksten Eindrücke, den ich heute
morgen gewonnen habe: diese Allgemeingültigkeit der Gesetze an
den Grenzen des Lebens - denn das ist auch eine Grenze des Lebens an den Küsten, eine Grenze des Lebens der Phanerogamen-
Gesellschaften gegen die völlige Phanerogamen-Leere des Milieus,
genau wie hier. Ich kann also nur die letzten Schlußworte von
Herrn Dr. WURCH durchaus unterstreichen, daß hier die gleichen
Gesetzmäßigkeiten bei den Bakterien-Gesellschaften gelten wie
bei unseren Phanerogamen-Gesellschaften.
Das zweite, was mich sehr beeindruckt hat, war, daß hier nicht
nur beschreibende, und auch nicht nur kausale Soziologie, sondern experimentelle Soziologie getrieben worden ist, und getrieben werden kann und sicher auch weiter getrieben werden wird,
mit einer außerordentlichen praktischen Bedeutung. Ich brauche
ja nur an die letzten Ausführungen der pathogenen Bildungen zu
erinnern. Ich bin erfreut über diese bahnbrechenden und grundlegenden Erweiterungen unserer schönen Wissenschaft hier in das
Gebiet der Human-Medizin.

R.CARBIENER:

Wir verwenden in unserem Wissenszweig sehr viel Analogie-Ver-
gleiche, und da Herr Professor TÜXEN eben gezeigt hat, wie sehr
ähnlich die Probleme an den Lebensgrenzen sind, möchte ich auf
die Hydrobiologie hinweisen, auf zwei Grenzprobleme, wo wir
ganz ähnliche Verhältnisse treffen. Im Brackwasser kommen nämlich auch plötzlich Substrat-Wechsel vor mit dem Wandel des
Salzgehaltes, wo wir dann ähnliche Artenarmut der Gesellschaften mit zugleich großem Individuen-Reichtum vorfinden, wo wir

auch sehr plötzliche Wechsel finden zwischen diesen Gesellschaften als Folge des Wechsels des Salzgehaltes.Ähnlich finden wir
auch in verseuchten Gewässern explosionsartige Entwicklungen
einzelner Mikroorganismen, die äußerst individuenreiche und artenarme Gesellschaften bilden mit allen Saprobität-Stufen,
die in der angewandten Hydrobiologie gut bekannt sind.Auch dort
zeigt sich eine schöne Analogie im Zusammenhang mit den an den
Grenzen der Lebensmöglichkeiten vorkommenden Gesellschaften,
die auf Substrat-Wechsel sehr stark und schnell reagieren.

A.APINIS:
Wir müssen doch manche Dinge auseinanderhalten. Es handelt sich
hier um ganz andere Typen der biologischen Einheiten, die wir
als heterotroph bezeichnen, oder als Parasiten-Saprophyten,
oder als Symbionten betrachten. Ökologisch sind sie grundsätzlich verschieden von einer autotrophen Vegetation. Im Verhältnis zu den Milieufaktoren sind sie sich sehr ähnlich, da verhält sich die autotrophe Welt wie die heterotrophe. Sie reagieren unter gleichen Temperatur-, Feuchtigkeitsverhältnissen usw.
Aber der Unterschied liegt in den Energie-Quellen. Alle Heterotrophen benützen die Energie-Quellen, die von anderen Organismen erzeugt werden. Das ist der grundsätzliche ökologische Unterschied. Wir müssen uns fragen,ob wir diese weitverbreiteten
Begleitgesellschaften auch als selbständige oder gemeinsame
Vergesellschaftungen betrachten dürfen? Diese Frage ist natürlich wichtig. Wenn wir diese Gesellschaften oder Populationen
von der autotrophen Pflanze oder von einem anderen Organismus
loslösen, würde das zu einem ökologischen Widerspruch führen.
Wir müssen dies doch auseinanderhalten, weil wir sonst diese
Unterschiede nicht klar sehen können.
Nun möchte ich noch eine Frage stellen: Haben Sie irgendwelche
Änderungen in der Zusammensetzung dieser Körperhöhlen beobachtet, wenn sich die Temperatur des Organismus ändert? Wenn sich
die Temperatur z.B.für eine längere Zeit erhöht, stellt sich
dann eine Änderung in der Zusammensetzung dieser Vegetation
ein, oder nicht?

E.BURRICHTER:
Die erwähnte explosionsartige Entwicklung der Bakterienpopula-
tion kennen wir auch aus den Böden, wenn plötzlich bessere

Standortsverhältnisse einsetzen. Das wird z.B. deutlich, wenn
nach längerer Trockenperiode Regen einsetzt. Bereits nach einem
Tag kann sich dann der Bodenbakterien-Gehalt verdoppeln oder
sogar vervielfachen, und auch die Zusammensezung der Flora ändern.
Wie ist es mit den Bakterienkolonien, den sogenannten Zooglöen?
Entwickeln sie sich bei Optimalzuständen?

R.LINDER:
Zuerst eine Antwort auf das, was Herr APINIS vorgetragen hat.
Müssen wir zwischen Autotrophen und Heterotrophen einen Unterschied machen? Ich glaube nicht.Ich glaube nicht, daß der Unterschied so groß ist. Wie Herr Professor TÜXEN gesagt hat,
glaube ich, daß hier sich nur der Maßstab ändert. Denn wir haben jedesmal einen Boden und die Bakterienvegetation hängt von
diesem Epithel ab, diesem geschichteten, lebenden Boden. Das
ähnelt einem höheren Boden, es sieht aus wie eine Gesellschaft
höheren Ranges in einem Boden. Nur gehen diese Veränderungen
viel schneller vor sich und sind zyklisch.
Es gibt während des Zyklus eine kleine Temperaturschwankung
(1°),aber die ist nicht bedeutend. Jedoch kann bei dieser Tem-
peraturänderung ein Umsturz der Gesellschaft auftreten; z.B.
ein L a c t o b a c i l l e t u m kann in diesem Falle bei
der Ovulation zu einem H a e m o p h i l e t u m werden und
umgekehrt. Aber das hängt doch hauptsächlich mit dem pH zusammen.
Bei optimaler Gesellschaftsentwicklung sehen wir die Herde nie
so deutlich, höchstens die Granulärformen hier und da auf den
Zellen, aber es ist, als ob sich diese Zooglöen in einem gehemm-
ten Milieu bilden, und wenn sich dann das Milieu angepaßt hat -
denn das Milieu ist lebendig - wenn ein Gleichgewicht zustande-
kommt, dann ist eine vollkommene Entfaltung da, und es sind
kaum mehr Herde vorhanden. Die Vegetation breitet sich ganz
aus, entwickelt sich nicht in Herden. Im Mund bleibt es allerdings bei diesen Zooglöen, dort bildet sich nie mehr.

R.TÜXEN:
Ich möchte nur noch ein Wort hinzufügen zu dem Vergleich mit
den S a l i c o r n i e t e n. Das war vielleicht nicht der
beste Vergleich. Denken Sie einmal an Lemna-Decken; wenn Sie

ein unscharfes, schlecht photographiertes Bild haben von einer
Wolffia, oder einer Lemna trisulca oder Lemna minor-polyrrhiza-
Gesellschaft usw., so können Sie das kaum, wenn nicht scharf
eingestellt ist, von einer solchen Gesellschaft unterscheiden.
Hier sind die Vergleiche, wenigstens die äußerlichen, noch viel
eindringlicher als bei den S a l i c o r n i e t e n.

R.LINDER:
Ich hätte gern eine Meinung über Sukzessionsfragen gehabt.

Th.WURCH:
La notion de Climax n'a été utilisée que pour le L a c t o -
b a c i l l e t u m. Cette association est le stade ultime des
associations de la flore vaginale et est consideré comme l'expression de l'état de santé - de l'optimum (de l'équilibre)pos-
sible.-   En gynécologie on parle de degré du pureté ou "Rein-
heitsgrad". Le terme Climax n'est pas utilisé pour les éléments
de la Biocénose (Trichomonas etc.).

R.TÜXEN:
Auch bei den L e m n i o n -Gesellschaften finden wir ja die
gleichen zyklischen Veränderungen, nur in längeren Zeiträumen,
die gleiche Unfähigkeit zur Sukzession. Eine L e m n a -Decke
stellt sich mit wenigen Individuen ein, entwickelt sich explosionsartig in wenigen Wochen und vergeht wieder, wenn die Temperaturen zu niedrig werden, überdauert in irgendeiner Form,
wie es auch die Bakterien tun, und stellt sich im nächsten Sommer wieder ein, sei es, daß sie an Ort und Stelle geblieben ist,
sei es, daß sie von neuem eingeschleppt wird wie etwa unsere
W o l f f i a -Gesellschaften, die sich in den letzten Jahren
zusehends bei uns vermehren. Wir kannten vor wenigen Jahren
ein oder zwei Stellen, wir kennen jetzt ein Dutzend Stellen
mit W o l f f i a. Auch hier ist keinerlei Sukzession zu sehen.
Das ist vielleicht eine der interessantesten Feststellungen von
Herrn LINDER, daß er hier Gesellschaften studiert hat - wir können das Lemna-Beispiel bei uns heranziehen - die sich einstellen, die sich zur optimalen Entwicklung ausbreiten, und die wieder zugrunde gehen. Und dieser Vorgang wiederholt sich dauernd
beim Menschen in kürzeren, hier in längeren Zeiträumen. Sie
sprechen von Klimaxstadien, wir würden vielleicht lieber von

einer Optimal- oder Schlußphase sprechen. Unter Klimax verstehen wir eigentlich die letzte Stufe einer Sukzession und wenden
den Begriff nicht bei diesen Spezialisten-Gesellschaften an.Wir
sprechen von Klimax-Gesellschaften bei der letzten Schlußphase
mehrerer Entwicklungs-Serien, die auf den verschiedenen Aus-
gangssubstraten schließlich zu einer einheitlichen Endstufe von
regionaler Verbreitung im Gleichgewicht mit dem Regionalklima
führen. Wir sprechen bei den L e m n e t e a und noch weniger
bei diesen äußerst lokalen abhängigen Spezialisten-Gesellschaften nicht von Klimax-Gesellschaften. Wir könnten sie dagegen
als Dauer-Initial-Gesellschaften bezeichnen. Wir könnten in ihrem Falle auch sehr gut von der potentiell natürlichen Vegetation sprechen, worunter ja auch die Klimax-Gesellschaft fällt.
Ich kann immer nur wieder sagen, wie anregend die Vergleiche
dieser Studien von uns bisher völlig unbekannten Spezialisten-
Gesellschaften mit unseren Erfahrungen in den altbekannten Gesellschaften sind. Denn hier werden allgemeine Gesetzmäßigkeiten im soziologischen Leben der Organismen deutlich. Sie haben
uns eine neue Welt eröffnet.

ÜBER DIE GRENZBEREICHE DER PILZVEGETATION

A.E.A p i n i s

Die Pilze unterliegen grundsätzlich denselben Umweltsfaktoren
wie andere Lebewesen. Der Einfluß dieser Standortsfaktoren zeichnet sich aus durch eine spezifische, artenreiche Amplitude mit
einem Minimum, Optimum und Maximum, die jedoch nicht streng festgelegt ist, und durch eine veränderte Faktoren-Kombination oder
durch innere Ursachen, die mehr oder weniger variieren kann.
Auch steht fest, daß die Resistenz unter ungünstigen Umweltver-
hältnissen dieselben extremen Werte aufweisen kann, wie das bei
anderen Mikroorganismen der Fall ist. Doch der fundamentale Unterschied der Pilze anderen Pflanzen gegenüber liegt in ihrer
heterotrophen Lebensweise, die wir als Saprophytismus,Parasitis-
mus und Symbiose bezeichnen. Doch durch die zahlreichen mehr
oder weniger festgelegten biochemisch-morphologischen und onto-
genetischen Anpassungen sind sie mit Pflanzen, Tieren und sogar
mit Vertretern der eigenen Klasse mehr oder weniger innig verbunden,(vergl.BAKER & SNYDER 1965). So ist die Pilzvegetation
des Standortes,d.h. des Bodens und der Luftschicht ohne regelmäßige Zufuhr aus Substratenergiequellen kaum denkbar. Die Zufuhr
wird durch die autotrophe Vegetation besorgt, wobei auch Bodenfauna, Stoffwechsel-Endprodukte der Blatt- und Wurzeloberflächen
als auch verschiedene Abbauprodukte organischer Substanz in Frage kommen. Eine Pilzvegetation des Bodens ist demnach ein wesentliches Indiz für die auf einer Oberfläche lebenden Pflanzen und
Tiere (Phyllosphäre,Rhizosphäre und Spermosphäre). Auf altern-
den und toten Organ-Oberflächen entwickelt sich diese weiter zu
einer epiphytischen oder einer abbauenden saprophytischen Pilzvegetation. Solche Pilzpopulationen werden als Pilz-Gesellschaf-
ten oder Synusien bezeichnet, die bestimmte lebende oder tote
Substrate besiedeln. Diese sind jedoch grundsätzlich abhängig
vom gegebenen Substrat der autotrophen Pflanzen, weil die wechselnden Pilzpopulationen mehr oder weniger schnell einen völli-
gen Abbau eines Substrates, je nach dessen chemischer Zusammensetzung, herbeiführen. Mit einer Erschöpfung der Nährstoffquelle
hat die Pilzvegetation ihre Existenzgrenze erreicht und wird
durch andere Organismen schnell zerstört und weiter abgebaut.
Die entsprechende Pilzpopulation kann nur weiter bestehen, falls

frische Nährsubstrate neu zugeführt werden oder die Arten dieser durch Sporendispersion an einer anderen Stelle passende
Substrate finden. Doch die Neukolonisation organischer Substrate ist in der Luftschicht und in der Bodenschicht einer Pflanzengesellschaft grundsätzlich verschieden. Die Neukolonisation
der Besiedlungsflächen in der Luftschicht eines Pflanzenbestan-
des geschieht durch zahlreiche Sporen, die meistens durch Wasser, Wind und Tiere passiv (oder aktive Abschleuderung der Sporen bei vielen Ascomyceten) verstreut werden. Bei einer Neukolonisation von Pflanzen- und Tierresten im Boden spielt die Spo-
ren-Dispersion dagegen eine untergeordnete Rolle, weil die meisten Bodenmikromyceten in einer Sporen-Produktion stark unterdrückt sind. Bei höheren Ascomyceten und Hutpilzen ist neben
einem extensiven Bodenmyzelium auch ein effektiver Sporen-Dis-
persionsmechanismus durch oberirdische Fruchtkörper ausgebildet. Die im Boden beobachteten Sporen sind meistens aus der
Vegetations- und Streuschicht durch Bodenfauna und Wasser abge-
lagert. Aktive Dispersion durch Zoosporen sind nur bei Oomyceten
und anderen niederen Phycomyceten bedeutsame Fortpflanzungsar-
ten. Bei einer direkten Beobachtung der Bodenpilzvegetation
findet man meistens wenig Sporenbildung in Bodenhohlräumen,aber
einen relativ hohen Anteil verschiedener Pilzmyzelien. Durch
spezielle Methoden ist es jetzt möglich, diese aus dem Boden zu
isolieren, und genau zu bestimmen.Die gewonnenen Tatsachen bezeugen, daß bei den meisten eigentlichen Bodenpilzen die Kolonisation verschiedener Nährsubstrate durch Myzelwachstum geschieht. Somit sind die Bodenpilze in der Lage, die nährstoffreichen Bodenräume zu überwinden und neue Substrat-Energiequel-
len aufzuschließen. Dieses Verhalten ist bei Oomyceten, Mukorineen, Bodenascomycten, Deuteromyceten und Hymenomyceten in mehr
oder weniger abweichender Form zu finden. Besonders eindrucks-
volles Myzelwachstum kann man bei humus- und hexenringbildenden
Agaricaceen beobachten. Bei Mykorrhiza bildenden höheren Pilzen
ist eine enge Bindung mit dem Wurzelsystem der entsprechenden
Baumart fast immer vorhanden. Symbiotische und asymbiotische
Hutpilze unterscheiden sich in ihrer Ökologie, was sich auch in
ihrer Aspektfolge äußert (vgl. SOBOTKA & SOBOTKOVA 1966). Da
die ökologischen Gegebenheiten in der bodennahen Luftschicht
anders sind, als die im Assimilationsbereich der höheren Pflanzen, wird hier nur von terrestrischen Pilzen gesprochen. Das fin-

det seinen Grund ganz besonders darin, daß seit langem bekannt
ist, welch bedeutsame Aufgabe höhere, aber auch niedere Pilze
im Wurzelbereich von Phanerogamen haben. Ferner haben die Pilze
als "Begleitflora" in bestimmten Pflanzengesellschaften einen
hohen ökologisch-gesellschaftstaxonomischen Aussagewert,wie dies
z.B. aus der Arbeit von JAHN, NESPIAK, TÜXEN (1967) über die
Pilze in Buchenwäldern zu ersehen ist.

EXTREME FEUCHTIGKEITSVERHÄLTNISSE

Die Aktivität, wie auch die Zusammensetzung der Artenmenge der
Bodenpilze, sind sehr stark von Standortsfaktoren abhängig.
Feuchtigkeitsverhältnisse (Menge und Verteilung der Niederschläge, Wassergehalt, Wasserkapazität, Saugkraft und Verdunstung
des Bodens) spielen eine bedeutende Rolle (vgl.FRIEDRICH 1940),
LANGE 1957, MISHUSTIN et al.1961 u.a.) in der Dynamik und Verteilung der Pilze. Unter mäßig feuchten Klimaverhältnissen Mitteleuropas weisen die Böden der verschiedenen Pflanzengesellschaften, einschließlich der Kulturböden, eine fortdauernde und
reichhaltige Pilzvegetation auf. Böden der Trockengebiete (vgl.
KILIAN & FEHER 1935, SABET 1935, HARVEY 1942, GAERTNER 1955,
BORUT 1960, NICOT 1960, MISHUSTIN et al. 1961, LOUB 1963), in
Wüsten, Savannen und Steppen dagegen bergen eine viel ärmere,
öfters kaum feststellbare Pilz-Vegetation, die sehr starken Tem-
peratur- und Feuchtigkeitsschwankungen ausgesetzt ist. Die Pilz-
flora dieser fast vegetationslosen oder mit einem nur spärlich
ausgebildeten Oberflächenwuchs ausgestatteten Geländes ist verhältnismäßig reich an Aspergillus- und Penicillium-Arten, so
wie anderen Deuteromyceten; Oomyceten und Hymenomyceten sind
selten, oder fehlen gänzlich. Diese Dürreperioden werden durch
verschiedene Ruhezustände (Dauermyzelien, Gemmen, Sklerotien
und Sporen) überstanden. Eine durch starke Einstrahlung bewirk-
te Sterilität der obersten Bodenschichten wird durch Neukolonisation wieder hergestellt. Eine stark ausgeprägte Überdauerungs-
fähigkeit bei vorübergehender Bodentrockenheit ist bei Bodenpilzen der feuchten Klimagebiete generell verbreitet. REMY
(1949) und APINIS (1964) konnten niedere und höhere Bodenmykro-
myceten in lufttrockenen Bodenproben wiederfinden, die acht Jahre im Laboratorium aufbewahrt worden waren.

Die Entwicklung und Artenverteilung der Pilze in Böden von unterschiedlichem Feuchtigkeitsgehalt zeigen öfters große Unterschiede auf,auch in Gebieten mit gemäßigtem Klima. Diese Diffe-

renzen wurden in Böden einiger alluvialer Dauerweiden an einem
Altwasser unweit Nottingham (APINIS 1958,1964,1967) beobachtet.
Folgende Standorte wurden für diesen Vergleich ausgewählt:

1. Trockene Weide (L o l i o - C y n o s u r e t u m) auf sandigem Lehmboden mit einem A-C-Profil.
2. Zeitweilig überflutete Weide (R a n u n c u l u s   r e p e n s-
   A l o p e c u r u s   g e n i c u l a t u s - A s s.) auf einem
   A-G-Boden.
3. Eine nasse Stelle in der feuchten Weide (R a n u n c u l u s
   r e p e n s - A l o p e c u r u s   g e n i c u l a t u s -
   A s s., S u b a s s. v o n   A g r o s t i s   s t o l o n i-
   f e r a ebenfalls auf einem A-G-Boden.
4. Eine feuchte Variante des L o l i o - C y n o s u r e t u m.
5. Uferzone mit einem G l y c e r i e t u m   m a x i m a e   auf
   einem subaquatischen Schlammboden.
6. Röhricht (S c i r p o - P h r a g m i t e t u m) auf subaqua-
   tischem Torf.
7. Initialstadien in der Marschen.

Der Wassergehalt der Böden nimmt allmählich mit steigendem Grund-
wasserspiegel längs der Transekt-Linie vom L o l i o - C y n o-
s u r e t u m   t y p i c u m in Richtung auf die Ufervegetation
zu. Die unterschiedlichen Wasserverhältnisse bedingen nicht nur
eine verschiedene Bodenbildung, sondern bewirken auch eine cha- ·
rakteristische Artenzusammensetzung innerhalb der Pilzpopulation. So ergibt sich mit zunehmender Bodenfeuchtigkeit ein Absin-
ken der Hymenomyceten-Arten (Tab.1).

Tabelle 1. Zahl der höheren Boden-Hymenomyceten in alluvialen
Dauerweiden an einem Altwasser unweit Nottingham mit Angaben
über den Wassergehalt und Redox-Potential (Eh) in einer Tiefe
von 20-25 cm im Sommer 1951

Pflanzengesellschaft	Artenzahl	Wassergehalt %	Eh mv 20° C
Lolio-Cynosuretum	24	16	414
etwas feuchteres Lolio-Cynosur.	8	--	---
Ranunculus repens-Alopecurus geniculatus-Ass.	2	40	238
Desgl.Subass.v.Agrostis stolon.	1	65	130
Glycerietum maximae	1	70	185
Scirpo-Phragmitetum	O	484	234

Die Wachstumsfrequenz der Pilzmycelien nimmt ebenfalls mit
steigendem Wassergehalt des Bodens stark ab; gleichermaßen ver-

hält es sich mit der Artenzahl aktiver Mycelien. Die ermittelten Redox-Potentiale unterstreichen dieses Faktum und weisen auf die schlechter werdende Bodendurchlüftung hin (Tab.2).

Tabelle 2. Vergleich der Aktivität der Bodenmikromyceten in der der oberen Bodenschicht (15-25cm) der verschiedenen Dauergrünland-Gesellschaften an einem Altwasser, August 1950 (APINIS 1967)

Pflanzengesellschaft	Wachstums-frequenz	Artenzahl	Relative Charakterart
Lolio-Cynosuretum	2,5	28	8
Ranunculus repens-Alopecurus genicu-latus-Ass.	O,9	22	4
Glycerietum maximae	O,4	18	6
Scirpo-Phragmitetum	O,7	13	4
Initialstadien der Sumpfbildung	O,3	8	1

Das Vorkommen höherer Phykomyceten-Arten in Böden dieser Transekten-Linie wurde mit besonderer Aufmerksamkeit studiert (APINIS 1964). Tab.3. zeigt die Artenverteilung der Phykomyceten-Gruppen als Resultat dieser Untersuchungen.

Tabelle 3. Artenzahl der Phykomyceten in verschiedenen Dauergrünland-Böden an einem Altwasser unweit Nottingham (APINIS 1958)

Standort Pilzgruppe	Blasto-cladiales	Saproleg-niales	Perono-boerales & Zoopagales	Mucorales
Lolio-Cynosuretum	—	9	11	2G
Ranunculus repens-Alopecurus genicu-latus-Ass.	2	13	12	22
Desgl.Subass.v. Agrostis stolon.	1	11	9	12
Glycerietum maximae (terrestre-Zone)	7	17	11	18
Glycerietum maximae (Schwingrasen)	7	15	12	15
Scirpo-Phragmitetum	8	8	11	15
Vegetationslos	1	1	5	9
Glycerietum Initial-stadium	—	6	5	14
Phragmitetum "	—	2	5	5
Salicetum "	2	3	4	2
	9	24	18	36

Aus dieser Feststellung geht hervor, daß Vertreter der B l a-
s t o c l a d i a l e s und S a p r o l e g n i a l e s die
nassen Sumpfböden bevorzugen, die M u k o r i n e e n dagegen
in feuchten, oder nur teilweise überfluteten Böden sich entwickeln. Die Erklärung für dieses Verhalten ist in der Konsti-
tution der einzelnen Pilze zu suchen. WALTER (1931) gibt für
das Verhalten der Schimmelpilze zu bestimmten Hydraturverhält-
nissen eine allgemeine ökologische Definition derart, daß die
Pilze das Außenmedium einer bestimmten relativen Dampfspannung
(RD) bedürfen,um ein Wachstumsminimum zu gewährleisten. Diese
Postulation hat bislang zu einer brauchbaren ökologischen Charakterisierung der Mikromyceten-Arten beigetragen. Gemäß dieser
Definition haben die xerophilen Arten eim Mycel-Wachstumsmini-
mum von 85-90% RD; hierzu zählen bestimmte Aspergillus- und Penicillium-Arten, die vornehmlich in trockenen Standorten und in
Trockengebieten vorkommen. Bei den meisten mesophilen Arten
liegt diese untere Grenze bei 90-95% RD. Diese Artengruppe ist
durch die meisten Mukorineen, eine Anzahl Deuteromyceten,Hefen
und Agaricaceen vertreten, die auch in größerer Anzahl in den
oben erwähnten Weiden vorkommen. Bei den hygrophilen Arten liegt
das Minimum bei 95% RD, und noch höher bei den Pythium-Arten.
Die hydrophilen Vertreter zeigen ihr Minimum schon bei RD 99,6%.
Hierzu gehören die eigentlichen Wasserpilze, wie Achlya, Dictyu-
uchus und Saprolegnia. Auf Grund dieser Angaben ist es leicht,
das Vorkommen von Bodenmikromyceten in alluvialen Böden mit star-
schwankendem Wasserhaushalt zu deuten.
Analoge Bodenfeuchtigkeitsverhältnisse sind auch an den mehr
oder weniger flachen, sandigen Ostküsten Englands anzutreffen,
die eine ähnliche Verteilung der Pilzvegetation (APINIS 1968)
wie im oben erwähnten Süßwasserbereich aufweisen. Jedoch hat
hier der Salzgehalt der ufernahen Böden einen starken Einfluß
auf das Artengefüge der dortigen Pilzvegetation (HÖHNK 1939-
1959 in SPARROW & JOHNSON 1961). Die Vegetation der höheren Pilze
zeigt auf dem Niveau des mittleren Hochwassers ihr Existenzmi-
nimum und gleichermaßen finden wir eine weitere Verarmung der
Kleinpilzvegetation in Richtung des niederen Wasserstandes (vgl.
HARDER & ÜBELMESSER 1955).
Die subaquatischen Meeresböden beherbergen, wie aus den wenigen
Untersuchungsergebnissen zu entnehmen ist, eine artenreiche Mikromycetenvegetation mit niederen und höheren Phycomyceten, He-

fen,Deuteromyceten und Ascomyceten. Die westlichen, küstennahen
Schlickböden des Atlantischen Ozeans sind in einer Tiefe von 13
bis 220 m artenreicher als die Tiefsee-Böden und zeigen eine
Reihe von Boden-Mikromyceten wie: Alternaria-, Aspergillus-,
Cephalosporium-, Chaetomium-, Cladosporium-, Rhizopus- und
Trichoderma-Arten. ROTH et al. (1962) konnten aus den küstenna-
hen Sedimenten der Florida-Küste zahlreiche Hefen isolieren
(Candida-, Cryptococcus-, Debaryomyces-, Rhodotorula- und Tri-
chosporon-Arten. Inzwischen ist eine ganze Reihe von Arbeiten
erschienen, die eingehend die marine Pilzvegetation abhandeln
(JOHNSON & SPARROW 1961). So berichtet HÖHNK (1959) über die
Tiefsee-Kernproben aus den Schelfregionen Irlands, Islands
und Grönlands aus einer Tiefe von 1100 bis 3425 m, aus der er
ein regelmäßiges Vorkommen der Pilze nachweist.

DIE PILZVEGETATION IN TIEFEREN BODENSCHICHTEN
Die oberen Schichten der natürlichen oder kultivierten, unter
mitteleuropäischen Klimaverhältnissen entstandenen Böden, weisen gewöhnlich die größte Anzahl unterschiedlichster Mikroorganismen auf, welche mit zunehmender Tiefe mehr oder weniger rasch
abnimmt. So ein räumliches Vorkommen der Bodenpilze im natürlichen Boden ist immer an die Verteilung der Wurzelmasse höherer
Pflanzen gekoppelt, weil sie die notwendigen Substratquellen
liefert, einschließlich unentbehrlicher Wuchsstoffe. Daraus kann
man entnehmen, daß die Tiefenverbreitung der meisten Bodenpilze
mit der Tiefenverteilung des Wurzelsystems höherer Pflanzen syn-
chron ist, und unter günstigen Boden- und Klimaverhältnissen
eine beträchtliche Tiefe erreicht. Zur Zeit liegen über die
Pilz-Population im Wurzelbereich nur sporadische Angaben vor.
Dennoch deuten vereinzelte Untersuchungen an, daß Pilze auch in
nicht durchwurzelten Boden-Horizonten anzutreffen sind. HARDER &
ÜBELMESSER (1957) konnten in einer Baugrube bei Göttingen Muco-
rineen bis 1,20 m, niedere Phycomyceten noch in einer Tiefe von
1,60 m finden,und ein regelmäßiges Vorkommen von Pilzhyphen
wurde selbst in 5.20 m liegenden Schichten festgestellt.
Die Tiefenverteilung der Bodenpilze ist auch von klimatischen
Einflüssen abhängig. Unter trocken-warmen Klimaverhältnissen
sind die oberen Bodenschichten oftmals einer starken Isolation
und Austrocknung unterworfen. Infolgedessen findet sich das Maximum der Rhizosphäre in tieferen Bodenschichten, in der auch
die meisten Bodenpilze als Wurzelbegleiter eine höhere Frequenz

erreichen. Analog hierzu finden wir dieses Phänomen in Frostböden
der Hochgebirge und Polargebieten (EKELÖF 1908, MISHUSTIN et.al.
1961).

Von den edaphischen Faktoren kommt der Bodendurchlüftung eine
fundamentale Bedeutung zu, weil diese nicht nur am Bodenbildungs-
prozeß beteiligt ist, sondern weitgehend die Pflanzenwurzeln
und somit auch die Entwicklung der Bodenmikroorganismen beeinflußt. Bei einer guten Gare und einem normalen Feuchtigkeitsgehalt ist auch in den tiefen Bodenschichten ein günstiger Gas-
austausch gewährleistet. Dagegen erleiden mittelschwere und
schwere Böden bei einer vorübergehenden oder andauernden Stau-
nässe schnell ein Sauerstoffdefizit, so daß aerobe mikrobiolo-
gische Prozesse zum Stillstand kommen. So finden wir bei hohem
Angebot organischen Materials unter solchen Bedingungen extreme
Erscheinungen (APINIS 1960,1967). MOORE (1954) fand, daß Boden-
Mikromyceten eines G e n i s t o - C a l l u n e t u m - Torf-
bodens, der eine lockere Struktur mit einer nur geringen Wasser-
sättigungskapazität besitzt, in dessen Gesamtmächtigkeit gleichermaßen verteilt waren. In einem Trichophorum caespitosum-
Torf mit hohem Wassergehalt, waren dagegen die Bodenpilze auf
die oberste 15 - 20 cm starke Auflage beschränkt. Bei ungenü-
gender Durchlüftung des Bodens wird die Aktivität der Bodenpilze merklich reduziert, wie dies an zwei alluvialen Weideböden
aufgezeichnet wurde (APINIS 1967).

Tabelle 4. Mykromycetenaktivität in verschiedener Bodentiefe
einer trockenen (Lolio-Cynosuretum auf A-C-Boden) und einer
feuchten Ranunculus repens-Alopecurus geniculatus-Ass. auf A-G-
Weide im Sommer 1950-1952 charakterisiert durch Frequenz und
Artenzahl der entsprechenden Pilze (vgl.APINIS 1967)

Standort	Bodentiefe cm	Frequenz	Artenzahl
Trockene Weide	2 - 5	4.2	24
" "	20 - 25	2.6	21
" "	50 - 60	1.9	10
" "	100 - 105	1.1	6
Feuchte Weide	2 - 5	3.1	27
" "	20 - 25	2.0	19
" "	50 - 60	0.4	9

In Tabelle 4 wird die Frequenz der wachsenden Pilzmycelien aus
unterschiedlicher Bodentiefe aufgezeigt, die in einem trockenen,
gut drainierten und in einem feuchten Weideboden gefunden wurden. Aus diesen Aufzeichnungen geht hervor, daß das Mycel-Wachs-

tum mit steigender Tiefe in beiden Substraten abnimmt. Dagegen
ist in einem frischen, gut durchlüfteten Weideboden die Wachs-
tumsabnahme von Pilzmycelien so kontinuierlich, daß selbst in
einer Tiefe von 1,05 m Frequenzgrenzwerte zu beobachten waren.
Dagegen sind die Pilzmycelien und ihr Wachstum im feucht-nassen
Weideboden nur im oberen, 25 cm starken A-Horizont zu finden. In
einer Tiefe von 50 - 60 cm, im Gleyhorizont des feinsandigen
Lehms, findet nur ein geringes Mycel-Wachstum statt. Auch die
Arten-Zusammensetzung der wachsenden Mycelien zeigt eine ähnliche Reduktion. Angestellte Versuche und Beobachtungen sind Beweise dafür, daß eine geringe Boden-Durchlüftung, die zu Reduk-
tionserscheinungen und Gleybildung führt, sich keineswegs günstig auf die Bodenpilz-Vegetation auswirken kann.

DER EINFLUSS HOHER UND NIEDRIGER TEMPERATUREN

Das Pilz-Wachstum läßt sich zur Temperatur in ein gewisses Verhältnis setzen. Für die meisten mesophilen Arten liegt die Temperatur zwischen 35° und 37° C mit einem Optimum von ca. 18°
bis 30° C. Bei weniger thermophilen Formen wurde eine Temperatur
von 45° bis 60° C als Maximum gefunden, mit dem Optimum zwischen
25° und 50° C. Diese maximale Temperaturgrenze der wenigen, käl-
teliebenden (psychrophilen) Pilzarten beträgt 20° bis 30° C mit
Optima zwischen 10° und 20° C und Minima bei ca. 0° C und darunter. Ökologisch bedeutsam ist auch die Überdauerungs-Fähigkeit
bei extrem niederen und hohen Temperaturen, worüber wir leider
verhältnismäßig wenig wissen. Im allgemeinen hat eine Temperatur
von über 60° C auf die zarten vegetativen Entwicklungsstadien
einen vernichtenden Einfluß. In Sonderfällen zeigten Sklerotien
eine Hitzeresistenz von 120°C für mehrere Stunden. Eine andau-
ernde oder kurzfristige niedere Temperatur wirkt kaum schädigend
auf Dauermycelien, Gemmen oder Sporen ein. Solch ein Temperatur-
einfluß hat seine Wirkung vielmehr in einer mehr oder weniger
starken Hemmung des Wachstums der Pilz-Mycelien. Für die Pilze
der Hochgebirge und Polargebiete sind Überdauerungsfähigkeit und
Wachstumsminima bei äußerst niedrigen Temperaturen ein kaum zu
verwunderndes Faktum, da diese Organismen an den sehr langsam
ablaufenden mikrobiellen Abbauprozessen in polaren Böden beteiligt sind (DOUGLAS & FEDOROW 1959). Schon SCHMIDT-NIELSEN (1902)
berichtete, daß die psychrophilen Schimmelpilze zu wachsen ver-
mögen. Weitere Beobachtungen wurden in Frost- und Kühlanlagen
gemacht, wo es nicht selten zu Schäden an den dort gelagerten

Lebensmitteln durch bestimmte psychrophile Schimmelpilze kommt.
BROOKS & HANSFORD (1923) fanden, daß Mucor-, Penicillium- und
Thamnidium-Arten auf Fleisch bei minus $2,2^{\circ}C$ zu wachsen vermö-
gen und Cladosporium herbarum und Chrysosporium pannorum sich
bei $-7,8^{\circ}C$ noch relativ gut entwickelten. CHRISTIAKOV & BOCHA-
ROVA (1938) konnten ein noch langsames Wachstum mehrerer, allgemein verbreiteter Bodenschimmelpilze (Aspergillus-, Botrytis-,
Cladosporium-, Fusarium-Arten, Hefen, als auch Mucor- und Penicillium-Arten in einem Zeitraum von einigen Wochen bis Monaten
noch bei -2° bis $-5^{\circ}C$ feststellen; nicht mehr dagegen bei -8°
bis $-18^{\circ}C$. Psychrophile Hefen sind in den letzten Jahren aus
antarktischen (DI MENNA 1960) und arktischen (FLINT & STOUT
1960, HAGEM & ROSE 1961, BOYD & BOYD 1963, SINCLAIR & STOKES
1965) Böden isoliert worden, die zwischen $0^{\circ}C$ (auch darunter)
und $20^{\circ}C$, aber nicht mehr bei $30^{\circ}C$ wachsen. Experimentelle Untersuchungen und deren Ergebnisse an borealen psychrophilen
Mikromyceten und sterile Mycelien der H y m e n o m y c e t e n
liegen vor (WARD et al.1961, WARD 1966). Diese bezeugen, daß
in borealen und polaren Böden eine allgegenwärtige Pilz-Vegetation existiert, was auch durch mehrere floristische Beiträge
belegt wird: HARDER (1954) hat uns den Bericht über die niederen Phycomyceten Lapplands und Spitzbergens überlassen, aus dem
hervorgeht, daß sich im ersten Gebiet vier Arten, im zweiten
aber nur zwei dieser Familie befinden. Dies bedeutet, daß die
Pilzdichte in borealen Gebieten, sofern es diese Gruppe betrifft
kleiner ist als in Gegenden mit gemäßigtem oder gar tropischem
Klima. Auch HÖHNK (1960) hat in einem kurzen Bericht über die
niederen Pilze Islands und Ostgrönlands (63°n.Br.) hierzu beigetragen. Diese und andere Arbeiten zeigen, daß die Literatur
über höhere Boden-Mikromyceten sehr umfangreich ist.
Auch vor 15 Jahren stellte mir Herr Dr. A.MELDERIS (British Mu-
seum of Natural History London) 35 Bodenproben verschiedener
Standorte Westgrönlands (Insel Disko, ca.70°n.Br.) zur Verfügung, die ich auf ihren Pilzgehalt untersuchen konnte (APINIS
1968 Mskr.). Die Pilzfrequenz dieser Böden war bedeutend kleiner, als das in mitteleuropäischen der Fall ist. Ungeachtet dessen, waren alle Bodenmikromyceten-Gruppen vertreten,einschließ-
lich der Ascomyceten und Oomyceten, so auch zwei Isolate thermo
philer Pilze (Tab.5).

Tabelle 5. Zahl der Arten und Gattungen der Bodenmikromyceten
aus Westgrönland (Disko)

Pilzgruppe	Zahl der Gattungen	Artenzahl	Domin.Gattungen
Oomyceten	6	8	Pythium
Zygomyceten	7	25	Mortierella
			Mucor
Ascomyceten	5	10	Arachniotus
			Gymnoascus
Hefen	4	5	Trichosporon
Deuteromyceten	28	70	Cephalosporium
			Chrysosporium
			Fusarium
			Penicillium
			Stemphylium
			Sterile Mycelia
			Trichoderma

Insgesamt wurden 118 Arten aus 60 Gattungen angetroffen. Die
Mehrzahl der Gattungen gehört zur Famile der D e u t e r o -
m y c e t e n, wie Chrysosporium, Cylindrocarpon, Fusarium,
Gliocladium, Penicillium, Stemphylium und Trichoderma. Auch die
Bodenascomyceten, wie Arachniotus, Gymnoascus und Talaromyces
wurden isoliert. Auffallend hoch war das Vorkommen steriler Mycelien, die wahrscheinlich zu den A g a r i c a c e e n und
anderen H y m e n o m y c e t e n gehören. Dieser Befund
stimmt gut mit der Schätzung der Arten-Zahl der höheren Pilze
Grönlands überein, die bei etwa 900 Arten liegen soll (SINGER
1954, LANGE 1957). Die höheren unterirdischen Pilze fehlen,
doch die A g a r i c a c e e n: Agaricus, Amanita, Cortinarius,
Hebeloma, Inocybe, Laccaria, Melanoleuca, Omphalina, Russula
u.a., sowie auch die Bauchpilse Bovista, Calvatia, Crucibulum
und Lycoperdon sind zahlreich. Nach SINGER (1952-1954) ist die
antarktische, höhere Pilzflora der arktischen sehr ähnlich.

Pilz-Symbiose mit höheren Pflanzen ist weit verbreitet (LANGE
1957) und infolgedessen finden sich Cortinarius, Hebeloma, Inocybe, Lactarius und Russula unter Birken und Weiden. Ferner
nimmt die Pilz-Symbiose mit Algen unter boral-arktischen Verhältnissen dominierende Ausmaße an, die der Vegetation ein cha-
rakteristisches Gepräge verleihen. Nach RÄSÄNEN (KLEMENT 1958)
sind einige nordische Flechtengesellschaften kälteliebend.Glei-
ches gilt für die höheren Pilze; so finden wir nach LANGE (1946/
1957) in Grönland (82° n.Br.) Laccaria laccata. So ist auch die
Vegetationsgrenze der Bodenmikromyceten, mit Ausnahme einiger

exponierter Felsflächen, durch die Schneegrenze festgelegt.Die
edaphischen und klimatischen Gegebenheiten sind in den Hochge-
birgen denen der Polargebiete ähnlich. Relief und Modulation
des Geländes bieten im ersteren Fall jedoch mit variablerer Exposition ungleich andere ökologische Verhältnisse, als das in
Polar-Gebieten der Fall ist. Nur wenige niedere Phycomyceten
sind aus der Schneeflora bekannt (KOL 1957, GARRIC 1965). Auch
hier ist die Bodenmikroflora durch die Schneegrenze bedingt.
Nach (KREHL-NIEFER 1951) weist das Vorkommen von Bodenmikromyceten in den Obersdorfer Alpen mit abnehmender Höhenlage (von
2656 auf 800 m) eine Populationszunahme auf. RIPPEL (1940) zeigte, daß die Exposition des Standortes die Entwicklung von Aspergillus niger stark beeinflußt. Nach MISHUSTIN et.al.(1961) findet in den höheren Gebirgsregionen des Altai-Gebirges eine Abnahme des Boden-Mikromyceten-Anteils in einer Höhe von 2600 m
statt, was durch die unterschiedlichen Boden- und Vegetations-
verhältnisse zu erklären ist. Doch wird auch dort eine allmähliche Reduktion des Pilzgehaltes in der Stufe von 1200 m bis
550 m angegeben, was mit der allgemeinen zonalen Verbreitung
der Bodenmikromyceten im Einklang stehen soll (MISHUSTIN 1961).
Vor einigen Jahren sandte mir Herr Dr.D.MULDER (Tea Reseach Institute of Ceylon) einige Bodenproben aus Tee-Pflanzungen verschiedener Höhenlagen aus Ceylon zu, die ich auf ihren Boden-
Mikromycetengehalt untersuchen konnte (Tab.6).

Tabelle 6. Zahl der Pilzkeime in 1g Kulturboden verschiedener
Höhenlagen in Ceylon (St.Coombs) bei 20°C und 38°C.

Höhe Gruppe	150 m	1060 m	1370 m	2000 m
Mesophile Pilze	366.000	316.200	372.400	957.600
Thermophile Pilze	3.520	3.960	3.800	4.790
insgesamt	369.520	320.160	376.200	962.390

Die Keimzahl der meso- und thermophilen Pilze nahm von 150 m
(Regenwald) bis ca.2000 m ständig ab. Auch die Anzahl der festgestellten Bodenpilzgattungen zeigte mit Ausnahme der O o m y-
c e t e n mit zunehmender Höhe eine ständige Abnahme. Eine
qualitative zonale Verbreitung in den vier Höhenstufen wird in
der Tab.7 deutlich (APINIS 1968).

Tabelle 7.   Zahl der Bodenmikromycetengattungen in Kulturböden
verschiedener Höhenlagen in Ceylon (St.Coombs)

Pilzgruppe	Höhe 150 m	1060 m	1370 m	2000 m
Oomyceten (10)[1]	2	2	5	5
Zygomyceten (9)	7	7	6	7
Ascomyceten (16)	11	6	6	6
Deuteromyceten (58)	27	27	24	22
	47	42	41	40

1) In () Gesamtzahl der Gattungen in einer Pilzgruppe.

Aus diesem Beispiel ist zu entnehmen, daß bei der Bestimmung
von Bodenpilzen eine Reihe von bodenmikrobiologischen, vegetationskundlichen und auch methodischen Betrachtungen in Frage
kommen.

Über die Verbreitung der höheren Pilze in europäischen Gebirgen
hat HEIM (1928/1947), FRIEDRICH (1940), FAVRE (1948/1955) für
die Alpen und COOKE (1955) für die USA berichtet. Die symbion-
tischen Großpilze, wie Amanita, Collybia, Cortinarius, Hebeloma,
Lactarius, Russula und Tricholoma sind an Baum- und Strauch-Ge-
sellschaften, andere Gattungen dagegen an Dung, Moose und andere Pilze, sowie verschiedenste Streu- und Humusformen gebunden.
FRIEDRICH (1940) nimmt eine geschlossene Pilzvegetation in den
Alpen bis ca.1700 m (Tirol) und bis 2000 m (Engadin) an; vereinzelt finden sich allerdings noch höhere Pilze unmittelbar an
der Schneegrenze, wo sie an windgeschützten Südhängen noch eine
Entfaltungsmöglichkeit haben.

DIE PILZVEGETATION DER PRIMÄR-SUKZESSIONEN

Die Entwicklung von Bodenpilzen ist mit der Initialphase einer
Bodenbildung eng verknüpft. Dieser Vorgang wird vom Boden-Aus-
gangsmaterial durch Mikroorganismen eingeleitet, was sich in
einer bestimmten Typifizierung der langsam oder schnell sich
vollziehenden Vegetationsentwicklung äußert. Die Mikromycetenvegetation spielt hierbei eine bedeutsame Rolle und ihre quantitative (Keimzahl) und qualitative (Arten-) Zusammensetzung, wie
auch der Anteil bestimmter ökologischer Artengruppen, kann als
zuverlässiger Zeiger für die Bodenfruchtbarkeit in einer Sukzessionsreihe dienen. Die primäre Flechtenbesiedlung der Felsen
und des Gesteins kann in Polar-Gebieten mehrere Jahre oder gar
Jahrhunderte beanspruchen (DAHL 1954), was sich in tropischen

oder subtropischen Gebieten bedeutend rascher vollzieht. Nach
der Krakatau-Eruption (1883) waren die Inseln zunächst noch steril, nach einer mykologischen Untersuchung in den Jahren 1933/
1934 wurden schon 310 Pilzarten gefunden, die 150 Gattungen um-
faßten (BOEDIJN 1940). Gefunden wurden 28 M y x o m y c e t e n
4 Mucoraceen, 3 D i s c o m y c e t e n, 56 P y r e n o m y c e-
t e n, 2 Rostpilze, 169 H y m e n o m y c e t e n und 51 D e u-
t e r o m y c e t e n, aber nur 13 epilithische Flechten. Leider wurde die Mikromycetenvegetation des Bodens nicht berücksichtigt. COOKE & LAWRENCE (1959) hatten in ihrer grundlegenden
Arbeit erstmalig die primäre, artenreiche Sukzessionsstufe der
Bodenmikromyceten am Rande rückschreitender Gletscher in Alaska
ausführlich beschrieben, die parallel mit der Sukzessionsreihe
der Kryptogamen- und Phanerogamenvegetation verlief. Ähnliche,
wenn auch nicht so anschauliche Beispiele findet man auch in
hochentwickelten Industriegebieten, in denen durch Bergbau oder
dergl. große Flächen (nach KNABE 1965) in Deutschland 85 000 ha
und in Großbritannien ca. 60 000 ha von Abraum und Kippen bedeckt wurden. Es wird angestrebt, diese verwüsteten Landstriche
durch Rekultivierung der Land- und Forstwirtschaft wieder zu-
gängig zu machen. Eine biologische Lösung dieser Problematik
wird auch an der Mikrobiologie nicht vorbeigehen können. Wissenschaftliche Grundvoraussetzung wird bei jeglichen Rekulti-
vierungsplänen das experimentelle Studium der Initialen einer
Primärsukzession (WESTON et.al.1965), nicht ohne Berücksichtigung der Bodenpilze, sein und bleiben müssen.

ZUSAMMENFASSUNG

Die Pilze sind als eine heterotrophe Organismengruppe auf orga-
nische Substrat- (Energie-) Quellen angewiesen, die durch autotrophe Vegetation oder andere Organismen (Algen, andere Mikro-
Organismen und Tiere) erzeugt werden. Reichhaltigkeit und Zufuhr von solchen Energiequellen bestimmen den quantitativen und
qualitativen Charakter der Groß- und Kleinpilz-Flora eines
Standortes. Doch ist die Kolonisation und Nutzung solcher Substrate im Boden und in der Vegetationsschicht auch stark durch
die chemisch-physikalischen und biotischen Standortsfaktoren
bedingt, wie z.B.durch Feuchtigkeit, Temperatur und Durchlüftung des Bodens, so wie durch periodische Änderungen (Schwankungen) der Standortsfaktoren. Extrem hohe oder niedrige Tempera-

turen, Trockenheit und schlechte Durchlüftung des Bodens wirken
hemmend auf die Aktivität der Pilzvegetation. Jedoch werden bestimmte ökologische Pilzarten-Gruppen wenig von extremen Standortsverhältnissen beeinflußt und durch diese sogar stark geför-
dert, wie manche hydrophile Wasserpilze, Trockenheit vorziehen-
de Schimmelpilze, thermophile Arten und Kälte liebende psychrophile Arten der Polargebiete und Hochgebirge. Deshalb sind weder extreme Temperaturverhältnisse, noch ein Überschuß an Feuchte, oder die Trockenheit des Bodens absolute Grenzen für die
Pilzvegetation des Standortes, vorausgesetzt, daß eine autotrophe Vegetation und andere Lebewesen anwesend sind.

SUMMARY

Certain ecological features of growth, development and dispersal of soil fungi are mentioned in relation to edaphic and climatic factors as well as to their host vegetation.
In general, the fungi show a very wide range of distribution so
that hardly any substratum or environment is devoid of their
presence. However, the wide range of distribution on various
substrata under extreme edaphic and climatic conditions is reflected in species composition of the particular population
which possesses specific ecological adaptation.
Extreme humidity conditions influence all the groups of soil
fungi as it was found along the permanent grassland transects
at a river arm near Nottingham (England) where the alluvial
pasture soils change from well-drained A-C type over A-G type
to subaquatic mud or peat soil of the swamps. Clearly marked
zones of occurence of larger and microscopic fungi were found
which coincided with the type of the particular grasslandvege-
tation. Water-logging combined with the formation of G-horizon
and the excess of water are the main factors responsible for
the occurrence of particular groups of fungi because of their
specific ecological adaptation (cf.Tab.1,2,3).
A similar zonation of fungi is found on the North Sea coast
(Lincolnshire,England) where the limit of occurrence of higher
fungi is at a medium tide level. At this sea level also the vegetation of microfungi diminish and this decrease continues in
sublitoral region deposits without reaching an absolute limits
even in the deep sea at the depth of 3425 m where presence of
fungi in the deep sea core samples (borings of the sea bed) has
been reported.

The depth distribution of fungi in soil depends upon climatic
and edaphic conditions. The main factor, however, determining
their occurence in soil is the distribution of the root mass
of the particular vegetation which regularly supply the necessary substrata to soil fungi. Because of this the depth limit of
soil fungi is determined by the depth of the roots but for certain lower fungi this limit may extend beyond this depth. Also
certain physical factors, such as inadequate soil aeration, excess in water in water logged soils or in a subaquatic mud may
cause (cf.Tab.4) limitations in occurence of lower and higher
fungi.

The wide range of various adaptions of fungi is the main cause
for their occurrence under extreme temperature in their natural
environments. Thus the growth limits of psychrophilic fungi,
which are reported from boreal and polar regions, may be as low
as 0°C or even -5.0^{o} to -7.0^{o}C while maxima as hight as 50° to
60°C have been found for thermophilous species. In contrast to
such extreme adaptions the majority of fungi are confined to
mesophilic or temperate and humid conditions. A general decrease
in fungi populations is reported from temperate towards the polar regions (zonal distributions). Despite this a relatively
rich flora of higher and the microscopic fungi (cf.Tab.5) is
reported from subarctic regions, where there is a short summer
with temperatures above 0°C in the top soil. A similar pattern
of occurrence of fungi is reported from high mountains where
even at lower altitudes (cf.Tab.6 & 7) a diminishing pattern of
occurrence is detectable. In general, the upper limit of occurrence at a high altitude is fixed by the snow line but even this
is not absolute while few lower fungi are found also amongst
the cryoflora of the mountain snow.Furthermore, soil fungi play
an important part in the primary succession of soil formation,
at it has been reported from the marginal areas of the retreating glaciers.

LITERATURVERZEICHNIS

APINIS,A.E.   -1960- Über das Vorkommen niederer Pilze in alluvialen Böden bestimmter Pflanzengesellschaften.- Mitt.flor.-soz.Arbeitsgem.N.F.8: 110-117. Stolzenau/Weser.
--   -1964- Concerning occurrence of Phycomycetes in alluvial soils of certain pastures, marshes and swamps.- Nova Hedwigia 8: 103-126. Weinheim.
--   -1967- Growth patterns of soil fungi in alluvial pastures and marshes.- In: Graff,O.& Satchell (Edit.): Progress in Soil Biology. pp.211-232. Braunschweig.
--   -1968- Bodenmikromyceten einiger Kulturböden verschiedener Höhenstufen in Ceylon.- Mskr.
--   -1968- Bodenmikromyceten aus der Umgebung von Disko (Grönland).- Mskr.
--   -1968- Das Verhalten der Pilze in bestimmten Grasland-Gesellschaften.- In: Tüxen,R.(Edit.): Gesellschaftsmorphologie. Ber.Intern.Sympos.1966 in Rinteln. Den Haag.
--   -1968/1969- (Bibliographie) Pflanzensoziologie und Mikrobiologie.- Excerpta Botanica B.Sociologica 9. Stuttgart.
BAKER,F.K.& SNYDER,W.C. (Edit.) -1965- Ecology of Soil-Borne Pathogens.- London.
BERRY,J.A.& MAGOON,C.A. -1934- Growth of microorganisms at and below 0°C.- Phytopathology 24: 780-796. Lancaster,Pa.
BISBY,G.R. -1933- Geographical distribution of fungi.- Bot. Rev.9: 466-482. Lancaster,Pa.
BOEDIJN,K.B. -1940- The Mycetozoa, Fungi and Lichenes of the Krakatau group.- Bull.Jard.Buitenzorg III,16: 358-420. Buitenzorg.
BORUT,S. -1960- An ecological and physiological study on soil fungi of the Northern Negew (Israel).- Bull.Res.Council of Israel 8B: 65-80. Jerusalem.
BOYD,W.L.& BOYD,J.W. -1963- Soil microorganisms of the McMurdo Sound area,Antartica.- Appl.Microbiol.11: 116-121. Baltimore.
BROOKS,F.E.& HANSFORD,C.G. -1923- Mould growth upon cold-store meat.- Trans.Brit.mycol.Soc.8: 113-142. London.
COOKE,W.B. -1955- Fungi of Mount Shasta.- Sydowia 9: 94-215. Horn,N.-Ö.
Cooke,Wm.B.& LAWRENCE,D.B. -1959- Soil mould fungi isolated from recently glaciated soils in South-Eastern Alaska.- J.Ecol.47: 529-549. Oxford.
CHISTIAKOV,F.M.& BOCHAROVA,Z. -1938- The influence of low temperature on the development of moulds.- Mikrobiologia 7: 498-573. Moskow.
DAHL,E. -1954- Crytogamic flora of the Arctic.-VII.Lichens.- Bot.Rev.20: 463-476. Lancaster,Pa.
DIETRICH,R.& HÖHNK,W. -1960- Studie zur Chemie ozeanischer Bodenproben I. Über die Kohlenstoff- Stickstoff- und Phosphor-Verhältnisse in Hinsicht auf die Mykoflora.- Veröff. Inst.Meeresforsch.7: 15-35. Bremerhaven.
DOUGLAS,L.A.& FEDROW,J.C.F. -1959- Organic matter decomposition rates in arctic soils.- Soil Sci.88: 305-312. Madison/Wisc.
EKELÖF,E. -1908- Bakteriologische Studien während der Schwedischen Südpolarexpedition 1901-1903.- In: Nordenskjöld,O. (Edit.): Wiss.Ergebnisse der Schwedischen Südpolar-Expedition 1901-1903. VI (1). Stockholm.
FAVRE,J. -1955- Les champignons supérieurs de la zone alpine du Parc National Suisse.- Ergebn.wiss.Untersuch.schweiz.

Nationalparks N.F.5: 1-212. Liestal.
FLINT,E.& STOUT,J.D. -1960- Microbiology of some soils from
    Antartica.- Nature 188: 767-768. London.
FRIEDRICH,K. -1940- Untersuchungen zur Ökologie der höheren
    Pilze.- Pflanzenforschung 22. Jena.
GAERTNER,A. -1955- Über das Vorkommen einiger niederer Erd-
    phycomyceten in Afrika, Schweden und an einigen mitteleu-
    ropäischen Standorten.- Arch.Mikrobiol.21: 4-56. Berlin.
GARRIC,R.K. -1965- On the cryoflora of the Pacific North-
    West.- Am.J.Bot.52: 1-8. Bloomingtown/Ill.
HAGEM,P.O.& ROSE,A.H. -1961- A psychrophilic Cryptococcus.-
    Can.J.Mikrobiol.7: 287-294. Ottawa.
HARDER,R. -1944- Über die arktische Vegetation niederer Phy-
    comyceten.- Nachr.Akad.Wiss.Göttingen, math.-physik.Kl.III,
    1: 1-9. Berlin.
--   & ÜBELMESSER,E. -1955- Über marine saprophytische Chy-
    tridiales und einige andere Pilze vom Meeresboden und Mee-
    resstrand.- Arch.Mikrobiol.22: 87-114. Berlin.
--   -- -1957- Notiz zur Frage des Vorkommens von Chytridineen
    und anderen Pilzen in tiefen Bodenschichten.- Arch.Mikrobi-
    ol.26: 353-357. Berlin.
--   & PERSIEL,I. -1962- Notiz über das Vorkommen niederer
    Erdphycomyceten in der Antarktis.- Arch.Mikrobiol.41: 44-5O.
    Berlin.
HARVEY,J.V. -1922- A study of western water moulds.- J.El.
    Mitchell Sci.Soc.58: 16-42. Chapel Hill, N.-Carolina.
HEAL,O.W., BAILEY,A.D.& LATTER,P.M. -1967- Bacteria, fungi
    and protozoa in Signy Island soils compared with those
    from a temperate moorland.- Phil.Trans.Roy.Soc.London B,
    252: 191-197. London.
HEIM,R. -1947- Sur quelques espèces nivales de Macromycetes
    des Alpes françaises.- Rev.Myc.12: 69-78. Paris.
HÖHNK,W. -1959- Ein Beitrag zur ozeanischen Mykologie.-
    Dt.hydrog.Z.Erg.-Heft 13 (3): 81. Hamburg.
-- -1960- Phycomyceten von Island und Grönland.- Veröff.Inst.
    Meeresforsch.Bremerhaven 7: 633-667. Bremerhaven.
JAHN,H., NESPIAK,A.& TÜXEN,R. -1967- Pilzsoziologische Unter-
    suchungen in Buchenwäldern (Carici-Fagetum, Melico-Fagetum
    und Luzulo-Fagetum)des Wesergebirges.- Mitt.flor.-soz.
    Arbeitsgem.N.F.11/12: 159-197. Stolzenau/Weser.
JOHNSON,T.W.& SPARROW,F.K. -1961- Fungi in Oceans and Estua-
    ries.- J.Cramer,Weinheim.
KILLIAN,C.& FEHER,D. -1935- Recherches sur les phénomènes
    microbiologiques des soils sahariens.- Ann.Inst.Pasteur 55:
    573-622. Paris.
KLEMENT,O. -1958- Die Stellung der Flechten in der Pflanzen-
    soziologie.-Vegetatio 8: 93-156. Den Haag.
KNABE,W. -1965- Observations on world-wide efforts to reclaim
    industrial wasteland.- In: Goodman,G.J., Edwards,R.W. and
    Lambert,J.M.(Edit.): Ecology and the Industrial Society
    p.263-296. Oxford.
KOL,E. -1957- Über die Verbreitung der schnee- und eisbewoh-
    nenden Mikroorganismen in Europa I.- Arch.Hydrobiol.52 (4):
    574-582. Stuttgart.
KREHL-NIEFFER,R.M. -1951- Verbreitung und Physiologie mikro-
    skopischer Bodenpilze.- Arch.Mikrobiol.15: 389-402. Berlin.
LANGE,M. -1957- Micromycetes III. Medd.om Grønland 148 (2):
    1-125. København.
LAUB,W. -1963- Untersuchungen zur Mikrobiologie afrikanischer
    Böden.- Die Bodenkultur 14: 189-208. Wien.

LIND,J.   -1934-   Studies on the geographical distribution of
    arctic circumpolar micromycetes.- Det kgl.Danske Videnskab.
    Selskab.Biol.Medd.11 (2): 1-152.   København.
MENNA,di,M.E.   -1960- Yeasts from Antartica.- J.gen.Mikrobiol.
    23: 295-300.   Cambridge.
MISTHUSTIN,E.N., PUSHKINSKAJA,O.J.& TEPLEKOVA,Z.F.   -1961-   Eko-
    logo-geografičeskoe raspzostranenie počvennych gribov.-
    Trudy Inst.Pačvoved.Akad.Nauk.Kazachskci SSR 12: 1-63.
MOORE,J.J.   -1954-   Some observations on the microflora of two
    peat profiles in the Dublin Mountains.- Scient.Proc.Dubl.
    Soc.26 (22): 379-395.   Dublin.
Muskat,G.   -1955-   Untersuchungen über Schimmelpilze bayerischer
    und tunesischer Böden. Floristisch-ökologischer Teil.-
    Arch.Mikrobiol.22 (1): 20.   Berlin.
NICOT,J.   -1960-   Some characteristics of microflora in desert
    sands.- In: Parkinson,D.& Waid,J.E.(Edit.): The Ecology
    of Soil Fungi p.94-97.   Liverpool.
PERSIEL.I.   -1960-   Über die Verbreitung niederer Phycomyten
    in Böden aus verschiedenen Höhenstufen der Alpen und an
    einigen Standorten subtropischer Gebirge.- Arch.Mikrobiol.
    36: 257-282.   Berlin.
REMY,E.   -1949-   Über niedere Bodenphycomyceten.- Arch.Mikro-
    biol.14: 212-239.   Berlin.
RIPPEL,A.   -1940-   Über die Verbreitung von Aspergillus niger,
    insbesondere in Deutschland.- Arch.Mikrobiol.2: 1-32.Berlin.
ROTH,F.J., AHEARM,D.G., FELL,J.W., MEYERS,S.P.& MEYER,S.A.
    -1962-   Ecology and taxonomy of yeasts isolated from various
    marine substrates.- Limnology and Oceanography 7 (2):
    178-185.   Baltimore.
SABET,Y.S.   -1935-   A preliminary study of Egyptian soil fungi.-
    Bull.Fac.Sci.Egyptian Univ.Cairo 5: 1-29.   Cairo.
SCHMIDT-NIELSEN,S.   -1902-   Über einige psychrophile Mikroorga-
    nismen und ihr Vorkommen.- Zbl.Bakt.Abt.2, 9: 145-147. Jena.
SINGER,R.   -1952/1953-   The agarics of the Argentine sector of
    Tierra del Fuego and limitrophous regions of the Magellanes
    area. I & II.- Sydowia 6: 165-226;7: 206-265.   Horn,N.-Ö.
SINCLAIR,N.A.& STOKES,J.L.   -1965-   Obligately psychrophilic
    yeasts from the polar region.- Can.J.Microbiol.11: 259-269.
    Ottawa.
SOBOTKA,A.& SOBOTKOVA,M.   -1966-   A contribution to the relati-
    ons of the growth of fruiting bodies of some hymenomycetes
    and of the ecological factors.- Česka Mykologia 20 (1):
    54-61.   Praha.
SÖRGEL,G.   -1957-   Vorkommen und Verbreitung epiphyller Pilze
    in China.-Dt.Ges.f.Pilzkunde 23: 100-117. Bad Heilbrunn.
SPARROW,F.K.   -1937-   The occurrence of saprophytic fungi in
    marine muds.- Biol.Bull.72: 242-246.   Lancaster,Pa.
SCHNEIDER,R.   -1954-   Untersuchungen über Feuchtigkeitsansprüche
    parasitischer Pilze.- Phytopath.Z.21: 63-78. Lancaster,Pa.
THOMAS,M.D.   -1965-   The effects of air pollution on plants and
    animals.- In: Goodman,G.T., Edwards,R.W.& Lambert,J.M.
    (Edit.): Ecology and the Industral Society p.11-33. Oxford.
WALTER,H.   -1931-   Die Hydratur der Pflanze.- Jena.
WARD,E.W.B., LEBAN,J.B.& CORMACK,M.W.   -1961-   The grouping of
    a low temperature basidiomycete isolated on the basis of
    cultural behaviour and pathogenicity.- Can.J.Bot.39: 297-
    306.   Ottawa.
--   --   -1966-   Preliminary studies of the physiology of Scle-
    rotinia borealis.- Can.J.Bot.44: 237-246.   Ottawa.

WESTON,R.L., GADGILL,P.D., SALTERS,B.R.& GOODMAN,G.T. -1965-
    Problems of revegetation in the Lower Swansea Valley, an
    area of extensive industrial dereliction.- In: Goodman,G.T.,
    Edwards,R.W.& Lambert,J.M. (Edit.): Ecology and the Indu-
    strial Society p.297-325. Oxford.

CYANOPHYTEN UNTER GRENZBEDINGUNGEN
DER PHOTOSYNTHESE

G.H. S c h w a b e
(Zusammenfassung)

Wo immer eine photoautotrophe Vegetation auftritt, sind auch
Cyanophyten als deren primitivste Vertreter, meistens in erheb-
licher Mannigfaltigkeit anzutreffen. Nur 2 Biotop-Gruppen scheinen dieser Regel nicht zu folgen: das ozeanische Pelagial ist
ähnlich blaualgenarm wie die von verschiedenen anderen Algen-
gruppen besiedelten Firn- und Eisflächen. Ihre sonstige "All-
gegenwart" wird nur deshalb kaum beachtet, weil sie unter günstigen Photosynthese-Bedingungen von anderen Vertretern des grünen Pflanzenreichs überlagert und unterdrückt, aber doch nicht
vollständig ausgeschlossen werden (relativ geringe Produktions-
leistung). Ihre Anwesenheit ist durch Kulturversuche leicht
nachweisbar.
Diesem allgemeine Verbreitungsbilde entsprechend treten Cyanophyten dort biotopprägend in den Vordergrund, wo andere Vertreter der Autotrophen sich nicht mehr behaupten können, d.h. unter Extrembedingungen für die grüne Pflanze schlechthin. Zu den
typischen Cyanophyten-Biotopen dieser Art gehören vor allem:
1. Thermal-Biotope unter Standort-Temperaturen zwischen 30°
   und 65° (bis 70°) bei alkalischer bis schwach saurer Reaktion, unabhängig von Elektrolyt-Konzentrationen.

2. Felsflächen im Hochgebirge, meistens mit hohem Strahlungsge-
   nuß und oft extremen Schwankungen von Temperatur und Wasser-
   versorgung.

3. Sonst vegetationslose Böden von Trockenwüsten (Tauwasser-Ver-
   sorgung, extreme Temperatur-Schwankungen).

4. Dysphotische Zonen im Gewässer (Schlammgrenzflächen), im Gestein (endolithische) und in Sandböden (endopedische Formationen). Zu letzteren gehört z.B. die Erscheinung des Farb-
   streifen-Sandwatts.

5. Grenzzonen der Anaerobiose. Diese finden sich, oft bei Anwesenheit von freiem Schwefelwasserstoff in manchen der vorgenannten Biotope (Thermal-Gewässer, Farbstreifen-Sandwatt,
   Faulschlamm).

Mit den genannten Sonderbedingungen für die biotopprägende Entfaltung von Cyanophyten-Beständen erweist dieser Pflanzenstamm auch in ökologischer Hinsicht seinen phylogenetischen Reliktcharakter (O_2-freie Ur-Atmospäre).
Was vielen Cyanophyten ermöglicht, sich in Grenz-Biotopen der angedeuteten Art üppig zu entfalten, sind gewisse physiologische Eigenarten, unter denen die folgenden besonders bedeutsam sein dürften:

1. Spezifische Widerstandsfähigkeit gegen bestimmte Faktoren (z.B. hohe Temperatur, Gegenwart von H_2S, extreme Austrocknung, hohe Strahlungsintensität).

2. Fähigkeit zur Kryptobiose: Phasen aktiver Lebenstätigkeit können ohne nachhaltige Schäden periodisch (etwa im Tages-Rhythmus) oder unperiodisch unterbrochen werden.

3. Fähigkeit zu autogener Melioration unter Grenzbedingungen.

Diskussionen zu den Referaten von G.H.SCHWABE und A.APINIS

R.TÜXEN:

Herr SCHWABE hat von seinen Farbbildern behauptet, sie wären Wiedergaben von Naturgegenständen. Man hätte glauben können, es wären Darstellungen abstrakter Malerei. Wenn es solche Parallelen gibt zwischen gegenstandsloser Malerei, wie sie 1960 mit mikroskopischen Bildern in Basel ausgestellt worden sind - A. PORTMANN hat ein Vorwort zu diesem bemerkenswerten Buch[1] geschrieben - und wie wir das in Hannover 1964 gemacht haben mit Bodenprofilen[2], so könnten Sie, Herr SCHWABE, noch eindrucksvollere Beispiele solcher Parallelitäten von herrlichen Naturformen mit den Erzeugnissen moderner Maler bringen, die sich gar nicht so naturfremd erweisen, wie behauptet wird, obwohl die Künstler die Natur-Bilder und -Formen gar nicht gesehen haben.
Ich bin glücklich darüber, daß wir allgemein über den ganzen Wissensbereich der Lebenskunde und über die ganze Erde reichende Erkenntnisse gewonnen haben. Wir sind alles andere als Spe-

[1] Schmidt,G.& Schenk,R.(Eds.) -1960- Kunst und Naturform.- Basel. 132 pp.

[2] Tüxen,R. -1964- Die Schrift des Bodens.- In: Schwind,M. (Edit): Schrift des Bodens-Sprache neuer Malerei. Das Gespräch 1: 11-29. Hannover.

zialisten, obwohl wir solche hören, die uns in die Tiefe und an
die Front ihrer Forschung geführt haben und ich hoffe, auch wei-
terführen werden. Ich möchte glauben, daß gerade dieses für unsere Symposien das erstrebenswerte Ziel und für sie auch bezeichnend sei.

Der Vortrag von Herrn SCHWABE stand den Bakterienvorträgen, die
wir zuerst gehört haben, näher, denn hier wurden selbständige
Gesellschaften, eigene Assoziationen aufgezeigt, und ich möchte gern anregen, Ihre Ergebnisse noch zu vertiefen. Das müßte
zu wissenswerten Ergebnissen führen.Herr APINIS hat - und das
fand ich besonders schätzenswert - die Pilze in Beziehung zu
den höheren Pflanzengesellschaften, als Glieder der höheren
Pflanzengesellschaften geschildert. Wir haben also selbständige
Kryptogamen-Assoziationen kennengelernt, wir haben aber auch
die Pilze in ihrer Funktion in wohldefinierten Phanerogamen-Gesellschaften, als Glieder derselben gesehen.

W.LÖTSCHERT:

Herr SCHWABE hat davon gesprochen, daß in den Thermal-Gebieten
und in heißen Quellen als Obergrenze eine Temperatur von 70°
festgestellt worden ist. Ich habe derartige Temperatur-Messun-
gen in Solfataren und Thermal-Feldern in Mittelamerika durchgeführt und kann seine Ergebnisse bestätigen. Es ist so, daß dort
überall als Obergrenze die Temperatur von 70° im allgemeinen
nicht erreicht wird, als Obergrenze konnte ich dort für eine
Synechococcus-Art 69° finden. Das stimmt ziemlich genau überein
mit den Angaben, die auch MOLISCH macht, der viele Jahre lang
in Japan solche Messungen angestellt hat und der ebenfalls als
obere Temperaturgrenze 69° gefunden hat. Der ganze fermentative
Apparat wird, wenn es über 70° hinausgeht, temperaturgeschädigt,
so daß auch in dieser Hinsicht ein Vorkommen unter höheren Temperaturen wohl ausgeschlossen erscheint.

Herr SCHWABE sprach davon, daß die C y a n o p h y c e e n im
allgemeinen bei extremeren Säuregraden zurücktreten und in diesem Zusammenhang ist es vielleicht interessant, daß diese Temperatur für Synechococcus bei 69°, die ich erwähnt habe, in einem Gebiet gefunden wurde, in dem das Wasser stark alkalisch
war. Die anderen Faktoren, von denen die Rede war, also z.B.
Licht (im Zusammenhang mit den Messungen von BÜNING),wirken ja
auch in etwa ausgleichend, oder ermöglichen es, daß diese C y-
a n o p h y c e e n bei höheren Temperaturen auftreten. Viel-

leicht kann Herr SCHWABE uns zu diesem Punkt noch etwas sagen?

Herr APINIS gab an, daß Dauerzustände bestimmter Pilze bei 120° aushalten. Dazu wäre die Frage zu stellen, inwieweit die Temperatur in das Innere dieser Sklerotien oder sklerotienartigen Zustände - denn um solche handelt es sich wohl - eingedrungen ist? Das wäre wohl noch zu prüfen.

Herr APINIS gab an, daß bestimmte Penicillium- und Thamnidium-Arten noch bei -2,2° Wachstumserscheinungen zeigen. Das kann wohl als richtig gelten, wenn man das Wachstum im üblichen Sinne definiert. Man weiß, das erste Ansätze wachstumsähnlicher Zustände schon bei -7 bis -8° gefunden wurden. In diesem Zusammenhang wäre es wissenswert, ob man schon einmal den osmotischen Wert bei Pilzen bestimmt hat. Vielleicht kann er uns dazu etwas sagen.

Dann fiel mir auf, daß die Anzahl der Pilzkeime mit der Höhe über Normal Null in den Tropen zugenommen hat. In diesem Zusammenhang wäre die Frage aufzuwerfen, inwieweit die Zahl der Pilze in einem normalen Boden bei gleichen Höhenstufen in den tropischen und den gemäßigten Breiten übereinstimmt, oder welche Unterschiede sich dabei ergeben.

Der Wasser-Gehalt des L o l i o - C y n o s u r e t u m und des P h r a g m i t e t u m wurde in Beziehung gesetzt zur Anzahl der Pilze. Inwieweit wurde der Sauerstoff-Gehalt, der wohl auch als limitierender Faktor in diesem Zusammenhang eine Rolle spielt, mitgemessen?

An eine vorhergegangene Diskussionsbemerkung von Herrn Professor TÜXEN über die Klimax-Gesellschaften anknüpfend, möchte ich anregen, die optimalen Entwicklungsstadien, die uns Herr SCHWABE von den Cyanophyceen im Bild vorgeführt hat, als Optimal-Gesellschaften zu bezeichnen; denn es gibt ganz ohne Zweifel ein Optimum, oder eine schmale Optimal-Zone, in der sie sich so üppig entwickeln.

E.BURRICHTER:

Zu dem Vortrag von Herrn Professor APINIS liegen in vieler Hinsicht analoge Verhältnisse bei den Boden-Bakterien vor. Das bezieht sich sowohl auf die Abhängigkeit von der Verteilung des Wurzel-Systems, als auch auf die Boden-Durchlüftung. Die erstgenannte Tatsache zeigt sich u.a. zwischen wenig durchwurzelten Acker- und intensiv durchwurzelten Grünland-Böden. In den Wur-

zelzonen der Grünland-Böden liegen die Bakterien-Zahlen um ein
Vielfaches höher als in vergleichbaren Acker-Böden. Der Einfluß
der Boden-Durchlüftung zeigt sich z.B. bei den verschiedenen
Bodenarten. In den feindispersen und wenig durchlüfteten schwe-
ren Lehm- und Tonböden setzt eine starke Reduktion des Bakteri-
enlebens ein. Das gilt auch für verdichtete Zonen strukturkran-
ker Böden im Bodenprofil.

R. TÜXEN:

Es ergeben sich noch weitere Vergleichsmöglichkeiten. Man sollte die Epiphyten-Gesellschaften an unseren Buchenstämmen im
F a g i o n erneut studieren. Wir sehen, wie diese in einer
klaren und auffälligen Zonierung um die Buchen-Stämme herum verteilt sind, indem an der Regen-Rinne, dort wo an der geneigten
Buche der Regen abfließt, ein dunkler Strich herabläuft, in dem,
wie ich mir habe sagen lassen, Blaualgen vorkommen.

G.H. SCHWABE

Wahrscheinlich sind hauptsächlich grüne Algen beteiligt. Auf
organischem Substrat pflegen die Blaualgen weitgehend zurückzu-
gehen. Immerhin können welche dabei sein.

R. TÜXEN:

Neben reinen Grünalgen-Überzügen von C h r o c o c c u s-Arten,
die als lebendige Hygrometer auf die Luft-Feuchte reagieren,
sind an den trockenen Seiten Flechten-Überzüge da, und am Stamm-
Fuß auch Moose und der Pilz Dichaena faginea. Diese Epiphyten-
Gesellschaften wechseln in ihrem Erscheinungsbild im Laufe des
Tages mit dem Licht, mit der Feuchtigkeit usw. Sie haben noch
weit stärkere Saison-Unterschiede. So trat in diesem Herbst bei
uns zum ersten Mal ein hellgrauer Pilz (Corticium centrifugum)
auf, der nach wenigen Wochen wieder verschwunden war. Er ist
vor kurzem von EBERLE[1] als selten und vorübergehend beschrieben. Er trat hier ganz plötzlich in diesem Herbst zu Tausenden,
fast auf jeder Buche in vielen Exemplaren auf, und verschwand
dann.(Nach etwa. 3 Jahren klang die Invasion ab).
Diese Kryptogamen-Gesellschaft, einschließlich der Bakterien-
Gesellschaften der menschlichen Körperhöhlen, sind m.E. eigene,
aber abhängige Assoziationen. Die Bakterien-Gesellschaften sind
ja nur möglich in Verbindung mit dem Menschen oder mit dem Tier,

---

[1] Eberle,G. -1967- Myzelkreise an Bäumen im herbstlichen
Wald.- H.Nass.Ver.f.Naturkde 97. Wiesbaden.

die Baum-Epiphyten-Gesellschaften nur in Verbindung mit dem
Wald. Sie sind Glieder einer höheren Gesellschaftsorganisation,
während andererseits die Blaualgen völlig selbständige, unabhängige Gesellschaften bilden. Viele Pilze, die uns Herr APINIS
vorführte, sind dagegen, wie zahllose andere Kryptogamen, Bestandteile einer Makrophyten-Gesellschaft.

R. CARBIENER:

Ich möchte mir nur erlauben, kurz zu den Grenz-Problemen bei
Pilzen etwas zu bemerken. Ich habe selbst nur wenig Erfahrung
mit höheren Pilzen, d.h. mit den im Gelände zu beobachtenden
Pilz-Karpophoren. Bei den heterotroph abhängigen Pilz-Gesell-
schaften sind natürlich die Grenzprobleme schwieriger, weil zu
den allgemeinen ökologischen, physikalischen, chemischen und
biotischen Konkurrenz-Faktoren, die wir bei der Phanerogamen-
Vegegation kennen, noch eine ganze Reihe anderer Faktoren hinzukommt. Da ist es die Frage der Bakterien-Konkurrenz, die Frage
der Verflechtung mit der höheren Vegetation durch Mycorrhiza,
die Frage der Antibiose, die Frage der Abhängigkeit der verschiedenen Variationen von der höheren Vegetation. Es gibt da
also ganz verschiedene Arten von Grenzproblemen. Eines ist bekannt und schon öfters festgestellt worden, daß unabhängige Pilz-
Gesellschaften auch mit den Grenzen der Phanerogamen-Gesellschaften zusammenfallen. Ich möchte nur kurz ein Beispiel aus
den Vogesen bringen. Im Tannen-Buchenwald der montanen Stufe
haben wir eine ziemlich reiche höhere Pilz-Flora, die in den
verschiedenen Gesellschaften dieser Stufe auch verschieden aus-
sieht. In den eutrophen sind mehr Bakterien, weniger höhere
Pilze, das ist ja allgemein bekannt, in oligotrophen sind mehr
höhere Pilze, also eine ziemlich artenreiche Pilz-Flora, die
zum Teil aus Mycorrhiza-Pilzen, zum Teil aus echten Saprophyten
besteht. Kommen wir nun in den subalpinen Krüppel-Buchenwald
der oberen Stufe, dann stellen wir eine deutliche Verarmung der
Pilzvegetation fest: Die höheren Pilze sind viel weniger arten-
reich, aber zugleich tritt - wie es oft in Grenzfällen vorkommt-
eine Massen-Entwicklung von gewissen Arten auf. Es scheint,
daß z.B. die Mycorrhiza-Arten im Buchenwald der höchsten Lagen
stark zurückgehen, so fehlt z.B. Boletus chrysenteron, die wir
überall in Buchenwäldern und Buchen-Tannenwäldern finden. Dagegen tritt eine Massen-Entwicklung von saprophytischen Arten
auf, deswegen vielleicht, weil der Laubfall der Buchenwälder

in der höheren Stufe sehr plötzlich Ende September - Anfang Oktober eintritt, und sich dann eine sehr dicke Streuschicht bildet, die fast 10cm mächtig ist. Zudem ist das Wetter um diese
Jahreszeit sehr feucht. Da kann man feststellen, daß im Oktober
manchmal der Boden von ungeheuren Mengen von sehr eurytopen Bu-
chenlaub-Saprophyten, z.B. Laccaria laccata var. amethystina,
Marasmius alliaceus oder verwandten Arten, bedeckt ist, zu denen sich Russula ochroleuca und Russula-Arten aus der emetica-
Gruppe gesellen. Laccaria laccata und Marasmius fallen am meisten
durch ihre ungeheure Massen-Entwicklung auf. Wenn man das Bu-
chenlaub im November oder später etwas aufwühlt, dann riecht es
nach Knoblauch von den Marasmius alliaceus-Hyphen, das geht bis
in den Dezember hinein. Warum ist aber der Mycorrhiza-Pilz
Chrysenteron z.B. nicht mehr da? Das sind Grenz-Probleme, die
eingehender beobachtet werden müßten.
Es gibt aber auch andere Grenzprobleme bestimmterer Art. Es ist
z.B. allgemein bekannt, daß gewisse Pilze die Säume, also die
Grenze zwischen Wald und Wiese, bevorzugen. Das kommt besonders
vor bei einer ganzen Serie von Mycorrhiza-Pilzen. Das habe ich
im Elsaß schon sehr oft beobachtet, und es ist auch aus der Literatur bekannt. Aber es ist vielleicht erwähnenswert,weil wir
damit zum Saum-Problem kommen. In den thermophilen Gebüschen
und Waldsäumen der elsässischen Ebene auf kalkreichen Böden
kommen z.B. seltene, aber sehr auffallende Arten vor, die besonders an der Grenze zwischen Wald und Wiese als Mycorrhiza-Pilze
einige Meter vom Gebüsch-Rand entfernt ihre Karpophoren entwik-
keln, und zwar sind sie in diesem Falle weitaus schöner als
wenn sie, wie es zum Teil auch vorkommt, im Gebüsch oder im
Wald selbst wachsen. Hier zeigt sich also ein deutliches Optimum der Karpophoren-Bildung. Das kommt z.B. bei Boletus pur-
pureus, einer relativ seltenen thermophilen Art, die wir im Elsaß gefunden haben, bei Boletus queletii, auch relativ selten,
am Saum der thermophilen Eichenwälder des F r a x i n o -
C a r p i n i o n und des L i g u s t r o - P r u n e t u m
vor. Einige Meter vom Gebüsch entfernt, wachsen die Karpophore
schon im September. Man ist erstaunt, mit welcher Kraft diese
Karpophore die noch relativ trockenen Böden an diesen Stellen
durchbrechen. Diese zwei Boletus-Arten sind Mycorrhiza-Pilze
von Eichen und wahrscheinlich auch von anderen Laubbäumen. Eine
andere Art, die so auffallend ist, ist Amanita strobiliformis

oder Amanita ·solitaria, ein Mycorrhiza-Pilz auch der Eiche, der
Linde und der Hasel. Diese Art kommt z.B. in F r a x i n o -
C a r p i n i o n innerhalb des Waldes auf sandigen Böden vor.
Man findet sie aber viel schöner und kräftiger - ich spreche
natürlich immer vom Karpophor - an den Mänteln, also am L i g u-
s t r o - P r u n e t u m, einige Meter vom Gebüsch entfernt.
Das hängt natürlich damit zusammen, daß dieses Mycorrhiza-Arten
sind, die an den Wurzelenden der Sträucher, die sich sehr weit
in die Wiese ausbreiten, zu finden sind. Das Karpophor entwickel·
sich etwa im Bereich der Grenze der Sträucher und der Bäume,
die im benachbarten Wald wachsen.
Nun stellt sich die Frage der Kausalität dieses Vorhandenseins
von Saum-Pilzen. Natürlich ist sie nicht nur rein deduktiv zu
lösen. Es kann aber doch vermutet werden, daß an diesen Standorten der Wiese der Wechsel von Trockenheit und Feuchte der
begünstigende Faktor ist. Die Arten, die ich genannt habe, sind
bekannte thermophile submediterrane Arten. Sie fruchten nach
einer sommerlichen Trockenheit mit den ersten Herbstregen. Manch·
mal sieht man die Karpophore Anfang September bei relativ noch
sehr trockenen Böden und ist dann erstaunt, mit welcher Kraft
sie hervorbrechen.

R. LINDER:
Ein F a g e t u m ist eine selbständige Gesellschaft. Aber ein
F a g e t u m kann nur gebildet werden innerhalb des Areals
der Buche, ein L a c t o b a c i l l e t u m innerhalb des
Areals der Vagina. Sollen wir da Grenzen aufstellen? Das glaube
ich nicht, obwohl die L a c t o b a c i l l e n Saprophyten
sind. Aber auch die Buche hängt von verschiedenen Bedingungen
ab, die ihr Areal bedingen. Ich wollte sagen, daß der Unterschied nicht so groß ist, und wir tatsächlich Bakteriengesellschaften aufstellen können, obwohl man sie als endemische Gesellschaften eines gewissen begrenzten Bereiches aufzufassen
hat.

W. LÖTSCHERT:
Am ehesten gewinnen wir in diesem Punkt Klarheit, wenn wir den
Gesichtspunkt der Autotrophie und der Heterotrophie der Pflanze
berücksichtigen. Die autotrophe Pflanze ist auf irgendeine Weise imstande, sei es, daß sie nun ihre Energie vom Licht oder
aus chemischen Prozessen gewinnt, Substanz zu produzieren, während das bei der heterotrophen Pflanze bekanntlich nicht der

Fall ist. Und wenn wir diesen Gesichtspunkt ins Feld führen,
dann haben wir auf der einen Seite die autotrophen und auf der
anderen Seite die heterotrophen Gesellschaften. Im Sinne der
klassischen Phytosoziologie lassen sich sowohl Bakterien- als
auch Cyanophyceen-Gesellschaften aufstellen.

Th. WURCH:
Ich möchte nur betonen, daß ich den Begriff Klimax nur für eine
Gesellschaft genommen habe, und diese Gesellschaft ist das
L a c t o b a c i l l e t u m. Für die anderen Gesellschaften
habe ich den Begriff nicht gebraucht. Experimentell kann man
z.B. in der Menopause, wenn die Epithel-Decke abgebaut wird, immer wieder eine Entwicklung des Epithels bewirken. Durch die
Glykogen-Speicherung unter dem Einfluß der Hormone, ob es das
oestrogene Hormon ist, ob es progestagene oder androgene Hormone
sind, kann man immer eine Auswirkung der Decke bewirken, und
dann stellt sich keine andere Gesellschaft ein als das L a c -
t o b a c i l l e t u m. Das ist das Normale, und deshalb habe
ich vom Klimax gesprochen. Früher, als diese Begriffe noch nicht
bekannt waren, sind die Menschen an der Pathologie von Trichomonas, an einem T r i c h o m o n a d e t u m gestorben. Es
entstand dann eine Kachexie. In einem L a c t o b a c i l l e -
t u m können sich auch T r i c h o m o n a d e n einstellen,
dann gibt es natürlich Gesellschaften mit Trichomonas, aber man
spricht nicht von einer Klimax. Wenn man Trichomonas zerstört,
bleiben die Lactobacillen, es können sich andere Gesellschaften
bilden, H a e m o p h i l e t e n usw. Das hängt von den Milieubedingungen ab. Aber es kommt nie zu einer Klimax. Den Begriff
Klimax habe ich nur für das L a c t o b a c i l l e t u m gebraucht, weil er das Normale, das Endstadium, der Gesundheits-
zustand ist.

J.J. BARKMAN:
Eine zu starke Differenzierung in autotrophe und heterotrophe
Synusien scheint mir unzweckmäßig. Unter den höheren Pflanzen
gibt es nur wenig rein autotrophe, die meisten sind mixotroph,
sowie rein heterotrophe Arten (Neottia nidus avis als Saprophyt,
oder Orobanche als Parasit). Alle diese gehören zu völlig verschiedenen Synusien, wie auch die Pilze, die übrigens selbst
wieder eigene Synusien eingehen können.
In den Wacholder-Gebüschen der Niederlande lassen sich folgende

Mikrogesellschaften zwischen höheren und niederen Pflanzen erkennen:

a. In den dichtesten Wacholder-Beständen ist der Boden völlig vegetationsfrei, er bildet ein Abiotop.

b. An der Stammbasis finden wir in lichteren Beständen Moos-Gesellschaften, die eine eindeutige Nord-Exposition bevorzugen.

c. Eine andere Moos-Gesellschaft finden wir in den Kronen der Juniperus-Gehölze.

d. Auf der Südseite der Stämme finden sich auf dichter Nadelstreu-Auflage keine Kryptogamen, jedoch Phanerogamen.

e. Weiter nach Süden gehend, gedeihen an der Grenze der E m p e t r u m oder C a l l u n a-Heiden Moos-Gesellschaften anderer Arten-Kombination.

Die beiden letzten Abschnitte (d+e) zeigen eine anders geartete Großpilzvegetation als die von b und c; die Grenzen von Moos-Synusien decken sich somit nur teilweise mit denen der Pilzgesellschaften.

A. APINIS:

Zur Frage der Resistenz gegenüber höheren Temperaturen habe ich nur die Literaturangaben erwähnt, habe aber auch persönliche praktische Erfahrung mit Materialien, die wir sterilisieren wollen, Grashalme, Graswurzeln usw. Diese enthalten viele parasitische und saprophytische Pilze in Sklerotium-Form oder als Dauer-Myzel. Die Behandlung dieses Materials während einer Stunde bei 115° im Autoklaven führt nicht zur Sterilität; man muß das dreimal wiederholen. Das ist ein Beweis, daß die Resistenz gegen hohe Temperaturen dieser verschiedenen Stadien sehr hoch sein kann, besonders bei den Pilzen, die in der Vegetationsschicht vorkommen. Denn wir wissen, daß manche Pflanzenoberflächen sehr stark erwärmt werden, und daß die Pilze dann doch noch existieren können.

Über die niederen Temperatur-Grenzen können Sie das selbst nachprüfen. Wenn Sie zuhause einen Kühlschrank haben, und Sie vergessen, ihn gut zu säubern, dann tritt nach einigen Wochen eine Pilz-Vegetation bei -3 bis -4°C auf. Die ersten Angaben über kälteliebende Pilze wurden von Kühlanlagen gemacht. Nur in der letzten Zeit sind Angaben über das Vorkommen dieser Pilze aus arktischen und subarktischen Böden bekannt geworden. Diese Daten sind experimentell wiederholt geprüft worden.

Über die Verbreitung der Pilze in den Tropen und in den gemäßig-

ten Zonen im Gebirge haben wir sehr wenig Material, um gültige
Schlüsse ziehen zu können. Es gibt von FRIEDRICHS eine Arbeit
über höhere Pilze in den Alpen. Ich habe beobachtet, daß die
Vegetation der höheren Pilze mit zunehmender Höhe zurückgeht.
In Tirol ist die Grenze bei 1700 m, im Engadin bei 2000 m fest-
gestellt worden.
Ich habe versucht, den Sauerstoff-Gehalt im Gley-Horizont zu
messen. Ich konnte keine Spur von Sauerstoff finden. Bei einem
Reduktionspotential von 80 mV gibt es keinen Sauerstoff mehr.

Herr Dr. CARBIENER hat interessante Fragen über die Grenzprobleme bei höheren Pilzen angeschnitten. Ich möchte nur erwähnen, daß die Pilz-Symbiose im arktischen Gebiet eine sehr ausgeprägte Erscheinung ist. Man findet z.B. dort Boletus-, Russu-
la-, Inocybe-, Hebeloma-Arten nur in Salix polaris- oder Betula nana-Gebüschen aber nirgendwo außerhalb derselben. Die Zah-
len-Angabe über die Pilze in Wiesen und Wäldern in einem alluvialen Gebiet bezieht sich auf die Wiese. Da gibt es keine
Gattung,die einen Mycorrhiza-Symbionten stellt. Da gibt es keine Russula-, Boletus-, Inocybe-Arten usw. Sie fehlen vollständig, so daß das L o l i o - C y n o s u r e t u m völlig sym-
biontenfrei ist. Bei den Wiesen-Gebüschen - wir haben in Nottingham diese Wälder noch erhalten - gibt es noch manche symbionti-
sche Pilze.

G.H. SCHWABE:
Wir müssen unterscheiden: Grenz-Temperatur aktiven Lebens und
Grenz-Temperatur der Dauerform. Die Grenz-Temperatur des akti-
ven Lebens liegt bei den Blaualgen nach allem, was wir wissen,
niemals über 70°, die Grenz-Temperatur der Dauerform liegt zum
Teil erheblich höher als 100°.

DIE FORM DER BÄUME AN DER BAUMGRENZE IN DER
SUBALPINEN STUFE IN ZENTRAL-JAPAN UND IHRE
AUSWERTUNG ALS LOKALKLIMATISCHES KENNZEICHEN

M.M. Y o s h i n o

1. Einleitung

Die windgeformten Bäume wurden von verschiedenen Standpunkten
aus diskutiert und klassifiziert. Nach den in Deutschland, England, Frankreich, Japan, der Schweiz, Italien, Tschechoslowakei
und den USA durchgeführten Untersuchungen unterscheidet man
vier Haupttypen (YOSHINO 1961,1966). Bei der Betrachtung eines
Typs, nämlich des Typs der Bäume mit vertikalem Stamm, deren
Zweige auf der Luv-Seite wie abgeschnitten erscheinen, wurde
versucht, die lokalen Eigenschaften ihrer Verbreitung in der subalpinen Stufe in Zentral-Japan zu untersuchen. Dieser Typ entsteht durch Schnee- und Eis-Gebläse während des Winter-Monsuns.
Nach der Klassifikation von WALTER (1951) kann man sie als"wind-
gescherte Bäume" bezeichnen.

2. Die Verbreitung der windgeformten Bäume auf Mt.Neko und Mt. Azuma.

(1) Überblick des Gebietes und der windgeformten Bäume.
Zuerst wird ein geographischer und klimatologischer Überblick
dieses Gebietes gegeben. Das Untersuchungsgebiet liegt in 36°
35'N, $138^{\circ}25$'E, rund 20 km südlich der Stadt Nagano in Zentral
Japan. In diesem Gebiet finden sich zwei alte Vulkane, Mt.Neko
(2.195 m) und Mt.Azuma (2.333 m). (Abb.1).
Die Beobachtungen wurden auf Hänge oberhalb von 1.900 m über NN
eingeschränkt. Dieses Gebiet gehört zur subalpinen Stufe.Obwohl
keine klimatologischen Werte für dieses Gebiet vorliegen, kann
man die Werte von einer Station am südwestlichen Bergfuß (Suga-
daira 1.301 m) heranziehen. Die Jahres-Mitteltemperatur liegt
dort bei 6.3°C, die Monats-Mitteltemperaturen betragen im Dezember, Januar, Februar und März, -3.1, -6.7, -6.5 und -2.6°C,
die mittlere Schneehöhe in diesen Monaten erreicht: 26,58,78
und 61 cm. Die wichtigsten Werte für diese Art windgeformter
Bäume, d.h. die Windstärke bei Schneefällen, wurde durch Werte
erlangt, die jeweils um 10 h (Ortszeit) vom Herbst 1940 bis
Frühjahr 1947 in derselben Station abgelesen wurden. Die Mo-
natsmittel der Windstärke waren 3.1 im Januar, 3.0 im Februar,
3.3 im März, 2.8 im April, 2.0 im Mai, 2.6 im November und 2.8

im Dezember.

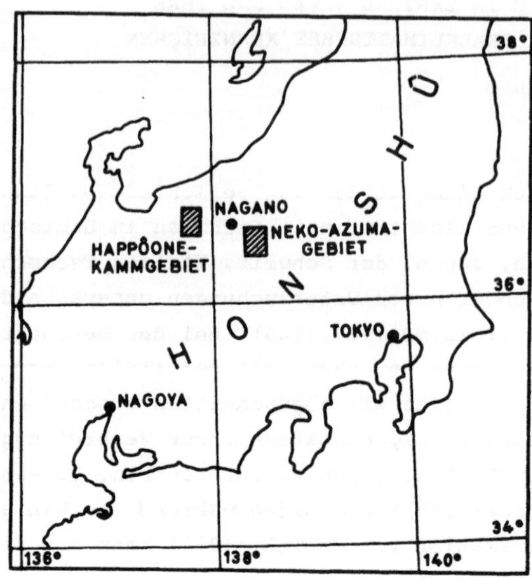

Abb.1. Die Untersuchungsgebiete in Zentral-Japan.

Als vorläufiges Ergebnis bei der Untersuchung windgeformter
Bäume in diesem Gebiet wurde folgendes festgestellt. Der mit
(a) bezeichnete Teil ist den Winterstürmen ausgesetzt und seine
Zweige auf der Luv-Seite werden durch Schnee und Raureif bela-
stet. Die Länge des Teils (a) schwankt sehr von Ort zu Ort
(0.5 - 3.0 m). Die Länge der Zweige im Lee der vorherrschenden
Windrichtung ändert sich auch stark (0.3 - 1.7 m) von Wuchsort
zu Wuchsort. Dies bedeutet, daß die kleinräumige Veränderung
der Windverhältnisse relativ groß ist. Dagegen zeigt die Länge
des Stammes und der Zweige, die vor den winterlichen Stürmen
durch die Schnee-Höhe geschützt werden, relativ kleine Änderungen: Die Länge des Stammes unter Schnee 0.5 - 1.4 m, die Länge
der Zweige unter Schnee 1.4 - 2.0 m. Um die Windverhältnisse
festzustellen, betrachtet man am besten Bäume derselben Art, die
etwa gleich alt sind. Allerdings war es im Untersuchungsgebiet
schwierig, diese Forderung zu erfüllen, da Bäume derselben Species nur vereinzelt auftreten. Daher wurde entschieden, alle
etwa 100-jährigen Nadelbäume (etwa 30 cm Durchmesser auf der
Höhe 30 - 50 cm auf der Erdoberfläche) von Abies veitchii, A.
mariesii, A.firma, A.homolepis, Picea jezoensis var. hondoensis

zu untersuchen.

Bei diesen Untersuchungen wurden nicht nur die windgeformten Bäume vermessen, sondern auch die Verteilung der Flechten, Usnea
diffracta und U.longissima, wurde berücksichtigt. Usnea wächst
im allgemeinen auf Zweigen und Stämmen von Bäumen in den nebeli-
gen intramontanen Regionen Zentral Japans in Höhen zwischen
1.000 und 2.500 m. Ihre Verbreitung ist ungleichmäßig durch die
Wirkung mikroklimatischer Windverhältnisse, die offensichtlich
ihr Wachstum in windigen, weniger feuchten Gebieten ausschließt.
Daher wurde angenommen, daß ihre Verbreitung windgeschützte Gebiete anzeigt. Wenn man nun die Verbreitung windgeschützter und
den Winden ausgesetzter Gebiete zusammenstellt (angezeigt durch
windgeformte Bäume), so kann man die Verteilung der herrschenden
Windrichtung und -stärke in diesem Gebiet kartieren.

(2) Beobachtungsergebnisse.

Abb.2 zeigt die Verbreitung der Gebiete mit windgeformten Bäumen und die Gebiete, wo keine windgeformten Gehölze auftreten,
und wo sich Usnea findet. Aus dieser Karte läßt sich folgendes
entnehmen: Windgeformte Bäume treten vor allem in Gebieten von
rund 2.020 m über NN auf, die dem Wind besonders ausgesetzt sind.
Einige Wuchsorte fanden sich in 2.000 m, andere in 2.050 m.

An den Hängen der Bergkämme in einer Höhe von 2.300 - 2.340 m
und 2.000 - 2.030 m, die sich vom Gipfel des Mt.Azuma in SW-Rich-
tung erstrecken, kann eine sehr aufschlußreiche Verteilung beobachtet werden, d.h. die Bäume, die sich direkt hinter dem Kamm
befinden, sind durch Wirbelwirkung deformiert und zwar entgegen-
gesetzt zu der vorherrschenden Wind-Richtung. Auf der dem Wind
zugekehrten Seite des Kammes sind windgeformte Bäume am Hang
vom Kamm bis in 40 - 70 m (vertikale Entfernung) unterhalb des
Kamms zu finden. Auf der Lee-Seite dagegen finden sie sich 20 -
70 m unterhalb des Kammes.

Die Verbreitungsgebiete von Usnea finden sich in einem Teil des
kleinen Daimyosinsawa-Talendes und auf dem lee-seitigen Hang des
Kamms, der sich vom Mt.Azuma in nordwestlicher Richtung erstreckt.
Es wird angenommen, daß sie nur in diesen beiden kleinen Gebie-
ten gedeihen, dank der hohen Luftfeuchtigkeit, die durch das
Nachlassen der Windstärke in diesen Gebieten bedingt ist.

Gemäß der Beobachtungen, die gerade nach dem Ende einer zwei Tage dauernden heftigen Wintermonsun-Wetterlage gemacht wurden,
war der Stand des Rauhreifs (englisch: 'silver thaw') folgender:

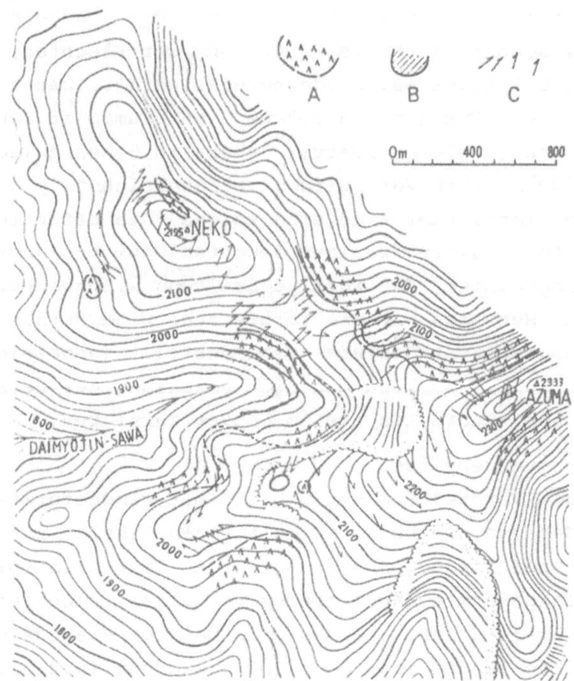

Abb.2. Verbreitung der windgeformten Bäume, der symmetrisch
gewachsenen Bäume und der von Usnea bedeckten Bäume in der
Gipfel-Region des Mt.Azuma und des Mt.Neko. A Gebiet, in dem
sich symmetrisch gewachsene Bäume finden. B Gebiet mit Usnea.
C windgeformte Bäume. Die Pfeile geben die Wind-Richtungen an,
die sich aus der Deformation dieser Bäume ablesen läßt.

Der Stand der Adhäsion von Schnee und Eis hatte sich entsprechend der Höhe über dem Boden verändert, wenn wir die einzelnen Bäume betrachten. Im ganzen Untersuchungsgebiet hatte eine
Veränderung entsprechend der Höhe über NN stattgefunden. Auf
der windwärtigen Seite der Bäume in den oberen 2/3 des Stammes
vom Teil (a) haftete der gefrorene Schnee, der sich aus torpe-
do-förmigen Teilen zusammensetzte, fest und kompakt. Das untere
Drittel von (a) ähnelte einem mit Zucker bestreuten Kuchen. Der
Durchmesser und die Länge jedes "Torpedos" der oberen 2/3 von
(a) betrugen 1 - 2 cm und 5 - 10 cm auf den Bäumen in einer Han‹
höhe von 2.140 - 2.250 m am Mt.Azuma. In tieferen Hanglagen
(2.030 m) erreichten sie 0.5 cm und 3 - 4 cm. Auf der Oberfläche der Schneedecke auf der Lee-Seite dieser Bäume wurden abge-

schlagene Blätter und Zweige beobachtet. Ihre Länge betrug in
der Regel 5 - 6 cm, teilweise jedoch 20 cm. Aus dieser Tatsache
läßt sich leicht der mechanische Effekt des Windes entnehmen.

(3) Lokal- und mikroklimatische Auswertung.
Die vorherrschende Windrichtung während des Winter-Monsuns in
2.000 m Höhe in der freien Atmosphäre über diesem Gebiet ist
West auf Grund von Ballon-Windmessungen in Nagano (36°40'N,
138°12'E).Daher sind die verschiedenen Wind-Richtungen durch
die Mikrotopographie des Gebietes bedingt.
Die lokal- und mikroklimatologischen Windverhältnisse lassen
sich folgendermaßen zusammenfassen: Die Westwinde passieren die
Paßlage zwischen Mt.Azuma und Mt.Neko mit hoher Geschwindigkeit
mit SW-Wind-Richtung. Die Kämme überweht der Wind meist rechtwinklig zur Richtung des Kamms. Hinter dem Kamm, gerade unterhalb des Kamms auf der Lee-Seite, findet sich fast die ent-
gegengesetzte Wind-Richtung. Dies bedeutet, daß hinter dem Kamm
ständig Wirbel gebildet werden. Das Gebiet, in dem die Luftströmungen durch den Einfluß der Mikrotopographie verstärkt werden,
findet sich hangabwärts vom Paß 30 - 40 m und vom Kamm 40 - 70m
auf der westexponierten Seite und vom Kamm 20 - 70 m auf der
Lee-Seite.

3. Die Verbreitung der windgeformten Bäume auf dem Kamm,
   Happôone.
(1) Geographische und klimatische Skizze des Gebietes.
Happôone liegt in den Nord-Japanischen-Alpen in Zentral-Japan.
Er läuft südlich vom Mt.Shirouma (2.933 m, 36°45'N, 137°12'E)
und Mt.Karamatsu (2.696 m, 36°41'N, 137°11'E) in Richtung von
WSW nach ENE. Die Länge des Kammes beträgt fast 7 km. (Abb.1
und Abb.3).

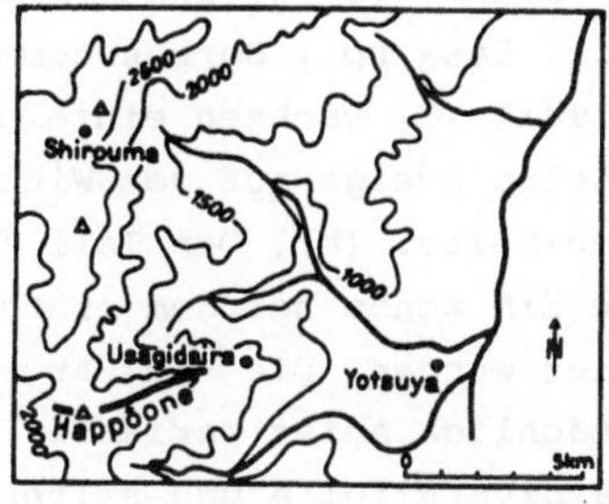

Abb.3. Das Beobachtungsgebiet und die Lage des Happôone-Kamms.

Die Beobachtungsergebnisse an den meteorologischen Stationen
Shirouma-dake (2.730 m), Usagidaira (1.650 m) und Yotsuya (800
m) während jeder zwei Sommermonate 1964 und 1965 zeigten folgende klimatische Verhältnisse: 1. Mit der Meereshöhe nimmt die
Windstärke zu und im Gegensatz dazu die Häufigkeit von Windstil-
len ab. 2. Die vorherrschende Windrichtung in Shirouma-dake
war W-NW, die mit der Strömungsrichtung in der unteren Troposphä-
re in Zentral-Japan übereinstimmt. 3. Im Gegensatz zu dieser
Windrichtung im Gipfelgebiet war sie in Usagidaira am Hang und
in Yotsuya am Talboden Süd, der Hauptrichtung des Haupttales
entsprechend. 4. Der Wind, der in Usagidaira stärker als 11 m/s
im Winter ist, weht aus W. Die Häufigkeiten nach der Windge-
schwindigkeit sind: 6-10 m/s 1.8%, 11-15 m/s 1.5%, 16-20 m/s
0.5%, 21-25 m/s 0.4% und 26-30 m/s 0.2%. Sie sind wichtig zur
Bildung der windgeformten Bäume. 5. Die Dauer der Schnee-Decke
ist sehr lang: die längste Dauer betrug 189 Tage im Winter 1964-
1965 während der Beobachtungsperiode von 1959-1965. 6. Die maximale Schnee-Höhe war 250 - 390 cm in Usagidaira. Wenn man als
Dichte der Schnee-Decke 0.3 nimmt, entspricht die Schnee-Decke
von 1 m Mächtigkeit einem Gewicht von 300 kg/m^2; d.h. auf der
Vegetation unter dem Schnee lastet im Winter ein Gewicht von
fast 1 ton/m^2 für 3 m Schneehöhe durchschnittlich.

(2) Beobachtungsergebnisse

Die windgeformten Bäume im Happôone-Kammgebiet zwischen 1.400 m
- 2.500 m wurden ausführlich untersucht, wobei sich folgende
Ergebnisse zeigten. Die allgemeinen Formen der Vegetation in
diesem Gebiet sind in Abb.4 dargestellt. Der Teil (c) erscheint
wie ein Teppich oder Kissen. In diesem Gebiet (c) wird die Vegetation hauptsächlich von Pinus pumila, Thuja standishii, Tsuga
diversifolia und anderen Arten wie Vaccinium smallii, Acer sp.,
Rhododendron tschonoskii, Sasa sp., Sorbus commixta var. glabrum
etc. gebildet. Aus dem Teil (c) wachsen einzelne Baumstämme hoch.
(a). Der Teil (a) hat keine Zweige auf der Wind-Seite und nur
kurze Zweige auf der Lee-Seite. (b). Den Teil (a) nennen wir hier
windgeformte Bäume, wie sie schon bei dem ersten Beispiel im
Azuma-Neko Gebiet gezeigt wurden. Die Baum-Arten , die den Teil
(a) bilden, sind hauptsächlich Abies mariesii, Abies veitchii,
Thuja standishii, Tsuga diversifolia und selten Larix leptolepis.

Die Teile (a), (b) und (c) der einzelnen Bäume wurden an 39 Stellen gemessen.

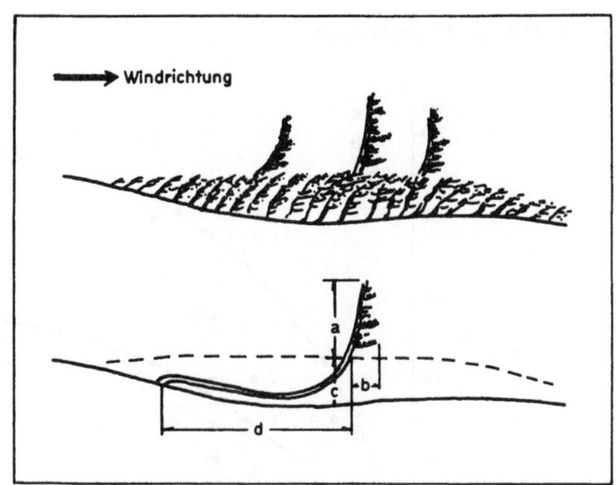

Abb.4. Skizze der windgeformten Bäume im Happôone-Kammgebiet.

Die Eigentümlichkeit, die besonders in diesem Gebiet gefunden
wurde, ist der lange Teil des Stammes, der Teil (d). Eine Thuja
zeigte zum Beispiel an einem Ort (1.995 m) (a) = 1 m, (b) = 1 m
und (d) = 4 m. Der Durchmeser des Teils (d) war 20 cm. Bei einem
anderen Beispiel von Thuja in einem anderen Ort (1.850 m) haben
wir (a) = 1.2 m und (d) = 4.7 m gemessen. Die Formen dieser
Bäume sind daher ganz ungewöhnlich.
Die Höhe des Teils (a) und die Länge des Teils (b) sind abhängig von der Wind-Geschwindigkeit in der Wintermonsun-Zeit unter
dem Einfluß von Schnee- und Eisgebläse, wie bereits erwähnt.
Der Teil (d) bildet sich durch den Druck der Schnee-Decke, die
sich am Hang während des Winters ganz langsam nach unten bewegt.
Die Kausalität des langen Teils (d) in diesem Gebiet scheint
mir folgende: Charakter der Thuja in Übereinstimmung mit der
Richtung der Kamm-Neigung und der vorherrschenden Windrichtung.

Wie erwähnt, beträgt die Dauer der Schnee-Decke etwa 6 Monate
und die maximalen Schnee-Höhen bis 3 m. Der Druck des Schnees
kann eine große Rolle bei der Bildung dieser Baumform spielen.

Die durch die windgeformten Bäume gezeigte Haupt-Windrichtung
ist West. Nur wenige Punkte zeigen WSW oder WNW.
Nun soll auf die Länge des Teils (a) in diesem Gebiet eingegangen werden. Die Beziehung zwischen (a) und (b) ergibt Kurven,

wie in Abb.5 dargestellt.

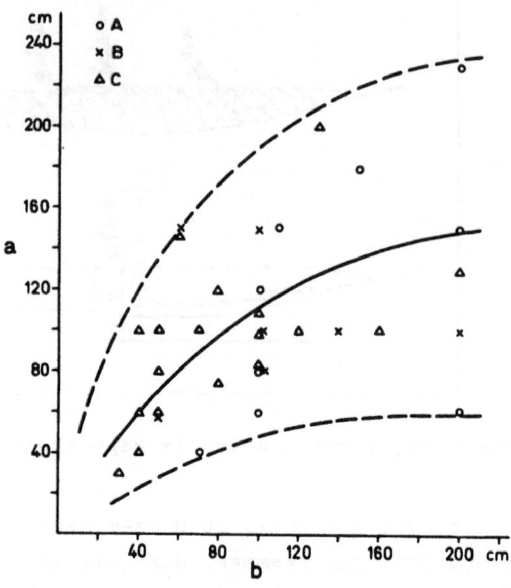

Abb.5.   Beziehung zwischen (a) und (b). (a und b: siehe Abb.4).
A: Abies mariesii und A.veitchii.   B: Tsuga diversifolia.
C: Thuja standishii.

Die Kurven von Abies mariesii und A.veitchii sind etwas anders
als die Kurven von Thuja standishii und Tsuga diversifolia. Die
Ursache dafür liegt darin, daß in diesem Gebiet Abies im höheren (meistens über 2.250 m) und Thuja und Tsuga im niedrigen
Teil (unterhalb 2.115 m) verbreitet sind. Das heißt mit anderen
Worten, daß der Stamm von Abies wegen der starken Wind-Geschwindigkeit relativ wenig nach oben wächst. Dabei muß man noch den
Unterschied in der Widerstandskraft zwischen den Baum-Arten berücksichtigen.

Die Beziehung zwischen (a) und Meereshöhe wurde dann in Abb.6
gezeigt. Maxima treten in den Standorten 2.250 m und 1.770 m
und das Minimum im Standort 1.900 m NN auf. In Abb.6 wurde auch
die Baumhöhe von Birken (Betula ermanii) dargestellt. Obwohl
die beobachteten Punkte von Betula nur wenige sind, stimmen die
Kurven gut miteinander überein.

Die Wuchs-Orte, an denen Maxima auftreten, liegen mikrotopogra-
phisch im allgemeinen dort, wo die Neigung des Happôone Kamms

in der laufenden Richtung relativ schwach ist - mit anderen Worten: Der Wind ist in dieser Lage relativ schwach und die Schnee-Decke ist hoch. Im Gegensatz hierzu hat das Gebiet zwischen 1.900 - 1.930 m in dem das Minimum auftrat, die mikrotopographische Situation mit einer schätzungsweise sehr starken Wind-Geschwindigkeit.

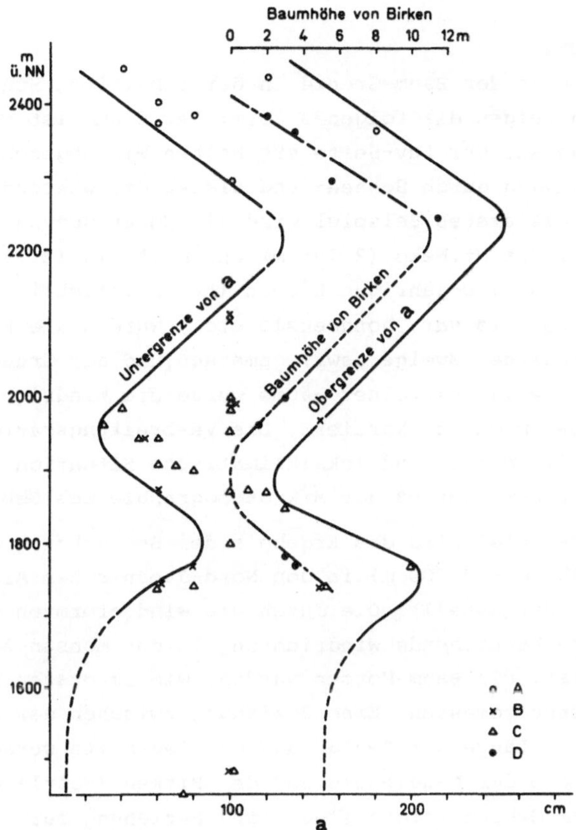

Abb.6. Beziehung zwischen (a) und Meereshöhe im Happôone-Kamm-gebiet für:
A: Abies mariesii und A.veitchii.
B: Tsuga diversifolia.
C: Thuja standishii.
D: Betula ermanii.

(3) Lokal- und mikroklimatische Auswertung.
Im Happôone Kamm-Gebiet stimmt die laufende Richtung des Kamms mit der Richtung der vorherrschenden Winde im Winter fast über-

ein. Die Dauer der hohen Schnee-Decke erreicht fast ein halbes
Jahr. Westwinde mit einer Geschwindigkeit von 10-20 m/s kommen
im Winter an über 40 Tagen im ganzen Kamm-Gebiet vor.
In dieser klimatischen Lage konnte man noch eine kleinräumige
Verbreitung des Windes und der Schnee-Decke aus der Verbreitung
der Höhe (a) (Baumhöhe) folgern.

ZUSAMMENFASSUNG

Die Nadelbäume an der Baum-Grenze in der subalpinen Stufe in
Zentral-Japan zeigen die folgende Form: Der Stamm ist vertikal,
und die Zweige auf der Luv-Seite erscheinen wie abgeschnitten.
Dies ist verursacht durch Schnee- und Eisgebläse während des Winter-Monsuns. Als erstes Beispiel wird die Untersuchung im subalpinen Gebiet des Mt.Neko (2.195 m) und Mt.Azuma (2.333 m) diskutiert. Bei etwa 100-jährigen Bäumen (Abies veitchii, A.marie-
sii, Picea jezoensis var. hondoensis etc.) wurden die Formen
(Baumhöhe, Länge der Zweige usw.) gemessen, und auf Grund der
windgeformten Zweige einzelner Bäume wurde die Windrichtung des
Winter-Monsuns im Gebiet kartiert. Die Verbreitungskarte ver-
anschaulicht die mikro- und lokalklimatische Situation der Windrichtung unter dem Einfluß der Mikrotopographie des Gebietes.

Als zweites Beispiel wird das Ergebnis der Beobachtung am Kamm
Happôone (1.450 m - 2.300 m) in den Nord-Japanischen Alpen in
Zentral-Japan dargestellt. Die durch die windgeformten Nadelbäume gezeigte herrschende Windrichtung in der Monsun-Zeit im
Winter war West. Die Baum-Formen wurden, wie im ersten Beispiel,
an 39 Standorten gemessen. Eine Beziehung zwischen der Höhe des
Stammes und der Länge der Zweige auf der Lee-Seite wurde dargestellt. Die Höhe der Nadelbäume und der Birken (Betula ermanii)
im ganzen Kamm-Gebiet zeigte eine klare Beziehung zur mikroto-
pographischen Lage, die den lokalen Unterschied der Wind-Geschwindigkeit und der Schnee-Tiefe verursacht. Bei der Beobachtung dieser Baum-Formen wird es möglich sein, die lokalklimati-
schen Verhältnisse in der subalpinen Stufe zu kartieren.

SUMMARY

" The form of trees on the tree limit in the subalpine zone in
Central Japan and their significance as local climatic indicators".

The needle trees on the tree limit in the subalpine zone in Central-Japan show the following forms: The trunk is vertical,but
the branches on the windward side of the trees are severed by
the effects of strong winter monsoonal wind carrying snow and
frozen rain. The results of investigations on the local distributions of the trees on the summit area of Mt.Neko (2.195 m)
and Mt.Azuma (2.333 m) are described first and secondly, those
on the ridge Happôone (1.450 m - 2.300 m) in the North-Alps in
Central-Japan. It was shown that the observed forms of the
trees(heiqhtof trunk, length of branches etc.) had close relations to their microtopographical situations, which control the
local distribution of wind and snow. On this account, in turn,
these wind-shaped trees can be utilized as indicators, which
sharply reflect the local climatic conditions in the subalpine
zone.

LITERATUR

WALTER,H. -1951- Einführung in die Phytologie III. Grundlagen
    der Pflanzenverbreitung. p.493-498. Stuttgart.
WEISCHET,W. -1963- Grundvoraussetzungen, Bestimmungsmerkmale
    und klimatologische Aussagemöglichkeit von Baumkronendefor-
    mationen.- Freiburger Geogr.Hefte 1: 5-19. Freiburg.
YOSHINO,M.M. -1961- Shôkiko (Geländeklima).- Tokyo pp.274.
-- -- -1964- Some local characteristics of the winds as re-
    vealed by wind-shaped trees in the Rhone valley in Switzer-
    land.- Erdkunde 18: 28-39. Bonn.
-- -- -1966- Wind-shaped trees as indicators of micro and
    local climatic wind situation.- Biometeorology 2: 997-1005.
    (Proc. 3rd International Biometeorological Congress in Sept.
    1963). Oxford.

E. OBERDORFER:

Als ich das erste Bild sah, wurde ich an das erinnert, was wir
im Schwarzwald am Feldberg Tännchen-Friedhof nennen. Auch bei
uns', vor allem im windgefegten Mittelgebirge, gibt es ja dieses
Phänomen. Es ist auch sicher manches bei uns in Mitteleuropa
darüber gearbeitet worden, selten aber so minutiös und erschöp-
fend, wie Sie das getan haben.

F. RUNGE:

Die mechanischen Faktoren Schneedruck und Eis spielen auch bei
uns eine Rolle, daneben jedoch auch die austrocknende Wirkung
des Windes. Wenn wir beispielsweise im Wesertal bei Rinteln
windgefegte Bäume antreffen, ist die Ursache hierfür sicherlich

nicht nur die Wirkung des Schnee- und Eisgebläses. Gibt es für
Ihr Gebiet Untersuchungen über die austrocknende Wirkung des
Windes auf die Gehölze?

M.M. YOSHINO:
Wie ich anfangs gesagt habe, kann man die windgeformten Bäume
in 4 Typen klassifizieren. Den einen Typus habe ich beschrieben.
Ein anderer Typus sind die Laubbäume, wie Apfel und Birne, die
nicht durch die Wirkung von Schnee und Eis, sondern durch die
Verdunstung auf der Luv-Seite verformt werden.Dort sterben die
Neu-Zweige ab, sie wachsen an der Lee-Seite neu. Solche windge-
formten Bäume gibt es auch bei uns.

E. OBERDORFER:
Ich darf vielleicht noch ergänzend dazu bemerken, daß bei uns
ja die Wind-Formung auch im Mittelgebirge eine Trockenerschei-
nung ist: die sog.Frosttrocknis. Das Absterben erfolgt nicht nur
mechanisch durch den Wind, sondern auch durch Austrocknung.

J.J. BARKMAN:
Es wäre wünschenswert, wenn eine solch gründliche Arbeit wie die
von Herrn Prof. YOSHINO auch einmal an der Meeresküste angestellt würde; hier wäre sicherlich der Salzgehalt der vom Meer
kommenden Winde als Faktum einzukalkulieren. In den Niederlanden sind an der Westküste die Bäume meist nach Osten gefegt,
aber in Medemblik, einer Stadt im nordwestlichen Abschnitt der
ehemaligen Zuider-See, waren die Bäume früher nach Südost gefegt. Seit der Eindeichung und Trockenlegung der Zuidersee und
der Änderung des Salzgehaltes (Süßwasser) entwickelten die Bäume Äste auch an der NW-Seite, dem Haupt-Einfallgebiet des hier
wirksam werdenden Windes. Eine offene Frage ist es bislang noch
immer, wie es sich mit dem Jod-Gehalt der zweifelsohne salztra-
genden Luftströmungen verhält, und welchen positiven oder negativen Wachstumsfaktor dieses Element hat.

F. RUNGE:
Ich glaube nicht, daß es so sehr das Salz ist. Ich glaube vielmehr, daß der Wind von der Seite am stärksten einwirkt, wo er
am meisten Zutritt hat. Auf Ihrer Insel Terschelling habe ich
windgeformte Bäume kartiert und habe festgestellt, daß auf der
Nordseite der Insel, also nördlich des Dünenwalles, Winde aus
nordwestlicher Richtung einwirkten; auf der Südseite, der Watt-
seite, dagegen Winde von Südwesten. Ich habe das damals auf an-

dere klimatische Verhältnisse geschoben, heute ist mir ganz
klar: Der Wind hat auf der Nordseite am besten von Nord Zutritt
und er wirkt von dort auf der Südseite von Süd und entfaltet
dort seine größte Kraft.

W.E. FÜRST KNYPHAUSEN:
Auch ich komme von der Nordseeküste,und auch bei uns werden die
Bäume vom Winde verformt. Erstmals im Jahre 1962 bei dem starken Sturm vom 16.Februar, erneut aber auch im vergangenen Frühjahr, auch wieder im Februar und im Oktober, haben wir an Koniferen Schäden aller Art feststellen können, soweit sie dem Winde direkt ausgesetzt waren.Diese Schädigungen waren dergestalt,
daß die windzugekehrten Seiten innerhalb von 14 Tagen bis 3 Wochen nach dem Sturm braun wurden und bei jungen Trieben, vor
allen Dingen bei Taxus aber auch bei Abies, von einer Stärke
bis zu 2, vielleicht 2 1/2 mm, die Zweige abstarben, und nicht
wieder austrieben. Mir ist damals gesagt worden, es wäre nicht,
wie wir vermutet haben, das Salz der Luft, sondern der Jod-Gehalt. Diese Frage ist bislang wohl noch nicht zufriedenstellend
beantwortet worden. Weiter wurde behauptet: Alle Koniferen sei-
en salz-resistent (nicht allerdings gegen die Überdosierung,wie
sie auf den Straßen in Wintermonaten angewandt wird). Gelände-
Beobachtungen haben bislang nur gezeigt, daß "salzträchtige",
vom Meer kommende Luftströmungen, auf das Wachstum von Koniferen einen größeren Einfluß nehmen, als dies bei Laubgehölzen
der Fall ist.

R. CARBIENER:
Die Erfahrungen zeigen, daß Koniferen windempfindlicher erscheinen als Laubbäume. An den Waldgrenzen, die von Laubbäumen
gebildet werden, da wo die Waldgrenze natürlich ist, sind es
ganz geschlossene Fronten von Bäumen, die wenig windempfindlich
sind. Koniferen erscheinen dagegen windempfindlich, wenn sie an
ihrer ökologischen Grenze wachsen. So ist es an der oberen Wald-
grenze; dort sind sie stark windempfindlich. So ist es aber
auch,wenn sie z.B. in maritimen Gebieten gepflanzt werden, wo
sie überhaupt standortsfremd sind, z.B. in Dänemark oder in
Nordwestdeutschland. Sehen Sie sich diese Pinus sylvestris oder
Abies- oder Picea- Pflanzungen an. Sie sehen jämmerlich aus!
Die Triebe sind schlecht, die Nadeln sind nicht ausgereift usw.,
weil sie nicht klimatisch am Platze sind, und in diesem Falle

sind sie windempfindlich, weil sie ökologisch unterlegen sind.
Ich glaube nicht, daß Koniferen allein durch den Wind am Wachstum verhindert werden können, wenn man bedenkt, wie sie in äußerst winterkalten Gebieten den trockenen Winterwinden bei -30
bis -40°C unbeschädigt widerstehen. In Alaska, in Zentral-Asien,
finden wir Koniferen von 30 m Höhe bei -40°C in Gebieten starker
trockener Winde.

M.M. YOSHINO:
Die vier Typen windgeformter Bäume sind: 1. Bäume mit vertikalem Stamm und zur Lee-Seite gebogenen Zweigen, verursacht durch
Winde während der Wachstums-Periode. 2. Bäume mit vertikalem
Stamm, deren Zweige auf der Luv-Seite wie abgeschnitten erscheinen, verursacht durch Schnee- und Eisgebläse während des Winters. 3.Bäume, deren Stamm und Zweige durch die während der
Wachstums-Periode wehenden Winde, entsprechend der Haupt-Windrichtung deformiert sind. 4. Bäume, deren Stamm sich zur Lee-
Seite der jeweils stärksten Winde neigt, deren Kronen jedoch
nicht sehr stark deformiert sind. Die Ursache dafür sind möglicherweise Taifune oder Hurrikane.
Die besprochenen windgeformten Bäume in Holland, die wir auch
an den Nordsee- bis Atlantischen Ozean-Küsten Nordwest-Europas
finden können, gehören zum Typ 3. Sie finden sich nicht nur an
diesen Küsten, sondern auch in anderen windig-reichen Gebieten
der Erde. Sommergrüne und immergrüne Laub- und auch Nadelbäume
zeigen diesen Typ.

Ergänzung:
Eine weitere Arbeit über die Klassifikation, Verbreitung und
Auswertung der windgeformten Bäume ist in der Zwischenzeit in
"Studies on windshaped trees", von M.M. YOSHINO, Climatological
Notes, Hosei University, Tokyo, Nr.12,1972, 52 S. erschienen.
Wir haben die von den windgeformten Bäumen abgeschätzte Wind-
stärke und -richtung mit denen von Instrumenten beobachteten
Resultaten imAjdovščina-Gebiet in Slowenien, Jugoslawien, und
im Ishikari-Gebiet in Hokkaido, Japan verglichen. Die Ergebnis-
se waren zufriedenstellend. Das Heft enthält die Bibliographie
über die windgeformten Bäume in 178 Aufsätzen, Monographien,
Büchern usw.

BEMERKUNGEN ZUR SAUM-MANTEL-FRAGE

P. J a k u c s [1]

SCHLUSSFOLGERUNGEN,VORSCHLÄGE

a. Die Zone des dynamischen Kampfes zwischen Wäldern und wald-
freien Teilen ist ein Übergangsstreifen. dessen eigenartiger
Charakter meistens nur eine Resultante der Faktormischungen
der zwei gegebenen und entgegengesetzten Bioräume ist. Die
künstlichen Waldgrenzen können sich am verschiedenartigsten,
teilweise vom Maße der anthropogenen Wirkung und ihrer Wie-
holung,teilweise aber von den ökologischen und floristisch-
soziologischen Gegebenheiten der Landschaft (Synökosystem)
abhängig herausbilden.

b. Die Mäntel und Säume bilden (auch im Sinne von Th.MÜLLER)
physiognomisch und struktuell tatsächlich existierende und
manchmal gut trennbare Einheiten. Ihre Einreihung in das zönologische System als Lokal-Assoziationen ist manchmal berechtigt.

c. Die wärmeliebenden Saum-Kennarten ("T r i f o l i o - G e -
r a n i e t a" - "O r i g a n e t a l i a" -Arten) sind größtenteils charakteristische Arten der submediterran-subkonti-
nentalen xerothermen Wälder (Q u e r c e t e a p u b e s -
c e n t i - p e t r a e a e -Klassenkennarten), sie dürfen
aus diesen nicht weggerissen werden. Diese Arten erreichen
im zentralen Raum ihrer Areale - unter natürlichen Umständen - in Wald-Phytozönosen ihre optimale Entwicklung und Vitalität, ferner die höchsten Stetigkeitsprozente und besten
Treuegrade. Es kann im allgemeinen vorgeschlagen werden, daß
in den zentralen Arealen die Treue- und Soziabilitäts-Ver-
hältnisse der Kennarten jedesmal beachtet werden sollten,
wenn man ihre zönologische Wertung bei der Aufstellung höherer Einheiten berücksichtigen will. (Das muß auch geschehen,
wenn der Forscher an Areal-Grenzgebieten arbeitet). Sollte
man das außer Acht lassen,so würden die neuen Verbände, Ordnungen und Klassen wie Pilze hervorschießen und das System
verlöre seine Reinheit und Realität.

[1] Der vollständige Vortrag wurde unter dem gleichen Titel in
Vegetatio 21 (1-3) The Hague 1970 veröffentlicht. Wir geben
darum hier nur die Zusammenfassung wieder.

d. Im floristisch-soziologischen System gehören die west- und
   zentraleuropäischen wärmeliebenden Saum-Mantel-Gesellschaf-
   ten in die Klasse der Wald-Gesellschaften. Es kann im allgemeinen festgestellt werden, daß sich die Säume aus den Kennarten der Plakor-Gesellschaften einer benachbarten Vegetati-
   onszone (Gürtel) lokal-klimatisch oder lokal-edaphisch bedingt zusammensetzen.

e. Auch bei der Einführung von neuen höheren floristisch-soziologischen systematischen Einheiten (Verband, Ordnung, Klasse,
   Division) darf man die Feststellung von BRAUN-BLANQUET über
   das Synökosystem, - neben der Priorität der Grundprinzipien
   der Treueverhältnisse der einzelnen Arten - nicht außer Acht
   lassen (1964,p.101: "Bei der Umschreibung des Synökosystems
   treffen sich Biosoziologie, Ökologie und Biogeographie").

f. Die Teilangaben sollten nach Möglichkeit in einer Form veröffentlicht werden, welche auch die Berechnungen mit bioma-
   thematischen Methoden ermöglicht.Das fördert Objektivität
   der Klassifizierung und so wird diese kontrollierbar.

g. Bei einigen Pflanzen tritt in einer extremen oder aufgelok-
   kerten ökologischen Lage, infolge des gesteigerten Daseins-
   kampfes, neben der generativen Vermehrung auch die Sproßkolonien-Bildung auf. Unter solchen Umständen kann ein Indivi-
   duum seine physiognomischen Dominanz-Verhältnisse in kurzer
   Zeit vervielfachen. Bei der Besprechung der Erscheinung der
   Sproß-Kolonien (Polykormie)-Bildung wünschten wir auf den
   ökologische Extreme anzeigenden Wert dieser aufmerksam zu
   machen. Unsere Meinung ist, daß diese Erscheinung bei der
   Wertung der Kennarten (besonders des Verbands-, der Ordnungs-
   und Klassen-Kennarten) als negatives Merkmal beachtet werden
   sollte.

h. Das wichtigste Ergebnis der mathematisch-statistischen Auswertung des pannonischen Hang-Waldsteppen-Komplexes,(geschlossener Hochwald-Buschwald-Mantel-Saum-Rasen): In den Präsenz-
   und Dominanz-Verhältnissen der Arten, ferner in den Faktoren
   des Standortes und Mikroklimas zeigt sich die schärfste signi-
   fikante Differenz zwischen Saum und Rasen.

i. Im jeweiligen Sukzessionsrhytmus sind die Saum-Mantel-Ökosy-
   steme in der produktionssteigernden Rolle der Wälder von
   großer Bedeutung.

E. OBERDORFER:

Ich danke Herrn JAKUCS für seine umfassende Darstellung der Auffassung der Probleme, wie sie von den ungarischen Erfahrungen
her gesehen werden. Nun ist es ja natürlich in erster Linie eine Frage der Vegetationsanalyse und ich meine, nach meinen be-
scheidenen Anschauungen, die ich selbst schon in Südost-Europa
gewinnen konnte, gibt es auch dort jene Bilder, die wir hier
als Mosaiken kennen, einen Wald, eine Busch-Gesellschaft, daran
anschließend die Saum-Gesellschaft, oft als Mosaik-Komplexe und
als Zonations-Komplexe,wenn wir die Ausdrücke, wie sie Herr
SEIBERT gestern geprägt hat, verwenden wollen. Ich erinnere mich
z.B., was wir erst im letzten Jahr gesehen haben - da waren sicher einige von Ihnen dabei - bei Wiel an der ungarischen Grenze
bei Koggendorf hatten wir ein sehr schönes Beispiel: das A c e -
r i - Q u e r c e t u m. Das war ein Hochwald, in dem ganz wenige Pflanzen wuchsen, nur Lithospermum purpureo und Polygonatum latifolium. Als wir aus dem Wald herausgetreten sind, fanden wir - ich habe selten eine so schöne offene Stauden-Gesell-
schaft gesehen - Peucedanum alsaticum, Geranium sanguineum,
Bupleurum falcatum u.a. Es gab eine ganz scharfe Grenze zwischen
diesem blumenreichen breiten Saum und dem Wald, in dem kaum ein
paar Arten waren. Und nun sagt Herr JAKUCS, daß es natürlich,-
und das ist ihm zuzugestehen - im pannonischen Raum Trockenwäl-
der gibt, in denen diese Arten auch mehr, als wir das vielleicht
von uns her gewohnt sind, in die Wälder hineingehen, ins Q u e r-
c e t u m c e r r i s,also in einen hohen und lichten Wald, in
dem nun auch die Saumarten stehen und er meint, man könne deshalb nicht die Klasse der T r i f o l i o - G e r a n i e t e a,
von diesem Verhalten aus gesehen, so beurteilen, wie wir das
hier tun. Herr JAKUCS hat mir, als er mich im letzten Herbst
besuchte, in das Gästebuch geschrieben: " Ein Gruß aus dem Zentrum des Flaumeichengebietes." Es war etwas gezielt, hat aber,
glaube ich, doch nicht ganz richtig getroffen. Das Zentrum des
Flaumeichen-Gebietes sehe ich mehr im südlichen Balkan, in Ser-
bien bis Nordgriechenland, in Italien, in Spanien, in Südfrank-
reich. Es ist nun erstaunlich, wie selten man in diesen Flaum-
eichen-Gebieten Geranium sanguineum und andere Arten dieser
Saum-T r i f o l i o - G e r a n i e t e a-Klasse sieht.Wenn man
nämlich nun das Areal dieser unserer T r i f o l i o-G e r a-
n i e t e a-Arten studiert, sieht man, daß es ganz anders ge-

formt ist als das Areal aller richtigen Flaumeichen - Wald-Arten,
Quercus pubescens, Fraxinus ornus,Ostrya, Acer monspessulanus.
Es sind nämlich Waldsteppen-Pflanzen, sie haben eine eurasia-
tisch, subkontinentale Verbreitung, Geranium sanguineum geht ja
bis nach Finnland hinauf und weit nach Rußland hinein in Verbindung mit den subkontinentalen Waldsteppen-Pflanzen. Es ist deshalb verständlich, daß eben gerade in Ungarn, wo wir ein Kon-
takt-Gebiet haben zwischen den submediterranen Flaumeichen-Ge-
sellschaften und den Waldsteppen-Gesellschaften, ein stärkeres
Eindringen der T r i f o l i o - G e r a n i e t e a im subkontinentalen Klima in ganz bestimmte Gesellschaften, eben die
Gesellschaften mit Quercus cerris, vorliegen. Wenn wir dann
weiter nach Süden kommen, wo die richtigen Quercus pubescens-
Wälder kommen, dann werden wir sehen, daß diese Arten überhaupt
sehr selten werden. Das können hier sicher die Herren aus den
submediterranen Gebieten bestätigen,daß man in Italien ganz selten einmal Geranium sanguineum oder auch verwandte Arten in einer der Waldgesellschaften sieht, wie wir das so häufig, vor
allem, wenn wir mehr in das subkontinentale Gebiet hineinkommen,
finden. Ich will damit nur sagen, daß es vielleicht doch vom
Gesamtverhalten her berechtigt ist, eine eigene Klasse aufrecht
zu erhalten, und daß wir dieses Übergreifen, wie wir es tatsächlich im subkontinentalen Kontakt-Raum haben, doch eben nur als
eine mehr regionale Erscheinung dieser subkontinentalen Wald-
steppen-Säume, die ja nun durch ganz Europa hindurch zu beobachten sind, verstehen können.

EINIGE BEMERKUNGEN ZU DEN
SAUMGESELLSCHAFTEN

Th. M ü l l e r

Schon in der Urlandschaft und nicht nur erst in der durch Menschen geformten Kulturlandschaft gab es Wald-Ränder, also Stellen, wo geschlossene Waldgesellschaften an andere, offene Pflanzengesellschaften angrenzten. Diese Wald-Ränder konnten durch
verschiedene Ursachen bedingt sein, durch lokale Waldgrenzen
wie Flüsse, Seen, Felsen usw., durch klimatische Verbreitungs-
grenzen des Waldes, die sowohl groß- als auch lokalklimatisch
bedingt sein können, oder dadurch, daß durch eine Natur-Kata-
strophe wie Brand, Windwurf, Schneebruch, Schädlings-Kalamität
usw. ein Teil des Waldes vernichtet wurde und damit waldoffene
Räume entstanden. Diese waldoffenen Räume sind durch die menschliche Tätigkeit gegenüber der Urlandschaft in ungeheurem Maße
vergrößert worden, wodurch auch die Wald-Ränder in entsprechen-
dem Umfange vermehrt wurden.
Daß der Wald am Wald-Rand nicht nur einfach aufhört, sondern
sich nach außen einen Abschluß gibt, ist eine schon lang bekannte Tatsache, die nur von den Pflanzensoziologen lange nicht
erfaßt wurde. Nicht umsonst gaben oder geben sich teilweise
noch die Forstleute so große Mühe, gerade den Waldrand oder
Waldrauf zu erhalten oder ihn entsprechend zu pflegen. Denn
hier wird der Stammraum des Waldes nach außen hin wie von einem
Mantel umhüllt, der die dem Waldes-Inneren ungünstigen Außen-
Einflüsse abhält. Dieser Waldmantel wird teilweise von den be-
standesbildenden Holzarten selbst gebildet durch eine bis weit
nach unten reichende Belastung, z.B. bei vielen Buchen- und
Fichtenwäldern oder aber von Gehölzen in Buschform, die nur die
Höhe des freien Stammraumes des Waldes erreichen und eigene
Waldmantel-Gesellschaften bilden, in denen vorwiegend lichtlie-
bende Sträucher vorkommen, deren Samen hauptsächlich durch Vö-
gel verbreitet werden. Diese Waldmantel-Gesellschaften sind in
der Kulturlandschaft weitgehend selbständig geworden und treten
unabhängig von Waldgesellschaften als Hecken und Gebüsche auf.

Zu Füßen der Mantel-Gesellschaften folgt ein mehr oder weniger
breiter Streifen mit höheren Kräutern oder Stauden, der nach
Herrn Professor TÜXEN treffend als Saum bezeichnet wird. Bis

jetzt wurden zwei Klassen mit Saum-Gesellschaften gefunden, nämlich einmal die thermophilen Gesellschaften der Klasse T r i -
f o l i o - G e r a n i e t e a, zum anderen die mehr hygro-
und nitrophilen Gesellschaften der Klasse A r t e m i s i e t a,
wobei alle Saum-Gesellschaften mehr oder weniger heliophil sind.
Genau wie die Mantel-Gesellschaften treten auch die Saum-Gesellschaften in der Kulturlandschaft oftmals selbständig und unabhängig von Hecken und Wäldern auf und haben sekundär eine wesentlich weitere Verbreitung erlangt, als sie ursprünglich in-
nehatten.
Die Saum-Gesellschaften sind nun physiognomisch ganz deutlich
verschieden von den Mantel- und Waldgesellschaften, indem zunächst einmal alle Holzarten fehlen, wodurch ihre Struktur und
ihr Aufbau ganz anders ist. Wie steht es aber, wenn man die Arten-Zusammensetzung der Waldbodenschicht mit der der Saum-Gesellschaften vergleicht? Es wäre wunderschön, wenn sich diese gegenseitig ausschließen würden; dann wären die Verhältnisse ganz
klar und einfach. Aber leider ist dem nicht so. Es läßt sich
nicht verheimlichen, daß in bestimmten Wald-Gesellschaften -
auch geschlossenen - neben den Wald-Arten Saum-Arten vorhanden
sind. Sie kommen hier jedoch meist nur mit geschwächter Vitalität vor, blühen und fruchten selten oder überhaupt nicht, bleiben oft klein oder schwächlich. Wenn z.B. Geranium sanguineum,
Trifolium rubens, T.alpestre und T.medium sowie andere Arten
aus der Klasse der thermophilen, Urtica dioica, Aegopodium po-
dagraria, Glechoma hederacea, Lamium maculatum und andere Arten aus der Klasse der nitrophilen Saum-Gesellschaften zwar in
geschlossenen Wäldern mit geschwächter Lebenskraft als Einzelpflanzen oder höchstens in kleineren Trupps vorhanden sind und
vielleicht so gar einmal schwach blühen, aber in dem Augenblick,
da sie genügend Licht erhalten, wie das gewöhnlich am Saum-Stand-
ort der Fall ist, sich sofort mit Hilfe ihrer kriechenden Wur-
zelstöcke höchst vital ausbreiten und größere Gruppen und Herden bilden sowie reichlich blühen und fruchten, dann kann ich
bei diesem Verhalten diese Arten einfach nicht als Wald-Arten
ansprechen. Man muß daher bei der soziologischen Beurteilung
dieser Arten unbedingt auch ihr ökologisches Verhalten unter
den jeweils gegebenen Standorts- und Konkurrenz-Verhältnissen
beachten und sie dann den Gesellschaften zuordnen, in denen sie
tatsächlich ihre größte Lebenskraft entfalten. Dies ist im Prin-

zip weder eine neue Erkenntnis noch eine neue Forderung, sondern
im Bereich anderer Klassen eine schon lang geübte Praxis. So
z.B. störte es bis jetzt kaum jemanden, daß zahlreiche M o l i -
n i o - A r r h e n a t h e r e t e a - Arten auch in Wäldern
vorkommen. Da es meist mehr oder weniger heliophile Arten sind,
zeigen sie in den Wiesen gewöhnlich größere Lebenskraft als in
den Wäldern und sind deshalb berechtigterweise als Kenn-Arten
zu der Klasse M o l i n i o - A r r h e n a t h e r e t e a
gestellt worden. Nach unseren heutigen Kenntnissen müssen wir
aber annehmen, daß sie zwar mit zurückgesetzter Lebenskraft in
Wäldern vorkommen, aber als heliophile Arten ihr ursprüngliches
Vorkommen weitgehend in A r t e m i s i e t e a -Saumgesell-
schaften haben und erst von da aus in die von Menschen geschaf-
fenen Wiesen-Gesellschaften eingewandert sind. Wie soll man solche Arten wie etwa Heracleum sphondylium, Anthriscus silvestris,
Geranium pratense, Melandrium dioicum, Angelica silvestris,
vielleicht auch Cirsium oleraceum sowie andere Arten soziologisch bewerten? Wenn man sie ihrem natürlichen, ursprünglichen
Vorkommen nach einordnet, dann muß man sie ohne Zweifel weithin
als A r t e m i s i e t e a - Saumarten ansprechen. Da dies
aber nicht unbedingt ein soziologisches Argument ist, die Arten
aber außerdem mit gleicher Vitalität in den Wiesen vorkommen,so
müßte man sie entweder in beiden Klassen nur als Begleiter werten, denn die Festsetzung, daß sie in der einen Kenn- in der
anderen Trenn-Arten seien, wäre willkürlich, oder man würde sie
zu einer der meistens so verschwommenen "soziologischen Artengruppen" zusammenfassen, die in beiden Klassen auftritt. Ein
weiterer Weg wäre der, daß sie sowohl als Arten der primären
Saum-Klasse als auch der sekundären, abgeleiteten Wiesen-Klasse
zu bewerten sind, wobei sie in den beiden Klassen in jeweils
verschiedener Arten-Zusammensetzung in verschiedener Gesell-
schafts-Struktur sowie bei verschiedenen ökologischen Bedingungen auftreten. Ich wollte diese Frage hier einmal anschneiden,
da bei den verfeinerten Analysen immer wieder das Problem auf-
taucht, wie Arten soziologisch zu bewerten sind, die mit gleicher Lebenskraft sowohl in primären als auch in abgeleiteten Gesellschaften vorkommen und in beiden struktuell von Bedeutung
sind.
Sowohl die thermophilen als auch die nitrophilen Saum-Arten treten in den Wäldern um so stärker auf, je lichter diese sind.Die

Lichtstellung der Wälder kann dabei menschlich bedingt sein,a-
ber auch ganz natürliche Ursachen haben,wie Alterung der Waldbestände,oder daß die ökologischen Verhältnisse derart sind,daß
sie einfach keinen richtig geschlossenen Wald mehr zulassen.Je
lichter der Wald wird,umso weniger lassen sich Wald und Saum
trennen. Dies ist vor allem bei den Flaumeichen-Wäldern der Ordnung Q u e r c e t a l i a p u b e s c e n t i s und den Säumen der Klasse T r i f o l i o - G e r a n i e t e a an den
Stellen öfters der Fall, wo aus ökologischen Gründen nur noch
ein sehr lockerer Wald möglich ist. Solche Fälle gibt es schon
an extremen Stellen im immer noch subatlantisch getönten Süd-
west-Deutschland. Sie nehmen aber gegen den kontinentalen Bereich
wesentlich zu. Aus dieser Tatsache kommen zunächst einmal die
verschiedenen Anschauungen über die Flaumeichen-Wälder und Saum-
Gesellschaften vom Kollegen JAKUCS und mir. Es ergibt sich
zwangsläufig die Frage, von welchen Typen der Flaumeichen-Wälder
aus man die Soziologie dieser Wälder beurteilen soll und muß.
Geht man dabei von den Busch-Wäldern oder jenen lichten, aus
einzelnen weit auseinanderstehenden Bäumen bestehenden, vom Kollegen JAKUCS treffend als "obstbaumplantagen-ähnlich" bezeichneten Wälder aus, dann sind ganz klar alle T r i f o l i o-G e-
r a n i e t e a-Arten darin enthalten und zwar nicht einmal unbedingt mit zurückgesetzter Lebenskraft.Die Schlußfolgerung daraus ist, daß es keine eigene, selbständige T r i f o l i o -
G e r a n i e t e a -Gesellschaften gibt. Daß es derartige Wälder gibt, bei denen die Trennung in Saum-, Mantel- und Wald-Gesellschaften nicht durchführbar ist und daß diese Fälle im kontinentalen Europa verbreitet sind, ist eine unbestreitbare, na-
turgegebene Tatsache. Ich möchte aber doch zu bedenken geben,
daß es sich bei den Flaumeichen-Wäldern - gemäß der Verbreitung
von Quercus pubescens selbst - ja nicht um kontinentale oder
subkontinentale, sondern im wesentlichen um submediterrane Wälder handelt. In diesem Bereich gibt es aber durchaus Flaumeichen
Wälder, in denen die T r i f o l i o - G e r a n i e t e a -Ar-
ten keine Rolle spielen und nur im Saum auftreten. Meiner Meinung nach muß man bei der soziologischen Beurteilung von diesen
Waldtypen im Submediterran-Gebiet ausgehen und die kontinentalen
Trockenwald-Typen, die in Süddeutschland lokalklimatisch, weiter
östlich großklimatisch bedingt sind, als Extrem-Ausbildungen betrachten, in denen die dort stet und durchaus vital auftreten-

den T r i f o l i o - G e r a n i e t e a -Arten als entsprechende Trennarten zu werten sind.

M. FÖRSTER:

Um die floristische Zusammensetzung von Waldgesellschaften eingehend beurteilen zu können, sind Betrachtungen über Struktur
und Dynamismus besonders wesentlich. Zudem haben wir zu prüfen,
in welchem Maße bestimmte Arten unter den gegebenen Standortsbedingungen konkurrenzkräftig genug sind, sich im gesamten Artengefüge durchzusetzen. Die im subatlantisch-atlantischen Gebiet als Saum-Arten erscheinenden xerothermen Arten sind unter
den gegebenen mesophilen Bedingungen gegenüber den mesophilen
Wald-Arten unseres Gebietes nicht konkurrenzfähig. Sie weichen
daher auf die Säume aus, in denen sie sich bei relativ optimalen Bedingungen durchsetzen können. Es ist aber darauf hinzu-
weisen, daß der deutsche Mittelgebirgsraum ein buntes Mosaik,
eine Verzahnungs- und Übergangszone darstellt.Auch hier stehen
besonders in Trockengebieten die gleichen T r i f o l i o-
G e r a n i e t e a - Arten bereits in den Wäldern. Die Vitalität der mesophilen Arten ist soweit reduziert, daß die Xerothermen sich im Konkurrenz-Gefüge durchsetzen können. Besonders
deutlich wird dies im Pannonicum. Die Bedingungen sind hier im
Saum derart, daß die zu betrachtende Artengruppe nur noch im
Wald optimal zu gedeihen vermag. Sofern sie außerhalb auftritt,
werden ausgesprochen xerische und reduzierte Formen gebildet.
Mit der Verschiebung in der Dominanz sind es Steppen-Arten,die
hier die Säume bilden. Das soziologische Verhalten ist also
durch eine klimatische Verschiebung bedingt. Die floristische
Struktur ist in den Wäldern, auch in den xerothermen, in verschiedenen Altersstufen unterschiedlich. Unter Altersstufen verstehen wir: Dickungs-, Stangenholz- und Baum-Phase. Nur in der
Optimalphase des Baumholz-Alters finden sich typische Bestände.
Demgegenüber finden wir gut entwickelte Säume, in denen auch ein
Teil der mesophilen Wald-Arten steht, auch in Dickungen. In diesen dichten Beständen bildet nur, sich vegetativ vermehrend, Lithospermum purpureo-coeruleum dichte Rasen. Der von Theo MÜLLER
vorgeführte "Obstplantagen-Wald" ist untypisch. Ihm fehlen vor
allem Sträucher; allenfalls könnte hier noch von einer Zerfalls-
Phase gesprochen werden. Auch hier wird der Unterschied zum Pannonicum, in dessen Wäldern F e s t u c o - B r o m e t a -Arten
erscheinen würden, deutlich. Relativ ist gleichermaßen der Licht-

Faktor. Im mitteldeutschen Trockengebiet wurde von mir ein xe-
rothermer Eichenwald aufgenommen, der der Masse nach mit 1,2,
das entspricht einem Ertagstafelwert von 240 fm, bestockt war.
In den geschlossenen Beständen standen in geschlossener Verzah-
nung: Dictamnus scoparius, Euphorbia cyparissias, Geranium sanguineum, Inula salicina u.a.m. Ferner muß noch auf die Bildung
von Sproß-Kolonien hingewiesen werden. Die Polykorm-Bildung
spielt für die Physiognomie der Säume eine wesentliche Rolle.
Da die Konkurrenzkraft anderer Pflanzen an solchen Standorten
minimal ist, ist auch der Abstand zwischen den einzelnen Rhi-
zomen so gering, daß ein Rasen-Aspekt entsteht.Im Wald dagegen
ist die Konkurrenzkraft anderer Pflanzen dermaßen groß, daß beispielsweise von Dictamnus nur noch Einzel-Exemplare und keine
Herdenbildung mehr vorkommt, die im Saum zu beobachtende extreme Vitalität dieser Pflanze also nicht überbewertet werden darf.
Das Fazit ist und bleibt, daß Saum-Arten, wie die oben genannten
der Konkurrenzkraft des Waldes unterliegen müssen. Den hier de-
monstrierten Eichenwald auf der Parndorfer Platte halte ich für
einen hoffnungslos verhauenen "Wirtschaftswald", der zur Diskussion des anstehenden Problems nicht dienlich sein kann.

BEMERKUNGEN ZUR PROBLEMATIK WALD-MANTEL-SAUM IN ÖKOLOGISCHER
SICHT AM BEISPIEL EINES INTRAZONALEN THERMO- UND HYGROPHILEN
VEGETATIONSKOMPLEXES: DIE AUENWÄLDER DES OBERRHEINGRABENS.

R. C a r b i e n e r

Problematik: Ist der Reichtum und die Vitalität der Holzarten,
Gebüsche und Lianen, der den Rhein-Auenwald des Oberrhein-Gra-
bens (F r a x i n o - U l m e t u m : Sommer-Hochwässer, starke
Grundwasser-Schwankungen, leichte gut belüftete Schluff-Sand-
Böden) vor allen andern temperierten Waldtypen (außer ähnlichen
Auenwäldern wie z.B. an der Donau) auszeichnet, nur eine sekun-
däre Folge der Mittelwald-Wirtschaft, oder wieweit ist diese
eigenartige vielschichtige und heterogene Struktur und die für
temperierte Verhältnisse äußerst holzartenreiche Zusammensetzung
des Rheinwaldes natürlich? Wenn man den Rheinwald mit dem in
völlig gleichartiger Weise bewirtschafteten und nur einige km
weiter westlich gelegenen Riedwäldern: P r u n o - F r a x i -
n e t u m, und thermophil-feuchtes ultra-eutrophes"U l m o -
Q u e r c o - C a r p i n e t u m" (noch unbeschrieben, schwache
Grundwasserschwankungen, Hochwasser im Winter, schwerere Böden),
vergleicht, so stellt man fest, daß die Gebüsche und Lianen weniger arten-reich und weniger vital sind, und oft von sehr kräf-
tigem Jungwuchs von Fraxinus oder Carpinus von vornherein unterdrückt werden. In der Rhein-Aue, sind im Gegenteil die Gebüsche
und Lianen in den häufigen natürlichen Lücken auch der Hochwäl-
der, und nach einem Schlag, so vital, daß sie den Baumwuchs lange Zeit einschränken und nur lückenhaft hochkommen lassen. Das
war auch früher im Naturwald und nach Hochwasserwüstungen der
Fall und darf also als natürlich angesehen werden.Vitis sylvestris ist fossil aus dem altpleistozänen, damals schon von Ulmus
campestris beherrschten Auenwald des Elsaß, also lange vor Ein-
griff des Menschen häufig gewesen, und erst die Mittelwald- und
Niederwaldwirtschaft hat diese Liane ausgerottet. Mir scheint,
daß diejenigen thermophilen Wälder, die aus Lichtholzarten zusammengesetzt sind, (Hauptbestandteil des U l m o - F r a x i -
n e t u m: Ulmus campestris-carpinifolia, Fraxinus, Quercus robur, Populus alba, Acer campestre) und auf nährstoffreichen aber
sehr labilen durchlässigen jährlich überschwemmten - im H o c h -
s o m m e r (!) wechseltrockenen (sehr kalkreichen) Böden stehen, als spezifische Merkmale den Baum-, Gebüsch- und Lianen-

Reichtum, die Heterogeneität der Struktur, die Unterdrückung der
Krautschicht und die Mehrschichtigkeit aufweisen. Ich definierte
diesen Wald (1966/67) als kalttemperierten Dschungel (Forêt
dense tempéré froide), unter Hinweis auf die ökologischen Ursachen dieser Eigenart. (Maximale Feuchtigkeit von Luft und Boden
in der Haupt-Vegetationsperiode bei gleichzeitig bleibender guter Belüftung der Böden, sehr große Bakterien-Tätigkeit). Andere warmtemperierte Auen-Urwälder z.B. der Kaspischen Region
(EMBERGER 1962) haben eine ähnliche Struktur und floristische
Zusammensetzung, und sind sehr nah mit den strauch-und lianen-
reichen altpleistozänen Auen-Wäldern des Elsaß (von GEISSERT
1967 beschrieben) verwandt.
Natürlich verstärkt die Mittelwald-Wirtschaft sehr den Mengen-
Anteil und die Vitalität der Gebüsche, und man weiß manchmal
nicht, was man noch als Wald oder auch als Gebüsch ansprechen
soll. Aber selbst in alten Hochwald-Beständen ist festzustellen,
daß die Formation sehr offen bleibt. Gebüsche und Lianen-, und
auch neophytenfreundlich ist (z.B. natürlicher Aufwuchs ange-
schwemmter Samen von thermophilen Arten wie Juglans, Aesculus,
Robinia). So verzeichnet man in den künstlichen Buchen-Pflan-
zungen (mit denen man heute, sowie mit Fichten, den Rhein-Wald
nach Abdämmung verheert), überhaupt keinen Unterwuchs von Gebüschen,(wenn,dann nur sehr spärliche und belanglose F a g e t a -
l i a - Arten in der Krautschicht) obwohl die Bestandesdichte
und das Alter der Bäume ungefähr die gleichen sind wie in be-
nachbarten-natürlichen hochwaldartigen Beständen des F r a x i-
n o - U l m e t u m, die sehr gebüschreich sind , und in denen
bis 25 cm hohe und armdicke C l e m a t i s und H e d e r a
wuchern.
So steht fest, daß, wie auch in andern stark thermophilen Eichen
wäldern des Elsaß (so z.B. in den wenig genützten Hardtrocken-
wäldern der Adonis vernalis, Dictamnus, Lithospermum purpureo-
coeruleum, Potentilla alba und P.rupestris) die Grenze zwischen
Wald, Mantel und Saum gegenüber nördlicheren Verhältnissen ihre
Schärfe einbüßt. Dies weil eben die thermophilen Wälder meist
aus Lichtholz-Arten bestehen.Die begrenzenden Faktoren, Licht
und Wärme, gewinnen gegen Norden Bedeutung und bewirken ein Aus-
wandern vieler Arten aus dem Innern des Waldes in die Mäntel
und Säume. Gegen Süden dagegen bewirkt ein anderer Grenz-Fak-
tor, die Feuchtigkeit, das Gegenteil, und das Fehlen eines ei-

gentlichen Mantels und Saumes ist typisch für die mediterrane
Hartlaub-Region.

Es wird interessant sein, diese Grenz-Problematik an den soeben
geschilderten Objekten im Elsaß eingehend zu studieren. Die Frage berührt auch die in den vorigen Symposien behandelten Struk-
tur- und Dynamik-Probleme.

Wenn auch die Herausarbeitung der Eigenart von Mantel- Schlei-
er- und Saumgesellschaften einen wesentlichen soziologischen
und ökologischen Fortschritt darstellt, so sollte andererseits
nicht die Personalität der Ganzheit, nämlich die landschaftsöko-
logisch so charakteristische Struktur und Dynamik eines Wald-
Komplexes übersehen werden.

Für den Rhein-Auenwald (F r a x i n o - U l m e t u m) scheint
ein Ligustrum-Cornus sanguinea und C.mas-Viburnum lantana, Prunus fruticans, Rhamnus cathartica- (um nur einige bezeichnende-
re wärmeliebende Arten zu nennen) Mantel mit einem Rubus caesi-
us-, Lithospermum officinale-, Solidago serotina-Saum mit Clematis, Humulus und Tamus communis (Schleier) bezeichnend zu sein,
(cf. P a d o - C o r y l e t u m tametosum, Moor 1958), während
im Ried eine banalere weniger thermophile doch nahe verwandte
Mantel-Gesellschaft (Corylus, Prunus padus, Ligustrum, Cornus,
Rosa arvensis) die vom P a d o - C o r y l e t u m Moor zum
L i g u s t r o - P r u n e t u m Tx. überzuleiten scheint,
und ein Dactylis aschersoniana-Festuca gigantea-Angelica syl-
vestris-Heracleum sphondylium australe-Saum (mit Humulus und
stellenweise Vicia dumetorum-Schleier) die Ränder eines noch
nicht in die Soziologie eingegangenen, zugleich sehr wärmeliebenden feuchten und eutrophen Eschen-Stieleichen-Hainbuchenwal-
des einnimmt (mit Carpinus und Quercus dominierend, viel Fraxinus, Prunus avium, Tilia cordata, Ulmus effusa und U.scabra,
Pirus piraster, Alnus glutinosa, Acer pseudoplatanus) in welchem
weder Stellaria nemorum noch Galium sylvaticum (cf.Oberdorfer)
vorkommen, Tamus fehlt vollständig, Clematis und Hedera sind
seltener und weniger vital.

ÜBER MARITIME WALDGRENZEN IN EUROPA UND JAPAN

R. Tüxen

Mit Hilfe von Farb-Dias wurde versucht, die verschiedenen durch
Regional-Klima, Wind, Salzspray, Relief, Gesteins-Unterlage und
Boden-Bildung natürlich bedingten maritimen Waldgrenzen der
west- und nordeuropäischen und einiger japanischer Küsten in
ihrer gesetzmäßigen Ausbildung aufzuzeigen. Ganz allgemein läßt
sich an europäischen Fels-Küsten eine Zonen-Folge von Gezeiten-
Assoziationen aus Kryptogamen (Algen, Flechten) - auf Geröllzo-
nen mit Spülsäumen mit C a k i l e t e a und H o n c k e n y o-
E l y m e t e a - von halophilen Felsspalten-Gesellschaften,
natürlichen halophilen Spray-Wiesen der A s t e r e t e a
t r i p o l i i, von Vorwald- (Mantel-) Beständen der R h a m-
n o - P r u n e t e a, (vor denen sich stellenweise Säume der
T r i f o l i o - G e r a n i e t e a ansiedeln können), und
stark windgeschorenen Wald-Gesellschaften unterscheiden. Alle
diese Gesellschaften wandeln sich mit dem Allgemeinklima von
Süden nach Norden. Die Wälder von Q u e r c e t e a i l i -
c i s- über F a g o - Q u e r c e t e a - Q u e r c e t e a
r o b o r i - p e t r a e a e- (auf quarzreichen Gesteinen) bis
B e t u l a - Gesellschaften (verschiedener Arten). An diesen
Fels-Küsten wird die Vegetation nur geduldet.
An Dünen-Küsten baut dagegen die Vegetation mit Hilfe des Windes aktiv neue geomorphologische Formen auf. Hier zeigt sich
eine Zonierung von C a k i l e t e a -Gesellschaften auf Spülsäumen, A m m o p h i l e t e a- bis H o n c k e n y o - E l y-
m e t e a - Dünen, S a l i c i o n a r e n a r i a e -(R h a m-
n o - P r u n e t e a - Gebüschen auf alten Dünen und Wald (Pinus, Quercus, Betula), die sich mit dem Allgemeinklima von Süden
nach Norden wandelt.
Schlick-Küsten werden ebenfalls durch die Vegetation unter der
Herrschaft des Wassers aktiv gestaltet. Die Zonation geht hier
von den Z o s t e r e t e a- über S a l i c o r n i e t e a-
oder S p a r t i n e t e a- zu A s t e r e t e a t r i p o-
l i i - Gesellschaften, die mit dem Ausklingen des Salz-Einflus-
ses in potentielle Fraxinus-Wälder übergehen, deren Mantel- und
Saum-Gesellschaften infolge der Bewirtschaftung durch den Menschen nicht mehr erkennbar sind.
Dünen- und Schlick-Küsten werden als 'schwache' Landschaften

den 'starken' Fels-Küsten gegenüber gestellt.

In Japan lassen sich ganz entsprechende Gesetzmäßigkeiten erkennen.

Weil der Vortrag ganz auf die Anschauung von Farb-Bildern abgestellt war, begnügen wir uns hier mit diesen kurzen Andeutungen und verweisen auf zwei Arbeiten, in denen einige dieser Fragen ausführlicher dargestellt sind.

LITERATUR

OHBA,T., MIYAWAKI,A.& TÜXEN,R. -1973- Pflanzengesellschaften
    der Japanischen Dünen-Küsten.- Vegetatio 26 (1-3):3-143.
    The Hague.

TÜXEN,R. -1973- Die westeuropäische Küste als Kampf- und
    Lebensraum.- Mitt.flor.-soz.Arbeitsgem.N.F.15/16:210-223.
    Todenmann/Göttingen.

E.OBERDORFER:

Ich danke Herrn Professor TÜXEN für diesen großartigen Vortrag, der uns auf dem Wege der Anschauung, nicht der Mathematik, unmittelbar natürliche Grenzen der Vegetation in einem umfassenden Bereich der Küste gezeigt hat. Wir dürfen jetzt in die Gesamtdiskussion über das Problem Saum, Mantel und Wald eintreten, wie wir es ja nun von verschiedenen, mehr theoretischen, mehr anschaulichen Seiten gehört haben.

P.JAKUCS:

Es ist eine bekannte aber bislang nicht genügend analysierte Erscheinung, daß einige Arten in ihren Areal-Grenzen größere Massenverhältnisse bzw. eine anscheinend größere Vitalität zeigen als in ihrem zentralen Areal. Die Ursache hierfür kann mit dem Kampf ums Dasein der Pflanzen erklärt werden. Zum Fortbestehen benötigt die Art unter ungünstigen Standortsverhältnissen eine größere Kraft. Sehr viele Pflanzen stellen sich unter solchen Bedingungen neben der generativen auf die vegetative Fortpflanzung ein, Geranium sanguineum, Polygonatum odoratum, Brachypodium pinnatum, Carex humilis und viele andere wärmeliebende Wald-Arten bilden Sproß-Kolonien (Polykorme) im Falle des Eintritts einer ungünstigen Änderung der ökologischen Bedingungen (Temperatur-Rückgang, Abnahme des Lichtes usw.). Sie kämpfen um ihr Fortbestehen nicht nur auf generativem, sondern auch auf ve-

getativem Wege. Diese beiden Formen kommen in der Physiognomie
und in den Abundanz-Dominanz-Werten sowie in der Soziabilität
sehr stark zum Ausdruck. Anthropogene Einflüsse wie verlassene
Weinberge, Waldrodungen, künstliche Waldränder werfen das ökologische Gleichgewicht um, was sofort zur Bildung einer Sproß-
kolonie führt. Einige Beispiele: Dictamnus albus findet sich im
Pannonicum in geschlossenen Wäldern nur spärlich. Hier vermehrt
er sich größtenteils nur generativ und lebt mit vielen Q u e r-
c e t e a - Arten in dispersem Vorkommen. Wie ich selbst es beobachtet habe und auch durch Wurzel-Profil-Untersuchungen von
MANFRED FÖRSTER aus dem Main-Tauber-Gebiet belegt wurde, erscheint Dictamnus in sekundären Waldlichtungen in regelmäßigen
Sproß-Kolonien und bildet große Gruppen. THEO MÜLLER hat dies
gleichermaßen demonstriert. In der Süd-Zone der sowjetischen
Wald-Steppe ist im zentralen Verbreitungsgebiet das Vorkommen
von Peucedanum officinale charakteristisch. Dieselbe Pflanze
entwickelt in den Lichtungen der Alkali-Steppen des pannonischen
Tieflandes wegen der Einengung der für sie günstigen ökologischen Verhältnisse riesige Sproß-Kolonien. Auch in den bisher
beschriebenen Saum- und Mantel-Gesellschaften kommt Polykormie
häufig vor, welche die hervorragende Dominanz eines Individuums
zeigt. . Die Polykormus-Bildung liefert eine physiognomische Dominanz, täuscht eine große Vitalität vor und beweist damit aber
auch für sie günstige, synökologische Umstände. Sie zeigt, daß
die Pflanze in eine extreme ökologische Lage gekommen ist, wo
das Konkurrenz-Gleichgewicht der konsoziierenden Pflanzen auf-
gelöst ist.
Zuletzt möchte ich die Parallele zwischen Geselligkeitsgrad, Soziabilität (PFEIFFER) und unseren Feststellungen bezüglich der
Sproßkolonien-Bildung hervorheben. PFEIFFER hat ausgeführt, daß
die Geselligkeit sehr stark vom Wettbewerb, sowie dem Entwicklungszustand der Gesellschaft abhängt und meistens unter extremen Lebensbedingungen zunimmt. Unsere eigenen Untersuchungen
der Jahres-Rhytmen der Polykormie, ferner ihres Verhaltens bei
Entziehung verschiedener ökologischer Faktoren stehen vor ihrem
Abschluß.
Bei der Abhandlung über Sproß-Kolonien-Bildung möchte ich auf .
das ökologische Extrem bzw. dessen soziologischen Wert aufmerksam machen. Unserer Meinung nach sollten diese Erscheinungen
bei der Wertung als Verbands-, Ordnungs- oder Klassen-Kennarten

als negatives Merkmal betrachtet werden.

A.NOIRFALISE:

Je n'ai pas étudié en Europe les associations des lisières (die
Mantel- und Saum-Gesellschaften) mais j'ai beaucoup observé ce
phénomène dans les régions tropicales, et notamment dans les
grandes zones forestières qui sont en Afrique, la forêt équatoriale d'une part et d'autre part de la forêt tropicale sèche.
Dans la forêt équatoriale il y a deux grands types de formation, avec des associations régionales, la forêt que l'on appelle
homogène et la forêt que l'on appelle hétérogène. La forêt homogène est une forêt comme la hétraie avec un couvert tout à
fait régulier extrèmement dense. Lorsqu'on traverse cette forêt
et qu'on arrive à une lisière, même si cette lisière n'est qu'un
très petit chemin, 3-4 m de large, on arrive à une Saum-Gesellschaft, qui est extraordinairement différente de tout le reste
et qui est caractérisée par des monocotylées surtout, Gingibé-
racées, Marantacées etc. Il y a donc là une difference très
nette entre la flore qui se trouve dans l'intérieur de la forêt
et de la flore de la Saum-Gesellschaft et de la Mantel-Gesellschaft. La Mantel-Gesellschaft ce sont surtout de lianes et la
Saum-Gesellschaft des monocotylées.
Lorsque maintenant on se trouve dans la forêt hétérogène, qui
est une forêt aussi fermée et aussi dense, mais hétérogène, avec
20,30 ou 40% d'espèces qui perdent leurs feuilles chaque année,
caducifoliées tropophiles, et avec un couvert tout à fait irré-
gulier, là la distinction devient beaucoup plus difficile parce
que les espèces du Saum se trouvent aussi dans la forêt. Partout
il y a un peu plus de lumière qui arrive au sol. En moyenne dans
la forêt dense équatoriale homogène il y a 1% de luminosité et
dans la forêt hétérogène 3 à 4%, et cela suffit pour avoir tout
de suite des pénétrations de la Saum-Gesellschaft dans la forêt
elle-même.Pour le phytosociologue qui travaille dans la forêt
homogène il y a une distinction très nette entre les deux formations,mais pour le phytosociologue qui travaille dans la forêt hétérogène,les espèces du Saum font partie de la forêt. Il
faut alors, ce qu'est très difficile en raison de la complexi-
té de la flore tropicale, des études innombrables pour faire
l'apart des deux catégories,si l'on n'avait pas comme référence
la forêt équatoriale homogène.

Maintenant je veux passer à une autre zone de végétation, parce
que là le phénomène se rapproche un peu de ce dont Monsieur JA-
KUCS a parlé. Ce sont les forêts tropicales sèches. Ce sont des
forêts de 10 ou 15 m de haut, avec seulement un étage (einstu-
fige Wälder),mais avec des cimes qui sont étalées, Bradystegia,
Isoberlinia etc.,des légumineuses. La luminosité dans cette forêt d'après les mesures que j'ai pu faire est de l'ordre de 10
à 15%. Donc ce n'est pas du tout sombre. Il y a beaucoup de lumière. Là aussi il y a une différence extrêmement nette entre
la végétation du sousbois et la végétation de la lisière et il
y a donc une Mantel-Gesellschaft et une Saum-Gesellschaft qui
sont tout à fait différentes.Ce qu'il y a de curieux c'est que la
Mantel-Gesellschaft et la Saum-Gesellschaft sont formées par
les plantes que l'on trouve dans la savane. La végétation de la
savane est une végétation de Mantel- und Saum-Gesellschaften. A
l'intérieur de la forêt la flore, qui est la flore herbacée ou
la flore buissonante, est totalement différente et cela peut
être verifié sur des centaines de kilomètres.Bien que les forêts
sèches tropicales soient parcourrues par les feus, les savanes
aussi, la forêt tropicale sèche et la savane ont une flore sensiblement différente. Donc là aussi ce phénomène s'observe mais
il disparait à partir du moment où la forêt tropicale sèche devient un "Obstbaumwald", c.à.d. où les arbres sont minces et
isolés les uns des autres, là il y a plus de distinction entre
la flore de la savane et la flore de la forêt. C'est la même
chose, et on les appelle non plus des "Wälder" mais des savanes
boisées.
Encore maintenant pour les forêts alluviales dans les régions
équatoriales. Dans les vallées les rivières et les fleuves sont
sauvages et par conséquent les forêts alluviales sont extrêmement hétérogènes. Il est très difficile, de faire un relevé sur
une surface minimum.Et là les vallés et les forêts alluviales sont
parcourrues par des chenaux et tout lelong des chenaux il y a
un Saum, et derrière le chenal, pour autant de 50 m de forêt,
il y a beaucoup moins d'espèces que lelong les chenaux. Là aussi
donc il y a une différence entre Saum- und Mantel-Gesellschaft
et la forêt proprement dite, la forêt qui comporte très peu
d'espèces d'après ce que j'ai pu voire, très peu d'espèces
d'herbacées et même très peu lianes.Là aussi il y a une différence, mais en raison de la structure extrêmement compliquée

il est impossible d'exprimer cela par des relevés, c.à.d. où il
y a des étendues homogènes, ce qui est relativement rare dans
ces vallées. Ces forêts de vallée sont toujours extrêmement hétérogènes et il y a souvent de petites clairières naturelles,là
où les eaux restent p.e. 15 jours, cela est une matière naturel-
le que les eaux, les fleuves varient en façon énormes, avec des
crues de 7,8,10 m,p.e. dans le fleuve Congo; là où il y a de
zones d'expansion des eaux jusque pendant 15 jours, il y a des
associations totalement différentes de celle qui se trouve à
côté exactement sous les arbres. Donc ce phénomène de Saum- et
Mantelgesellschaft se retrouve en Afrique, mais il devient in-
existant lorsqu'on se trouve dans l'état limite des "Obstbaum-
Wälder".

R.TÜXEN:

Um ein Fazit zu ziehen, damit wir auch sehen, daß sich die Reise von Herrn JAKUCS gelohnt hat, möchte ich noch ein Wort sagen.
Ich glaube, daß die Aufnahme-Methodik besonders in den älteren
Arbeiten, sei es bei Ihnen in Ungarn, sei es bei uns,nicht ganz
dieselbe ist, wie heute bei uns. Man kann also wahrscheinlich
ältere Aufnahmen weder in Ungarn noch hier mit derselben Sicherheit verwenden wie die neueren.
Andererseits haben mich Ihre Ausführungen überzeugt, Herr JAKUCS
daß die thermophilen Pflanzen unserer Säume, der T r i f o l i o
G e r a n i e t e a, in Ihrem Lande wirklich in Wäldern wachsen
können. Das Verhalten dieser Arten in verschiedenen Synökosyste-
men ist vielleicht verschieden.
Wir dürfen andererseits nicht überhören, (vergl.p.191) daß Geranium sanguineum und soziologisch ihm nahestehende Arten auch
in Ost-Europa auf Lichtungen ihre Polykormie entfalten. Sind
das nicht gerade unsere Säume?
Aber bitte, hören Sie noch ein Drittes. Als wir hier auf Grund
von SISSINGH's Anregungen die Ackerunkraut-Gesellschaften studierten, da haben wir eine Ordnung V i o l e t a l i a   a r-
v e n s i s gehabt. Die bestand aus den Gesellschaften der Hack
früchte (Kartoffeln und Rüben) einerseits, und der Halmfrüchte
(Winter-Roggen und -Weizen) andererseits. In dieser Ordnung wurden Arten vereinigt, die im Mediterrangebiet zwei Klassen, den
S e c a l i n e t e a und den C h e n o p o d i e t e a   a l-
b a e angehören, die jedoch bei uns nicht zu trennen sind in·

folge der Rotation, so nennen wir den wechselnden Anbau von Som-
mer- und Winterfrucht. Aus anthropogenen Gründen sind hier also
zwei Klassen so miteinander verschmolzen, daß man hier nie auf
den Gedanken hätte kommen können, zwei Klassen aus diesem sehr
homogen gemischten Gesellschaften zu machen. Und dennoch gibt
es z.B. in Spanien gewaltige Gebiete, wo seit 100 Jahren oder
mehr nur Weizen kultiviert wird, wo reine S e c a l i n e t e a-
Gesellschaften wachsen, und es finden sich in Süd-Frankreich
oder in Italien oder auch in Spanien große Gebiete, wo nur Wein-
bau getrieben wird, wo im Sommer gehackt wird und nie S e c a -
l i n e t e a-Arten anzutreffen sind, und wo es nur C h e n o-
p o d i e t e a-Gesellschaften gibt. Dort sind zwei Klassen
scharf getrennt, und niemand zweifelt daran, daß sie dort berechtigt sind.
Bei uns sind T r i f o l i o - G e r a n i e t e a , R h a m -
n o - P r u n e t e a und Wald-Gesellschaften als Klassen getrennt. Es gibt Gebiete wo sich ihre Arten anders verhalten.
Dort kann man diese Trennung nicht vornehmen. Das spricht aber
nicht gegen die Existenz dieser Klassen, besonders wenn wir die
Wahl der Aufnahme-Flächen bedenken.

AKTUELLE UND POTENTIELLE GRENZEN DES LATSCHENGÜRTELS IM QUELLGEBIET DES LECH (VORARLBERG)

Otti Wilmanns, J. Ebert

Zu den eindrucksvollsten Vegetationsgrenzen gehören jene,welche
die Höhenstufen der Gebirge voneinander absetzen: Waldgrenzen,
Baumgrenzen, Krummholzgrenzen und Grenzen der zwergstrauchrei-
chen Gebiete. Sie sind hier physiognomisch auffällig und daher
seit langem bis in die jüngste Zeit oft diskutiert worden (5,16).
Unsere eigene Arbeit wurde zunächst einfach durch häufige Exkursionen und Wanderungen von Studenten und Alpinisten im Raum
um die Freiburger Hütte veranlaßt. Beim Aufstieg schließt sich
dort an den oft stark durchweideten Fichtenwald eine Zone sehr
lockeren Baumwuchses mit etwas Krummholz an; weiter gelangt man
durch Krummholz von Pinus mugo, auch dieses oft stark aufgelich-
tet und von Rhododendron-Beständen durchsetzt, in die Stufe der Ra-
sen- und Schuttgesellschaften. Beim Vergleich unserer Beobachtungen mit der Literatur und verschiedener Arbeiten untereinander (2,4,5,6,8,10,13,14,15) ergaben sich indessen über einige
Punkte Unklarheiten oder gar kontroverse Darstellungen, die sich
kurz in zwei Fragen fassen lassen: 1. Fällt die natürliche, klimatische Wald-Grenze (im üblichen Sinne) mit der Baum-Grenze (im
üblichen Sinne) zusammen, ist der oft beobachtbare Gürtel zwischen ihnen beiden also wirtschaftsbedingt oder nicht? 2. Gibt
es einen eigenen, natürlichen Zwergstrauch-Gürtel, der durch die
Dominanz von Ericaceen,vor allem Rhododendren, gekennzeichnet
ist? Das im wesentlichen von J.EBERT im Rahmen seiner Staats-
examensarbeit erhobene Material sei daher unter folgenden Aspek-
ten betrachtet: Wo liegen <u>aktuelle</u> Grenzen des Latschengürtels?
Wo liegen <u>potentielle</u> Grenzen, wobei also vom menschlichen Einfluß abstrahiert wird? Und wo liegen klimatische Grenzen, wobei
außerdem von der Auswirkung lokaler orographischer Besonderheiten wie Schutthalden und Steilwänden abgesehen wird?
Das Gebiet, für welches unsere Schlüsse gelten sollen, liegt in
den Klostertaler Alpen südlich von Bregenzer Wald und Allgäu,
nördlich der Straße Bludenz-Arlbergpaß. Geologisch bildet es,
aufgebaut aus Schichten der ober-ostalpinen Lechtal-Decke, ein
Stück Südrand der Nördlichen Kalk-Alpen. Die größte geschlossene Ausdehnung besitzen Hauptdolomit und Plattenkalk des Nor und
ihre Schuttmassen; und auch wir beschränken uns auf karbonati-

tisches Muttergestein. Sich dort, wo Schiefer oder Mergel anste-
hen, an Fragen der Gehölz-Grenzen zu wagen, ist wenig erfolgver-
sprechend; dazu sind siese günstigen, frischen Standorte seit
allzu langer Zeit der mit intensivem Weidebetrieb verknüpften
Rodung unterworfen. Außerdem bildet hier die Grünerle Krummholz,
und die stärker als die Fichte bedrohte, aber höher steigende
Arve greift schon von den Zentral-Alpen herein. Sie weist auf
entsprechende klimatische Anklänge hin: Die Niederschläge liegen im Gebiet bei rund 2 m/Jahr, ebenso die mittlere Schnee-
höhe.
Die Schwierigkeit, von menschlichen Eingriffen zu abstrahieren,
ist in unserm Raum nicht geringer als in anderen Teilen der
Kalk-Alpen; schon für das 14.Jahrhundert ist die Einwanderung
von Walsern in Vorarlberg nachgewiesen, welche oberhalb der von
Rätoromanen besiedelten Täler als Handelsleute ihre Saumpfade
anlegten und als waldfeindliche Viehzüchter von oben her talwärts rodeten (9).Das Ausmaß der Vegetations-Veränderung auf
feinerdereichen Böden und die Bodenerosion mit Rasenabschälung
durch Viehtritt und Kammeisbildung auf den Blößen zeigt, daß
der menschliche Einfluß selbst in der subalpinen und alpinen
Stufe dem in den Tieflagen stellenweise nicht nachsteht.
Versuchen wir zuerst, die aktuellen Grenzen und Stufen nach
rein physiognomischen Kriterien festzulegen. Wir beginnen mit
dem gipfelwärtigen <u>Ausklang von Pinus mugo</u>. In dem auf der Karte dargestellten Gelände ließen sich 7 Stellen auffinden, wo es
sich offenbar auch um die klimatische Obergrenze handelt (Beispiele: In den Bänken, am Mehlsack); sie lag dabei fast stets
bei 2150 m (einmal, in SW-Exposition, bei 2180 m).(Bei allen
Höhen-Angaben ist mit einer Abweichung von ±10 m infolge Unge-
nauigkeit des Lufft-Höhenmessers und Luftdruck-Schwankungen zu
rechnen). Es ist wohl unnötig, näher auszuführen, daß die aktuellen Grenzen nicht geschlossen durchlaufen, sondern zerrissen
sind. Die Latschen haben hier eine Höhe von etwa 1/2 m, müssen
nach stichprobenhaften Jahrring-Zählungen aber dennoch mehrere
Jahrzehnte alt sein; ich fand an einer Stelle in dieser Höhe an
einem Ast 1/2 mm Durchmesser-Zuwachs pro Jahr, bei 2100 m waren
es 2 mm. Ganz vereinzelt lassen sich, selbst noch in 2275 m Höhe, Pinus-Krüppel beobachten. Von tiefer stehenden, charakteristisch leewärts niederliegenden Kampfformen darf man sich nicht
täuschen lassen und sie etwa als einzelnstehende Vorposten an-

sprechen; schon die dicken Tangelhumus-Reste,welche ihr Astwerk
vor Abtragung geschützt hat, erweisen ihre Relikt-Natur. In ihrem Schutz findet man auch verschiedentlich noch Jungfichten.

In tieferen Lagen erreichen die <u>Latschen</u> bald <u>2 m</u>; dieser Bereich
liegt, wie unser Aufnahmematerial zeigt, bei 2000 m NN. Unter
noch günstigeren Bedingungen schließlich werden sie 3 - 4 m hoch
und übertreffen damit bereits die mittlere Schneehöhe des Gebietes. Bei 2000 m liegt eine weitere physiognomische Grenze: Die
Fichte macht sich nun deutlich bemerkbar, indem sie nicht nur
im Krummholz verborgen vegetiert, sondern es stellenweise baumförmig überragt.- Das charakteristische "getupfte" Bild dieser
Stufe ist nicht völlig zufallsbestimmt; die jahrhundertelange
Rodung zur Gewinnung von Weide und Brennholz war vielmehr deutlich selektiv: die frischeren Senken und ebenen Flächen sind
krummholzfrei; geringer war der Anreiz zum Hieb an trockenen
Hängen mit minderem Graswuchs; an Steilkanten aber mußte ein
Latschensaum das Vieh vor dem Absturz schützen.
Der Beginn von <u>Fichtenwäldern</u> bewirkt die nächste physiognomische Gürtelobergrenze: die aktuelle obere Fichtenwaldgrenze erreicht im Gebiet rund <u>1850</u> m und steigt lechabwärts um 20 - 30 m
an. Beobachtbar ist sie an 5 Stellen (z.B. südlich der Freiburger Hütte und um die Tannleger Alpe). Dies ist zu wenig, um et-
waige Höhendifferenzen zwischen den Expositionen sichern zu können. Die aktuelle Grenze des Fichtenwaldes selbst ist ebenso wie
die der Latschenbestände aus orographischen und anthropogenen
Gründen zerrissen: Krummholz greift noch weit in diesen Gürtel
hinein, ja stellenweise noch bis in die Laubwaldstufe hinunter.

Gehen wir von den teilweise zeitbedingten aktuellen und physiognomischen Grenzen über zur Frage nach den potentiellen Grenzen der Vegetation. Nach dem bisher Gesagten ist folgende Gliederung wahrscheinlich: bis 1850 m ist die regionale potentiell
natürliche Vegetation Fichtenwald und zwar, wie unsere Aufnahmen zeigen, das P i c e e t u m   s u b a l p i n u m; darüber
folgen bis 2000 m regional Latschen-Bestände mit eingesprengter,
das Krummholz überragender Fichte, soziologisch also das R h o-
d o d e n d r o - M u g e t u m = E r i c o - P i n e t u m,
Schneeheide-Knieholz, und, nur kleinflächig, das R h o d o -
d e n d r o - V a c c i n i e t u m; darüber bis 2150 m das Gleiche mit niederwüchsigen Fichtenkrüppeln; oberhalb der 2150 m beginnt das natürliche Reich alpiner Rasengesellschaften, auf Kar-

bonat-Gestein meist des C a r i c e t u m f i r m a e,des Pol-
sterseggen-Rasens.

Lassen sich nun auch die soziologischen Kriterien, also Kriterien der Gesellschafts-Qualität und solche der Konkurrenz zwischen Fichte und Latsche, für diese Grenzen finden?- Im Bereich
unterhalb 1850 m läßt die Wüchsigkeit der Fichte im Piceetum
subalpinum keinen Zweifel an ihrer dominierenden Rolle, auch
wenn in den fast überall stark durchweideten Beständen nicht
nur krautige Weidezeiger aus den Rasen-Gesellschaften, sondern
an lichten Stellen so gar Latschen auftreten. (s.Tab.) Im Bereich episodisch befahrener Lawinenbahnen des Lechtals, die an
den dauernd umgelagerten Schutt der kombinierten Wildbach- und
Lawinenrinnen seitlich anschließen, ist jedoch das R h o d o -
d e n d r o - M u g e t u m aktuelle und potentiell natürliche
Lokalgesellschaft, also Dauergesellschaft. Hier hält sich das
P i c e e t u m nicht. Im einzelnen ist freilich schwer abzugrenzen, wo es wirklich nur aus natürlichen Grenzen Halt macht.-
Schließlich kann das R h o d o d e n d r o - M u g e t u m
auch noch kurzfristig natürliches Sukkzessionsglied sein: auf
ruhendem, nicht mehr überflutetem Schotter war bei 1550 m NN
diese Entwicklung sehr klar zu erschließen: Zuerst stellt sich ein
C a r i c e t u m f i r m a e mit gut entwickelter Kryptogamen-
Schicht ein; auch Kiefer fliegt früh an. Die Humusbildung geht
rasch vor sich, vor allem durch Erica carnea. Dadurch siedeln
sich im Schutz der Sträucher bald Rasen von Waldboden-Moosen wie
Hylocomium splendens und Rhytiadiadelphus triquetrus an.Dies ist
der bevorzugte Keimungsort für die Fichte; man findet sie prak-
tisch nie ohne Pinus mugo, während das Umgekehrte oft der Fall
ist. Unter schätzungsweise 40- bis 60- jährigen Bäumen wachsen
auf dem mittlerweile 15 - 20 cm mächtigen Tangel-Humus reichlich
Vaccinien.- Das R h o d o d e n d r o - M u g e t u m erscheint
auch als Folgegesellschaft des T h l a s p i e t u m r o t u n-
d i f o l i i auf Hangschutt. Diese Beobachtungen stimmen recht
gut mit denen von LÜDI(10) und MAYER et al.(11,12) überein. Letztere konnten kürzlich (12) sogar Datierungen ganz ähnlicher Be-
siedlungsstadien in den Berchtesgadener Alpen durchführen.- Daß
das Schneeheide-Knieholz endlich als anthropogene Gesellschaft
auch in der Fichtenstufe bei Schlag und extensiver Beweidung
auftreten kann, ergibt sich schon aus dem unmittelbaren Kontakt
zum P i c e e t u m.

Vor der Besprechung des eigentlichen Krummholzgürtels sei kurz
auf seine Obergrenze eingegangen. Daß die beobachtete aktuelle
Maximal-Höhe gleich der potentiellen ist, läßt sich auch soziologisch begründen: Oberhalb 2150 m fehlen die an sich ja häufigen Charakterarten des Rhododendro-Mugetum fast ganz: Rhododendron hirsutum, Sorbus chamaemespilus, Daphne striata und Erica
carnea sind dort sehr selten und niederwüchsig, verhalten sich
also wie Pinus mugo. Dagegen sind sie regelmäßig in den Kalk-
stein-Rasen darunter zu finden: man kann sie hier als syngene-
tische Differential-Arten von S e s l e r i e t a l i a -Gesell-
schaften werten, diese wiederum sind Ersatzgesellschaften des
Schneeheide-Knieholzes. Ein Aufnahme-Beispiel stehe für viele:
C a r i c e t u m   f i r m a e. 2000 m NN; 20°SW, getreppt;
Hauptdolomit. Aufnahmefläche 6 qm; Vegetationsdeckung 85%.
Kennarten der E l y n o - S e s l e r i e t e a:
Carex sempervirens 2a, Carex firma 2m, Globularia cordifolia 2m,
Alchemilla coniuncta 1, Androsace chamaejasme 1, Anthyllis vul-
nearia ssp.alpestris 1, Crepis alpestris 1, Dryas octopetala 1,
Galium anisophyllum 1, Gentiana clusii 1, Helianthemum alpestre
1, Phyteuma orbiculare 1, Sesleria varia 1, Pedicularis rostra-
to-capitata +, Festuca pumila r, Hieracium bifidium r, Hieracium villosum r, Sedum atratum r;
Kennarten der E r i c o - P i n e t e a:
Daphne striata 2m, Polygala chamaebuxus 1, Erica carnea +, Rhododendron hirsutum +, Gymnadenia odoratissima r;
Sonstige:
Bellidiastrum michelii 1, Campanula scheuchzeri 1, Chrysanthe-
mum leucanthemum fo.1, Cladonia-Primärthallus 1, Hippocrepis
comosa 1, Thymus polytrichus 1, Tortella tortuosa 1, Carlina
acaulis +, Euphrasia minima r, Gentiana campestris r°, Homogyne
alpina r, Pinus mugo jg. r, Poa alpina r, Primula auricula r.
Je nach Weide-Intensität gibt es alle Übergänge vom latschen-
freien R h o d o d e n d r o - M u g e t u m bis zu Rasengesellschaften, welche ihre Herkunft noch durch ihre Arten-Kombination verraten oder gar - als P o i o n   a l p i n a e - den
ähnlich unspezifischen Fettweiden der Tieflagen korrespondieren.
Die Rhododendron hirsutum-Bestände sind jedenfalls als Fragment-
Gesellschaften aufzufassen; natürlich im soziologischen Sinne
sind sie allenfalls kleinflächig am Rande von Lawinenbahnen.
Sie bilden aber im Gebiet auf Karbonat-Gestein sicher keinen

eigenen Gürtel oberhalb des Krummholzes. Dies ist die Antwort
auf unsere Frage 2 der Einleitung.
Läßt sich die 2000 m-Grenze, die Obergrenze jener Stufe, in der
die Fichte noch, wenn auch selten, baumförmig vorkommt und die
Latschen wenigstens überragt, soziologisch untermauern?- Die
beigegebene Stetigkeitstabelle ist gekürzt; nicht aufgeführt
sind die Moos- Flechten-Synusien; denn Fichtenwald wie Krummholz bieten so viele Habitate und sind daher so kryptogamenreich,
daß man leicht 25 - 30 Arten pro Aufnahme erhält; so haben wir
diese einer eigenen Studie vorbehalten.
Zwei Assoziationen werden von Latschen beherrscht: Das R h o -
d o d e n d r o - V a c c i n i e t u m,als Bodensaures Alpenro-
sen-Gebüsch in den Zentralalpen weit verbreitet, kommt kleinflächig auch über Karbonat-Gestein vor, wenn der Standort boden-
frisch ist und eine genügend mächtige Humusschicht für ein oberflächlich saures Substrat sorgt. Zur näheren Diskussion steht
hier aber nur das R h o d o d e n d r o - M u g e t u m, das
Schneeheide-Knieholz, mit einer typischen, überwiegend südexpo-
nierten, und einer frischeren, meist nordexponierten Subassoziation, nach Rhododendron ferrugineum benannt, mit stärkerer Hu-
mus-Auflage; diese Einheit steht dem R h o d o d e n d r o -
V a c c i n i e t u m nahe, enthält aber die E r i c o - P i-
n i o n-Kennarten noch mit guter Stetigkeit. Hingewiesen sei
auf das allgemein reichliche Vorkommen von V a c c i n i o -
P i c e e t e a-Kennarten; es belegt die Flügelstellung des
R h o d o d e n d r o - M u g e t u m innerhalb der neuen Klasse E r i c o - P i n e t e a (wenn man nicht überhaupt vorzieht,
es bei den Fichtenwäldern und Beerstrauch-Gestrüppen zubelassen).
Für unsere Fragestellung wichtig ist die Tatsache, daß man in
allen Latschengesellschaften zwanglos Hoch- und Tieflagen-Formen
herausschälen kann; besonders letztere sind durch eine Fülle
von Arten, die freilich wenig stet sind, differenziert (s.Tab.).
Der Tabellenkopf aber zeigt, daß die kritische Grenzzone zwischen ihnen bei rund 2000 m liegt, die Überschneidungen sind
ganz gering.
Wie verhält sich die Fichte? In beiden Formen des R h o d o -
d e n d r o - M u g e t u m verjüngt sie sich, wenn auch nur
vereinzelt. In der Hochlagenform ist sie jedoch keinesfalls in
der Lage, als Konkurrent der Latsche eine nennenswerte Rolle zu
spielen. Das R h o d o d e n d r o - M u g e t u m ist dort

unzweifelbare regionale potentiell natürliche Vegetation.-Es
bleibt noch der Gürtel zwischen 1850 und 2000 m zu besprechen.
Die Verjüngung der Fichte ist hier, verglichen mit Gebieten voller Vitalität, sehr gering, vor allem an trockenen Standorten.
An frischen, z.B. im N a r d i o n, sieht man auch freien Jungwuchs; an trockenen sind die Jungpflanzen dagegen auf den Humus
vorwüchsiger Zwergsträucher angewiesen. Im geschlossenen R h o-
d o d e n d r o - M u g e t u m jedoch ist die Feld- und Bodenschicht so dicht, daß die spärlichen Keimlinge starker Konkurrenz begegnen. Der Schluß auf das R h o d o d e n d r o - M u-
g e t u m ist also auch hier so gut wie sicher.
Für andere geologische Substrate als Karbonat-Gestein sollte
man freilich in geeigneten Gebieten noch Beobachtungen sammeln;
denn da muß die Wettbewerbssituation etwas verschoben sein.Dies
möge eine kleine Beobachtung beleuchten: Auf geologisch "bunter"
Moräne oberhalb des Formarin-Sees beobachtete ich um 1950 m NN
eigenartig dicht stehende Gruppen von Fichten, umgeben von Latschen und N a r d i o n; ungewöhnlich nahe bei den älteren
Stämmen, noch im Schatten ihrer Kronen, wuchsen kräftige, jünge-
re Pflanzen; die unteren Äste der älteren Bäume waren stark zu
Boden geneigt. Es erinnerte mich an ein Bild, wie es uns in Süd-
Norwegen 1965 an der Baumgrenze vorgeführt wurde: Vegetative
Fortpflanzung der Fichte. In der Tat: auch hier in den Alpen be-
wurzeln sich die früh und anhaltend vom Schnee zu Boden gedrück-
ten Äste, so daß ganze Gruppen von Fichten-Individuen aus einem
einzigen Keimling entstehen können.
Nun zuletzt zur ersten Frage, der Natürlichkeit separater Wald-
und Baumgrenzen.- ELLENBERG (5) berichtete über Baumgrenzen in
tropischen Gebirgen unter ungestörten Verhältnissen; dort fallen Wald- und Baumgrenze zusammen, d.h. ein geschlossener, zunächst hochstämmiger Wald geht allmählich ohne Gruppenbildung
in krüppelwüchsigen "Krummholzwald" über und endet schließlich
abrupt in geschlossener Front; oberhalb derselben wachsen nur
noch niedere, unter der Schneedecke geschützte Büsche. ELLENBERG
stellte überzeugend dar, daß es sich bei den Waldgrenzen um absolute, d.h. nicht konkurrenzbedingte Grenzen handle und daher
bei gleichmäßigen Gelände dort, wo ein einziger Baum stehe,auch
mehrere wachsen könnten. Wie läßt sich dies mit unseren Ergebnissen in Einklang bringen? Der Schlüssel liegt in einer Fußnote
jener Arbeit: Ökologisch, wenn auch nicht unbedingt morpholo-

gisch, sei als "Baum" ein Holzgewächs dann zu bezeichnen, wenn
es die normale Schneehöhe des Gebietes überrage.Nach dieser zweifellos sinnvollen Definition ist also unser R h o d o d e n -
d r o - M u g e t u m der tieferen Lagen als "Wald" zu bezeichnen; dann wird die Waldgrenze in den Lechtaler Alpen auf Karbonat-Gestein nicht von der Fichte, sondern von der Latsche gebildet; dann liegt sie recht hoch, bei 2000 m. Gewiß: um hochstäm-
migen, Bauholz-liefernden Wald handelt es sich nicht; aber die
Bildung einer geschlossenen Gehölz-Synusie, damit verbunden das
Eindringen von Waldpflanzen, auch die reiche Aufgliederung in
Habitate haben diese Tieflagen-Formen mit dem angrenzenden Fichtenwald gemeinsam.
Der Gedanke, den (allerdings wohl gesamten,nicht nur unteren)Lat-
schen-Bereich als Wald-Bereich aufzufassen, ist nicht neu, wenn
auch wenig beherzigt. GRADMANN (7) wies in seiner so klaren Ge-
dankenführung darauf hin, daß es willkürlich und anthropozen-
trisch sei, just 4-5 m Höhe von einem "Baum" zu verlangen; zu
plausiblen Höhen-Vergleichen zwischen den einzelnen Alpenteilen
komme man nur dann, wenn man die nordalpine Krummholzgrenze mit
der zentralen Lärchen- und Arvengrenze vergleiche. In der Tat
fügen sich unsere 2000 m dann sehr gut ein. Für das Allgäu kommt
OBERDORFER (13) auf eine Alpenrosen-Grenze von 2050 m NN; das
ergibt (wenn man ebenso wie für die Lechtaler Alpen 150 m ab-
zieht) für die dortige Waldgrenze in unserm Sinne eine Höhe von
1900 m NN; die Zentral-Alpen mit 2300 bis 2400 m NN schließen
sich dann sinnvoll an.
So läßt sich trotz der starken Veränderungen durch den wirt-
schaftenden Menschen auch in unserer alpinen Vegetation die allgemeine, weltweite Gesetzmäßigkeit in der Struktur der Waldgrenze aufweisen.

SUMMARY

We mapped the actual limits of Picea abies forests and stands of
Pinus mugo and found the following heights in the district: The
actual upper limit of the P i c e e t u m s u b a l p i n u m
at 1850 m, that of the R h o d o d e n d r o - M u g e t u m
with 2 m growing height and bigger spruce trees at 2000 m, the
actual upper limit of the R h o d o d e n d r o - M u g e t u m
without woody plants of more than 2 m lying at 2150 m.
These limits are phytosociological limits toc: Above 2150 m the

characteristic species of the R h o d o d e n d r o - M u g e-
t u m lack nearly, whilst they are frequent in the S e s l e-
r i e t a l i a - "Ersatzgesellschaften" downwards. As to the
R h o d o d e n d r o - M u g e t u m one can distinguish two
forms, one below, one above a line of about 2000 m.
Above 1850 m it is difficult for spruce to grow from seed on limestone and dolomite. On morain we observed vegetative reproduc-
tion by rooting of downbended twigs.
From an ecological viewpoint it is reasonable to draw the timber line and the tree line there, where the woody plants exceed
the middle snow cover, here in the northern Alps 2 m. Consequently it is Pinus mugo, which forms the timber line in this
district. One can show, that it forms a continuous front under
natural conditions like in tropical mountains and it is not dis-
solved into patches and single trees.

LITERATUR

1. AMPFERER,O., BENZINGER,Th.& REITHOFER,O. -1932- Geologische
      Karte der Lechtaler Alpen: Klostertaler Alpen; 1:25000.
      Wien.
2. CZELL,Anna, SCHIECHTL,H.M., STAUDER,S.& STERN,R. -1966-
      Erhaltung des Naturschutzgebietes "Großer Ahornboden"
      durch technische und biologische Maßnahmen.- Jb.Ver.
      Schutze Alpenpflanzen u.-tiere 31: 33-56. München.
3. Ebert,J. -1965- Die aktuellen und potentiellen Grenzen des
      Krummholzgürtels im Gebiet der Freiburger Hütte.- Staats-
      examensarbeit Freiburg i.Br.
4. EBLIN,B. -1901- Die Vegetationsgrenzen der Alpenrosen als
      unmittelbare Anhalte zur Festsetzung früherer bzw. mög-
      licher Waldgrenzen in den Alpen.- Schweiz.Z.Forstw.52:
      133-138, 157-162. Zürich.
5. ELLENBERG,H. -1966- Leben und Kampf an den Baumgrenzen der
      Erde.- Naturwiss.Rundschau 19: 133-139. Stuttgart.
6. FRIEDEL,H. -1956- Die alpine Vegetation des oberen Möllta-
      les (Hohe Tauern).Erläuterungen zur Vegetationskarte der
      Umgebung der Pasterze (Großglockner).- Wiss.Alpenvereins-
      hefte 16: 153 pp. Innsbruck.
7. GRADMANN,R. -1931- Süddeutschland 2.- Stuttgart.
8. LANDOLT,E. -1960- Unsere Alpenflora.- Zollikon-Zürich.
9. LIEHL,E. -1958- Vorarlberg - Exkursion des Geographischen
      Institus der Universität Freiburg i.Br. 1.-10.8.1958.
      (Mskr.der Teilnehmer-Berichte; Geogr.Inst.Freiburg).
10.LÜDI,W. -1921- Pflanzengesellschaften des Lauterbrunnenta-
      les und ihre Sukzession.- Beitr.geobot.Landesaufn.
      Schweiz 9: 350 pp. Bern.
11.MAYER,H., FELDNER,R.& GRÖBL,W. -1967- Montane Fichtenwäl-
      der auf Hauptdolomit im Naturschutzgebiet "Ammergauer
      Berge".- Jb.Ver.Schutze Alpenpflanzen und -tiere 32:20-
      43. München.
12.MAYER,H., SCHLESINGER,B.& THIELE,K. -1967- Dynamik der Wald-
      entstehung und Waldzerstörung auf den Dolomitschuttflä-

chen im Wimbachgries (Berchtesgadener Alpen).- Jb.Ver.
Schutze Alpenpflanzen und -tiere 32: 132-160. München.
13.OBERDORFER,E.   Beitrag zur Vegetationskunde des Allgäu.-
Beitr.naturkdl.Forsch.Südwestdeutschl.9: 29-98. Karlsruhe.
14.SCHARFETTER,R. -1938- Das Pflanzenleben der Ostalpen.- Wien.
15.SCHROETER,C. -1926- Das Pflanzenleben der Alpen. 2.Aufl.-
Zürich.
16.TRANQUILLINI,W. -1967- Über die physiologischen Ursachen
der Wald- und Baumgrenze.- Mitt.Forstl.Bundesvers.anst.
Wien 75: 457-487. Wien.

V.WESTHOFF:

Sie haben uns gezeigt, wie man eine natürliche Vegetationsgrenze erfassen kann, auch wenn man nicht über Komputer und sonstige Apparate verfügt, sondern vor allem seine Augen verwendet,
seinen common sense und circumstantial evidence. Wenn man alles
mit dem richtigen Verstand anwendet, wie Sie es aus langer und
guter Erfahrung im Gelände kennen, kann man damit sehr viel erreichen. Sie haben uns das ganz klar angezeigt. Ich möchte zur
Erweiterung des Baum-Begriffes - wenn ich so sagen darf - noch
ergänzen, daß die nordischen Forscher ungefähr dieselbe Auffassung vertreten. Wenn man den Birken-Gürtel der subarktischen
Zone betrachtet, wo, wie Herr Professor TÜXEN uns das gestern
so schön gezeigt hat, die Birke nicht die maritime, sondern die
subarktische Baumgrenze bildet, so sieht man, daß auch da die
Frage "Was ist ein Baum?" so beantwortet werden kann: Es ist die
Lebensform, die den Schnee überragen kann, und daß man mit dieser Definition besser auskommt, als mit der Angabe einer absolu-
ten Höhe.

R.TÜXEN:

Sie zeigten uns die Verjüngung oder die vegetative Vermehrung
der Fichte und erinnerten an ein Beispiel in Norwegen. In der
Lüneburger Heide zwischen Celle und Ülzen sind zahlreiche Fichten bekannt, die sich genau in der gleichen Weise benehmen, die
weite Zweige nach außen auf den Boden legen und vielfach dort
Wurzel schlagen.

M.WRABER:

Der Vergleich zwischen den Lechtaler und den südöstlichen Alpen (Julischen und Savinja-Alpen, Karawanken) ist interessant,
denn nach Südosten hin kommt es zu einer bedeutenden Grenz-Ver-
schiebung. Die obere Wald-Grenze wird dort nicht vom P i c e e-

t u m   s u b a l p i n u m, sondern vom A d e n o s t y l o -
P i c e e t u m gebildet. Die südöstlichen Alpen stehen unter
einem ziemlich starken Einfluß des mediterranen und ozeanischen
Klimas, und da stellt sich der subalpine Fichtenwald als Höhen-
Gürtel nicht mehr ein, sondern ein anderer Fichtenwald, der etwas mit Buche gemischt ist. Die obere Wald-Grenze verläuft in
der Höhe von 1600 - 1700 m, ist aber nicht klimatisch, sondern
menschlich bedingt. Man vermutet, daß sie durch die hundert- und
so gar tausendjährige Alp-Wirtschaft um 200 - 400 m herabgedrückt
wurde, und anstelle des Waldes hat sich die Latsche eingestellt,
natürlich mit anderen Strauch-Arten. Da kommen die verschiedenen
Alpen-Weiden vor, mit Arten wir Lonicera nigra, Salix appendicu-
lata, Salix glabra, S.waldsteiniana und viele andere. Die obere
Grenze des Latschen-Gürtels liegt bei 2100 - 2200 m. Fichten,
Lärchen und stellenweise sogar noch Buche kommen einzeln oder
horstweise nur im unteren Teil des Latschen-Gürtels (bis ungefähr 1900 - 2000 m) vor, höher oben gibt es nur reine Pinus mu-
go-Bestände.
Eine subalpine Grünland-Gesellschaft, die bei uns sehr oft vorkommt und sehr ausgebreitet ist, ist das sog. S e s l e r i o -
S e m p e r v i r e t u m, oder einfach S e m p e r v i r e -
t u m. Es ist festgestellt worden, daß diese subalpine Grasflur
nur in der ehemaligen Waldstufe vorkommt. Wenn wir aber da mit
einem C a r i c e t u m f i r m a e oder mit einem F i r -
m e t u m zu tun haben, ist das ein Zeichen, daß wir uns da
schon oberhalb der natürlichen oder klimatischen Wald-Grenze befinden.

Otti WILMANNS:
Eine derartige Höhen-Differenzierung innerhalb des S e s l e -
r i o n, also S e s l e r i o - S e m p e r v i r e t u m unten, C a r i c e t u m oben mit scharfem Ausschluß ist in den
Lechtaler Alpen sicherlich nicht durchführbar. Das C a r i c e-
t u m geht zweifellos in den Krummholz-Gürtel über, wohl nicht
in den Wald-Gürtel im klassischen Sinne, also unterhalb 1850 m.
Wie weit dort das S e s l e r i o - S e m p e r v i r e t u m
vorkommt, ist nicht sicher; denn nach unten hin kommt eben schon
C a r l i n o - S e m p e r v i r e t u m als eine doch sehr
viel wärmeliebendere Ausbildung. Zur Breite des Krummholz-Gür-
tels kann ich eigentlich nicht viel sagen. Dazu sind eben die

klimatischen Verhältnisse in Jugoslawien doch zu verschieden.

Man sollte zu einem Wald die Holz-Bestände rechnen, die die
mittlere Schnee-Höhe überragen, die also in der Lage. sind, der
Frost-Trocknis zu widerstehen. Das trifft nicht für den gesamten Latschen-Gürtel zu, oberhalb der 2000 m ist er etwa ständig
unter der mittleren Schnee-Decke und wird dann sukzessive kleiner, genauso wie es in der Skizze von ELLENBERG auch für die
tropischen Gebirge gezeigt wird.

WALD- UND BAUMGRENZE IN DEN VOGESEN

R. C a r b i e n e r

Zusammenfassung des französischen Vortrags.
I. Besprochenes Gebiet. Hauptkamm der südichen Hochvogesen
   ("Grande Crête"), höchster Teil des Gebirges und einziger,
   in welchem sich die Frage der klimatischen Waldgrenze stellt.
   Morphologie: sehr wichtig! Schmales (nur 0,2 - 1 km breites!)
   aber sehr ausgedehntes (40 km Länge) und massives, wenig ein-
   gekerbtes Kamm-Gebiet. Nördliche Hälfte Granit, NNO-SSW ori-
   entiert, südliche Hälfte Grauwacke, allmählich nach SO und
   schließlich mit einem W-O gerichteten Abschnitt mit dem Großen Belchen 1420 m steil zur Rheinebene abfallend. Mittlere
   Höhe des gesamten Kamm-Streifens 1250 m, mittlere Höhe der
   Hauptgipfel 1350 m so z.B. im zentral gelegenen Hohneck-Mas-
   siv und am süd-östlichsten Abschnitt (Belchen-Massiv). Mittlere Höhe der Paß-Scharten über 1200 m. An der Ostseite der
   nördlichen N-S gerichteten Abschnitte fast ununterbrochene
   Kette aneinandergereihter Kare. An der West-Flanke tief ein-
   geschnittene Gletscher-Tröge parallel zum Kamm (steile West-
   hänge!) mit Ausnahme des Hohneck-Massivs aus dem Ost-West ge-
   richtete Täler nach Westen ausstrahlen.
   Klima: Niederschlags- und Bewölkungs-Klima eu-ozeanisch:
   Über 2 m Gesamtniederschlag, stärkste Dezember-Niederschlag
   ganz West- und Mittel-Europas (300 mm). Winter-Semester/Som-
   mer-Semester = 1,3 (vergl. Feldberg: 0,8) starke Frostwech-
   sel-Häufigkeit (Naß-Schnee, Reif): nivometrischer Koeffizient im Winter-Halbjahr (Schnee/Regen) = 50% (vergl.Feldberg
   75%). Westwinde (meist stark) 7 Tage/10.
   Thermisches Klima subozeanisch: Juli-Mittel 11° (10,5° vor
   50 Jahren), Januar -4°.
II.Problematik.
   1. Ökologie der Wald-Grenze (W.G.) Wird die klimatische W.G.
   in den Vogesen erreicht? Alle älteren Autoren (1900 - 1940):
   nein (FLAHAULT, GAMS, AICHINGER, ISSLER 1924 usw.) Aktuelle
   W.G.sei im gesamten Kamm-Gebiet auf Weide-Betrieb zurückzuführen. Wenn W.G. lokal als natürlich anerkannt wird (ISS-
   LER 1942 Hohneck), dann wäre sie ursprünglich von "lichtungs-
   reichem" Buchen-Gebüsch, das die ganze Kamm-Fläche und die
   Kare bedeckt hätte, überlagert gewesen, und auf geomorpholo-

gische Prozesse in den Karen, oder "Wind-Wirkung" auf dem
Kamm (mechanische Prozesse oder winterliche Frost-Trocknis)
zurückzuführen.

2. Probleme der Unterscheidung zwischen Wald- und Baum-Grenze
(bzw.Krüppel-Grenze). Veränderungen der Morphologie der Wald-
und Baum- (Krüppel)-Grenzen je nach Sonnen- und Wind-Exposi-
tion.

3. Problem der West-Grenze der europäischen Coniferen (Picea
abies, Pinus montana). Wurde bisher besonders aus histori-
scher Sicht gedeutet.

4. Problem der Grenz-Verschiebungen während der Wärme-Zeit,
und des Überlebens der an größere waldfreie Gebiete gebundenen subalpinen und alpinen Floren-Elemente.

III. Eigene Stellungnahme (cf.CARBIENER 1962/1963)

1. Die klimatische (thermische) Wald-Grenze wird auf mehreren zusammenhängenden Kamm-Partien erreicht (z.B. Hohneck-
Massiv), ist aber relativ unabhängig von der Höhen-Lage. Sie
schwankt zwischen 1250 und 1370 m je nach Orientierung des
Kammes zur Haupt-Windrichtung, Exposition, Morphologie des
Gebietes, d.h. den Lokal-Klimaten. Die Vogesen bilden ein
Schulbeispiel einer maritimen klimatischen Waldgrenze eines
ausgesprochen i n s e l a r t i g e n Gebirges mit orogra-
phischem Barriere-Charakter im west-europäischen Haupt-Zir-
kulations-Gebiet der aus dem isländisch-schottischen Zyklo-
nentätigkeits-Zentrum gesteuerten allochtonen feuchten Luft-
massen. Begrenzender ökologischer Faktor = ungenügende Som-
mer-Temperaturen, wobei die physiologische Wertung einer
$10^{\circ}5 - 11^{\circ}$ Juli-Isotherme bei dem strahlungsarmen Nebel-Kli-
ma (3 mal mehr Nebel als in Reykjavik auf Island!) viel
schlechter ist als bei kontinentalem strahlungsreichen Klima
(was BROCKMANN JEROSCH 1919) gegen KÖPPEN (1927 recht gibt).
Ökologische Wirkung des Windes indirekt ,d.h.thermetisch.Mechani
sche Wirkung, winterliche Unbilden sehr untergeordnet. Frost-
Trocknis unbedeutend als begrenzter Faktor des Baumwuchses
und diesbezügliche Untersuchungen (z.B. MÜLLER-STOLL 1954 am
Feldberg) führen zu Fehlschlüssen. (Bisherige anthropomorphe
Überbewertung der winterlichen Faktoren: Märchen des fehlen-
den Wasser-Nachschubs aus gefrorenem Boden, u.a. noch verbreitete grobe Irrtümer). Morphologie und Physiologie des
Waldes (Sterilität, allmählicher Krüppel-Wuchs, Bodenfrucht-

barkeits-Unterschiede nur in der Krautschicht bemerkbar) für
thermische W.G.charakteristisch.
Verschiebungen der Höhenlage der W.G. erklären sich
a) durch Variationen des "Kamm-Effektes" = Ausmaß der Strom-
linien-Verdichtungen, der Düsen-Effekte, die eine relative
Anpassung der Isothermen an das Profil der Kamm-Linie bei
senkrecht zur Wind-Richtung liegenden Kamm-Partien bewirken.
Bei parallel verlaufenden, Ausbleiben des Kamm-Effekts.
b) durch Vorhandensein von w i n d - k a n a l i s i e -
r e n d e n T ä l e r n auf der West-Flanke (Herabsetzung
der W.G.) (cf. anemo-orographische Systeme nach JENIK). Gradient der Baumhöhen-Abnahme mit der Höhe stark von Besonderheiten des Lokal-Klimas der Hänge abhäufig: Beispiel laminier-
ter oder unterdrückter Höhenstufen.
2. Morphologie der W.G.
Auf West-Flanken bildet der Krüppel-Wald an den erhalten ge-
bliebenen natürlichen Grenz-Bereichen als 3 m hohes dichtes
Gebüsch eine geschlossene Grenz-Front mit Zusammenfallen von
Wald, Gebüsch und Krüppel-Grenze. An Süd-Flanken der Kare
D e p r e s s i o n der Wald-Grenze mit steppenartigem Charakter: Auflösung des Waldes in Busch-Gruppen inmitten sub-
alpiner Hochgras-Steppen (C a l a m a g r o s t i o n).
An Nord- und Ost-Hängen lichtungsreiche W.G. durch nivale
Prozesse (Schneewehen, Firn-Massen, Wächten) und tundraarti-
ge W.G.-Aspekte durch Mosaik von Krüppelwald und klimatischen
Zwergstrauchheiden.Künstlich herabgesetzte Wald-Grenzen überall
morphologisch leicht von den natürlichen unterscheidbar.
3. Die West-Grenze der boreo-alpinen Coniferen hat klimatische und nicht vegetationsgeschichtliche Ursachen.(cf. FIRBAS
1949, 1952, CARBIENER 1962,1963,ELLENBERG 1963.)Argumente:Fehl-
schlagen von großen wiederholten Anpflanzungen im Zentral-Massiv;
maritime Fichten-Grenze in Norwegen; Allgemeinheit der Grenz-
Bildung durch Fagus (z.T. mit Sorbus aucuparia und Acer pseu-
doph.) in Westeuropa (Zentralmassiv, West-Pyrenäen, Korsika,
Apennin) oder durch morphologisch und biologisch sehr ähnliche Laubholz-Gesellschaften anderer maritimer Gebirge in NW-
Europa (Betula pubescens) Japan (Betula ermani).Chile (Notho-
fagus). Ursache: Komplex; vielleicht Bedürfnis kontinentaler
Thermo-Periodizität und schlechtes Fortkommen der im ozeanischen Naß-Schnee begrabenen Jungbäume (Pinus montana stirbt

in ozeanischen Schneewehen immer ab! NEGRE 1932).

4. Die Waldgrenz-Verschiebungen in der Wärme-Zeit waren sehr
gering. Dieser Befund gilt für alle ozeanischen Gebirge West-
Europas und ist den verschieden starken Verschiebungen der
Nadelwald-Grenze in kontinentalen Gebieten gegenüberzustel-
len. Beweise:

a) Bodenkunde: jahrtausend alte Humus-Silikatböden direkt
über der natürlichen heutigen W.G., Böden die sich von denen
der künstlich entwaldeten Teile, höchsten 1 Jahrtausend alten Weideflächen (Podsolige Böden oder Braunerden mit dünner
Rohhumus-Auflage, d.h. Konservierung der nur ganz oberflächlich veränderten alten Buchenwald-Böden) gut unterscheiden.

b) Pollen-Analyse: Andauernd starker Anteil der NBP in wald-
grenznahen Mooren und Rohhumus-Böden.

c) Floristische:Überleben eines ziemlich reichen alpinen,
nicht saxikolen und nicht an Moore gebundenen, hemerophilen
Floren-Elements (in scharf ökologisch differenzierten jeweils
verschiedenen Assoziationen zugehörigen Gruppen) mit selbst
zeitweiliger Krüppel-Bewaldung unvereinbar (ozeanische W.G.=
dicht geschlossener Buchen-Kriechwald; Überleben von Sibbal-
dia und Schneetälchen-Flora z.B. deutet auf Permanenz großer
vollständig kahler Kamm-Flächen (Kondition sine qua non der
großen Firnschnee-Ansammlungen) CARBIENER 1966.

Deutung: wird versucht; grundsätzliche Unterschiede in der
Auswirkung der Wärme-Zeit je nach Klima-Charakter. Größere
Wintermilde = Verlängerung der Vegetations-Periode in kontinentalen Gebieten. Beispiel: Nord-Wandern des Kiefernwaldes
in NO-Finnland und Kola in den letzten 50 Jahren, aber keine
bekannte Verschiebung der West-Norwegischen Birkenwald-Gren-
ze.

5. Soziologische Zuordnung des subalpinen Buchenwaldes sehr
schwierig: Problem der Grenze zwischen F a g e t a l i a
und B e t u l o - A d e n o s t y l e t a l i a. Muß wahrscheinlich von den F a g e t a l i a abgetrennt werden. Das
"B e t u l e t u m t o r t u o s a e" Norwegens zeigt sehr
ähnliche Strukturen und auch floristische Zusammensetzung.

VEGETATIONSGRENZEN IN VULKANGEBIETEN

E m i l i a   P o l i

Wir werden nur von tätigen Vulkanen sprechen. Die Vegetations-
Grenzen sind besonders dort eigenartig, wo der Vulkan - sei es
durch intensive oder beliebig abgeschwächte Tätigkeit - seine
Wirkung zeigt.

Auf den erloschenen Vulkanen, die jede Tätigkeit seit mehreren
Jahrhunderten eingestellt haben, ist die Vegetation und ihre
räumliche Anordnung ungefähr die gleiche wie in irgendwelchen
anderen Gebieten (vergl.auch unten). Ein klares Beispiel zeigen
die alten Vulkane Zentral-Afrikas (Kenya, Rouvenzori, Kilimand-
scharo), auf denen die Vegetation und ihre räumliche Anordnung
nicht verschieden von der anderen, nicht vulkanischer Erhebun-
gen ist.

Ein tätiger oder relativ junger Vulkan, auf dem verschiedene
Substrat-Typen (aus Pyroklastika oder verschiedenen Lava-Arten
gebildet), Standorte (wie Fumarolen, Thermal-Quellen, Solfataren, Geysere u.a.) und auch Mikro-Standorte (wie z.B. Spalten
und Höhlen) zu unterscheiden sind, zeigt hingegen verschiedene
örtliche ökologische Einheiten, von denen eine jede von einer
bestimmten Pflanzengesellschaft oder einem besonderen Gesellschafts-Komplex besiedelt ist. Die räumliche Gliederung der Vegetation ist deswegen auch in einem klimatisch und florenge-
schichtlich einheitlichen Gebiet sehr eigenartig, weil außer
den die Gesellschaften bedingenden und überall wirkenden abiotischen und biotischen Umwelt-Faktoren, in vulkanischen Gebieten, neue Einflüsse hinzukommen.

Folgende Faktoren spielen für die räumliche Anordnung der Vegetation eine maßgebliche Rolle:

    a. Oberflächliche Morphologie der Lavaströme
    b. Verschiedene Alter der Laven
    c. Bodentemperatur und Wärme-Gradient
    d. Chemische Zusammensetzung des Bodens
    e. Häufigkeit der Ausbrüche

Das Klima, das für die Besiedlung der sterilen vulkanischen Böden entscheidend ist(vgl.POLI 1970-71)spielt bei der Problematik
der Vegetations-Grenzen in Vulkan-Gebieten keine besondere Rolle.

Die Wirkung aller entscheidenden Faktoren für die Vegetation
und ihre Anordnung ist verschieden, weil sie sich je nach dem
Überwiegen mancher Erscheinungen ändert. Auch kleine Änderungen
in diesem Faktoren-Komplex verursachen Veränderungen des Vege-
tationsbildes und dessen Grenzen.

Es ist aber noch hervorzuheben, daß diese Faktoren im allgemeinen nicht alle zusammen gleich stark auf eine bestimmte Fläche
der Vegetation und ihre Grenzen einwirken. Ein Faktor allein
kann z.B. mit seinen qualitativen und quantitativen Einflüssen
mehrere Typen von Substraten, Standorten oder Mikro-Standorten
bilden. Es ist also wesentlich, die Wirkungsweise eines jeden
Faktors auf die Vegetations-Grenzen im einzelnen zu behandeln.

a. Oberflächliche Morphologie der Laven.
Nach der Oberflächen-Struktur der Laven, bedingt durch den Vulkanismus und das Relief vor dem Fließen der Lava, lassen sich
zwei Lava-Arten unterscheiden: Schollen-Laven mit einheitlich
zusammenhängender Oberfläche und Brocken-Laven mit schlackenar-
tiger Oberfläche.

Solche Haupt-Typen von Lava-Oberflächen bilden ökologisch unterschiedliche Standorte, die von entsprechenden Pflanzengesellschaften besiedelt werden.

Die Schollenlaven-Flächen liefern hauptsächlich zwei verschiedene Substrate: mehr oder weniger breite Schollen und dazwischen
mehr oder weniger tiefe Spalten, die oft als Mikro-Standorte zu
betrachten sind, da eigene Boden- und Mikroklima-Verhältnisse
vorherrschen.

Auf den platten Schollen hängt die Pflanzen-Besiedlung von der
Menge des abgelagerten Feinmaterials und deren Neigungswinkel
ab und schreitet langsam voran, so daß derartige Lavaflächen
lange Zeit "nur" von Flechten-Kolonien besiedelt sind. Die Risse und Sprünge zwischen den einzelnen Schollen - die mehr oder
weniger mit Feinmaterial zugeschüttet worden sind- tragen hingegen höheren Pflanzenwuchs (Farn- und Phanerogamen-Gesellschaften).

Unser Untersuchungsobjekt sind mehrere Lavaströme des Aetna.
Ein deutliches Beispiel zeigt der Lava-Strom der Jahre 1614 -
1624, der eine große Fläche des Nord-Abhanges des Vulkans in
der Hochgebirgs-Stufe (etwa zwischen 1700 und 2500 m) bedeckt

und zum Teil als "Lava dei Dammusi"[1] bezeichnet wird. Jede
"Dammuso" (= Scholle) wird von einer steilen kompakten Platte
gebildet, die von Flechten-Gesellschaften mit Rhizocarpon geo-
graphicum, Parmelia prolixa, Candelariella vitellina, Caloplaca
festiva u.a. besiedelt ist.
Die mehr oder weniger tiefen Spalten und Risse zwischen den
Schollen sind hauptsächlich von Pionier-Stadien des R u m i c i -
A s t r a g a l i o n besiedelt in welchen Rumex aetnensis
(J.et C.Presl), Ciferri et Giac., Senecio aetnensis (Jan.) Fiori, Festuca levis (Hack.) Richter, Achillea ligustica All. typi-
ca Fiori, Potentilla calabra (Ten.) Fiori, Satureja aetnensis
Fiori, Cerastium minus Presl, Hypochaeris robertia Fiori u.a.
vorherrschen. (vgl.POLI 1965 119-120).
Solche Pionier-Stadien, soziologisch wenig abgrenzbar, unterscheiden sich durch die Arten-Zusammensetzung, die in enger Beziehung mit der Tiefe und der Breite der Spalten steht. So findet man in benachbarten Spalten verschieden entwickelte Stadien
derselben Sukzessions-Serie. Das zeigt, wie variabel und damit
kaum unterscheidbar die Vegetations-Grenzen zwischen verschiedenen Pionier-Stadien der Spalten desselben Lava-Stromes sind.

Alle Pionier-Stadien verändern mit der Zeit ihre Arten-Zusammensetzung; die Vegetationsgrenzen dagegen erfahren jedoch in kurzen Zeiträumen keine bedeutenden Veränderungen.
Ein besonderes Beispiel (aus QUEZEL 1965) werden wir vom Emi-
Koussi (Tibesti) anführen. Auf diesem Vulkan, der seit langem
nicht mehr tätig ist, bilden die sehr kompakten Laven in der
Hochgebirgs-Stufe (zwischen 2800 und 3400 m) eine unentwirrba-
re Gesamtheit von 2 - 7 m tiefen und 0.50 - 2 m breiten Spalten, "Lappiaz" genannt. Sie bieten dem Pflanzenleben besondere
mikroklimatische und edaphische Bedingungen, weil sie das ganze Jahr im Schatten liegen, und die Luftfeuchtigkeit sehr hoch
ist. Der Boden ist naß oder sehr feucht, weil dauernd Wasser in
Form kleiner Bäche oder Pfützen vorhanden ist.

[1] "Dammusi" bedeutet in der Landes-Sprache Häuser-Decke. Dies
     erklärt sich daraus, daß dieser Lava-Strom aus verschiede-
     nen Decken, eine neben der anderen und eine auf der anderen,
     gebildet erscheint, und dadurch eine dachziegelartige Struk-
     tur aufweist.

Derartige Standorte werden hauptsächlich von zwei Pflanzen-Gesellschaften[1] H e l i c h r y s u m   m o n o d i a n u m -
D i c h r o c e p h a l a   t i b e s t i c a -Ass. und E r a -
g r o s t i s   k o h o r i c a - E r o d i u m   o r o p h i -
l u m -Ass. besiedelt, die reich an Moosen, Lebermoosen und Far-
nen sind.

Das übrige Gebiet des Hoch-Emi-Koussi ist, besonders auf sandigem und lockerem Boden[2], meist von einer steppen-artigen Vegetation besiedelt, in der die A r t e m i s i a   t i l h o a n a
- E p h e d r a   t i l h o a n a - Ass. vorherrscht.
Auf der Hochgebirgs-Stufe des Emi-Koussi wird die Vegetation
von bestimmten Dauer- und Schlußgesellschaften gebildet. Vegetations-Grenzen sind hier deswegen nicht nur häufig und sehr
scharf begrenzt, sondern vermutlich auch unveränderlich.
Der Vergleich mit dem vorherbeschriebenen Lava-Strom des Aetna
(von 1614-1624) bestätigt, daß die Vegetations-Grenzen in beiden Gebieten unabhängig von den örtlichen Gesellschaften und
dem Entwicklungszustand der Vegetation sind; sie hängen also
hauptsächlich von der Oberflächen-Struktur der Lava-Ströme ab.

Die Brockenlava-Flächen bilden mehr oder weniger große Brocken,
deren Zwischenräume mit Feinmaterial aufgefüllt sind. Auf den
Brocken selbst werden sich Flechten und Moose, in den Zwischenräumen hingegen - wenn sie mit Feinmaterial angefüllt sind -
höhere Pflanzen ansiedeln.
Diese Zwischenräume können je nach Tiefe und Höhe des eingebrach-
ten Materials weitere Standorte bilden, die von verschieden entwickelten Pflanzengesellschaften , und oft von ebensovielen Lebensformen, besiedelt sind.
Auf mehreren derartigen Lava-Strömen des Aetna kann man folgende
Vegetationseinheiten, die unter sich eine ökologische Einheit
bilden, unterscheiden:

[1] Sie sind biogeographisch sehr interessant, weil ihre floristische Zusammensetzung zum großen Teil von biogeographi-
   schen und palaeoklimatischen Relikten gebildet ist. Pflanzensoziologisch werden sie auch als Relikte angesehen (QUEZEL
   1965).

[2] Man weiß nicht genau, von welchen Gesellschaften die Flächen
   der kompakten Lava, welche die "Lappiaz" bildet, besiedelt
   sind. Man kann aber vermuten, daß sie von anderen Pflanzengesellschaften - verschieden von denen der "Lappiaz" und des
   sandigen lockeren Bodens - bewohnt werden.

a. Flechten- und Moos-Gesellschaften auf den Brocken-Oberflä-
   chen;

b. Therophytenreiche Gesellschaften, hauptsächlich aus der Klasse T h e r o - B r a c h y p o d i e t e a, die auf flachen,
   mit Feinerde überschichteten Brockenstellen gedeihen.

c. Kräuter-Gräser-Gesellschaften, meistens C i s t o - L a -
   v a n d u l e t e a - Gesellschaften, in Zwischenräumen mit
   dünner Feinerde-Schicht.

d. Halbsträucher-Gesellschaften, wie z.B. R u m e x a e t -
   n e n s i s - H e l i c h r y s u m i t a l i c u m - Stadium, das eine Garigue zu sein scheint, auf den mit Feinmaterial angefüllten Zwischenräumen.

e. Busch-Gesellschaften, wie verbreitete G e n i s t a a e t -
   n e n s i s - Ges., auf genügend mit Feinmaterial angefüllten
   Zwischenräumen;

f. Buschwald seltener Hochwald-Gesellschaften als mehr oder weniger weit entwickelte Vorstadien der entsprechenden Schlußphase der heutigen potentiell natürlichen Vegetation des Gebietes. Je nach Höhenlage und Standortsbeschaffenheit finden
   sich Gesellschaften der Q u e r c e t e a i l i c i s, der
   Q u e r c o - F a g e t e a u.a. in sehr tiefen Mulden mit
   genügend tiefgründigem Boden.

Die Brockenlava-Ströme, die im allgemeinen viel mehr Standorte
als die Schollenlava-Ströme bilden können, sind wie diese mosaikartig geordnet.

Die meisten Lava-Ströme aber, besonders jene, die sehr mächtig
sind und sich über große Gebiete erstrecken, gehören nicht einfach zur einen oder anderen Lava-Art, da ihre Oberfläche teilweise von Schollenlava- oder Brockenlava-Flächen gebildet wird.

Auf derartigen Lava-Strömen sind wegen ihrer wechselnden Ober-
flächen-Beschaffenheit mehrere Standorte und damit mehrere Pflanzengesellschaften zu unterscheiden.

Ein typisches Beispiel zeigt uns der große Lava-Strom des Aetna
aus dem Jahre 1669, von dem ein Querschnitt in der Abb.1 wiedergegeben wird, der auf einer etwa 100 qm großen Fläche in der
Nähe von Catania (Nesima inferiore) aufgenommen wurde.

Dieser enge Raum wird, den verschiedenen Standorten entsprechend, von folgenden Pflanzengesellschaften besiedelt:

A. Fels-Flächen und -Nasen mit Flechten-Gesellschaften, mit über-
   wiegendem Rhizocarpon geographicum oder Stereocaulon vesuvi-

anum.

B. Untere Fläche der Brocken, meist vom Moos-Gesellschaften besiedelt.

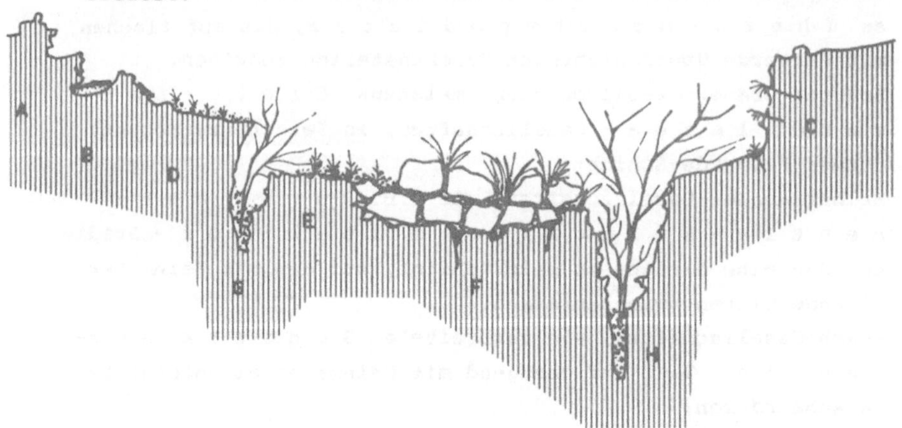

Abb.1 Querschnitt eines Lava-Stromes des Aetna (von 1669) bei
       Nesima inferiore (Catania), der einen kleinräumigen mo-
       saikartigen Gesellschafts-Komplex zeigt (Erklärungen im
       Text).

C. Steilwände von der C h e i l a n t e s   f r a g r a n s -
   P a r i e t a r i a   l u s i t a n i c a - Ges. (A s p l e -
   n i e t e a) besiedelt.

D. Kleine Schollen-Oberfläche, von einer dünnen Feinerde-Schicht
   überdeckt, die von S e d u m   c a e r u l e u m - Ges.
   (T u b e r a r i e t e a   g u t t a t i) besiedelt ist.

E. Breitere Schollen-Oberflächen mit stärkerer Feinerde-Auflage,
   von einer wenig stabilen therophyten-reichen Gesellschaft
   mit Vulpia ciliata Lk., Lamarkia aurea Moench., Arabis thali-
   ana L., Valerianella puberula DC.als vorherrschende Pflanzen
   besiedelt.

F. Mehr oder weniger große Brocken, in deren Zwischenräumen eine Halbstrauch-Vegetation mir vorherrschend Chamaephyten wie
   Rumex scutatus L., Centranthus ruber DC., Prasium majus L.
   und Hemikryptophyten wie Senecio cineraria DC., Scrophularia
   canina L. u.a. auftritt.

G. Wenig tiefe Spalten mit einer Busch-Vegetation mit überwiegend Pistacia therebinthus L. und Rhamnus alaternus L.

H. Sehr tiefe Spalten mit Buschwald-Vegetation mit vorherrschenden baumartigen Ficus carica L. caprificus Risse, Quercus
   pubescens Willd., Pistacia therebinthus L.

Auch in diesem Fall bilden die Pflanzengesellschaften - die
teils als Dauergesellschaften, teils als verschieden entwickelte Stadien der Q u e r c i o n   i l i c i s - Serie aufzufassen sind - auf sehr engem Raum ein regelloses mosaikartiges Vegetationsbild.

Wie es durch die oben beschriebenen Beispiele genügend erläutert
wurde, kann man feststellen, daß die Vegetationsgrenzen auf den
Lava-Strömen in enger Beziehung zu deren Oberflächen-Beschaffen-
heit stehen,und daß sie auf jeder beliebigen Lava-Art natürliche mosaikartige Gesellschafts-Komplexe bilden, die zum großen
Teil als syndynamische Gesellschafts-Komplexe (auch Besiedlungs-
Komplexe genannt) anzusehen sind.

Zu den Laven im weiteren Sinne gehören auch die Pyroklastika,
unter welchen man die Lockerstoffe versteht,die durch vulkanische Gase in festem oder flüssigem Zustand ausgeworfen wurden
(RITTMANN 1963). Je nach ihrer Größe, Form und Art werden sie
Aschen, Sande, Lapilli, Bomben, Wurf-Schlacken usw. genannt.
Ökologische Bedeutung haben hauptsächlich die Aschen und Sande,
die Hauptbestandteil der Pyroklastika sind.

Die von derartigem Locker-Material bedeckten Gebiete zeigen im
allgemeinen gleichartige Oberflächen, die der Vegetation mehr
oder weniger einheitliche Substrate bieten. Die Vegetation, die
sich darauf ansiedelt, ist - unabhängig von ihrem Entwicklungszustand - sehr homogen, und deswegen ist ihre Begrenzung einfach und leicht erkennbar, da sehr breite, klimatisch einheitliche Oberflächen meistens von einer einzigen Pionier-Gesell-
schaft besiedelt werden.

Als Beispiel hierfür diene die Hoyê-Lava des Fuji-san. Ein großer Teil des SO - Abhangs dieses Vulkans ist zwischen 1200 -
1300 m und 2600 - 2700 m von Aschen und Lapilli überdeckt, die
bei der letzten Eruptionstätigkeit im Jahre 1707 ausgeschleu-
dert wurden. Auf solchem Substrat kommen vor:

  bis 1350-1400 m ein sommergrüner Laubwald mit vorherrschenden
  Quercus mongolica Fischer var.grosseserrata Rhbd.et Wils.,
  Sorbus commixta Hedl., Alnus-Arten (A.tinctoria Sarg. var.ob-
  tusifolia Sarg., A.matsumurae Callier, A.firma Sieb.et Zucc.)
  und dazwischen Larix leptolepis Gordon, Pinus densiflora Sieb.
  et Zucc., u.a.

Von 1400 bis ca.1700 m eine Busch-Vegetation (Pionierstadien
  der A n g e l i c a   h a k o n e n s i s-M i s c a n t h u s

o l i g o s t a c h y s - Ass. mit vorherrschenden Polstern
von Rosa luciae Franch.et Rochebr. var. fujisanensis Makino
bis 1500 m und mit Zwergsträuchern von Salix reinii Fr.et Sav.
bis etwa 1700 m.
Von 1700-1800 m bis etwa 2300 m eine sehr spärliche Pionier-
Gesellschaft mit vorherrschendem Cirsium purpuratum Matsum.
und Polygonum weyrichii F.Schmidt var. alpinum Maxim. (ein
Pionier-Stadium der C a r e x s t e n a n t h a - S t e l -
l a r i a n i p p o n i c a - Ass.?), die einige hundert Meter unter dem Hoyé-Krater aufhört.

Geprägt wird diese Vegetation, auch in ihrem Grenzbereich, im
wesentlichen durch kleinklimatische und edaphische Einflüsse[1]
Das bedeutet, daß, obwohl die Vegetation sich noch weiter entwickeln wird, - in diesem Fall zum Teil zu Gesellschaften der
F a g e t e a c r e n a t a e - ihre räumliche Anordnung (vertikale und großräumig gürtelartig) keine bedeutende Veränderungen erfahren wird.

Aus diesen Überlegungen kann man ersehen, daß die Oberflächen-
Beschaffenheit der Pyroklastika keine besondere Bedeutung für
das Problem der Vegetationsgrenzen hat.

b. Verschiedenes Alter der Laven[2]

Bis jetzt sind nur die ziemlich jungen Lava-Ströme behandelt
worden. Diese fördern die Bildung kleinräumiger, mosaikartiger
Vegetations-Komplexe, die entweder von Schollen- oder von Brok-
ken-Oberflächen verursacht werden.

Auf sehr alten Lava-Strömen, die von höher entwickelter Vegetation besiedelt werden, herrscht großräumig eine den Klimaverhältnissen besser angepaßte Gesellschaft vor, oder eine solche,
die sich ihr am meisten nähert. Das bedeutet, daß die kleinen
Standortsunterschiede, die auf einem jungen Lava-Strom sehr von
einander abweichende Gesellschaften hauptsächlich als verschiedene Stadien derselben Sukzessions-Serie bedingen, mit der Zeit,
d.h. mit zunehmenden Alter der Laven immer weniger werden bis
zum Vorherrschen jener Gesellschaft, die sich auch physiognomisc

[1] Der vulkanische Ursprung des bodenbildenden Materials bewirkt in diesem Fall - nach unserer Meinung - in der klimatischen Höhen-Abstufung der Vegetation nur eine Verschiebung
nach unten.

[2] Dieser Faktor wird hier allein behandelt; d.h. in erster Linie von den Höhenlagen der Laven und daher auch von den Klimaverhältnissen unabhängig.

dem Klima besser anpaßt. Auf diese Weise zeigt sich, besonders
in Gebieten, die vom Menschen wenig beeinflußt und gestört sind,
ein ganzer Lavastrom von einer oder von nur sehr wenigen und
großräumigen Pflanzengesellschaften besiedelt. Gleichzeitig fallen entsprechende Änderungen an den verschiedenen Vegetations-
grenzen auf, welche die Pflanzengesellschaften als syndynamischen Gesellschafts-Komplex kennzeichnen, da sie mit der Zeit
verschoben und ausgelöscht werden.

Es genügt als deutliches Beispiel zwei Lavaströme des Fuji-san
zu erwähnen: den alten Takamarubi-Lava-Strom auf der NO-Seite
des Vulkans, nordwestlich des Yamanaka-Sees, der von einem reinen Picea polita-Bestand bedeckt ist; der Aogigahara-Lava-Strom
(von 860) auf der NW-Seite des Vulkans, der von einem immergrünen Nadelwald, mit vorherrschenden Tsuga sieboldii Carr., Picea
polita (Sieb.et Zucc.) Carr., Abies bicolor Maxim. besiedelt
wird. Dieser Wald bedeckt kilometer-weit (14 km entlang des
SW - Abhanges und 5 km breit) den ganzen Lava-Strom, so daß er
dort als "Baum-See" bezeichnet wird.

Beide Wälder, die vielleicht noch sehr weit vom letzten Stadium
der Vegetationsentwicklung entfernt sind, werden durch die entsprechenden Lava-Ströme, die sie besiedeln, genau begrenzt. Die
Grenzen sind in diesem Fall sehr auffallende, umso mehr als um
beide Lava-Ströme Gras-Gesellschaften mit einer vorherrschenden
M i s c a n t h u s   s i n e n s i s-Ges. (M i s c a n t h e -
t e a) sich ausbreiten.

Auf den alten Lava-Strömen werden die Vegetationsgrenzen nicht
durch kleinflächige Standort-Änderungen und die Morphologie der
Lava-Oberfläche bestimmt. Hier ist vielmehr das verschiedene
Alter benachbarter Lavaströme die Ursache. Solche Gebiete bilden meistens großräumige Vegetations-Komplexe und deren Grenzen
fallen als Formationsgrenzen im allgemeinen mit den Grenzen der
Lava-Ströme verschiedenen Alters zusammen.

Das verschiedene Alter nachbarter junger Lava-Ströme bewirkt daher auch sehr scharfe Grenzen zwischen ihnen. Solche Grenzen
trennen aber nicht Pflanzen-Formationen, wie bei den alten Lava-Strömen, sondern verschiedene kleinräumig, mosaikartig aus-
gebildete Vegetations-Komplexe, welche jeden Lava-Strom kennzeichnen. Mit der Zeit werden solche Grenzen verstärkt und innerhalb eines jeden kleinräumigen, mosaikartigen Vegetations-Komplexes werden, wie oben gesagt, die Vegetationsgrenzen verscho-

ben oder verwischt. Die Grenzen zwischen zwei kleinräumigen mosaikartigen Vegetations-Komplexen werden deswegen mit steigendem Alter Formations-Grenzen bilden. Der Vegetations-Komplex
ist nur großräumig erkennbar, wenn er ein großes Gebiet von Lava-Strömen verschiedenen Alters umfaßt.
Wenn zwischen den Lava-Strömen große Pyroklastika-Flächen liegen, wird sich das landschaftliche Vegetationsbild nicht verändern. Eine Pflanzen-Formation oder mehrere, welche die Pyroklastika-Flächen besiedeln, stellen ebensoviele Einheiten des großräumigen Vegetations-Komplexes dar.
Das landschaftliche Vegetationsbild wird sich hingegen bei jenen Lava-Strömen verändern, die mehrere verschieden große Flächen, von bedeutend älteren Böden gebildet, als Inseln (auf dem
Aetna "Dagale" und auf Hawai "Kipukas" genannt) umfassen. In
diesem Fall werden die mosaikartigen Besiedlungs- oder Formati-
ons-Komplexe der Lava-Ströme durch derartige Inseln höher entwickelter Vegetation unterbrochen.

c. Boden-Temperatur und Wärme-Gradient

Die höchsten Temperaturen kommen in der Natur in Vulkan-Gebieten und zwar bei tätigen Kratern während ihrer Ausbruchs-Peri-
ode vor. Die Temperatur ist so hoch (bis 1400°C), daß jedes
pflanzliche Leben unmöglich wird.
In Thermen, Fumarolen und Solfataren bleiben die Temperaturen
im allgemeinen viel niedriger und von einem bestimmten Wärmegrad
ab (im allgemeinen ab 80°C) erlauben sie pflanzliches Leben.
Organismen, die sich den höchsten Temperaturen in den Thermen
am leichtesten anpassen, sind Bakterien und Cyanophyceen. In
einer warmen Quelle sind bei 80°C lebende Spalt-Algen und bei
74° und 69°C gut entwickelte Confervaceen gefunden worden.
(SCHIMPER 1935). Faden-Algen sind in Thermen von 59°C (REIN
1896) und Diatomeen in solchen von $54-60^{\circ}$C (SCHNETZLER 1889) und
so gar von 94°C noch festgestellt worden. PEDICINO (1873) hat
als extremste Wärme-Grenze des Pflanzenlebens in den Fumarolen
von Ischia (Italien) 62°C angegeben. Er hat lebende Algen-Kolo-
nien (der Gattung Protococcus) bei $54-62,5^{\circ}$C gefunden und für nur
kurze Momente bis zu 67°C bemerkt.
Exakte Untersuchungen über die Verteilung solcher Organismen,
entsprechend dem Wärme-Gradienten, sind spärlich. Wir können
aber vermuten, daß sie sich in horizontaler Gürtelung nach den
abnehmenden Werten der Temperatur und der Feuchtigkeit verbrei-

ten, indem sie sich nach und nach von der Wärme-Quelle entfernen.

In den Fumarolen des Vesuvs, innerhalb des Haupt-Kraters, hat
MAINI (1962) an den Spalten und bis zur Tiefe von 15-20 cm die
Cyanophyceen: Hapalosyphon laminosus (Hass.)Kirch., Nostoc
sphaericus A.G. (bei 40°C) Chroococcus turgidus West und eine
Chlorophycee der Gattung Cosmarium gefunden. Wenige Meter von
den heißen Spalten entfernt und noch auf warmem Boden hat er
Bryophyten, wie Ditrichum subulatum (Br.eur.)Hampe, Bryum pro-
vinciale Phil., Leptobryum piriforme Schimp., Marchantia poly-
morpha L. und die Farne Adiantum capillus-veneris L. bemerkt,
welche die ersten erkennbaren Pionier-Gesellschaften: M a r -
c h a n t i a   p o l y m o r f a - B r y u m   p r o v i n c i -
a l e-Ges. und L e p t o b r y u m   p i r i f o r m e - D i -
t r y c u m   s u b u l a t u m-Ges. darstellen.

Eine interessante Gürtelung der Vegetation bei den Fumarolen
von Toussidé im Herzen der Sahara (Tibesti) haben DE MIRE et
QUEZEL (1959) nachgewiesen. (Abb.2) Zwischen 3000 und 3100 m

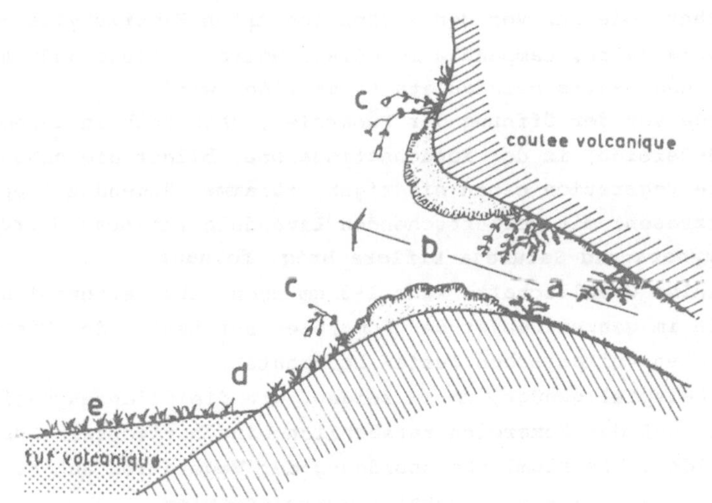

Abb.2. Querschnitt einer Fumarole von Toussidé, Tibesti (aus
       DE MIRE et QUEZEL 1959), der eine horizontale Gürtelung
       der Vegetation zeigt (Erklärungen im Text).

breitet sich eine fast ununterbrochene Zone von Fumarolen aus,
denen ausschließlich Wasserdampf mit einer Temperatur von maxi-

mal 45°C in den tiefsten Schichten, und durchschnittlich 38°C
an der Mündung, entströmt. (Nach MONOD kann er 60°C erreichen).
Dieser Dampf schlägt sich an den Wänden nieder, an denen sich
eine sehr eigenartige Vegetation findet, deren Arten-Kombination stark von ihrer Umgebung abweicht. Sie verteilt sich in 5
deutliche Zonen, wie folgt (vgl.Abb.2):

A. In der tiefen Schicht ist das Gestein mit einem kompakten
   Überzug von Cyanophyceen bedeckt (mit vorherrschenden Apha-
   nocapsa biformis, Schizotrix lardacea, Tolyopothryx tibesti-
   anensis,Arten der Gattungen Chroococcus und Stigonema u.a.),
   zu denen sich, ein wenig weiter oben - und zwar 1,5 m vor der
   Mündung - Farne (Apslenium adiantum-nigrum L., Adiantum capillus-veneris L., Cheilantes maderensis), Arten der Gattungen Selaginella und Oldenlandia und Moose der Gattung Fossombronia, gesellen.

B. Die folgende Zone - etwa 50 cm vor der Mündung bis zu ihr -
   ist durch einen 15 cm reich entwickelten Teppich von Moosen
   (23 verschiedene Arten) gekennzeichnet.

C. Die Mündung der Fumarolen wird von einer Mikro-Gesellschaft
   bewohnt, die nur von den xerophilen Arten Fimbristylis minu-
   tissima Maire, Campanula monodiana Maire, Mollugo nudicaulis
   Lam. und Oxalis corniculata L. gebildet wird.

D. Gerade vor der Öffnung der Fumarolen, aber noch in ihrem Einfluß-Bereich, in der Kondensationszone, bildet die schon rei-
   chere Vegetation einen niedrigen, zusammenhängenden Teppich
   (Kurzrasen) mit vorherrschenden Lavandula antineae Maire fo.
   platynota und Satureja biflora Briq. fo.nana.

E. Dieselbe Gesellschaft, kaum 2-3 cm hoch, aber artenreicher,
   wurde im ganzen Gebiet der Fumarolen auf den in der Tiefe
   feuchten vulkanischen Tuffen beobachtet.

Dieses Beispiel genügt, um zu zeigen, wie die Pflanzengesellschaften auf den Fumarolen tatsächlich eine horizontale Gürtelung bilden. Die räumliche Anordnung der Vegetation äußert sich
in diesem Falle als horizontaler Gürtel-Komplex.

d. Chemische Zusammensetzung des Bodens

Im Vulkan-Gebiet spielt dieser Faktor besonders bei den Solfataren eine wesentliche Rolle. An derartigen vulkanischen Standorten ist für das pflanzliche Leben und seine räumliche Anordnung vor allem die chemische Zusammensetzung des Bodens mit seinen hohen Temperaturen entscheidend und einschränkend.Eine wich-

tige Rolle kommt auch Gasen wie SO_2, H_2S u.a. zu, die dem Untergrund entströmen und für die meisten Pflanzen entwicklungshem-
mend sind.
Die auffällige Vegetation der Solfataren ist deswegen sehr eigenartig, und unterscheidet sich von der Vegetation der Umgebung. Diese Einzigartigkeit der Solfataren-Vegetation hängt
hauptsächlich von den Besonderheiten der Bodenbeschaffenheit ab.

Der Boden der Solfataren ist bekannt für seinen hohen Säuregrad.
Er ist sehr arm an Nährstoffen und Stickstoff und sehr reich an
Alaun und Kieselsäure. Auf derartigem Boden können sich keine
anderen Pflanzen ansiedeln als diejenigen, die im Bezug auf
Nährstofferwerb sehr genügsam sind (Oligotrophen), eine gewisse
Toleranz gegen Aluminium-Jon haben und zugleich hohe Temperatu-
ren ertragen können.
Das bewirkt eine sehr starke Auslese unter den Pflanzen (vgl.
POLI 1970-71,und eine bestimmte Verteilung der Arten entspre-
chend den sinkenden Werten von Temperatur und Säuregehalt im
Boden, sowie der giftigen Dämpfe des Luftraumes. Das verursacht
zugleich eine deutliche Abgrenzung in der Vegetation, die, wie
bei den Fumarolen, räumlich horizontale Gürtel-Komplexe bildet.
Gleiches konnten wir bei den verschiedenen japanischen Solfataren (Numayu, Pon-pon-yama, Kogen-Onsen, Io-san, Hakone u.a.)
feststellen. Die Pflanzen, die sich am nächsten an die Schwefel-
Quelle herandrängen, bestehen überwiegend aus Flechten und Moosen. Als erste Blütenpflanze findet sich in der nächsten Zone
eine Cyperus-Art, von der Miscanthus sinensis Anderss. gefolgt.
Bei der Numayu-Solfatare auf Hokkaido (Japan) sind folgende Gürtel zu unterscheiden (Abb.3):

a. Nackte Zone unmittelbar an der Schwefel-Quelle, die aus zwei
   Öffnungen besteht. In solcher etwa 1 m breiten Zone fehlt die
   Vegetation vollkommen.
b. Flechten-Zone. Kaum 1 m von der Schwefel-Quelle entfernt, also in der lebensfeindlichsten Zone, breitet sich ein dichter
   Pflanzen-Teppich aus, dessen erster Gürtel von Flechten der
   Gattung Cladonia, und zwar Cladonia carassensis unter Vorherrschen von Cladonia macilenta Hoff. ssp. therophylla gebildet wird.
c. Moos-Zone. Es folgt eine Zone, in der die Boden-Oberfläche
   eine Temperatur von 37-38°C zeigt, und die Vegetation vor

allem durch Moose der Gattung Campylopus vertreten ist. Die
Flechte Cladonia macilenta Hoff.ssp.therophylla der vorliegenden Zone kommt noch vereinzelt vor. Auf der äußeren, der,
der Solfatare abgewandten Seite, erscheint die erste Blütenpflanze: eine Cyperus-Art[1] .

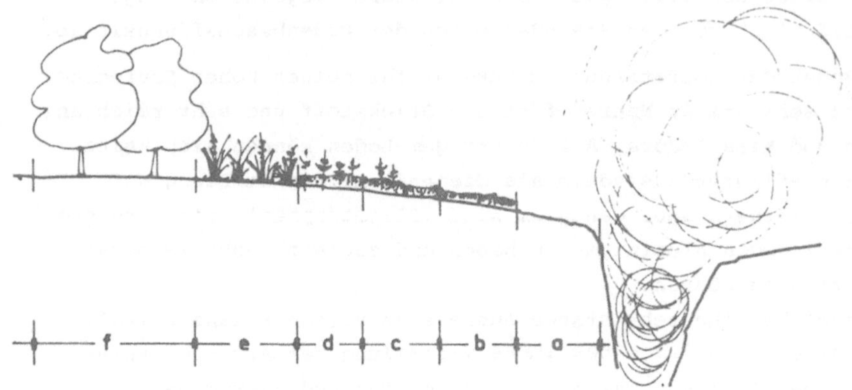

Abb.3. Horizontale Vegetations-Gürtelung bei der Numayu-Solfa-
        tare, Japan (Erklrärungen im Text).

d. Moose-Gräser-Zone. Von der Schwefel-Quelle mindestens 3 m
   entfernt, wird diese Zone durch Moose der Gattung Hypnum,fer-
   ner durch denselben Cyperus und Deschampsia flexuosa (L.)
   Trin. gekennzeichnet.
e. Gräser-Zone. Diese letzte Zone ist durch Miscanthus sinensis
   Anderss., ein typisches Gras der japanischen Solfataren, gekennzeichnet.
Außerhalb des M i s c a n t h u s-Gürtels beginnt der Q u e r-
c u s   c r i s p a-Wald, der die benachbarte, nicht mehr von
den Solfataren bedingte Vegetation beherrscht.
YOSHIOKA, SAITO und TACHIBANA (1965) haben die Vegetation der
japanischen Solfatare des Osoreyama (N-Honshu) untersucht und
beschrieben. Sie haben außerhalb des nackten Bodens, der am
nächsten an der Schwefel-Quelle liegt, 6 für Solfatare charakteristische Pflanzengesellschaften (C a r e x   a n g u s t i -
s q u a m a-Ges., P h r a g m i t e s   c o m m u n i s - S c i r-

[1] Ihre Bestimmung war wegen des schlechten Zustandes der Pflanze unmöglich.

p u s   t a b e r n e m o n t a n a e - Ges., D e s c h a m p -
s i a   f l e x u o s a - Ges., L e d u m   p a l u s t r e ssp.
d i v e r s i p i l o s u m var. n i p p o n i c u m - Ges.,
R h o d o d e n d r o n   f a u r i a e var. r o s e u m -
I l e x   s u g e r o g i ssp. b r e v i p e d u n c u l a t a -
Ges., S a s a   p a n i c u l a t a - S a s a   k u r i l e n -
s i s - Ges.),sowie eine Q u e r c u s m o n g o l i c a var.
g r o s s e s e r r a t a - Ges.unterschieden, wobei die letzte
den Übergang zwischen der Solfataren-Vegetation und der Klimax-
Waldgesellschaft des Gebietes (F a g u s   c r e n a t a - T h u -
j o p s i s   d o l o b r a t a var. h o n d a i - Ges.) bildet.

Die vorgenannten Verfasser haben die räumliche Verbreitung der
Gesellschaften nicht zu den Schwefel-Quellen in Beziehung gesetzt. Sie haben dagegen Daten über die chemische Boden-Beschaf-
fenheit ermittelt. In der nackten Zone des Bodens wurden pH-Werte von 1,4 - 1,8, in der C a r e x   a n g u s t i s q u a m a-
Ges. von 2,8 - 3,8 gefunden.Der Boden der anderen Gesellschaften zeigt pH-Werte von 2,5 - 3,0.
Das bedeutet, daß sich in einem Solfatare-Gebiet, das von verschiedenen Schwefel-Quellen durchflossen wird, die Vegetation
durch das gegenseitige Einwirken der beieinander liegenden
Schwefel-Quellen keine deutlich horizontale Gürtelung bilden
kann, weil es keine regelmäßige Abwandlung der Werte für die
wichtigen Boden- und Luft-Faktoren gibt. Die verschiedenen Pflanzengesellschaften sind jedoch erkennbar, weil sie sehr deutlich
gegeneinander abgegrenzt sind. Im wesentlichen ist es die chemische Beschaffenheit des Bodens, die keine vollkommene, sondern
nur eine unterbrochene Gürtelung der Vegetation bewirkt.
Etwas Gleiches haben wir auch an den japanischen Io-san Solfataren (Hokkaido) bemerkt.
Die Vegegations-Grenzen sind um Solfataren und in Solfataren-
Gebieten überall sehr deutlich erkennbar. Im allgemeinen konnte
beobachtet werden, daß horizontale Gürtel-Komplexe in Solfata-
ren-Gebieten (die reich an Schwefel-Quellen sind) unvollständig
ausgebildet sind.

e. Häufigkeit der Ausbrüche
Die teilweise oder ganz ausgefüllten Krater der nicht mehr tä-
tigen Vulkane gleichen im allgemeinen - wie schon gesagt - den
Gipfeln anderer Gebirge. Die Vegetation dieser Krater-Böden

ist je nach Breitegrad, Höhenlage und den dadurch bedingten klimatischen Faktoren verschieden.

Dies konnten wir auf dem Fuji-san feststellen, dessen Haupt-Krater seit mehreren Jahrhunderten nicht mehr tätig ist (vgl.POLI 1964).

Etwas anderes wurde aber bei den erloschenen Vulkanen auf Java und Sumatra von JUNGHUHNS (1854 u.1865), ERNST (1910), FABER (1927) nachgewiesen. In diesen Gebieten sind die Innenwände der erloschenen Krater-Kegel stets durch vollständiges Fehlen von Baum-Gesellschaften gekennzeichnet, während die Außenhänge, soweit sie nicht zur alpinen Stufe gehören, meistens dicht bewaldet sind.

Die Vegetation der nicht mehr tätigen Krater-Kegel hat in diesem Fall ihre eigene Anordnung, die hauptsächlich durch vulkanische Tätigkeit bedingt ist. Die Vegetations-Grenzen, zumindest die Grenzen der Formationen, sind deshalb verschieden von jenen der anderen, nicht vulkanischen Gipfel. Das zeigt, wie schwer eine Typisierung der Vegetations-Grenzen auf erloschenen Krater-Kegeln ist.

Was wir hier jedoch besonders behandeln wollen, ist die Besiedlung und die Anordnung der Pflanzengesellschaften bei tätigen Kratern,des Pflanzenlebens, welches durch die vulkanische Tätigkeit und Häufigkeit der Ausbrüche der Krater bedingt ist.

Derartige Krater-Kegel bieten pflanzlichen Organismen extreme oder gar keine Lebensbedingungen. Sie werfen oft festes Material, Asche, Bimsstein, Lavabrocken aus, mit denen giftige Gase, vermischt mit Wasserdampf, aufsteigen.

An der inneren Seite der Krater und an den stets neu mit solchen Auswurf-Stoffen überschütteten Abhängen der Kegel ist jedes Pflanzenleben unmöglich.

Das völlige Fehlen einer Vegetation auf den tätigen Krater-Kegeln ist aber nicht auf die klimatischen Bedingungen des Standorts - wie manche Verfasser vermuten - zurückzuführen. Es beruht vielmehr auf der eruptiven Tätigkeit des Kraters, wenn auch in manchen Fällen mehrere Faktoren zusammentreffen (das jähe Abfallen der Seitenwände des Kraters und ihre durch Wind und Regenwasser begünstigte Verwitterung, die ungünstigen klimatischen Verhältnisse bei den höheren Vulkanen usw.). Denn auch die Gipfel und Krater-Felder tätiger Vulkane von mäßiger Höhe (wie z.B. Stromboli, Komagatake, Aso u.a.) zeigen auf den durch

Eruptionen stets gefährdeten, obersten Hängen keine Vegetation.

An den tätigen Vulkanen folgt unterhalb einer je nach der Intensität der vulkanischen Tätigkeit vom vegetationslosen Krater tiefer reichende pflanzenfreie Zone - von uns als "Vulkan-Wüste" (POLI 1970-71 und ined.)bezeichnet - ein sehr spärlich und unregelmäßig von Pflanzen besiedelter Abschnitt.

Die Ansiedlung von Pflanzen in dieser zweiten Zone wird aber oft durch die Häufigkeit der vulkanischen Tätigkeit erschwert, und zum Teil verhindert. Während der mehr oder weniger langen Ruhezeiten der Vulkane versuchen die widerstandfähigsten Pionierpflanzen das neue Substrat zu besiedeln, werden jedoch durch die nächste Eruption wieder vernichtet. Das zeigt, wie schwer das Leben für die Pflanzen an den Abhängen der tätigen Krater ist, auch wenn sie sich den besonderen Standortsbedingungen angepaßt haben.

Eine derartige Zone kann als Kampf-Gürtel zwischen dem Pflanzenleben und der vulkanischen, das Leben vernichtenden Wirkungskraft bezeichnet werden.

Unterhalb dieser Zone, die besonders nach oben schwer zu begrenzen ist, haben wir im allgemeinen jene Vegetation, die normalerweise den örtlichen trockenen Geröll-Halden der tätigen Vulkane entspricht, jeweils abhängig jedoch von den Klima-Verhältnissen, welche durch den Breitengrad und Höhenlage bedingt sind.

Diese, von der Höhenlage unabhängige, Zonierung der tätigen Kraterkegel ist von uns auf vielen Vulkanen (Aetna, Stromboli, Aso, Asama, Komagatake, Tokachi u.a.) beobachtet worden.

Als typisches Beispiel sei hier der Zentral-Kegel des Aetna mit seinem subterminalen NO-Kegel angeführt. Der dauernde Anfall an Pyroklastika, sowie die häufigen Lava-Effusionen[1] , bewirken auf dem Aetna-Gipfel laufende Oberflächen-Veränderungen, so daß keine Ansiedlung von Vegetation und damit auch keine Bodenbildung erfolgen kann. Hinzu kommen die vulkanischen Gase, die lange Kälte- und Schnee-Periode und die fast ständige Umlagerung des Locker-Materials durch Wind und Wasser.

Die vulkanische Wüsten-Zone auf dem Aetna ist deshalb sehr breit. Die stets von neuem mit Auswurf und ausfließendem Material überschütteten Abhänge sind wenigstens bis zu 300 m weit talwärts

[1] Im Jahre 1964 sind z.B. sechs Lava-Ströme aus dem Zentral-Krater herausgeflossen und haben große Flächen bedeckt.

völlig vegetationslos.

Die Kampfgürtel-Zone beginnt bei 3050 m auf dem NW-Abhang und
bei 2950 m auf dem S-SO-Abhang. Ihre obere Grenze wechselt aber
sehr. Als Folge der letzten Lava-Ströme wird sie manchmal unter
2600-2500 m herabgedrückt.

Die untere Zone, d.h. die R u m i c i - A n t h e m i d e t u m
a e t n e n s i s -Zone, in der die Vegetation hauptsächlich
von klimatischen Faktoren bedingt wird[1] , erreicht eine maximale Höhe von 2850-2900 m.

Das zeigt sehr deutlich, wie die vulkanische Tätigkeit ein entscheidender Faktor für die Pflanzen-Ansiedlung und überhaupt
der Lebensmöglichkeiten für Pflanzen auf den tätigen Krater-Kegeln darstellt. Die von der vulkanischen Tätigkeit abhängige
räumliche Höhen-Zonierung der Vegetation, oder besser des Pflanzenlebens, ist hauptsächlich eine vertikale edaphische Gürtelung, in welcher das Pflanzenleben talwärts immer bessere Lebensmöglichkeiten findet.

Ein schönes Beispiel für räumlich vertikale Grenzen, die vor allem edaphische Grenzen sind, gibt die vulkanische Oshima-Insel,
die von TEZUKA (1961) beschrieben wurde. Der Autor hat auf der
ganzen Insel, deren einziger Gipfel der tätige Krater-Kegel (M.
Mihara,755 m) ist, vom Meeresspiegel bis zum Gipfel folgende
Zonen unterschieden:

a. Wald mit vorherrschenden Alnus sieboldiana Matsum. Prunus
   lannesiana (Carr.)Wils. var.speciosa (Koidz.)Makino fo. sim-
   plicifolia, Camellia japonica L. Pinus thunbergii Parl.

b. Immergrüne Laubwald-Zone mit vorherrschenden Shiia sieboldii
   (Makino) Makino und Machilus thunbergii Sieb.et Zucc., als
   Klimax-Vegetation für das ganze Gebiet bezeichnet.

c. Immergrüner und sommergrüner Laub-Mischwald mit vorherrschenden Alnus sieboldiana Matsum, Eurya japonica Thunb. var.mon-
   tana Bl., Camellia japonica L., Cornus controversa Hems.u.a.

d. Gebüsch-Zone mit vorherrschend Alnus sieboldiana Matsum. und
   Weigela grandiflora (Sieb.et Zucc.)K.Koch.

e. Wüsten-Zone mit spärlichen Polygonum cuspidatum Sieb.et Zucc.
   und Carex okuboi Frank.

f. Nackte Zone in der Nähe der Krater-Gipfel völlig vegetations-

---

[1] Diese Pioniergesellschaft, das extremste Beispiel von
    Pflanzengesellschaften in der höheren Lage des Aetna,
    ist von uns (vgl.POLI 1965, p.70 u.217) als Klimax-Ge-
    sellschaft bezeichnet worden.

los.

Die Pflanzengesellschaften dieser Zonen sind in einer progressiven primären natürlichen Sukzession wie folgt miteinander verbunden:

Nackter Boden "Wüste" Gestrauch immergrüner-sommergrüner Mischwald immergrüner Laubwald (als Klimax bezeichnet).

Das zeigt, daß die jetzige Höhenstufen-Gliederung eine räumlich vertikale Grenz-Betrachtung erlaubt, und nur edaphische Bedeutung hat und durch die Zeit begrenzt ist. D.h., daß sie mit der Zeit, sofern es die vulkanische Tätigkeit gestattet, verschwinden werden. Derartige Grenzen sind also auch als zeitlich begrenzte, edaphische zu betrachten.

SCHLUSSFOLGERUNGEN

Aus diesen Überlegungen geht hervor, das das Problem der Vegetations-Grenzen in Vulkan-Gebieten komplex und schwer zu fassen ist, da in derartigen, sehr veränderlichen Biotopen mehrere, einander überschneidende Faktoren zur Wirkung kommen.

Das Ziel dieser Arbeit ist es, die wichtigsten Aspekte dieses vielseitigen Problems in seiner Gesamtheit darzulegen. Wir hoffen, daß wir mit diesem kurzen Bericht Anregungen für jene, die das Problem weiter zu vertiefen wünschen,gegeben haben.

Aus diesem Grund wurde hier eine allgemeine Typisierung der wichtigsten entscheidenden Faktoren aufzuzeigen versucht, die - vor allem in Verbindung mit dem Vulkanismus - nicht nur die Vegetation als solche,sondern auch ihre räumliche Anordnung bedingen.

Eine zusammenfassende Übersicht gibt das folgende Schema:

RÄUMLICHE ANORDNUNG DER VEGETATION (Vegetationsbild)	ENTSCHEIDENDE VULKANISCHE FAKTOREN	STANDORT (oder Gebiete)
kleinräumige mosaikartige Vegetationskomplexe	oberflächliche Morphologie der Lava-Ströme	Schollenlava-Ströme Brockenlava -Ströme
großräumige Vegetations-Komplexe (Formations-Komplexe)	verschiedenes Alter der Laven	nicht sehr junge,klimatische einheitlich vulkanische Gebiete
horizontale Vegetations-Gürtel	Temperatur und chemische Zusammensetzung des Bodens	Fumarolen, Thermal-Quellen, Solfataren

vertikale edaphische Vegetations-Stufen	Häufigkeit der Ausbrüche	tätige Krater-Kegel manche klimatisch einheitlich vulkanische Gebiete

Das Schema zeigt, wie sich das Problem der Vegetations-Grenzen
in den Vulkan-Gebieten ganz eigentümlich darstellt und wie deren
räumliche Anordnung in der Vegetation, je nach Übergewicht
des einen oder anderen der entscheidenden vulkanischen Faktoren
und je nach deren Einfluß auf die Beschaffenheit des Standortes
oder des ganzen Gebietes,in der Landschaft verschiedene Vegeta-
tionsbilder zeigen kann.

Am Ende möchten wir vor allem hervorheben, daß wir hier nicht
alle Umwelt-Faktoren der Vegetations-Grenzen in Vulkan-Gebieten,
sondern nur jene, die hauptsächlich mit dem Vulkanismus verbunden sind, behandelt haben; denn die in jeder beliebigen Zone
wirksam werdenden Umwelt-Faktoren[1] äußern sich gleichermaßen
in vulkanischen, wie in allen anderen Gebieten. Klimatische Einflüsse bewirken in allen Gebieten mit gleicher geographischer
Lage dieselben vertikalen Höhenstufen[2].

[1] Der menschliche Einfluß mitgerechnet.

[2] Sie haben mit vertikalen edaphischen Vegetations-Gürteln
(vgl.oben) nichts zu tun.

ZUSAMMENFASSUNG

Es wird vor allem erklärt, daß in tätigen Vulkanen, seien sie
intensiv oder abgeschwächt tätig, Vegetations-Grenzen häufiger
und schärfer als in irgendeinem anderen Gebiet sind. In solchen
Vulkan-Gebieten wird die Vegetation von verschiedenen Faktoren
beeinflußt, und zwar spielen bei der Verbreitung so wie der Begrenzung der Pflanzengesellschaften folgende eine maßgebliche
Rolle:

1. Die physikalische Struktur des vulkanischen Material und außerdem, wie bei jedem physikalischen Typ, die Verschiedenheit des Substrates. Dieses bedingt oft im Vulkan-Gebiet verschiedene Substrate und Standorte, die durch- und nebeneinander als kleine Flecken eines Komplex-Mosaiks gemischt sind.
2. Das unterschiedliche Alter des vulkanischen Materials, das
   bei nicht sehr jungen Substraten noch eine große Variations-

breite für eine sich entwickelnde Vegetation erlaubt.

3. Die Boden-Wärme und zwar der Wärme-Gradient des Terrains in
   bestimmten Flächen wie z.B. bei Fumarolen und Thermal-Quellen.

4. Die chemische Zusammensetzung des Bodens, besonders in Solfataren, die durch ihren hohen Säuregehalt bekannt sind.

5. Die Häufigkeit der Ausbrüche, besonders bei den Kratern, aus
   denen Gase und immer neues festes Material entweichen.

Die Wirkung dieser Faktoren auf die Vegetations-Grenzen ist verschieden, weil sie sich je nach dem Überwiegen des einen oder
des anderen ändern. Das wird durch Beispiele aus verschiedenen
Vulkan-Gebieten der Erde erläutert.

RIASSUNTO

Limiti della vegetazione nei vulcani

Nei vulcani attivi, quelli qui in prevalenza trattati, gli aggruppamenti vegetali hanno nei diversi ambienti una disposizio-
ne singolare. I limiti fra i diversi aggruppamenti possono
essere infatti più frequenti e più evidenti che non nei territori non vulcanici. Ciò dipende soprattutto dai fattori legati
al vulcanismo quali ad esempio:
- morfologia superficiale delle colate laviche
- diversa età delle colate laviche
- calore del suolo e gradiente termico di esso
- composizione chimica del suolo
- attività vulcanica presso i crateri.
Questi fattori si aggiungo a tutti gli altri fattori biotici e
abiotici dell'ambiente.
I fattori legati al vulcanismo, poiché vengono ritenuti determi-
nanti per i limiti della vegetazione, vengono trattati separata-
mente e indipendentemente da tutti gli altri, con esemplificazio-
ni tratte anche dai vulcani che l'A. ha visitato. I risultati,
sintetizzati in uno schema conclusivo, sono i seguenti:
1) La morfologia superficiale delle colate laviche, diversa nei
   due principali tipi di lava distinti: lava a lastroni e lava
   a blocchi scoriacei, determina sia nell'uno che nell'altro
   tipo di lava diversi e numerosi substrati, variamente inter-
   ferenti e colonizzati da altrettanti aggruppamenti vegetali.
   Ne consegue che in superfici relativamente piccole (100-200

mq) coesistono tutti i tipi di substrato e quindi tutti gli
aggruppamenti vegetali distinti. I limiti fra questi sono
quindi frequenti e non sempre sociologicamente chiaramente
definibili trattandosi il più delle volte dei elementi (stadi)
della stessa serie evolutiva. Il paesaggio vegetale risulta
costituito da un mosaico, i cui pezzi, piccoli, si interse-
cano ripetutamente.

2. La diversa età delle colate laviche ha un significato nel de-
   terminare i limiti della vegetazione specialmente se si trat-
   ta di territori vulcanici non molto recenti. In questo caso
   intere colate laviche sono colonizzate da una o poche formazioni vegetali, più o meno vicine alla fase evolutiva fina-
   le della vegetazione. Tra colate laviche di diversa età quindi, colonizzate da formazioni vegetali differenti, i limiti
   sono molto evidenti. In questi casi il paesaggio vegetale
   viene ad essere costituito ancora da un mosaico, ma i cui
   pezzi sono rappresentati da formazioni vegetali molto estese.
   Il complesso degli aggruppamenti vegetali delle colate laviche recenti tende col tempo a costituire questo tipo di mosaico.

3. Il calore del suolo, e precisamente il gradiente termico di
   esso, dertermínana una zonizzazione orizzontale degli aggruppamenti vegetali per valori decrescenti della temperatura,
   man mano che ci si allontana dalla fonte di calore.
   Ciò si verifica soprattutto nelle sorgenti termali e nelle
   fumarole.

4. La composizione chimica del suolo, in particolare il grado
   di acidità di esso, dertermina pure una zonizzazione orizzontale degli aggruppamenti vegetali, cosa che si verifica soprattutto presso le solfatare.

5. L'attività vulcanica presso i crateri, in particolare la fre-
   quenza di questa, determina una particolare zonizzazione ver-
   ticale sui fianchi dei crateri e nel territorio ad essi più
   prossimo qualsiasi altitudine essi raggiungano. Questa zonizzazione è dovuta principalmente a fattori edafici ed è li-
   mitata nel tempo essendo in stretta dipendenza con l'attività
   del vulcano.

Esempi di questo tipo di zonizzazione sono stati riscontrati
anche in territori vulcanici molto estesi.
Si conclude precisando che il problema qui affrontato è stato

trattato nelle linee generali allo scopo di metterne in eviden-
za i principali aspetti, onde poter fornire ben precisi punti
di partenza a chi volesse ulteriormente approfondire l'argomen-
to.
La tipologia tentata vuole avere solo questo significato.

SUMMARY

The limits of the Vegetation on Volcanoes

On active volcanoes, which are the ones principally dealt with
here, the plant-communities and their distribution are quite
singular. The limits between the various plant communities are,
in fact, more frequent and clearer than in non-volcanic regions.

This depends, above all, on the factors connected with volcanoism:
    the superficial morphology of the lava streams
    the different ages of the lava streams
    the heat of the soil and especially its thermic gradient
    the chemical composition of the soil
    the volcanic activity near the craters
These factors act together with all the other ecological biotic
and abiotic factors.
Since the factors connected with volcanoism are considered determining as to the limits of the vegetation, they are dealt
with separately with numerous examples taken from various volcanoes that the Author has visited. The results synthesized in
a conclusive table are as follows:
1) The superficial morphology of the lava streams, different in
   the two principal types of lava distiguished, (lava in slabs
   and lava in blocks), determines, in both types of lava, different and numerous substrata interwoven with each other and
   colonized by different plant-communities. These, therefore,
   have a mosaic like arrangement in the space, constituting, as
   a whole, a small group of mosaic-like vegetation, which remains independant of the degree of vegetation evolution.
2) The different ages of the lava streams is significant in determining the limits of the vegetation only if it concerns
   not very recent volcanic regions. In this case, entire lava
   streams are colonized by one, or few plant formations, more
   or less close to the final stage of the evolution of the ve-

tation.

Between contiguous lava streams of different ages, the limits are very evident, since they were colonized by different plant-communities. In this case, the plant landscape
is composed of vast groups of plant formations.

3) The heat of the soil and to be exactly, its thermic gradient,
   determine a horizontal zonation of the plant-communities according to the debreasing temperature values, as one gradually
   moves away from the source of heat.
   This is noticable above all in thermal springs and fumaroles.

4) The chemical composition of the soil, and in particular its
   degree of acidity, also determines a horizontal zonation of the
   plant-communities, something which is particularly noticable
   Solfataras.

5) The volcanic activity in the craters and in particular its
   frequency, determines a particular vertical zonation on the
   sides of the craters and in the regions nearest to them.

However this zonation has only edaphic significance since it
depends directly on the activity of the volcano.
Examples of this type of zonation have also been found in large volcanic regions.

LITERATUR

ASAI,T.   -1952a-   Die Vegetation auf den Kegeln des Aso-Vul-
     kans.-Kumamoto J.of Science 1: 1-31.   Kumamoto.
-- -- -1952b-   Zur Ökologie der Vulkanpflanzen von Asosan.-
     Kumamoto J.of Science 1: 33-81.   Kumamoto.
AUBERT DE LA RUE,E.   -1958-   L'homme et les volcans.- Geographie
     Humaine 30. Collection dirigée par Pierre Deffontaines.
     Paris.
BRAUN-BLANQUET,J.   -1964-   Pflanzensoziologie.- Wien-New York.
BRUNEAU DE MIRE,P.& QUEZEL,P.   -1959-   Sur quelques aspects de
     la flore résiduelle du Tibesti: les fumerolles du Toussi-
     dé et les lappiaz volcaniques culminaux de l'Emi Koussi.-
     Bull.Soc.Hist.Nat.de l'Afrique du Nord 50: 126-145. Alger.
BURKILL,J.H.   -1926-   Vegetation on lava surfaces of various
     ages in the crater of Kilauea,Hawai.- Proc.Linn.Soc.Lon-
     don 138: 23.   London.
EGGLER,W.A.   -1959- . Manner of invasion of volcanic deposits by
     plants, witk further evidence from Paricutin and Jorullo.-
     Ecol.Monogr.29 (3): 267-284.   Durham,N.C.
ERNST,A. -1910-   Die Besiedlung vulkanischen Bodens auf Java
     und Sumatra.- Vegetationsbilder, Reihe 7 (1-2). Jena
FABER,F.C.   -1927-   Die Kraterpflanzen Javas in physiologisch-
     ökologischer Beziehung.- Arbeiten aus dem Treub-Laborato-
     rium: 1-119. Buitenzorg.
FOSBERG,F.R.   -1959-   Upper limits of Vegetation of Mauna Loa,

Hawai.- Ecology 40 (1): 144-146. Lancaster,Pa.
GHIGI,A. -1961- Tra i vulcani di Hawaii.- Natura e Montagna,
    se.2 (1): 23-31.Bologna.
HAYATA,B. -1926- Succession in the vegetation of Mt.Fuji and
    the formulation of a new theory, the succession theory,
    in opposition to the natural selection theory.- Proc.of
    third Pan-Pacific Science Congress,Tokyo: 1867-1868. Tokyo.
KLAUSING,O. -1958- Untersuchungen über Vegetation und Wasser-
    haushalt am Volcan de San Salvador.- Ber.deutsch.bot.Ges.
    71 (10): 439-452. Stuttgart.
KUMAGAI,S.& TSUKADA,T. -1960- Succession of plant community
    on the lava flow in the Tockachi.- Hoppo-Ringyo 140-141:
    384-422.
KUWABARA,Y. -1957ā- On the vegetation of volcanic groups of
    Iwaonupuri in Hokkaido.- Hokuriku J.of Botany 6 (3): 81-
    84.
-- -- -1957b- On the vegetation of volcanic groups of Iwao-
    nupuri in Hokkaido.- Hokuriku J.of Botany 6 (4): 94-98.
LEONARD,A. -1958-1959- Contribution à l'étude de la colonisa-
    tion des laves du volcan Nyamuragira par les végétaux.-
    Vegetatio 8: 250-258. Den Haag.
LÖTSCHERT,W. -1959- Vgetation und Standortsklima in El Salva-
    dor.- Botan.Studien 10: 1-79. Jena.
LOJACONO,M. -1878- Le isole Eolie e loro vegetazione.- Palermo.
MAINI,S. -1963- Recenti insediamenti Crittogamici nei pressi
    delle Fumarole Vesuviane.- Giorn.Bot.It.,n.s.70 (5-6):
    547-522. Firenze.
MATSUURA,S. -1961- Ecological study of solfatara on Owakidani,
    Hakone.(Part.I).- Bull.Nat.Hist.Hakone 1: 1-6. Hakone.
MEROLA,A. -1957- Osservazioni sull'ecologia e sulla biologia
    dei vegetali viventi presso le fumarole.Nota I. Termotro-
    pismo radicale e riscaldamento del terreno in Erica arbo-
    rea L.- Delpinoa n.s.10: 5-20. Napoli.
-- -- -1951- Osservazioni sull'ecologia e sulla biologia dei
    vegetali viventi presso le fumarole.Nota II. Lo sviluppo
    dell'apparato radicale in Myrtus communis L. cresciuto su
    terreno fumarolico.- Boll.Soc.Natur.Napoli 66: 3-6- Napoli.
MÜLLER-DUMBOIS,D. -1974- Aspekte der Sukzessionsforschung
    auf der Insel Hawai. In: Tüxen,R.(Edit.): Fragen der Ge-
    sellschaftsentwicklung, Syndynamik.- Ber.Intern.Sympos.
    1967 in Rinteln. Lehre.
OHBA,T. -1969- Eine pflanzensoziologische Gliederung über die
    Wüstenpflanzengesellschaften in der alpinen Stufe Japans.-
    Bull.Kanagawa Pref.Mus.1(2): 24-70. Kanagawa.
PEDICINO,N. -1873- Poche osservazioni presso le terme.- Rend.
    R,Accad.Sci.Fis.Mat.5. Napoli.
POLI,Emilia -1964- L'Etna e il Fujiyama.Cenni geobotanici com-
    parativi.- Ann.Botanica 28 (1): 125-148. Roma.
-- -- -1965- La vegetazione altomontana dell'Etna.- Flora et
    Vegetatio Italica 5: 1-251. Sondrio.
-- -- -1970- Überblick über die Vegetation der Hochgebirgs-
    Stufe des Asama-Vulkans (Japan).- Vegetatio 120 (1-4):
    74-96. The Hague.
-- -- -1970-1971- Aspetti della vita vegetale in ambienti vul-
    canici.- Ann.Botanica 30: 47-80. Roma.
RIPPA,A. -1936- Cenni sulla vegetazione del cratere della Sol-
    fatara di Pozzuoli.- Bull.Orto Bot.Napoli 13: 11-20.Napoli.
RITTMANN,A. -1963- Les volcans et leur activité.- Paris.
ROBIJNS,W.& LAMB,S.H. -1939- Preliminary ecoligical survey
    on the island of Hawaii.- Bull.Jard.Bot.de l'Etat 15:

241-344. Bruxelles.
ROBYNS,W. -1932- La colonisation végétale des laves récentes
     du volcan Rumoka (Laves de Kateruzi).- Mem.Inst.Roy.Colo-
     nial Belge,Sect.Sc.Nat.Med.1(1): 3-29. Bruxelles.
SCHIMPER,A.F.W. -1935- Pflanzengeographie auf physiologischer
     Grundlage.- Dritte Auflage v.F.C.von Faber.Jena.
SCHMÜCKER,T. -1927- Beiträge zur Kenntnis der Hochgebirgsflo-
     ra Javas und zur Theorie der Pflanzenausbreitung.- Beih.
     Bot.Zblatt 5,43,II,34-68. Jena
SKOTTSBERG,C. -1930- Remarks on the flora of the high Hawaii-
     an volcanoes.- Medd.bot.Trädgärd 6: 47-65.
-- -- -1941- Plant succession on recent lava flows in the
     island of Hawaii.- Göteb.Kungl.Vet.och Vitterhets-Sam-
     hälles Handl.,Sjätte Följden,Ser.B 1(8): 4-32. Göteborg.
STEINDORSSON,S. -1957- Um groour i Reykjaneshraunum.- The
     Vegetation in the Lava-field of the Reykjanes peninsula
     South-West Iceland.- Arsrit Raektunarfelags Norourlands,
     54: 137-150. Akureyri.
TAGAWA,H. -1964- A study of the volcanic vegetation in Saku-
     rajima South-west Japan. I.Dynamics of vegetation.- Mem.
     Faculty of Science, Kyushu University, Ser.E(Biology) 3
     (3-4): 166-225. Fukuoka.
-- -- -1965- A study of the volcanic vegetation in Sakurajima,
     southwest Japan.II. Distributional pattern and successi-
     on.- Jap.J.Bot.19 (6): 127-148. Tokyo.
-- -- -1966- A study of the volcanic vegetation in Sakurajima,
     South-west Japan. III.Trap sampling of disseminules on the
     lava flow and the culture experiment of some pioneer
     mosses.- Science Reports Kagoshima University 15: 63-83.
     Kagoshima.
TEZUKA,Y. -1961- Development of vegetation in relation to soil
     formation in the volcanic Island of Oshima,Izu,Japan.-
     Jap.J.Botany 17 (3): 371-402. Tokyo.
TOBLER-WOLFF,Gertrud & TOBLER,F. -1914-1915- Vegetationsbil-
     der vom Kilimandscharo.- Vegetationsbilder v.Karsten u.
     Schenck 12: 2-3. Jena.
TOHYAMA,M. -1962- Forest Vegetation in the Central Cone, Mt.
     Hakone,Central Honshu, Japan.- Mem.Faculty Agriculture,
     Hokkaido University 4 (1): 7-23. Sapporo.
TURNER,E.P. -1928- A brief account of the re-establishement
     of vegetation on Tarawera Mountain since the eruption of
     1886.- Transact.and Proceed.N.Z.Inst.59: 60-66.Wellington.
WERNER,D. -1968- Naturräumliche Gliederung des Ätna. Land-
     schaftsökologische Untersuchungen an einem tätigen Vul-
     kan.- Göttinger Bodenkundl.Ber. 3: 1-171. Göttingen.
YOSHII,Y. -1932- Revegation of Volcano Komagatake after the
     great eruption in 1929.-Botan.Magazine 46 (544): 208-219.
     Tokyo.
-- -- -1937- Aluminium Requirements of Solfatara-plants.- Bo-
     tan.Magazine 51 (605): 262-269.(Shibata Commemoration N.1)
     Tokyo.
-- -- -1939- Untersuchungen über vulkanische Pflanzengesell-
     schaften.- Ecological Study 67:203-290. Berlin,Heidelberg,
     New York.
-- -- -1940 Untersuchungen über vulkanischen Pflanzengesell-
     schaften.- Ecological Study 74-76: 59-160. Berlin,New
     York,Heidelberg.
-- -- -1942- Pflanzengesellschaften nach Ausbruch auf dem Ko-
     magatake. Ecological study 170-220.
YOSHIOKA,K. -1950- Studies on the vegetation of Mt.Zao.- Sci.

Rep.Tohoku Univ.4th.Ser.(Biology)18 (3): 342-350.
YOSHIOKA,K. -1958- The vegetation on the lava flow in comparison with those of surrounding areas.- Sci.Rep.Faculty Art and Sci.Fukushima Univ.7: 45-57. Fukushima.
-- -- -1965- Development and recovery of vegetation since the 1929 eruption of Mt.Komagatake,Hokkaiodo.- Ecol.Rev.16 (4): 271-292. Sendai.
-- -- SAITO,K.& TACHIBANA,H. -1965- Solfatara vegetation at Osoreyama.- Ecol.Rev.16 (3): 137-151. Sendai.

LEGENDE ZU DEN TAFELN

Abb.1. Kleinflächiger mosaikartiger Gesellschafts-Komplex auf
einem alten Schollenlava-Strom des Aetna (etwa 2000 m). Auf den
Fels-Flächen R h i z o c a r p o n   g e o g r a p h i c u m-
Ges., in den weniger tiefen Spalten F e s t u c a' l e v i s-
P o a   a e t n e n s i s-Ges., in tieferen Spalten Zwergsträu-
cher mit Berberis aetnensis.
Abb.2. Großräumiger mosaikartiger Vegetations-Komplex auf dem
SO-Abhang des Aetna (etwa 1900-2000 m). Das A s t r a g a l e-
t u m   s i c u l i, die am weitesten verbreitete Gesellschaft
des Gebietes, ist von den Kratern und vom Lava-Strom von 1910
an mehreren Stellen zerstört worden.
Abb.3. Aetna, O-Abhang etwa um 1000 m. Das Bild zeigt, wie
scharf die Vegetations-Grenze zwischen zwei verschiedenaltrigen
Lava-Strömen ist.Links die G e n i s t a   a e t n e n s i s -
Ges. auf einem sehr alten Lava-Strom, rechts ein Lava-Strom mit
typischer Brocken-Oberfläche (vom Jahre 1852), hier von Flech-
ten- und Moos-Gesellschaften besiedelt.
Abb.4. Solfataren-Gebiet in Hakone (Japan). Die Gürtelung der
Vegetation ist hier sehr klar. Von vorne nach hinten sind zu
unterscheiden: nackte Zone, Gräser-Zone mit Miscanthus sinensis,
und Busch-Zone vermutlich von Pieris japonica.
Abb.5. Teilansicht der Miscanthus sinensis-Zone. Dieses Gras
ist sehr häufig an den japanischen Solfataren, weil es den hohen Säuregehalt des Bodens (pH 2,5) verträgt.
Abb.6. Ledum palustre ssp. decumbens-Zone bei der Solfatare Io-
san (=Schwefelberg) in Japan. Die zwei Ledum-Gürtel (im Vordergrund und im Hintergrund) sind durch die äußerst saure nackte
Zone getrennt. Auch diese E r i c a c e e paßt sich dem hohen
Säure-Gehalt des Bodens sehr gut an.
Abb.7. Vulkan-Wüsten-Zone bei dem tätigen Vulkan Aetna, auf dem
sie mindestens 300 m hoch ist.
Abb.8. Auf dem sich seit langem in Ruhe befindlichen Vulkan
Fuji-san erstreckt sich bis zum Kraterkegel die klimatisch bedingte Kryptogamen-Stufe. Auf dem Bild eine S t e r e o c a u-
l o n   v e s u v i a n u m-Ges.

Abb.1

Abb.2

Abb.3

Abb.4

Abb. 5

Abb. 6

Abb.7

Abb.8

Abb.9

Abb.9. R h a c o m i t r i u m-Ges. am Haupt-Krater des Fuji-
san (3776 m), zur klima-bedingten Kryptogamen-Stufe gehörig.

Alle Aufnahmen von der Autorin.

GRENZPROBLEME IN AUSTRALISCHEN LANDSCHAFTEN

H. D o i n g

Aus der Fülle von Problemen, welche die noch so wenig untersuchte australische Vegetation darbietet, sollen einige herausge-
griffen werden, welche sich auf das Thema dieses Symposium beziehen, um damit die Vegetation dieses Erdteils mit zur Diskussion zu bringen.
Die hier zu behandelnden Beispiele, welche sich fast alle auf
konkrete Grenzen beziehen, können folgenden Haupt-Themen zugerechnet werden:
1. Natürliche(Klimax-)Vegetation, beherrscht von Eucalyptus-Arten.
2. Menschlich bedingte Vegetation (hauptsächlich Grünland- und
   Unkraut-Gesellschaften).
3. Küsten-Vegetation (größtenteils natürliche Dauergesellschaften).
4. Allgemeine Bemerkungen über die wichtigsten Formationen.

1. Die Eucalyptus-Wälder und -Gebüsche sind in den meisten Gebieten stabile Gesellschaften. Je dichter die Vegetation, desto
wichtiger werden aber Brände als katastrophaler Faktor. Es gibt
Gebiete, wo Vernichtung des Waldes durch Brand eine Sukzession
zur Folge hat, wobei verschiedene Baum-Arten (Acacia, Eucalyp-
tus oder Regenwald-Arten) miteinander abwechseln. Dabei können
sehr scharfe natürliche Grenzen entstehen, z.B. zwischen Regenwald (welcher im Klimax-Stadium keine Eucalyptus-Arten enthält)
und Eucalyptus-Wald. Umgekehrt liegt beim Auftreten solch scharfer Grenzen der Verdacht nahe, daß Brand die Hauptursache dafür
gewesen sein könnte. Im allgemeinen werden die Grenzen zwischen
Regen- und Eucalyptus-Wald hauptsächlich bestimmt von: a.Länge
der Trockenperioden (kontinentale Regenwald-Grenze), b.Tempera-
tur, besonders Frostvorkommen (kontinentale, montane und polare
Regenwald-Grenze), c.Boden-Fruchtbarkeit (WEBB 1963), d.Brand.

Es ist aufschlußreich, die Eigenschaften des tropischen Regen-
waldes (der temperierte Regenwald ist in mancher Hinsicht sehr
verschieden), des Eucalyptus-Waldes (COSTIN 1952) und der europäischen Fallaub-Wälder in Bezug auf Grenz-Erscheinungen in der
Landschaft miteinander zu vergleichen.
Bei dem tropischen Regenwald hat man es in der Regel mit ver-

hältnismäßig kleinen Art-Arealen zu tun.Auf einer gewissen Land-
oberfläche ist die Zahl der Baum-Arten sehr groß, sowohl innerhalb eines geographischen Gebietes, als auch in einer einzelnen
Gesellschaft. Die verschiedenen Arten kommen meistens nicht in
großen Mengen vor. Es gibt in der Regel nur wenige Kräuter und
Zwergsträucher, die außerdem diagnostisch nicht sehr wichtig
sind (WEBB c.s.1967). Die Kenn-Arten sind hauptsächlich hoch-
wüchsige Baum-Arten. Die verschiedenen Gesellschaften sind meistens unscharf gegeneinander begrenzt.
Auch im Eucalyptus-Wald sind die Art-Areale oft ziemlich klein,
die Zahl der Baum-Arten innerhalb eines Gebietes ist groß, und
die Krautschicht,die auch diagnostisch nur geringen Aussagewert
hat,ist artenarm. Die einzelnen Gesellschaften sind dagegen in
der Regel arm an Baum-Arten, die jedoch in großer Individuen-
Zahl auftreten. Die einzelnen Baum-Arten können Kenn-Arten sein,
öfter aber sind es die Baumarten-Kombinationen, welche die Assoziationen charakterisieren.In manchen Gesellschaften spielen außerdem die Sträucher und Zwergstrauch-Arten eine wichtige Rolle,
sowohl in der Arten-Anzahl als auch diagnostisch. Im Savannen-
Wald dominieren grasartige Pflanzen im Unterwuchs, welche für
diesen als höhere Einheit großen, für die einzelnen Assoziationen aber geringen diagnostischen Wert haben. Die Eucalyptus-Ge-
sellschaften sind, abgesehen von Sukzessions-Reihen, in der Regel recht scharf gegeneinander begrenzt.
In den europäischen Fallaub-Wäldern haben im Verhältnis dazu
die Bäume und Sträucher große Art-Areale. Einzelne Gebiete sind
arm an Baum-Arten. Die Gesellschaften sind es meistens auch, wobei oft dominierende Baum-Arten im Unterwuchs sehr verschiedene
Arten-Kombinationen enthalten können.Kenn-Arten befinden sich
darum besonders unter den Kräutern und Gräsern. Die Gesellschaften können scharf oder unscharf gegeneinander abgegrenzt sein.

Der tropische Regenwald hat also am stärksten den Charakter eines "Kontinuums". Im europäischen Fallaub-Wald sind die Grenzen
manchmal unscharf. Es lassen sich aber doch deutliche Typen unterscheiden, die gewöhnlich an sehr vielen Stellen wiederzufin-
den sind, und die Kartierung erleichtern. Im Eucalyptus-Wald
kehren die Typen nicht so oft wieder, und die Landschaft zeigt
ein buntes Mosaik von Gesellschaften, welche sich durch die
scharfen konkreten Grenzen leicht kartieren lassen.
2. Bei der Entwicklung von natürlichen zu menschlich bedingten

Gesellschaften,unter dem Einfluß von Beweidung und der Zerstö-
rung der Baumschicht, lassen sich drei Hauptstadien unterscheiden:
a. Extensive Beweidung. Die meisten Bäume werden (oft allmählich) entfernt oder getötet (durch Ringeln), viele Arten der ursprünglichen Vegetation verschwinden, einige andere breiten sich
dagegen stark aus. Es entsteht größere Einförmigkeit, manche
Grenzen werden verwischt.
b. Zunehmende Beweidung, keine Düngung oder Ansaat. Ein neues
Gleichgewicht stellt sich ein. Viele neue einheimische Arten,
welche in der ursprünglichen Landschaft eine untergeordnete Rolle spielten, und einige importierte Arten siedeln sich an. Die
Variation nimmt stark zu und ist oft größer, als die in der natürlichen Landschaft. Die Zahl der Grenzen nimmt damit auch zu,
aber sie werden oft weniger scharf.
c. Mit weiterer Intensivierung der Beweidung, Düngung (Superphos-
phat) und Ansaat (Trifolium subterraneum) stellen sich ziemlich plötzlich importierte Gesellschaften ein, welche hauptsächlich aus europäischen, zum Teil auch aus amerikanischen, afri-
kanischen und australischen Arten bestehen. Die Artenzahl und
Variation nimmt immer stärker ab, die Schärfe der Grenzen nimmt
vorübergehend zu, kann aber in extrem artenarmen Parzellen
schließlich wieder abnehmen.
Gegenüber den räumlichen, kann man auch von "zeitlichen Grenzen" sprechen. Das Entstehen des Stadiums"a"geschieht katastro-
phal, stellt also eine scharfe zeitliche Grenze dar. Danach
kommt eine stabile Periode, mit unscharfen zeitlichen Grenzen,
bis zu dem Zeitpunkt, an dem die Beweidung eine gewisse Stärke
überschreitet. Diesem ziemlich schnellen Übergang zum Stadium"b"
folgt eine stabile Periode von langer Dauer (in Australien jetzt
schon über ein Jahrhundert. In Europa können die damit überein-
stimmenden Stadien über ein Jahrtausend existiert haben). Mit
starker Phosphat- und Nitrat-Anreicherung kommt dann eine zweite Katastrophe, welche zum Stadium"c"führt, während dessen ist
noch eine Reihe allmählicher Veränderungen möglich.
Die Existenz dieser Stadien ist eine vielfach bestätigte Erfah-
rungs-Tatsache (TIVER & CROCKER 1952, DOING 1972). Durch die
oft sehr großen und uneinheitlichen Parzellen und durch das Nebeneinander von Stadien in der wirtschaftlich genutzten Landschaft, die in Europa gewöhnlich weit von einander entfernt wa-

ren, sind auch räumlich oft nicht klar getrennt. Die Landschaft
ist in dieser Hinsicht voller Übergangs- und Regenerationspha-
sen, Resten ursprünglicher Vegetation und durchzogen von Erosi-
ons-Rinnen, versalzten oder durch Dammbau vernäßten Stellen,
Vieh-Lagerplätzen usw. Erst durch sorgfältige Vegetations-Aufnahmen von kleinen, homogenen Beständen, lernt man allmählich
die Typen zu unterscheiden. Bestimmte Differenzierungen zwischen einheimischer und importierter Vegetation lassen sich dabei überhaupt nicht lokalisieren. Die Gegensätze mit der europäischen Landschaft, mit ihren scharf begrenzten, kleinen Parzellen mit stabiler Nutzungsweise, wo die natürlichen räumlichen
Grenzen an manchen Stellen verwischt, und an anderen Stellen
durch jahrhundertlange menschliche Beeinflussung eher verschärft
sind, sind sehr groß (DOING 1966 c). Hier ist ein grobkörniges
Mosaik mit oft scharfen Grenzen entstanden - dort ist die Kultur-Landschaft ein Gewebe von vielen Fäden (Arten-Gruppen), wo-
rin sich erkennbare Figuren wiederholen, aber wo die räumlich
homogenen Flecken so klein, die Grenzen so unscharf und die Veränderungen so verschiedenartig und so schnell sind, daß z.B.
die Kartierung von Gesellschaften, auch wenn diese sich auf
Grund von Aufnahmen theoretisch oft noch recht gut unterscheiden lassen, für praktische Zwecke unausführbar wird.
3. Die Küsten-Gesellschaften sind meistens verhältnismäßig artenarm und scharf begrenzt, oft von wenigen, dominanten Arten
beherrscht und spontanen Veränderungen unterworfen (BIRD 1965).
Die Landschaften, in denen sie vorkommen, lassen sich mit denen
in Europa viel besser vergleichen als in den beiden vorhergehenden,hier diskutierten Fällen. Dies hängt wohl damit zusammen,
daß sich für Spezialisten-Gesellschaften im allgemeinen weltweite Vergleiche recht gut durchführen lassen, was z.B. auch für
Wasser- und Sumpf-Gesellschaften, Zwergstrauch-Heiden u.dgl.zu-
trifft. Die australischen Küsten-Gesellschaften sind oft nur
kleinflächig entwickelt, aber prinzipiell nicht schwer zu kartieren. Mangroven (z.B. Avicennia marina), Strand-Gesellschaf-
ten (z.B. Cakile edentula oder C.maritima), Weißdünen (z.B. Spinifex hirsutus), Gebüsch-Ränder (z.B. Lomandra longifolia) und
Salzsprüh-Gebüsche an dem Meer zugewandten Felsen (z.B. Westrin-
gia rosmariniformis) bilden gewöhnlich schmale Bänder, welche
sich auf die obengenannte Art charakterisieren lassen. Weltweite Vergleiche von parallelen Zonen zeigen ein fast universelles

Vorrücken der Gehölz-Grenze in immer größere Meeresnähe mit zunehmender Jahres-Temperatur, zum Teil auch mit zunehmenden Jah-
res-Niederschlägen und gleichmäßigerer Verteilung der Niederschläge.

4. Zu den australischen Formationen muß im allgemeinen noch bemerkt werden, daß diese, oft zu Unrecht den Formationen in anderen Erdteilen zugerechnet, falsch interpretiert oder zu nichts-
sagenden Sammel-Einheiten verallgemeinert werden (z.B. DANSERE-
AU 1957, FIRBAS 1967, POLUNIN 1960, SCHMITHÜSEN 1959, vgl.dazu
BEADLE 1951, DOING 1966b, WILLIAMS 1955). Die Regenwälder, die
C h e n o p o d i a c e e n-Halbwüsten, die alpinen Rasen-Gesellschaften, die meistens ausgesprochenen Spezialisten-Gesellschaften und vielleicht auch die von Akazien beherrschten "Baum-
wüsten" (DOING 1966a) und die tropischen und subtropischen Gras-
Savannen und Savannen-Wälder (vgl. BEARD 1967) lassen sich gut
vergleichen mit parallelen Formationen in anderen Erdteilen.
Die geschlossenen Eucalyptus-Wälder bilden aber eine Formation,
welche ausschließlich in Australien (und Tasmanien) vorkommt,
und es ist irreführend, diese teils als "Hartlaub-Wälder" teils
als "Regen-Wälder" zu interpretieren, wie dies auf manchen Welt-
Karten geschehen ist. "Steppe" im Sinne einer baumlosen, geschlossenen, kurzrasigen, sommertrockenen Gras-Formation,
"Dornbusch" und vegetationslose Wüste gibt es in Australien
nicht, oder wenigstens nicht in solchen Ausmaßen, die dargestellt werden können im üblichen Maßstab 1:100.000.000. Die
"Mallee" (hohe Eucalyptus-Gebüsch-Gesellschaften im trocken-me-
diterranen Klima) und die "Spinifex-Wüste" (Igelgras-Polster-Ge-
sellschaften der Halbwüste im Innern Australiens, besonders auf
Sanddünen) lassen sich physiognomisch kaum mit den Gesellschaften in parallelen Klima-Gebieten anderer Kontinente vergleichen.
Wie vorsichtig man hierbei, sogar dann, wenn die Lebensformen
oberflächlich übereinstimmen, sein muß, zeigt TROLLS (1958)
Zuordnung der australischen "Grasbäume" (Xanthorroea, Kingia
und Dasypogon) zum "Dracaena-Typ" der Trocken-Savannen. Diese
Pflanzengruppe zeigt in Australien eine ausgesprochene Vorliebe
für ärmere, schwer durchlässige Böden in küstennahen Gebieten
mit hohen Niederschlägen außerhalb der Tropen!

ZUSAMMENFASSUNG

Beispiele wichtiger Grenzen in der australischen Vegetation werden besprochen und durch Diapositive belegt.
1. Eucalyptus-Wälder werden verglichen mit tropischen Regen-
und europäischen Fallaub-Wäldern in Beziehung zu ihren inneren
und äußeren Grenzen.
2. Bei Grünland- und Unkraut-Gesellschaften wurden eine Reihe
von anthropogenen Stadien und ihre zeitlichen Grenzen diskutiert.
3. In der Küsten-Vegetation zeigte sich eine Situation, die mit
derjenigen in Küsten-Landschaften in Europa gut vergleichbar ist.
4. Schließlich wurden die australischen Formationen mit dem üblichen System der Formations-Klassen verglichen (siehe z.B. die
noch immer viel verwendete Welt-Karte von BROCKMANN-JEROSCH).
Für die tropischen und subtropischen Regen- und Savannen-Wälder
und für manche alpinen und ariden Gesellschaften ist eine Einordnung ohne besondere Schwierigkeiten möglich. Der geschlossene
Eucalyptus-Wald, die "Mallee" und die "Spinifex-Wüste" sollten
aber auch auf Welt-Karten kleinsten Maßstabs als eigene Formations-Klassen dargestellt werden. Die Unterscheidung von "Steppen", "Dornbusch" oder"Hartlaub-Wäldern" in Australien ist irreführend (BEADLE 1951, DOING 1966b).

SUMMARY

Problems of vegetation boundaries in Australian landscapes.

A brief review is given of some problems concerning boundaries
within the following types of Australian vegetation:
1. Eucalypt forests, as compared with tropical rainforests and
European deciduous forests;
2. grazed communities, mainly in relation to various degrees of
human influence;
3. coastal vegetation;
4. Australian vegetation formations in general, compared with
formations in similar climatic regions in other parts of the
world.
It is stressed that Eucalypt forests, Eucalypt shrubs ("mallee")
and Triodia-desert ("spinifex")should be distinguished on world
maps of vegetation (even those on an extremely small scale) as
separate formations.

LITERATUR

BEADLE,N.C.W.   -1951-   The misuse of climate as an indicator of
        vegetation and soils.- Ecology 32: 343-345. Durham,N.C.
BEARD,J.S.   -1967-   Some vegetation types of tropical Australia
        in relation to those of Africa and America.- J.Ecology
        55: 271-290. Oxford.
BIRD,E.C.F.   -1965-   A geomorphological study of the Gippsland
        Lakes. The Australian National University. Research
        School of Pacific Studies. Dept.of Geography Publication
        G/1. Canberra.
COSTIN,A.B.   -1952-   The ecosystems of the Monaro Region of New
        South Wales.- Sydney.
DANSEREAU,P.   -1957-   Biogeography.- New York.
DOING,H.   -1966a-   Enkele opmerkingen over het begrip "hoofd-
        formatie".- Gorteria 3: 5-11. Leiden.
-- --   -1966b-   Het landschap rondom Australie's hoofdstad.-
        De Levende Natuur 69: 171-178, 201-211, 230-239. Arnhem.
-- --   -1966c-   A comparison of pasture-weed problems in tempe-
        rate Australia and in Europe.- J.Australian Institute of
        Agriculture Science 32: 302-304. Sydney.
-- --   -1972- Botanical composition of pasture and weed commu-
        ties in the Southern Tablelands region.- C.S.I.R.O. Div.
        of Plant Industry Techn.Pap.No.30. Melbourne.
FIRBAS,F.   -1967-   Pflanzengeographie.- In: Strasburger: Lehr-
        buch der Botanik für Hochschulen.29.Aufl. Stuttgart.
POLUNIN,N.   -1960-   Introduction to Plant Geography.- London.
TIVER,N.S.& CROCKER,R.L.   -1951-   The grasslands of South-East
        South Australia in relation to climate, soils and deve-
        lopmental history.- J.British Grassland Society 6: 29-80.
        Aberystwyth.
TROLL,C.   -1958-   Zur Physiognomik der Tropengewächse.- Jahres-
        ber.Gesellsch.Freunden u.Förderern d.rheinischen Fr.-Wil-
        helms-Univ.Bonn e.V. Bonn.
WEBB.L.J.   -1963-   The influence of soil parent materials on the
        nature and distribution of rainforests in South Queens-
        land.- The symposium on ecological research in humid tro-
        pics vegetation. (Kuching, Sarawak). Tokyo.
-- --   TRACEY,J.G., WILLIAMS,W.T.& LANCE,G.N.   -1967-   Studies
        in the numerical analysis of complex rainforest communi-
        ties. II. The problem of species sampling.- J.Ecology 55:
        525-538. Oxford.
WILLIAMS,R.J.   -1955-   Vegetation regions.- In: Atlas of Austra-
        lian Resources. Dept.of National Development. Canberra.

I.ZONNEVELD:

Wir haben gesehen, daß die natürliche Waldgrenze scharf oder
auch aufgelockert und allmählich sein kann. Eine scharfe Baum-
Grenze kann das Resultat von Massen-Wirkung, also ein Massen-
Effekt sein, wie er bei WESTHOFF und auch von mir für Wasser-
Vegetationen beschrieben wurde. Das grundlegende Prinzip ist:
"Einigkeit macht stark". So etwas hilft gegen ungünstige Fakto-
ren, denen durch Zusammenarbeit (d.h. Zusammenschluß der Indivi-
duen),zu widerstehen ist. Auch mechanische Angriffe gehören da-

zu wie Wellenschlag und strömendes Wasser, aber auch Temperatur,
Sonnen-Strahlung, Austrocknung und Luftfeuchtigkeit. In dichten
geschlossenen Pflanzen-Beständen kann das durch die Vegetation
bedingte Mikro-Klima abgeändert werden. Die scharfe Buchenwald-
grenze, die uns aus den Vogesen gezeigt wurde, war temperatur-
bedingt. Innerhalb des Waldes wird die Temperatur nivelliert,
die großen Unterschiede sind nicht mehr da. Gegen die meisten
edaphischen Faktoren (auch hydrologisch-edaphische und durch
Klima beeinflußte Boden-Eigenschaften) hilft das Zusammenschließen nicht; gerade das Umgekehrte kann unter Umständen besser
sein. Wo edaphische Faktoren waldbegrenzend werden, wird in vielen Fällen das horizontale edaphische Gefüge das Vegetations-
Gefüge bestimmen, wenn das edaphische Gefüge aufgelockert ist
(Mikrorelief, Termiten-Aktivität usw.). Auf relativ kleinen Maßstab übertragen, kann der zuletzt genannte Zustand einen relativ kontinuierlichen Übergang der Wald-Grenze bewirken. Es besteht dann augenscheinlich ein Paradoxon: Ein an sich allmählich
fortschreitender Klima-Gradient bewirkt mit Hilfe des Massen-
Effektes eine scharfe Diskontinuität. An sich diskontinuierli-
che edaphische Faktoren bewirken aber (besonders bei kleinem
Maßstab gezeichnet), eine mehr allmähliche oder aufgelockerte
Wald-Grenze. Der endogene Einfluß der Vegetation ist in dem
ersten Fall ganz klar.

VEGETATIONSGRENZEN UND BODENREAKTION

W. L ö t s c h e r t

Einleitung

Die Erkenntnis, daß die Boden-Reaktion für die Verteilung von
Einzelpflanzen und Vegetation entscheidend ist, knüpft an die
Beobachtungen über die Boden-Stetigkeit alpiner Pflanzen an.
Sie führten nach der Einführung der pH-Meßtechnik zu dem Ergeb-
nis, daß der pH-Faktor die Verteilung bestimmter Arten und Ge-
sellschaften bestimmt und somit für die kausale Erklärung von
Vegetations-Grenzen entscheidend ist. Dabei ist der pH-Wert nur
ein Faktor im komplexen Reaktions-Ablauf des Bodens, der über-
dies beträchtlichen jahreszeitlichen Schwankungen unterliegt.
(LÖTSCHERT 1965).

Aufgrund eingehender Standorts-Analysen gelangt PEARSEAL (1952)
zu dem Ergebnis, daß im nordwestlichen Mitteleuropa wichtige
pH-Grenzen bei 3,9; 4,9 und 6,5 liegen. Die Böden über pH=6,5
sind basengesättigt und durch gute Krümel-Struktur ausgezeichnet. Die Böden mit pH=6,5-4,9 sind in der Regel kalkarm, aber
noch genügend basenhaltig, um gute Krümel-Struktur mit günstiger Wasser-Bilanz zu garantieren. Zwischen pH=4,9-3,9 sind die
Nährstoff-Verhältnisse mäßig, während die Böden mit pH<3,9 als
nährstoffarm angesehen werden müssen. (Vgl.auch ELLENBERG 1958).
Eine wichtige Grenze liegt bei pH=3,9. Während bei pH 3,9 in
der Regel Braunerden ausgebildet sind, herrschen bei pH<3,9 Pod-
sol-Böden mit Rohhumus- und Moder-Auflagen.

Die schwedischen Moore teilt SJÖRS (1950) anhand der durchgeführten pH-Messungen wie folgt ein:

Mosses	pH=3,7-4,2
Extreme poor fens	pH=3,8-5,0
Transitional poor fens	pH=4,8-5,7
Intermediate fens	pH=5,2-6,4
Transitional rich fens	pH=5,8-7,0
Extreme rich fens	pH=7,0-8,4

Diese beiden Beispiele zeigen, daß der Boden-Reaktion bei der
Analyse von Vegetations-Grenzen im Gelände entscheidende Bedeutung zukommt. Dabei ist die Frage, ob der pH-Faktor auch in den
Tropen zur Kausal-Analyse des Vegetations-Mosaiks verwendet werden kann, bisher wenig beachtet worden (vgl.LÖTSCHERT 1959).
Dies ist einerseits darauf zurückzuführen, daß die Frage der Boden-Reaktion ihren Ausgang von Europa genommen hat, andererseits

darauf, daß die Standortsfaktoren-Analyse in den Tropen wesentlich schwieriger ist.
Ich möchte die Bedeutung der Boden-Reaktion für das Zustande-
kommen der Vegetations-Grenzen anhand von drei Beispielen dar-
legen, die aus verschiedenen Arbeitsgebieten und verschiedenen
Vegetationskomplexen stammen.

1. C e p h a l a n t h e r o - und L u z u l o - F a g e - t u m an der nördlichen Bergstraße

Im Gebiet der nördlichen Bergstraße südlich Darmstadt besteht
der Westrand des Odenwaldes aus Melibokus-Granit. Er grenzt mit
einer Verwerfung an den nördlichen Oberrhein-Graben mit diluvialen kalkarmen oder kalkreichen Flugsanden. Am Fuße der Granit-
Berge, so z.B. am Darsberg, südöstlich von Jugenheim findet
sich vielfach eine Löß-Auflage,die mit zunehmender Höhe auskeilt (Abb.1). Während die diluvialen Flugsand-Flächen des Ober-
rhein-Grabens weithin mit D i c r a n o - P i n e t e n wech-
selnder Ausbildungsform bedeckt sind, stocken am West-Abfall
des Odenwaldes Buchenwälder. Sie sind auf den Lößböden als
C e p h a l a n t h e r o - F a g e t u m, auf den Granit-Ver-
witterungsböden als L u z u l o - F a g e t u m ausgebildet.
Beschreibungen des Boden-Profils und Vegetations-Aufnahmen der
jeweiligen Gesellschaft finden sich bei LÖTSCHERT (1952).

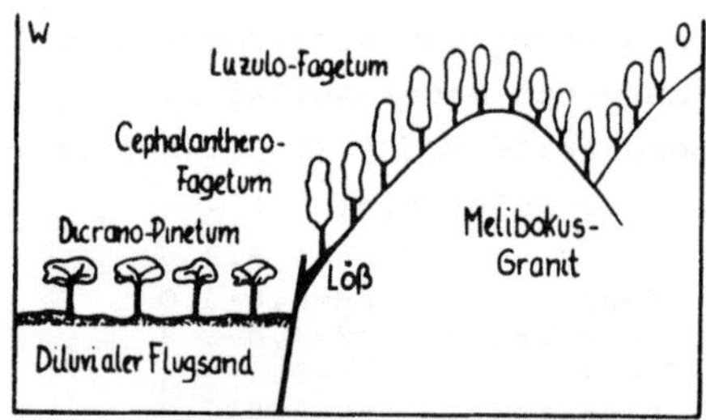

Abb.1. Schematisches Vegetations-Profil am Westrand des Oden-
       waldes südlich Jugenheim und Vegetations-Grenze zwischen
       C e p h a l a n t h e r o - und L u z u l o - F a g e -
       t u m auf Löß und Granit. Auf den diluvialen Flugsanden
       westlich der Verwerfung der Rhein-Ebene D i c r a n o -
       P i n e t u m.

Die Boden-Reaktionsverhältnisse in der Rhizosphäre der Krautschicht sind anhand eines pH-Vegetationsprofils in Abb.2 dar-

gestellt, die einige typische Krautschicht-Arten berücksichtigt[1)]

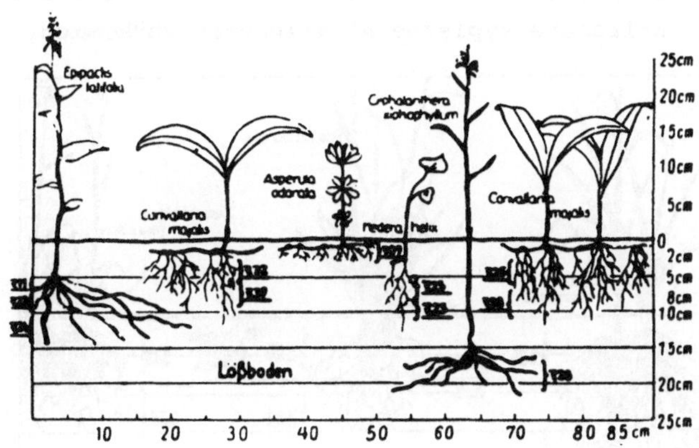

Abb.2. Schematisches pH-Vegetationsprofil aus dem C e p h a -
l a n t h e r o - F a g e t u m des Darsberges südlich
Jugenheim.

Sie zeigt, daß die pH-Werte in der Braun-Erde des Löß in der
Wurzel-Region alle über pH=7,0 liegen. Demgegenüber bewegen sich
die pH-Werte in der Wurzel-Region der Krautschicht-Arten Luzula
nemorosa, Deschampsia flexuosa und Poa nemoralis im L u z u l o -
F a g e t u m (Abb.3) mit Braunerde über Granit zwischen pH =
3,95-4,8.
Besondere Aufmerksamkeit verdient jene Zone am Berghang, in welcher der Löß auskeilt und beide F a g i o n -Gesellschaften
aneinandergrenzen. Die mikrotopographische pH-Analyse an dieser
Vegetations-Grenze ergibt beträchtliche pH-Unterschiede in der
Wurzel-Region der Krautschicht. Die entsprechenden pH-Werte von
Abb.4 zeigen, daß der pH-Unterschied an dieser Vegetations-Grenze bis 2,5 pH-Einheiten beträgt. Dabei ergibt sich gleichzeitig,
daß die an diesem Standort vorkommende Rasse von Convallaria
majalis keinesfalls als Alkaliphyt anzusehen ist. Sie dringt
vielmehr von der alkalischen Löß-Unterlage mit ihren Ausläufern
auf den sauren Granit-Verwitterungsboden vor. Dagegen gilt die im

---

[1)] Die Ermittlung der pH-Werte erfolgte in einer Boden-Suspen-
sion (Boden:Wasser = 2:5,Gewicht) mittels Chinhydron-Elek-
trode im Laboratorium. Die 2.Stelle hinter dem Komma bei den
pH-Angaben der Abb. 2-4 ist nur als Relativ-Zahl anzusehen.

L u z u l o - F a g e t u m vorkommende Deschampsia flexuosa
seit den klassischen Untersuchungen von OHLSEN (1923) als ste-
nöcischer Acidophyt, während im C e p h a l a n t h e r o - F a -
g e t u m zahlreiche typische Alkaliphyten vorkommen.

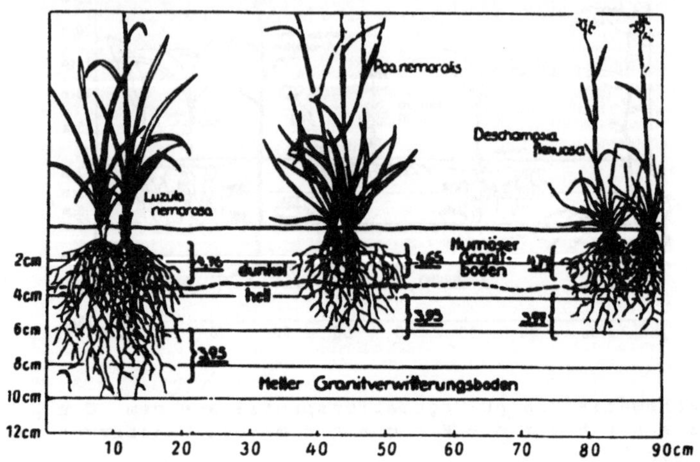

Abb.3. Schematisches pH-Vegetations-Profil aus dem L u z u l o -
F a g e t u m des Darsberges südlich Jugenheim.

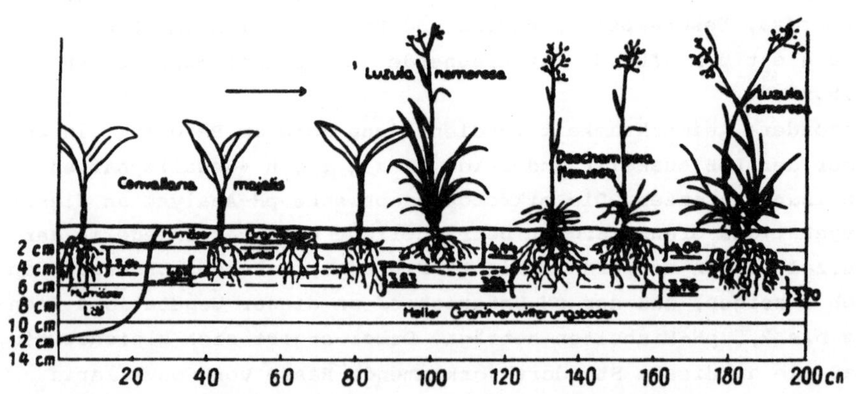

Abb.4. pH-Vegetations-Profil von der Vegetations-Grenze zwischen
C e p h a l a n t h e r o - und L u z u l o - F a g e -
t u m an der Auskeil-Grenze von Löß über Melibokus-Granit.

2. Hochmoor und Flachmoor im Gebiet der Baltischen
   Jungmoräne

Das zweite Beispiel stammt aus dem Hochmoor-Flachmoor-Komplex
des Salemer Moores. Dieses subkontinentale Baum-Hochmoor liegt
rund 4km südöstlich Ratzeburg im Winkel zwischen Salem, Ziethen
und der Glazial-Rinne vom Plötscher- und Garren-See auf dem Ter-
ritorium der weichsel-eiszeitlichen Jungmoräne. Das rund 2km
lange und im Mittel 500 m breite Hochmoor ist zum größten Teil
als Ledum palustre-reiches Betula pubescens-Pinus silvestris-
Hochmoor (B e t u l o - S p h a g n e t u m l e d e t o s u m
et v a c c i n i e t o s u m) entwickelt. Dieses wird an seinen
Rändern im Unterwuchs vielfach nicht mehr von Ledum palustre
sondern von Eriophorum vaginatum beherrscht und grenzt an seinem Ostrand über verschiedene Zwischenmoor-Gürtel an ein prachtvoll entwickeltes C a r i c e t u m e l a t a e (vgl.LÖTSCHERT
1964).
Aus dieser Übergangszone vom subkontinentalen Baum-Hochmoor zum
Flachmoor zeigt Abb.5 ein schematisches pH-Vegetationsprofil[1].

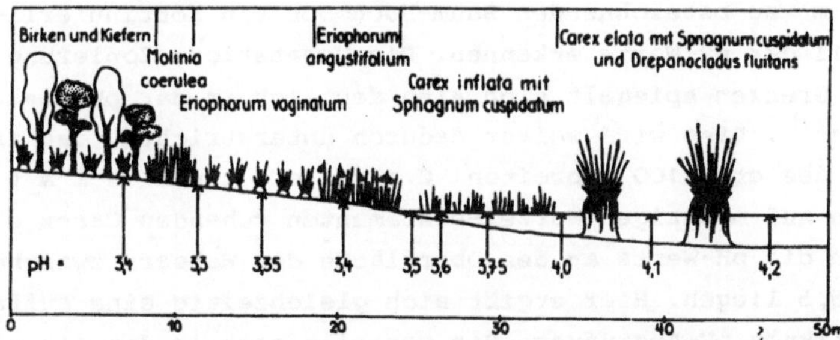

Abb.5. Schematisches pH-Vegetations-Profil vom Ostrand des Sa-
       lemer Moores bei Ratzeburg (subkontinentales Baum-Hoch-
       moor). Grenze zwischen Großseggen-Gürtel (C a r i c e -
       t u m e l a t a e) und Baum-Hochmoor (B e t u l o -
       S p h a g n e t u m) mit verschiedenen Zwischenmoor-Gür-
       teln. pH-Werte an den durch Pfeile gekennzeichneten Stel-
       len in 5cm Tiefe gemessen.

Dieses beginnt in der hochmoor-nahen Zone des C a r i c e t u m

[1] Die angegebenen pH-Werte wurden mit Hilfe einer Einstab-Meß-
    kette vom Typ 505 Agro in Verbindung mit einem tragbaren
    Batterie-pH-Meter E 280 der Firma Metrohm AG, Herisau,Schweiz
    unmittelbar am Standort in 3-5 cm Tiefe im stehenden Wasser
    gemessen. Die Agro-Sonde gestattet pH-Messungen bis zu einer
    Eintauch-Tiefe von 80 cm.

e l a t a e, in dem der Zwischenmoor-Charakter durch die flot-
tierenden Sphagnum cuspidatum- und Drepanocladus fluitans-Dek-
ken augenfällig ist. Die pH-Zahlen liegen hier bereits bei pH=
4,2 und sinken am Rand der Streifseggen-Zone auf pH=4,0. Sie
zeigen somit trotz der Anwesenheit der Crex elata-Horste bereits
deutliche Zwischenmoor-Verhältnisse an.
In der von Carex inflata (= C.rostrata) gebildeten Übergangszone ist ein rascher Abfall der pH-Werte bis pH=3,5 festzustellen. Dieser Anstieg der H^+-Konzentration setzt sich in den aus
Eriophorum angustifolium und Eriophorum vaginatum gebildeten
Zwischenmoor-Gürteln fort. Am Rande des Hochmoores selbst ist
zwischen den Sphagnum-Rasen der Eriophorum vaginatum-und der
Molinia coerulea-Zone ein pH-Wert von 3,3 zu verzeichnen. Im
Birken-Kiefern-Hochmoor selbst steigen die gemessenen pH-Werte
wieder geringfügig auf 3,4 an. Zusammenfassend läßt sich somit
vom hochmoor-nahen Rand des C a r i c e t u m   e l a t a e
über die einzelnen Zwischenmoor-Zonen bis zu dem als B e t u -
l o - S p h a g n e t u m   l e d e t o s u m   e t   v a c c i n i -
e t o s u m   zu bezeichnenden Baum-Hochmoor ein kontinuierli-
cher Abfall der pH-Werte erkennen. Die Vegetations-Zonierung
mit ihren Grenzen spiegelt sich also deutlich in der pH-Abstufung wieder[1]. Dies wird weiter dadurch unterstrichen, daß in
der Mitte des etwa 100 m breiten C a r i c e t u m   e l a t a e
mit seinen auf mächtigen Wurzel-Postamenten ruhenden Carex ela-
ta-Horsten die pH-Werte an der Oberfläche des Wassers zwischen
pH=6,2 - 6,5 liegen. Hier ergibt sich gleichzeitig eine auffallende vertikale pH-Abstufung. Sie verteilt sich in den einzelnen Stufen wie folgt:

          5 cm Tiefe: pH=6,2
         10 cm Tiefe: pH=5,8
         15 cm Tiefe: pH=5,7
         20 cm Tiefe: pH=5,5
         25 cm Tiefe: pH=5,4
         30 cm Tiefe: pH=5,3
         35 cm Tiefe: pH=5,2
         40 cm Tiefe: pH=5,1
         45 cm Tiefe: pH=5,1
         50 cm Tiefe: pH=5,1

[1] Zur Abgrenzung von Flach- und Zwischenmoor auf der einen und
     Hochmoor auf der anderen Seite wird auch vielfach der Ca-Ge-
     halt des Wassers verwendet. Der Grenzwert liegt nach den An-
     gaben von ELLENBERG (1963) bei 1 mg Ca/l. Dies deckt sich
     mit den Ergebnissen von unseren Untersuchungen im Schwarzen
     Moor der Rhön.

Eine ähnliche pH-Verteilung wurde mehrmals an verschiedenen
Stellen im Magnocaricetum festgestellt.
Wie die Zahlen zeigen, fallen die pH-Werte mit zunehmender Tiefe kontinuierlich ab. Sie sind in Abb.5 nicht mehr berücksichtigt, da die Meßstelle weit außerhalb des Bildes liegt. Dagegen
wurde am rechten Bildrand von Abb.5 ebenfalls die Vertikal-Abstu-
fung der H^+-Konzentration gemessen. Sie zeigte die folgenden
Werte:

```
 5 cm Tiefe: pH=4,2
10 cm Tiefe: pH=4,0
15 cm Tiefe: pH=3,9
20 cm Tiefe: pH=3,8
25 cm Tiefe: pH=3,6
30 cm Tiefe: pH=3,4
35 cm Tiefe: pH=4,1
40 cm Tiefe: pH=4,5
```

Im Gegensatz zum typischen Caricetum elatae ist
hier am Rand des Großseggen-Riedes in der Kontakt-Zone zum Zwischenmoor hin nicht ein gleichmäßiger Abfall, sondern ein er-
neuter Anstieg der pH-Werte in der Tiefe zu verzeichnen. Diese
Erscheinung geht wohl darauf zurück, daß hier noch eutrophe Reste des ehemaligen Großseggen-Riedes vorhanden sind, die inzwischen von den in Ausbreitung begriffenen mesotraphenten
Sphagneten der Hochmoor-Randzone überwachsen wurden.
Eine ähnliche vertikale pH-Verteilung an dieser Vegetations-
Grenze wurde wiederholt festgestellt (vgl.LÖTSCHERT 1963).

   3. Kegelkarst-Vegetation und immergrüner Kiefern-Eichen-
      Wald in West-Cuba

Das dritte Beispiel stammt aus dem Kegelkarst-Gebiet der Sierra
de los Organos ("Orgel-Gebirge") in der Provinz Pinar del Rio,
West-Cuba.In diesem klassischen Karst-Gebiet hat sich über mehrfach gestaffelten mit Sandsteinen und Tonschiefern abwechseln-
den mesozoischen Kalken eine typische tropische Kegelkarst-
Landschaft gebildet, eine geologische Karte zeigt Abb.7 (vgl.
LEHMANN, KRÖMMELBEIN,LÖTSCHERT 1956, LÖTSCHERT 1958).
Die aus Jura- bis Kreide-Kalken bestehenden Karst-Ketten und
turmförmigen Einzelberge tragen einen endemismen-reichen, völlig unveränderten Trocken-Wald (Mogoten-Vegetation), während
auf den Sandsteinen und Tonschiefern ein durch den Menschen
veränderter Kiefern- oder Eichenwald vorhanden ist. Die beiden
letzten Vegetationstypen sind durch markante Vegetations-Grenzen gegen den xerophytischen Primärwald abgesetzt (Abb.8). Sie

lassen sich nicht immer voneinander trennen, sondern gehen vielmehr stellenweise ineinander über.

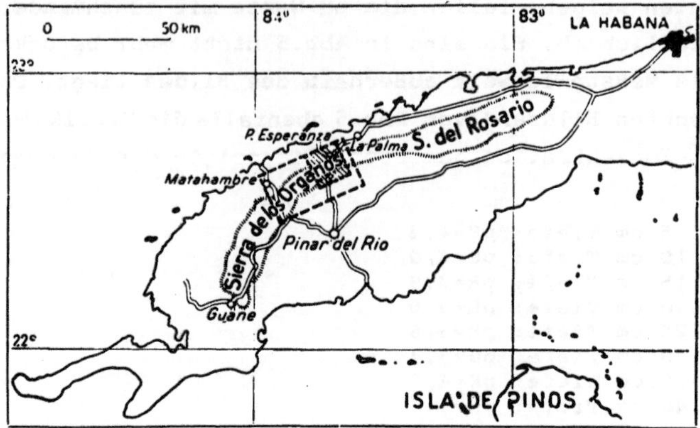

Abb.6.Übersichtskarte mit Untersuchungsgebiet in der Sierra de
los Organos, West-Cuba. Die geologische Karte von Abb.7
und die Vegetations-Karte von Abb.8 erfassen den schraf-
fierten Teil.

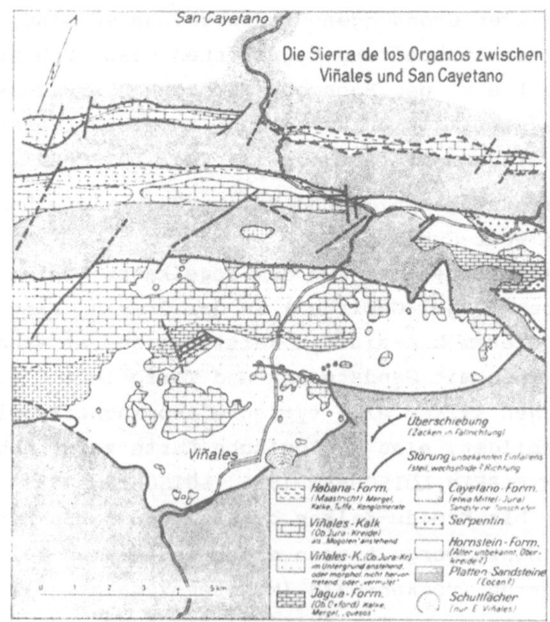

Abb.7. Vereinfachte geologische Karte des mittleren Teiles der
Sierra de los Organos (nach KRÖMMELBEIN)

Die auf den homogenen, viefach senkrecht aufragenden Kegelkarst-
bergen festgestellten Arten des xerophytischen Mogoten-Waldes
sind in Tabelle 1 zusammengestellt[1] . Unter den aufgeführten
Arten finden sich ohne Zweifel zahlreiche stenöcische Kalk-
pflanzen, für deren Gedeihen die geringe H^+-Konzentration be-

Tab.1. Liste der wichtigsten Pflanzenarten des Trocken-Urwal-
       des auf Kalk in West-Cuba. CH=Charakterpflanzen der Mo-
       goten-Vegetation, E=Endemismen Cubas.

```
E Ch  Gaussia princeps Wendl.
E Ch  Spathelia brittonii Wils.
E Ch  Omphalea hypoleuca Griseb.
E Ch  Portlandia pendula Wr.
E Ch  Lisiantus seleniflorus (Griseb.)Urban
E Ch  Eugenia galeata Urban
E Ch  Marcgravia calcicola Britt.
E Ch  Gesneria rupincola Urban
E Ch  Gesneria celsioides (Griseb.)Urban
E Ch  Leptocereus prostratus Br.& Rose
E Ch  Leptocereus assurgens (Wr.)Br.& Rose
E Ch  Erythrina cubensis Wr.
E Ch  Erythroxylon havanense Jacq.
E Ch  Thrinax microcarpa Sarg.(=T.punctulata Becc.)
E Ch  Anthurium venosum Griseb.
E Ch  Agave tubulata Trel.
  Ch  Bursera shaferi (Br.& Wils.)Urban
  Ch  Plumeria obtusa L.
  Ch  Adelia ricinella L.
  Ch  Bombax emarginatum (A.Rich.)Wr.
  Ch  Ekmanianthe actinophylla (Griseb.)Urban.
  Ch  Guettardia calcicola Br.
  Ch  Solandra grandiflora Sw.
  Ch  Zamia latifoliata Prenl.
  Ch  Philodendron scandens Koch
  Ch  Philodendron lacerum (Jacq.)Schott
  Ch  Adiantopsis rupincola Max.
E     Microcyas calocoma (Miq.)DC.
E     Anthurium recusatum Schott
E     Matelea oblongata (Griseb.)Woods.
E     Securidaca elliptica Turcz.
      Gymnanthes lucida Sw.
      Savia sessiliflora (Sw.)Willd.
      Casasia calophylla A.Rich.
      Rhipsalis cassutha Gaertner
      Selenicereus grandiflorus (L.)Br.& Rose
      Hohenbergia penduliflora (A.Rich.)Mez
      Tillandsia fasciculata Sw.
      Peperomia spec.
      Pilea spec.
      Dryopteris reptans (Gmel.)C.Chr.
```

[1] Die gesammelten Arten wurden zum größten Teil vom damaligen
    Bearbeiter der Flora von Cuba, Herrn H.ALAIN, im Herbarium
    zu Vedado-Habana bestimmt. Hierfür möchte ich Herrn ALAIN
    auch an dieser Stelle noch einmal herzlich danken.

        Dryopteris dentata (Forsk.)W.Chr.
        Asplenium veracundum Chapm.
        Adiantum tenerum Sw.
        Adiantum sericeum (DC.)Eaton
        Selaginella spec.

zeichnend ist.Zu ihnen gehören die durch aufgetriebene Stämme ausgezeichnete Palme Gaussia princeps, die nur auf den höchsten Mogoten-Gipfeln wachsende, prachtvoll rosa blühende, im Habitus an eine Palme erinnernde Rutacee Spathelia brittonii und der meist an den oberen Rundungen der Bergkuppen wachsende, durch flaschenförmige Stämme ausgezeichnete Bombax emarginatum. Ferner sind typisch Omphalea hypoleuca, Portlandia pendula, Lisianthus seleniflorus sowie die krautigen Fels-Besiedler Marcgravia calcicola, Gesneria rupincola und G. celsoides, Guettardia calcicola und Adiantopsis rupincola. Sie alle müssen als Kalk-Pflanzen angesehen werden.Es gibt also ohne Zweifel auch in den Tropen stenöcische Kalkpflanzen, obwohl der Wasser-Versorgung hier in der Regel größere Bedeutung zukommt. Die Boden-Reaktion des lateritischen Verwitterungs-Materials in Spalten des nackten Kalk-Gesteins, auf Steilabsätzen und in den Jamas betrug im Trocken-Wald pH=9,4.[1] Die mit zahllosen Korrosions-Rillen und -Gruben übersäten schroffen Kegelkarst-Berge sind an der Oberfläche völlig nackt und mit zahllosen endolithischen Flechten übersät. Sie zeigen wegen der unterirdischen Entwässerung eine sehr geringe Tendenz der Bodenbildung.

Durch scharfe Vegetations-Grenzen getrennt, siedelt auf den sauren Sandsteinen und Tonschiefern der Cayetano-Formation der aus der endemischen Pinus tropicalis und aus P.caribaea[2] aufgebaute Kiefernwald. Seine Zusammensetzung geht aus Tabelle 2 hervor.

Tab.2. Liste der in den Kiefernwäldern auf sauren Sandsteinen
     und Schiefern festgestellten Arten.

        E Ch Pinus tropicalis Morel
          Ch Pinus caribaea Morel
        E Ch Coperniciapauciflora Burret
          Ch Coccothrinax miraguana (H.B.K.)Becc.
          Ch Pachyanthus poiretii Griseb.
          Ch Miconia delicatula A.Rich.
          Ch Clidemia neglecta D.Don.
          Ch Clidemia strigillosa (Sw.)DC.

[1] Ermittlung mit Ionometer nach LAUTENSCHLÄGER Nr.1675 unter
    Verwendung der Kalomel-Chinhydron-Elektrode.

[2] Pinus caribaea bildet in Nicaragua die Südgrenze der Gattung
    Pinus (LAUER 1954).

```
Ch  Clidemia hirta D.Don.
Ch  Chaetolepis cubensis (A.Rich.)Tr.
Ch  Befaria cubensis Griseb.
Ch  Lyonia myrtilloides Griseb.
Ch  Rondeletia correifolia Griseb.
Ch  Cassytha americana Nees (=C.filiformis L.)
Ch  Tabebuia lepidophylla (A.Rich.)Greenm.
Ch  Declieuxia mexicana DC.
Ch  Chrysobalanus pellocarpus G.F.W.Meyer
Ch  Lycopodium cernuum L.
Ch  Odontosoria wrightiana Maxon
Ch  Dicranopteris flexuosa (Schrad.)Underw.
    Miconia prasina (Sw.)DC.
    Miconia ibaguensis (Bonpl.)Tr.
    Conostegia xalapensis (Bonpl.)D.Don.
    Roystonea regia (H.B.K.)O.F.Cook
    Polygala uncinata Wright
    Polygala leptocaulis T.& G.
    Cyrilla racemiflora L.
    Myrica cerifera L.
    Sauvagesia brownei Planch.
    Hypericum styphelioides A.Rich.
    Hypericum fasciculatum Lam.
    Eriosema violaceum (Aubl.)D.Don.
    Cassia hispidula (Vahl)Britt.& Rose
    Catopsis floribunda (Brongn.)L.B.Smith
    Tillandsia fasciculata Sw.
    Tillandsia bulbosa Hook.
E   Tillandsia flexuosa Sw.
E   Habenaria bicornis Lindl.
    Habenaria replicata A.Rich.
    Buchnera elongata Sw.
    Angelonia pilosella Kick.
    Xyris ambigua Beyr.
    Syngonanthus lagopodioides (Griseb.)Rutzl
    Burmannia bicolor Mart.
    Drosera capillaris Poir.
    Genlisea luteo-viridis Wright
    Blechnum serrulatum Rich.
    Dryopteris reticulata (L.)Urban
    Adiantum petiolatum Desv.
    Trichomanes pinnatum Hedw.
```

Sehr charakteristisch für diese P i n e t e n ist wie in den
aus Pinus oocarpa gebildeten Wäldern von El Salvador eine aus
Melastomataceen bestehende Strauchschicht. Neben den Melastomataceen bekunden Befaria cubensis und Lyonia myrtilloides, daß
die Ericaceen auch in den Tropen als mycotrophente Arten an saures Substrat gebunden sind.

Die Grenze zwischen Kiefern- und immergrünem Eichen-Wald aus
Quercus virginiana ist nicht immer deutlich, beide gehen vielmehr bei tiefer Verwitterung und erhöhter Feuchtigkeit des Bodens in eine Misch-Formation über. Auch werden Arten des Kiefernwaldes im Eichenwald angetroffen und umgekehrt. Die Zusammenset-

zung der Vegetation im Eichenwald geht aus Tabelle 3 hervor.
Sie berücksichtigt auch die epiphytischen Tillandsia-Arten und
zeigt zugleich, daß der berühmte Reliktendemit Microcycas calo-
coma nicht nur auf den Kalken der Kegelkarst-Berge, sondern
auch auf den sauren Pizarras-Sandsteinen vorkommt. Im übrigen
ist die Vegegation des Eichenwaldes sehr heterogen und wechselnd

Tab.3. Zusammensetzung der aus Quercus virginiana gebildeten
       immergrünen Eichenwälder unter Berücksichtigung eini-
       ger Sonder-Assoziationen auf saurem Sandstein.

```
        Ch Quercus virginiana Mill.
        Ch Clusia rosea Jacq.
        Ch Clusia minor L.
        Ch Hypericum styphelioides A.Rich.
        Ch Chiococca alba (L.)Hitchc.
        Ch Davilla rugosa Poir.
        Ch Miconia serrulata (DC.)Naud.
           Miconia prasina (Sw.)DC.
           Miconia ibaguensis (Bonpl.)Tr.
           Conostegia xalapensis (Bonpl.)D.Don.
           Clidemia divaricata (Wr.)Cogn.
           Tetracigia bicolor (Mill.)Cogn.
           Pachyanthus poiretii Griseb.

    E      Microcycas calocoma (Miq.)DC.
           Ouratea nitida (Sw.)
           Citharexylum caudatum L.
           Rondeletia correifolia Griseb.
           Chrysophyllum oliviforme L.
           Eugenia axillaris (Sw.)Willd.
           Ateleia cubensis Griseb.
           Allophyllus cominia (L.)Sw.
           Calophyllum brasiliense var.antillanum (Britt.)
           Casearia guianensis (Aubl.)Urban        Standl.
           Amaioua corymbosa H.B.K.
           Matayba oppositifolia (A.Rich.)Britt.
           Eugenia farameoides A.Rich.
           Palicourea crocea (Sw.)R.& S.
           Faramea occidentalis (L.)A.Rich.
           Alibertia edulis A.Rich.
           Coccocypselum guianense (Aubl.)Schum.
           Comocladia dentata Jacq.
           Bauhinia cumanensis H.B.K.
           Smilax havanensis Jacq.
           Clitoria rubiginosa Juss.
           Lygodium cubense H.B.K.
           Pteridium aquilinium var.caudatum (L.)Sadeb.
           Tillandsia bulbosa Hook.
           Tillandsia pruinosa Sw.

    E      Tillandsia flexuosa Sw.
           Tillandsia fasciculata Sw.
           Tillandsia usneoides L.

    Arten des "thicket shrub":
           Brya ebenus L.
```

> Buxus wrightii Muell.
> Rhus copallina var.leucantha DC.
> Tabebuia lepidophylla (A.Rich.)Greenm.

Pteridophyten-Assoziationen in Bach-Tälern:
> Lycopodium cernuum L.
> Odontosoria wrightiana Maxon
> Dicranopteris flexuosa (Schrad.)Underw.

Chaparral-Pflanzen:
> Curatella americana L.
> Byrsonima crassifolia (L.)H.B.K.

Hygrophile Arten in Tälern:
> Cecropia peltata L.
> Alsophila myosoroides Liebm.
> Andira inermis (Sw.)H.B.K.
> Genipa americana L.
> Xylopia aromatica (Lam.)Mart.
> Didymopanax morototonii (Aubl.)Dcne.& Planch.
> Pithecollobium obovale (A.Rich.)Wright

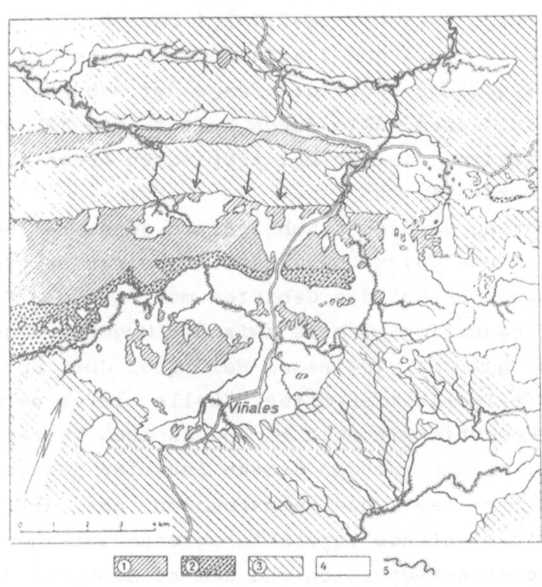

Abb. 8. Vegetations-Karte des zentralen Teiles der Sierra de los Organos auf
der Grundlage von Luft-Bildern 1:40 000 und Gelände-Untersuchung.-
(1) Mogoten-Vegetation mit Spathelia brittonii, Bombax emarginatum
und Gaussia princeps. (2) Mogoten-Vegetation mit Ficus spec. (3) Kie-
fern-Eichen-Wald mit Quercus virginiana, Pinus tropicalis, P.cariba-
ea und Melastomataceen-Strauchschicht. (4) Kulturlandschaft mit Nico-
tiana tabacum, Manihot utilissima, Xanthosoma sagittaefolia, Ipomoea
batatas und Saccharum officinarum. (5) Bachbegleitender Eugenia jam-
bos-Wald.- Die Pfeile kennzeichnen quer über Bergrücken verlaufende
Vegetations-Grenzen zwischen Mogoten-Trockenwald und Kiefern-Eichen-
wald.

Eine Vegetationskarte des untersuchten Gebietes zeigt Abb.8. In
dieser sind neben den behandelten Assoziations-Komplexen, eine
wenig xerophytische Mogoten-Mischvegetation mit verschiedenen
Ficus-Arten, Kulturland und bachbegleitender Sekundär-Wald mit
Eugenia jambos ausgeschieden. Letztere sollen jedoch in diesem
Zusammenhang unberücksichtigt bleiben. Vergleicht man die Vege-
tationskarte mit der unabhängig davon aufgenommenen geologischen
Karte, so ergibt sich eine auffallende Übereinstimmung. Die Ko-
inzidenz von geologischem Substrat und Vegetations-Formation
tritt somit in diesem klassischen Kegelkarst-Gebiet in selten
eindringlicher Weise zutage (vgl.LÖTSCHERT 1958).
Die Boden-Reaktion in den lateritisch verwitterten Pizarras-
(=Cayctano) Sandsteinen und -Schiefern beträgt im Kiefern- und
Eichenwald pH=5,0-5,5, während sie sich in der Rhizosphäre der
Kegelkarst-Pflanzen zwischen pH=7,7-9,4 bewegt. Es finden sich
somit unter den in Tab.2 und 3 aufgeführten Arten zahlreiche
Acidophyten. Allerdings müßte die Frage der stenöcischen Acido-
philie gerade in diesem Falle durch Kultur- und Nährstoff-Ver-
suche geklärt werden. Sie ist bei tropischen Arten bisher wohl
nicht geprüft worden.
Besondere Beachtung verdienen schließlich jene Stellen, an denen Vinales-Kalk und Cayetano-Sandstein aneinander grenzen.Dies
ist bei dem "Schuppenbau" des Gebietes mehrfach der Fall. Hier
verläuft eine Vegetations-Grenze zwischen Mogoten-Vegetation
und immergrünem Quercus virginiana-Wald quer über Bergrücken.
Zwar ist der Vinales-Kalk an dieser Stelle tiefer verwittert
und die Boden-Feuchtigkeit erhöht, so daß auch hygrophile Arten
aus den Tälern, so die auf den Großen Antillen beheimatete Kö-
nigspalme Roystonea regia hier einstrahlen, aber die quer über
den Bergrücken verlaufende Vegetationsgrenze zwischen der alka-
liphytischen Mogoten-Vegetation und dem acidophytischen Kiefern-
Eichenwald und ihrer unterschiedlichen Boden-Reaktion ist von
seltener Eindringlichkeit!

ZUSAMMENFASSUNG

Die Beziehungen zwischen Vegetations-Grenzen und Boden-Reaktion
werden anhand von drei verschiedenen Beispielen aus verschiedenen Vegetations-Gebieten dargestellt. Die Untersuchungen erstrek-
ken sich auf folgende Assoziationen bzw. Asssoziations-Komplexe:
    1. Grenze zwischen C e p h a l a n t h e r o und L u z u -

      l o - F a g e t u m auf LÖß bzw. Granit am Westrand des
Odenwaldes südlich Darmstadt. Dabei ergeben sich auf kleinstem Raum pH-Unterschiede von 2,5 pH-Einheiten in der Rhizosphäre der Krautschicht-Arten.

2. Übergangszone zwischen Flachmoor (C a r i c e t u m e -
l a t a e) und subkontinentalem Baummoor (B e t u l o -
S p h a g n e t u m l e d e t o s u m et v a c c i -
n i e t o s u m)mit verschiedenen Zwischenmoor-Zonen am Beispiel des Salemer Moores bei Ratzeburg in Holstein. Die
pH-Werte fallen vom Sphagnum-reichen Rand des Großseggen-
Riedes bis zum Rand des Baum-Hochmoores von pH=4,2 auf pH
=3,4 ab.Gleichzeitig ergibt sich ein pH-Wert-Abfall mit
zunehmender Tiefe im Wasser und zwischen den flottieren-
den Sphagnen des hochmoor-nahen Großseggen-Riedes.

3. Grenze zwischen primärem xerophytischem Urwald. (Mogoten-
Vegetation) auf reinen mesozoischen Kalken und Kiefern-
Eichenwald auf Pizarras-Sandstein und -Tonschiefer. Die
pH-Werte liegen auf den Kegelkarst-Bergen bei pH=7,7-9,4,
auf den Sandsteinen bei pH=5,0-5,5. Besonders auffallend
sind dort quer über Bergrücken verlaufende Vegetations-
Grenzen zwischen Mogoten-Vegetation und immergrünem Quercus virginiana-Wald. Pflanzen-Listen für die einzelnen
Vegetations-Formationen werden mitgeteilt.

Im Falle 1 und 3 sind die pH-Werte Ausdruck der verschiedenar-
tigen geologischen Unterlage. Sie werden durch Vegetation und
Edaphon modifiziert (allogene pH-Stufung). Im Falle des Hoch-
moor-Flachmoor-Komplexes bestimmt die Vegetation in erster Linie die Boden-Reaktion (autogene pH-Stufung).

SUMMARY

    Boundary lines of vegetation and reaction of the soil.

Correlations between boundary lines of vegetation and reaction
of the soil are demonstrated by 3 examples from various areas.
The investigations deal with the following associations or formations of vegetation:

1. Boundary line between C e p h a l a n t h e r o - and
L u z u l o - F a g e t u m on loess respectively granite
on the western slope of the Odenwald south of Darmstadt.

In this case differences of 2,5 pH units in the rhizosphe-
re of the herb layer were found within a small area.

2. Transitions between a bog (C a r i c e t u m   e l a t a e)
and a subcontinental fen covered by Betula pubescens and
Pinus silvestris (B e t u l o - S p h a g n e t u m   l e-
d e t o s u m   et   v a c c i n i e t o s u m) with various
zones (Salemer Moor near Ratzeburg in the north-western
part of Germany). The pH-values decrease from the border
of the   C a r i c e t u m   e l a t a e rich in Sphagnum
to the fen from pH=4,2 to pH=3,4. At the same time a decrease of the pH-values with increasing depth in the water of the bog and between the floting Sphagnum plants
from 5-50 cm has been found.

3. Boundary line between a xerophytic forest on tropical limestone sierras ("vegetation of mogotes") and a mesophy-
tic mixed forest of Pinus tropicalis, P.caribaea and Quercus virginiana on hilly Pizarras sandstones in the western
part of Cuba (Sierra de los Organos, Pinar del Rio). The
pH-values in the rhizospere of the limestone vegetation
range between pH=7,7 and 9,4, in the mixed pine oak forest on the sandstones between pH=5,0 and 5,5. Impressing
is the boarder line of vegetation between the 2 types of
vegetation on a single slope. A plant list of the various
forest types is presented.

In the case 1 and 3 pH-values are mainly a result of the different geological formations which are only little modified by
vegetation (allogene graduation of pH). In the case 2 the acidity of the substrate is mainly determined by rotting plants
(autogene graduation of pH).

LITERATUR

ELLENBERG,H. -1958- Bodenreaktion (Einschließlich Kalkfrage).-
     Handb.Pflanzenphysiol.IV: 638-708. Berlin,Göttingen,
     Heidelberg.
-- --   -1963- Vegetation Mitteleuropas mit den Alpen.- Stutt-
     gart 943 pp.
LAUER,W. -1954- Zentralamerika. Bericht über eine Forschungs-
     reise.- Erdkunde 8: 207-212. Bonn.
LÖTSCHERT,W. -1952- Vegetation und pH-Faktor auf kleinstem
     Raum auf Kalksand, Löß und Granit.- Biol.Zentralbl.71:
     327-348. Leipzig.
-- --   -1958- Die Übereinstimmung von geologischer Unterlage
     und Vegetation in der Sierra de los Organos (Westcuba).-
     Ber.deutsch.botan.Ges.71: 55-70. Stuttgart.

LÖTSCHERT,W. -1959- Kalkpflanzen auf saurem Untergrund. Ein
       Beitrag zur Frage der relativen Standortkonstanz.-
       Flora 147: 417-428. Jena.
-- -- -1963- Mikrotopographische pH-Messungen in Hoch- und
       Flachmooren.- Z.Bot.51: 452-467. Jena.
-- -- -1964- Vegetation, Trophiegrad und pflanzengeographische
       Stellung des Salemer Moores.- Beitr.Biol.Pfl.40: 65-111.
       Berlin.
-- -- -1965- Neuere Untersuchungen zur Frage jahreszeitlicher
       pH-Schwankungen. Ein zusammenfassender Überblick.- Angew.
       Bot.38: 255-268. Berlin.
LEHMANN,H., KRÖMMELBEIN,K.& LÖTSCHERT,W. -1956- Karstmorpho-
       logische, geologische und botanische Studien in der Sier-
       ra de los Organos auf Cuba.- Erdkunde 10: 185-204. Bonn.
OLSEN,C. -1923- Studies on the hydrogen-ion concentration of
       the soil and its signifiance to the vegetation, especial
       to the natural distribution of plants.- Comptes rendus
       trav.Lab.de Carlsberg 15 (1).Kopenhagen.
PEARSALL,W.H. -1952- The pH of natural soils and its ecologi-
       cal signifiance.- J.Soil Sci.3: 41-51. Oxford.
SJÖRS,H. -1950- On the relation between vegetation and elec-
       trolytes in north Swedish mire waters.- Oikos 2: 241-
       258. København.

R.TÜXEN:

Der pH-Wert kann im Laufe der Jahreszeiten und Jahre schwanken.
Ich möchte gern erfahren: wieweit wirken sich derartige Schwankungen im pH innerhalb eines Jahres auf die gezeigten Unterschiede in benachbarten Gesellschaften aus? Wann wurden diese Messungen gemacht?

W.LÖTSCHERT:

Die pH-Messungen im C e p a l a n t h e r o - F a g e t u m
und L u z u l o - F a g e t u m wurden während der Sommermonate Juli-August, die im Ratzeburger Moor im August durchgeführt.

Die Bestimmungen in West-Cuba erfolgten während der Regenzeit
(August-Oktober). Beide geben die Unterschiede an ähnlichem
Standort zu fast gleicher Zeit wieder.

M.W.TRENTEPOHL:

Eigene Befunde zeigen, daß die pH-Werte verschiedener Pflanzen
z.B. in einer 1 m^2 großen Wiesenfläche, in den Rhizosphären eine Abweichung von etwa 1-1,5 ergaben. Welcher Wert kommt solchen
Messungen jahreszeitlicher Schwankungen dann noch zu? Sind es
nicht Zufälligkeitsdaten, die man beim Erfassen kleiner und

kleinster Unterschiede dabei ermittelt?

W.LÖTSCHERT:
Wir haben den pH-Faktor als Indikator für edaphische Verhältnisse benutzt, um dadurch Unterschiede zwischen Vegetationseinheiten deutlich machen zu können.

A.APINIS:
Die Boden-Azidität ist ein guter Zeiger für bestimmte dynamische Vorgänge tropischer Böden, über die wir wenig wissen. Besteht ein Zusammenhang zwischen den verschiedenen pH-Werten
der oberen Bodenschicht und dem Abbau der durch die Vegetation
erzeugten Pflanzenreste (Streu), entweder in der hohen Abbau-
Geschwindigkeit oder in der Humus-Form?

W.LÖTSCHERT:
Die pH-Verhältnisse in den Tropen sind lediglich während dieses
Aufenthaltes in der Sierra de los Organos analysiert worden.
Darüber haben wir keine weiteren Untersuchungen ausgeführt. A-
ber wir haben seinerzeit die Abhängigkeit der Ammonifikation
und der Nitrifikation vom pH-Wert untersucht. Im allgemeinen
findet man im alkalischen oder im schwach sauren Bereich eine
Nitratation und eine geringe Ammonifikation, also Nitrat- und
Ammoniak-Bildung, dagegen im sauren Bereich nur noch Ammoniak-
Bildung. In dem Maße, in dem sich das pH innerhalb des Boden-
Profils ändert, wenn es also unmittelbar an der Oberfläche alkalischer ist und nach unten saurer wird, findet im allgemeinen
in dieser Abstufung auch eine Abnahme z.B. der Nitrat-Bildung
statt.

R.CARBIENER:
Ich möchte kurz beifügen, daß man dabei vorsichtig sein muß,
weil starke Nitrifizierung auch auf sehr sauren Böden wohl vorkommen kann. Es ist natürlich im allgemeinen, wie Sie es sagten.
Aber es gibt ziemlich viele Ausnahmen. Es sind viele solche Fälle bekannt, z.B. aus Schweden und aus den Hoch-Vogesen. Im subalpinen Buchenwald sind in der Humus-Schicht die pH-Bedingungen
sehr sauer, 4,5-4,8. Sie schwanken im Laufe des Jahres zwischen
ungefähr 4,2, (sie gehen manchmal bis 4 herunter), und ganz maximal 4,5-4,9. Und in diesen Böden ist da, wo sie feucht genug

sind, enorme Nitrifikation festzustellen. Es waren die tätigs-
ten Böden, die ich überhaupt in den Vogesen gefunden habe. In
den ganzen Hoch-Vogesen finden Sie nirgendwo so tätige Böden,
was die Nitrifizierung anbelangt, und die Nitrat-Mengen liegen
um das Fünffache höher als solche, die ich in den C a l a m a-
g r o s t i s -Hochgras-Steppen gefunden habe, die sonst auch
relativ tätige Böden haben. Also es gibt Ausnahmen, und man muß
mit einer Verallgemeinerung sehr vorsichtig sein.

W.LÖTSCHERT:
Der Wassergehalt des Standortes spielt für die Stickstoff-Bin-
dung eine maßgebliche Rolle; sinkt dieser unter ein bestimmtes
Minimum, arbeiten die Stickstoff-Bildner im allgemeinen nicht
mehr, oder nur noch in beschränktem Umfang. Deswegen stellt man
im Brutversuch den Wassergehalt auf einen standardisierten Wert,
meist 20-30 Gewichtsprozent ein.

R.TÜXEN:
Wenn auch meine Bemerkung vom pH hier abweicht, so erlaube ich
sie mir doch; denn es geht um die Nitrifizierung. Es ist noch
gar nicht so lange her, daß Dr.HÖLL berichtete über die Befunde
des Nitrat-Gehaltes in den Lysimeter-Abwässern, die wir um Stolzenau aufgefangen hatten. Er fand nicht nur hohe Nitrat-Werte
unter den Äckern, sondern auch unter dem Kiefern-Forst. Diese
Angaben wurden damals bezweifelt oder gar energisch bestritten.
Auf der anderen Seite war an den Befunden von HÖLL überhaupt
nicht zu zweifeln. Es blieb also ein gewisses Mißbehagen. Man
sagte z.B., wenn das so sei, dann müßte im tieferen Grundwasser
ja auch ein hoher Nitrat-Gehalt sein. Herr HÖLL konnte tatsächlich mit solchen Analysen aufwarten; das Grundwasser war in 18m
Tiefe erstaunlich reich an Nitraten. Es ist mir bemerkenswert,
daß Herr CARBIENER auch auf den sauren Böden eine sehr starke
Nitrifikation festgestellt hat.

J.MORAVEC:
Wir haben gefunden, daß nicht das pH ein limitierender Faktor
für die Nitrifikation ist, sondern das Erscheinen von Austausch-
Aluminium im Boden-Komplex. Und wir haben gefunden, daß es z.B.
in solchen Hochstauden-Buchen- und Ahorn-Buchenwäldern - noch
zur Nitrifizierung gekommen ist, obwohl die Böden pH 4,6 gehabt

haben. Das ist ein ziemlich niedriges pH. Aber dort war ein
ziemlich hoher Humus-Gehalt, der zwar nicht ganz durch Kalzium
gesättigt worden ist, aber die Differenz war mit Austausch-Wasserstoff gemacht worden, nicht mit Austausch-Aluminium. Wenn mar
nämlich bei einer alkalischen Titration die Austausch-Kapazität
bestimmt, dann stellt man nur die Summe von Austausch-Aluminium
und Austausch-Wasserstoff fest. Man muß dann das Aluminium näher analysieren, sonst bekommt man nicht die richtigen Resultate. Natürlich haben wir im Boden-Komplex noch nicht-dissoziier-
te Säuren, die man aber nur bestimmen kann, wenn man den Boden
mit einer alkalisch reagierenden Lösung versetzt, z.B. mit Kal-
zium-Azetat oder mit einer gepufferten Lösung wie z.B. NERICH's
Barium-Chlorid mit Äthanolamidol.
Vielleicht ist hier in Nordwest-Deutschland der Sand so ausge-
waschen und besteht nur aus Quarz, so daß der Humus kein Aluminium enthält. Dann ist es möglich, daß auch dort eine Nitrifika·
tion auftritt.

A.NOIRFALISE:

Dans le L u z u l o - F a g e t u m des Ardennes belges il
existe des variantes assez contrastées, allant du L u z u l o-
F a g e t u m f e s t u c e t o s u m au L u z u l o - F a-
g e t u m v a c c i n i e t o s u m. Les pH ne varient pas
d'un type à l'autre; ils sont souvent même plus bas dans la hêtraie à fetuque que dans la hêtraie à myrtille. Par contre, les
teneurs en nitrates diminuent sensiblement du type à fetuque au
type à myrtille. Parallèlement les teneurs en aluminium libre
augmentent fortement. Il n'est donc pas exclus que des teneurs
trop élevées en aluminium libre ne soient toxiques pour les or-
ganismes nitrifiants et que ce facteur soit beaucoup plus impor·
tant que la concentration en ion hydrogène.

W.LÖTSCHERT:

Ich bin mir klar darüber,daß der Boden ein überaus komplexes
System verschiedener Faktoren darstellt,in dem der pH-Wert zwar
vielfach aber nicht immer mit der Vegetation korreliert ist. Es
lag mir daran,Fälle aufzuzeigen,bei denen in Abhängigkeit von
der geologischen Unterlage Vegetations- und pH-Wert-Grenzen parallel gehen.

DIE VEGETATIONSKUNDE ALS ENTSCHEIDENDER GESICHTSPUNKT
BEI ERMITTLUNG PFLANZENGEOGRAPHISCHER GRENZEN

M. W r a b e r †

Beim Versuch einer pflanzengeographischen Gliederung Sloweniens
(M.WRABER 1968) stießen wir auf sehr große Schwierigkeiten. Den
Hauptgrund dafür dürfen wir wohl in der pflanzengeographischen
Lage des Landes erblicken, das den Knotenpunkt von drei Vegetations-Grenzen darstellt, nämlich der eurosibirischen, der mediterranen und der hochalpin-nordischen, und weiter in seiner klimatischen Übergangslage zwischen dem ozeanischen (atlantischen)
und dem kontinentalen (pannonischen) Klima-Regime mit den man-
nigfaltigsten lokalen klimatischen Einflüssen und Abänderungen,
dann in seiner außerordentlich bewegten Oberflächen-Gestaltung,
in seinen abwechslungsreichen geologischen Verhältnissen und
nicht zuletzt in seiner stürmischen Floren- und Vegetations-Ge-
schichte während der Eiszeit und in der Nacheiszeit.
Als Folge solcher natürlichen und vegetationsgeschichtlichen
Verhältnisse ergibt sich in der Vegetationsdecke des Landes ein
ungeheurer Floren-Reichtum und ein bunter Gesellschafts-Teppich
mit Übergängen nach allen Seiten. Versucht man nun eine pflanzengeographische Gliederung des Landes durchzuführen, so stellt
man fest, daß die Grenzen zwischen den Vegetations-Gebieten meistens sehr fließend sind, und daß nur in selteneren Fällen eine
klare, unbestrittene Grenzziehung möglich ist. Wir haben bei der
Grenzbestimmung mehrere Gesichtspunkte in Betracht gezogen,grund-
sätzlich aber wollten wir floristische und vegetationskundliche
Kriterien als entscheidend zu Rate ziehen. Aber in einem Lande,
wo sich die Floren-Gebiete vielfach berühren, überschneiden und
durchdringen, und wo dazu auch die arealkundlichen Kenntnisse
von den meisten Pflanzen-Sippen noch ungenügend sind, kommt man
auf dem Wege der rein floristischen Betrachtungsweise bei der
Grenzbestimmung nicht weit. Im Gegenteil hat sich die Vegetation mit dem ihr innewohnenden ökologischen Aussagewert, ihren wesentlichen Gesellschaftseinheiten und nicht zuletzt auch durch
ihre verhältnismäßig leicht erkennbare charakteristische Physiognomie bei der Bestimmung der vegetationskundlichen Areal-
bzw. Gebietsgrenzen als zuverlässiger und zweckdienlicher erwiesen.
Ich möchte nun zunächst auf Grund dieser Erkenntnisse und Erfah-

rungen eine grundlegende Überlegung vom Wesen und Wert der floristischen und der vegetationskundlichen Betrachtungsweise bezüglich der pflanzengeographischen Grenzbestimmung erörtern und
nachher am konkreten Beispiel einer klaren Grenzziehung die dies-
bezüglichen Vorteile der Vegetationskunde näher behandeln.
Flora oder Vegetation als grenzbestimmender Faktor?
Wie bereits erwähnt, läßt uns die rein floristische Betrachtungs·
weise bei der pflanzengeographischen Grenzbestimmung oft im
Stich, oder gibt uns wenigstens keine eindeutigen und sicheren
Stützpunkte. Diese Erkenntnis wurde in unserem Lande erworben,
das sich durch einen großen Floren-Reichtum auszeichnet, als
Folge seiner pflanzengeographischen Lage und seiner besonderen
natürlichen Gegebenheiten einerseits und als Ergebnis der plei-
stozänen und holozänen Floren-Wanderungen andererseits. Es treten natürlich noch andere Momente auf, die eine pflanzengeographische Gliederung des Landes erschweren, insbesondere noch die
sehr mangelhaften arealkundlichen Kenntnisse der Pflanzen-Sippen
unseres Landes. Trotz aller dieser negativen Momente, die in anderen europäischen Ländern wahrscheinlich nicht so schwer ins
Gewicht fallen, dürften jedoch unsere Erwägungen und Erfahrungen auch von allgemeinem Interesse sein. Sie haben uns nämlich
dazu bewogen, den Zeigerwert sowohl der floristischen wie auch
der vegetationskundlichen Argumente schärfer ins Auge zu fassen
und ihre Beweiskraft sorgfältig zu überprüfen.
Sollten auch die zuverlässigsten Areal-Karten der leitenden
Pflanzen-Sippen vorliegen, auf Grund derer man mit einer gewissen Sicherheit auf pflanzengeographische Grenzen schließen könnte, so würde man dennoch gar zu oft hoffnungslos dastehen und
keine sicheren Grenzen ziehen können. Versuchen wir uns diese
auf den ersten Blick unerwartete Feststellung zu erklären!
Man muß zunächst festhalten, was L.EMBERGER (1966: 155-156),der
hervorragende Kenner des Mediterran-Gebietes, über die Flora
und die Vegetation aussagt. Nach seiner Meinung ist die Flora
das Ergebnis der Phylogenese, also eine historische Tatsache,
während die Vegetation das Produkt des Milieus, also ein ökologisches Faktum ist. Es ist eine allgemein anerkannte Feststellung, daß derselbe Standort immer dieselbe Vegetationseinheit
ergibt, vorausgesetzt, daß es sich um ein mehr oder weniger ein-
heitliches Vegetations-Gebiet handelt, bzw. daß (physiognomisch
erfaßt) derselbe Vegetations-Typ entsteht, wenn es sich um ver-

schiedeneVegetations-Gebiete handelt, während die floristische
Zusammensetzung im letzten Fall auch etwas verschieden sein
kann. Um pflanzengeographische Gebiete mit größerer Sicherheit
abgrenzen zu können, muß man sich vorzugsweise auf ökologische
Faktoren stützen, die vor allem in der Vegetation und viel weniger in der Flora ihren komplexen Ausdruck finden. Es sind vor
allem das Klima und die Vegetation, die im Einklang stehen und
sich gewissermaßen auch gegenseitig beeinflussen. Während die
Flora je nach den entwicklungsgeschichtlichen Veränderungen auf
verschiedenen, aber ökologisch mehr oder minder gleichwertigen
Raum-Einheiten verschieden sein kann, stellt die Vegetation als
sicht- und greifbarer Ausdruck der Raum-Ökologie die einzig ra-
tionelle Grundlage zur Abgrenzung der territorialen pflanzengeographischen Einheiten. "Mit den floristischen Verwandtschafts-
verhältnissen sind auch ökologische, chorologische, sowie flo-
rengeschichtliche Beziehungen verbunden und bieten dafür Gewähr,
daß in mehrfacher Hinsicht Zusammengehöriges im System auch nebeneinander zu stehen kommt" J.BRAUN-BLANQUET 1951: 5.
Mehrfach unternommene ökologische Experimente an Pflanzen-Arten
mit bestimmtem ökologischen Zeigerwert, der ihnen in der pflanzensoziologischen Systematik einen gewissen Charakterarten-Rang
verleiht, haben eindeutig bewiesen, daß ihre ökologische Valenz
recht groß ist, und daß sie sich in Reinkulturen ökologisch anders verhalten als in der Natur, d.h. in bestimmten Pflanzengesellschaften wird ihre Reaktionsfähigkeit durch die jeweiligen
Konkurrenten wesentlich eingeschränkt. Der gewaltige regulieren-
de Faktor, der in den natürlichen Vegetationseinheiten über die
Reaktions-Amplitude einzelner Pflanzenarten entscheidet,ist der
Konkurrenzkampf, in welchem sich die Pflanzen-Individuen durch
ihre verschieden große Wettbewerbs-Kraft gegenseitig ökologische
Schranken stellen. Es ergibt sich aus dieser Feststellung die
klare und für gewisse Arbeitsziele sehr wertvolle Erkenntnis,daß
Pflanzen-Gesellschaften eine viel engere und bestimmtere ökologische Amplitude besitzen als die einzelnen, sie zusammensetzenden Pflanzen-Arten. Den Pflanzen-Gesellschaften fällt also
ein qualitativ viel höherer und sicherer ökologischer Zeigerwert zu als den Pflanzen-Sippen, die ihren floristischen Kern
bilden, es möge da entweder an die geläufigen Begriffe der charakteristischen Arten-Kombination (i.S.von J.BRAUN-BLANQUET)
oder der ökologischen Zeiger-Gruppen bzw. der soziologischen

Gruppen (i.S.von A.SCAMONI und H.PASSARGE) gedacht werden.(Vgl.
dazu H.ELLENBERG 1958: 14-16, 1960: 119-120!)
Wir machen in dieser Gedankenrichtung einen Schritt weiter. Es
ist allgemein bekannt, daß wissenschaftliche Methoden für die
Erfassung und Messung des Standortsfaktoren-Komplexes noch ausbleiben. Und so werden die Standorte einfach an ihrem Erzeugnis,
d.h. den Pflanzengesellschaften, "gemessen" und "geeicht". Nach
R.TÜXEN (1954:476) lassen sich "auch auf kleinstem Raume die
feinsten Veränderungen im Gewebe des Vegetationsteppichs nicht
nur zwanglos und objektiv beschreiben, sondern auch kausal verstehen". Wir erblicken darin die sicherste pflanzensoziologisch-
ökologische Grundlage zur Ermittlung der pflanzengeographischen
Grenzen, wobei dem Vegetationsforscher die sehr wichtige Aufgabe zufällt, die "Pflanzengesellschaften von den feinsten bis zu
den umfassendsten und ihre Grenzen zu finden" (R.TÜXEN,ibid.).

Es ist nicht zu leugnen, daß es in der Natur nur ausnahmsweise
scharfe Grenzen gibt, und daß im allgemeinen die Grenzen fließend sind. Man darf also eher von Übergangs-Zonen als von klaren Grenzen sprechen. In diesem Zusammenhang möchte ich wieder
R.TÜXEN (1961: 165) zitieren: "Entscheidend für die Grenzziehung
bleibt aber stets das Vorhandensein oder das Fehlen der diagnostisch maßgebenden Kenn- und Trennarten. Die Grenzen zu den Kon-
takt-Gesellschaften sind besonders als verschieden breite Über-
gangszonen ausgebildet".
In der zeitgemäßen Pflanzengeographie gilt als Hauptgesichts-
punkt für biogeographische Gliederung und Abgrenzung das soziologische Element oder das Coeno-Element, das Arten mit gleichen
bioökologischen Bedingungen vereinigt. Es "können dabei nur solche Arten herangezogen werden, die kein allzu großes Areal besitzen und eine stärkere großklimatische Abhängigkeit erkennen
lassen" (H.FREITAG 1962: 48-49). Da aber das ökologisch-sozio-
logische Verhalten der meisten für Abgrenzungszwecke in Betracht
kommenden Kenn- und Trennarten gerade in ihrer gesellschaftli-
chen Gebundenheit am klarsten zum Ausdruck kommt, so ist dies
ein Beweis mehr für die Zweckmäßigkeit und Richtigkeit unserer
Stellungnahme in Bezug auf die pflanzengeographische Gliederung
des Landes. Für die allgemeine, großräumige phytogeographische
Gliederung sind vor allem die großklimatisch gebundenen Vegeta-
tions-Einheiten (klimato-zonale Gesellschaften) entscheidend,
während die kleinklimatisch und relief-bedingten für die engere

oder genauere Grenzbestimmung maßgebend sind, ja, in einem Land
mit stark betontem Übergangsklima und mit unendlich abwechslungsreichem Groß- und Klein-Relief spielen die extrazonalen und
azonalen Pflanzengesellschaften sehr oft sogar die Hauptrolle.

Ausgehend von den Pflanzengesellschaften, falls deren floristische Zusammensetzung gut bekannt und vergleichsmäßig ausgearbeitet ist, kann man sich im Besitz guter Erfahrungen und zuverlässiger chorologischer Kenntnisse bei der Grenzbestimmung mit
Erfolg auch einzelner Pflanzen-Arten oder noch besser Arten-
Gruppen bedienen. Dieser Weg kann aber nur dann erfolgreich betreten werden, wenn man mit gründlichen Vegetations-Kenntnissen
ausgestattet ist; in Zweifelsfällen wird man jedoch immer wieder auf vegetationskundliche Merkmale zurückgreifen müssen.

Um diese grundsätzlichen Überlegungen zusammenzufassen, erklären wir die floristisch gut erfaßten und klar umschriebenen Vegetationseinheiten als unentbehrliche Grundlage für die pflanzengeographische Gliederung eines Landes, indem in der floristischen Arten-Zusammensetzung dieser Einheiten neben den qualitativen auch quantitative soziologisch-ökologische Merkmale zu erkennen sind.

DIE GRENZE ZWISCHEN DEM SUBMEDITERRANEN UND DEM DINARISCHEN
GEBIET
Wir wollen nun den Verlauf der Grenze zwischen dem Submediterranen und dem Dinarischen Gebiet Sloweniens als Beispiel einer vegetationskundlich sehr klaren Abgrenzung eingehender behandeln
und die Grenzziehung im Gebirgs-Massiv Nanos näher verfolgen.

Das slowenische Submediterrane Gebiet nimmt in verschiedener
Hinsicht eine eigenartige Stellung ein; es besitzt unter allen
pflanzengeographischen Gebieten Sloweniens die höchste Selbst-
ständigkeit, gleichzeitig aber auch die größte innere Gliederung. Seine Grenze gegen das Dinarische Gebiet tritt am schärf-
sten hervor, nicht nur hinsichtlich der Vegetation, sondern sie
trägt auch bezüglich der Landschafts-Ökologie und der Wirtschaftsgeographie einen ausgeprägten, charakteristischen Zug. Diese
Grenze verläuft längs des südwestlichen Randes des Dinarischen
Gebirgszuges und bedeutet auch in klimatischer Sicht eine ausgesprochene Scheidelinie.
Das Submediterrane Gebiet Sloweniens zeichnet sich in klimatischer Hinsicht durch starke Frühjahrs- und Herbstniederschläge,

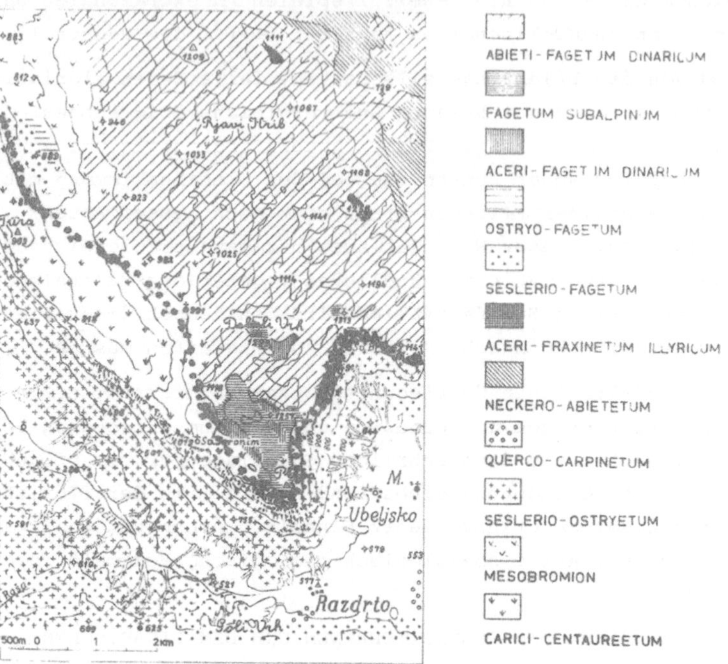

eine mäßig ausgeprägte Sommerdürre- und -hitze und durch eine
kältebedingte winterliche Vegetationsruhe aus, die durch den eis-
kalten Fallwind, die Bora (slowenisch "burja"), noch verschärft
wird. Die jahresdurchschnittliche Niederschlagsmenge beläuft
sich auf 1000/1100 mm an der adriatischen Meeresküste und steigt
am oberen Gebietsrande bis 2500/3000 mm an. Die Zunahme der
Niederschläge landeinwärts ist also außerordentlich groß und
rasch. Charakteristisch für den größten Teil dieses Gebietes
sind auch die großen und schroffen Temperatur-Unterschiede zwischen der wärmsten und kältesten Jahreszeit (absolute Tempera-
tur-Differenz bis 60°C), aber auch kurzfristige Temperatur-
Schwankungen sind ungewöhnlich groß. (Vgl. dazu M.WRABER 1954a,
1967). Klimatisch stellt das slowenische Submediterrane Gebiet
eine Übergangszone zwischen dem Einfluß-Bereich des atlantischen
und des mediterranen Klima-Regimes, etwas abgeändert durch den
teilweise abgeschwächten Einfluß des kontinentalen (pannonischen)

Klimas dar.

Vegetationskundlich gehört das slowenische Submediterrane Gebiet zum Mediterranen Vegetationskreis. Nach der Auffassung älterer pflanzengeographischer und geobotanischer Forscher der Balkan-Halbinsel (G.BECK v.MANAGETTA, L.ADAMOVIĆ, M.RIKLI, F.MARKGRAF, E.OBERDORFER u.a.m.) galt nur der Bereich der immergrünen Hart-laub-Vegetation (Q u e r c e t a l i a   i l i c i s) als medi-terran, oder es wurde ein großer Teil des kontinentalen Karst-Gebietes mit mitteleuropäisch-dinarischen Vegetationseinheiten in den Mediterranen Vegetationskreis einbezogen. Nach dem neue-sten Stand unserer Kenntnisse, die wir in erster Linie den kro-atischen Forschern I.HORVAT und S.HORVATIC verdanken, wird der ganze submediterrane Vegetations-Gürtel zum Mediterranen Gesell-schaftskreis gerechnet; die laubwerfende Wald- und Gebüsch-Ve-getation des Verbandes   O s t r y o - C a r p i n i o n   o r i -e n t a l i s   und insbesondere noch seine Degradationsstadien zeigen nämlich floristisch, ökologisch und genetisch engere Be-ziehungen zur mediterranen als zur mitteleuropäischen Vegetati-on.

Das slowenische Submediterrane Gebiet besitzt nur einen schma-len Küsten-Gürtel eumediterraner Vegetation (O r n o - Q u e r -c e t u m   i l i c i s   c o t i n e t o s u m), der sich an der Triester Bucht hinzieht und staatlich zu Italien gehört. Vom submediterranen Verband des   O s t r y o - C a r p i n i o n   ist das   C a r p i n e t u m   o r i e n t a l i s   c r o a t i -c u m   nur schwach vertreten, während das   S e s l e r i o -O s t r y e t u m   als der "Karstwald" schlechthin bezeichnet werden kann, da er als klimatozonale Gesellschaft den größten Teil des slowenischen Submediterranen Gebietes einnimmt und we-gen seiner großräumigen Ausdehnung systematisch reichlich ge-gliedert ist. Als azonale bzw. extrazonale Vegetations-Gebilde vom mitteleuropäischen Gepräge haben sich unter speziellen Re-lief- und Klima-Bedingungen Vegetations-Basen erhalten (Q u e r -c o - C a r p i n e t u m   s u b m e d i t e r r a n e u m, Q u e r c o - C a s t a n e e t u m   s u b m e d i t e r r a -n e u m, A s a r o - C a r p i n e t u m, L u z u l o - F a g e -t u m, L u z u l o - C a r p i n e t u m   u.a.). Von den Degra-dationsstadien des   S e s l e r i o - O s t r y e t u m   sind mehrere Wiesen-, Trockenrasen- und Steintrift-Gesellschaften ausgebildet, die den Ordnungen   S c o r z o n e r o - C h r y -

s o p o g o n e t a l i a , B r o m e t a l i a e r e c t i und
S e s l e r i e t a l i a j u n c i f o l i a e angehören.

Es sei vorübergehend bemerkt, daß der slowenische Teil des submediterranen klimatozonalen Vegetationsgürtels, der sich an der
Adria-Küste entlang zieht und nach S.HORVATIĆ (1963a, 1963b,
1964) zur eigenen Adriatischen Provinz des Mediterranen Vegetationskreises gehört,einen etwas abgeschwächten mediterranen Charakter besitzt, der durch das Ausbleiben oder das Seltenerwer-
den typischer mediterraner Elemente und durch den zunehmenden
Einschlag der mitteleuropäischen und alpin-hochdinarischen Flora zustande kommt. Man darf also das slowenische Submediterrane
Gebiet als eine nordwestlich auskeilende und ausklingende Übergangszone der Adriatischen Vegetations-Provinz zwischen dem Me-
diterran und Mitteleuropa betrachten.

Nach der floristischen Pflanzengeographie stellt das slowenische Submediterrane Gebiet eine Durchdringung von echten mediterranen und illyrisch-mediterranen Elementen der (i.S.von S.
HORVATIĆ 1963a:15), wobei letztere jedoch mit ihrer illyrisch-
südeuropäischen Gruppe (S.HORVATIĆ 1963a:125, 1964:16) vorherrschen.

LITERATUR

BRAUN-BLANQUET,J. -1951- Pflanzensoziologie.2.Aufl.- Berlin.
ELLENBERG,H. -1958- Über die Beziehungen zwischen Pflanzen-
        gesellschaft, Standort, Bodenprofil und Bodentyp.-
        Angew.Pflanzensoz.15. Stolzenau/Weser.
-- -- -1960- Ökologische Pflanzengeographie.- Fortschr.Bot.22.
        Berlin,Göttingen,Heidelberg.
EMBERGER,L. -1966- Place de la région méditerranéenne française
        se dans l'ensemble méditerranéen.- Ass.prof.Biol. et Gé-
        ol.de l'enseignement public 53.
FREITAG,H. -1962- Einführung in die Biogeographie von Mittel-
        europa mit besonderer Berücksichtigung von Deutschland.-
        Stuttgart.
HORVAT,I. -1954- Pflanzengeographische Gliederung Südosteuro-
        pas.- Vegetatio 5-6. Den Haag.
-- -- -1962- Die Grenze der mediterranen und mitteleuropäi-
        schen Vegetation in Südosteuropa im Lichte neuer pflan-
        zensoziologischer Forschungen.- Ber.Deutsch.bot.Ges.75.
        Stuttgart.
HORVATIĆ,S. -1963a- Vegetacijska karta otoka Paga s općim
        pregledom vegetacijskih jedinica Hrvatskog Primorja.
        (Carte des groupements végétaux de l'ile nord-adriatique
        de Pag avec un aperçu général des unités végétales du
        Littoral Groate).- Prirodosl.istraž.JAZU 33. Zagreb.
-- -- -1963b- Biljnogeografski položaj i rašćlanjenje našeg
        Primorja u svijetlu suvremenih fitocenoloških istraži-
        vanja. (Pflanzengeographische Stellung und Gliede-

rung des ostadriatischen Küstenlandes im Lichte der neu-
esten pflanzensoziologischen Untersuchungen).- Acta bot.
Croat.22. Zagreb.
HORVATIĆ,S. -1964- Fitocenološke jedinice vegetacije krškog
prodručja Jugoslavije kao osnova njegovog biljnogeograf-
skog rašćlanjavanja.(Phytocoenologische Vegetationseinheiten des Karstgebietes Jugoslawiens als Grundlage seiner pflanzengeographischen Gliederung).- Acta bot.Croat.,
vol.extraord. Zagreb.
ľÜXEN,R. -1954- Die räumliche, durch Relief und Gestein bedingte Ordnung der natürlichen Waldgesellschaften am nördlichen Rande des Harzes.- Vegetatio 5-6. Den Haag.
-- -- -1961- Vegetationskartierung.- Method.f.Heimatforsch.
in Niedersachsen. Hildesheim.
WRABER,M. -1954- Splošna ekološka in vegetacijska oznaka slo-
venskega krasa. (Caractéristique générale écologique et
végétative du Karst slovène).- Gozd.vestnik. Ljubljana.
-- -- -1967- Ökologische und pflanzensoziologische Charakte-
ristik der Vegetation des slowenischen küstenländischen
Karstgebietes.- Mitt.ostalp.-dinar.pflanzensoz.Arbeits-
gem.7. Trieste.
-- -- -1969- Pflanzengeographische Stellung und Gliederung
Sloweniens.- Vegetatio (Festschrift Tüxen) 17. The Hague.

ZUSAMMENFASSUNG

Beim ersten Versuch einer pflanzengeographischen Gliederung Sloweniens (M.WRABER 1968) hat sich herausgestellt, daß in diesem
ausgesprochenen, floristisch sehr reichen Übergangsgebiet, in
dem sich drei Floren-Regionen berühren und überschneiden, nämlich die eurosibirische, mediterrane und hochalpin-nordische,
reine floristische Gesichtspunkte für eine pflanzengeographische Einteilung nicht ausreichen. Außer der betonten Übergangslage ist vor allem die mangelhafte Kenntnis der Verbreitung
der bedeutsamen pflanzengeographischen Elemente der Grund für
die Schwierigkeiten bei der Ermittlung der pflanzengeographischen Grenzen.
Vegetationskundliche Gesichtspunkte haben sich als viel zuverlässiger erwiesen als floristische, besonders noch nach einer
gründlicheren Erforschung der Vegetation im ganzen Lande und
der Vegetations-Kartierung einiger Landesteile. Trotz guter Areal-Karten von pflanzengeographischen Leitpflanzen könnte man ohne Berücksichtigung der Vegetation in vielen Fällen die Abgrenzung zwischen den einzelnen pflanzengeographischen Gebieten
nicht durchführen.
Nach L.EMBERGER (1966:155-156) ist die Flora das Ergebnis der
Phylogenese, also eine historische Tatsache, während die Vegetation das Produkt der Umwelt, also ein oekologisches Faktum

ist. Da die Vegetation ein gut sichtbarer Ausdruck der komplexen Wirksamkeit der Standort-Faktoren ist und mit dem Wechsel
derselben sich auch die Vegetation ändert,können wir anhand der
Vegetationsgrenzen auch pflanzengeographische Grenzen feststellen, die sich in der Hauptsache nach ökologischen Gesichtspunkten richten, während die einzelnen floristischen Elemente viel
weniger von standörtlichen Verhältnisse abhängig sind und infol
gedessen auch die pflanzengeographischen Grenzen mit geringerer
Sicherheit andeuten.

Experimente an Pflanzen-Arten mit bestimmtem ökologischen Zeigerwert und mit gewissem diagnostischen Aussagewert in der
pflanzensoziologischen Systematik haben eindeutig bewiesen, daß
ihre ökologische Valenz recht groß ist, und daß sie sich in
Reinkultur anders verhalten als in der Pflanzen-Gesellschaft,
wo die Konkurrenz-Beziehungen ihre Reaktionsfähigkeit wesentlich einschränken und ihre ökologische Amplitude bestimmen. Es
folgt daraus die wertvolle Erkenntnis, daß die Pflanzen-Gesellschaften eine viel engere und näher bestimmte ökologische Spann
weite besitzen als die einzelnen, sie zusammensetzenden Pflanzen-Arten. Deswegen kommt den Pflanzen-Gesellschaften ein viel
höherer und sicherer ökologischer Zeigerwert für die Beurteilung der Standortsverhältnisse zu als den Pflanzen -Arten,die
das floristische Inventar einer Gesellschaft bilden (vgl.dazu
H.ELLENBERG 1958: 14-16, 1960: 119-12o!).

Da wissenschaftliche Methoden zur Erfassung und Messung des
Standortsfaktoren-Komplexes noch sehr mangelhaft und unzuläng-
lich sind, werden die Standorte nach den Pflanzen-Gesellschaften "gemessen" und "geeicht". Man kann "auch auf kleinstem Raum
die feinsten Veränderungen im Gewebe des Vegetationsteppichs
nicht nur zwanglos und objektiv beschreiben, sondern auch kausal verstehen" (R.TÜXEN 1954:476). Diese Feststellung ist die
beste Grundlage für eine sichere und objektive pflanzengeographische Grenz-Bestimmung. Nach demselben Autor bleibt für die
Grenzziehung entscheidend das Vorhandensein oder Fehlen der diagnostisch maßgebenden Pflanzenarten. Natürlich sind scharfe
Grenzen in der Natur sehr selten, so daß man kaum irgendwo von
Grenz-Linien, sondern vielmehr nur von Grenz-Zonen sprechen kan

Für eine allgemeine, großräumige pflanzengeographische Gliede-
runng sind vor allem die makroklimatisch gebundenen (klimatozonalen) Vegetationseinheiten entscheidend, während die mikrokli-

matisch und relief-bedingten für die engere und genauere Grenzbestimmung maßgebend sind, was besonders in einem Lande mit betontem Übergangsklima und mit unendlich abwechslungsreichem
Groß- und Kleinrelief, wie Slowenien es ist, eine hervorragende
Rolle spielt. Ein guter Kenner der Vegetation wird sich bei der
Grenzbestimmung auf Grund erprobter Erfahrungen auch einzelner
Pflanzen-Arten oder noch eher Arten-Gruppen mit Erfolg bedienen,
er wird jedoch in Zweifelsfällen und zur Bestätigung seines
richtigen Vorgehens immer wieder auf vegetationskundliche Methoden zurückgreifen. Wir dürfen also die floristisch gut erfaßten und ökologisch klar umgrenzten Vegetation-Einheiten als
unentbehrliche Grundlage für die pflanzengeographische Gliederung des Landes erklären.

Auf Grund eines Ausschnittes der Vegetationskarte erörtert der
Verfasser den Verlauf der Grenze zwischen dem submediterranen
und dem dinarischen Gebiet, die hier sehr klar ist und zwei Wel-
ten trennt, wesentlich verschieden nicht nur in vegetationskund-
licher und floristischer Hinsicht, sondern auch bezüglich der
Landschafts-Ökologie und der Wirtschaftsgeographie. Diese Grenze sondert das mediterrane Gebiet, das von Vegetationseinheiten
aus dem Verbande O s t r y o - C a r p i n i o n   o r i e n -
t a l i s   mit seinen zahlreichen Degradationsstadien beherrscht
wird, vom mitteleuropäischen, für das die Gesellschaften der
Verbände F a g i o n   i l l y r i c u m und F a g i o n   m e-
d i o e u r o p a e u m   besonders bezeichnend sind.

SUMMARY

At the first attempt to divide Slovenia into phytogeographical
regions (WRABER 1968) it became apparent that, in such a marked-
ly transitional and floristically rich country where three floristic provinces overlap (viz.Euro-Siberian, Mediterranean and
Arctic-Alpine), floristic aspects are not sufficient to determine the phytogeographical boundaries. In addition, an incom-
plete knowledge of the spread of the major phytogeographical
elements poses problems in boundary definition. Vegetation studies, especially mapping, have proved more reliable than floristic studies. Yet, despite good maps of the areal distribution
of the main phytogeographical units, the boundaries between several regions could not be reliably defined.
EMBERGER (1966:155-156) considers the flora of any region to be

the result of past phylogenetic development,whereas vegetation
is viewed as the product of its ecological environment. Since
vegetation is a visual expression of a complex of ecological
factors and, since with changes in these complexes there are
also vegetation changes, it is possible to determine both vegetation and phytogeographical boundaries. On the contrary, the
various floristic elements are much less dependent on environmental factors and are less reliable for such definition.
Experiments with plant species with a definite ecological prefe-
rence and which play a diagnostic part in the phytosociological
system have clearly shown that they behave differently in pure
culture than in plant societies, where competition limits their
reacting capability and determines their ecological amplitude.
This emphasises the fact that plant societies have a much narro-
wer ecological amplitude than the several species forming the
society. Hence, a much higher indicator value for the reflection of habitat conditions can be ascribed to plant societies
than to single plant species (cf.H.ELLENBERG 1958: 14-16, 1960:
119-120!).

It is well known that the scientific methods of determining and
measuring the complex of ecological factors are imperfect and
inadequate and it has become practic to "measure" and "gauge"
habitats on the basis of their product - their vegetation units.
R.TÜXEN (1954:476) maintains that by phytosociological methods
even the finest changes in vegetation can be objectively determined and causally explained. This is the most reliable basis
for the description of phytogeographical borders and according
to the same author (1961:165), the presence or absence of diag-
nostically significant species is decisive for delimitation.
Sharp border lines are rare in nature and transitional border
zones are more common.

On a broad geographical scale, societies delimited by macrocli-
matic conditions (climatozonal) are of importance, whilst societies defined by microclimatic or relief conditions(azonal)are of
importance in a region of transitional relief and climate types
such as Slovenia. Using a section from a vegetation map of Slovenia, the floristic, vegetational,ecological and economic difference between the Submediterranean and the Dinaric regions is
expounded and the boundary shown to be quite distinct. This boun
dary separates the Mediterranean region, dominated by the vege-

tation units of the alliance O s t r y o - C a r p i n i o n
o r i e n t a l i s and its degradation stages, from the central Europaean region, characterized by the forests of the alli-
ances F a g i o n i l l y r i c u m and F a g i o n m e -
d i o e u r o p a e u m.

G.LAVRENTIADES:
Bei uns in Griechenland ist Spartium junceum sehr häufig und mit
großer Vitalität in der Q u e r c i o n i l i c i s - Zone vom
Norden bis zu den südlichen Küsten Griechenlands vorhanden. Da
kann man sehr häufig ganz in der Nähe des Meeres in dem Hart-
laub-Gebüsch der Q u e r c i o n i l i c i s - Zone Spartium
junceum-Pflanzen von 2-3 m Höhe finden. In welcher Zone haben
Sie diese Art gesehen?

M.WRABER:
Wir haben in Slowenien keinen Q u e r c e t a l i a i l i -
c i s -, sondern nur den O s t r y o - C a r p i n i o n -Gür-
tel und darin kommt diese Pflanze bei uns, größtenteils in den
höheren Lagen, hauptsächlich im Gürtel der orientalischen Hain-
buche, also des C a r p i n e t u m o r i e n t a l i s vor.

R.TÜXEN:
Es hat mir große Freude gemacht, zu sehen, daß Quercus ilex und
Spartium junceum außerhalb des Q u e r c i o n i l i c i s -
Gebietes vorkommen. Ich kenne an der west-französischen Küste,
nördlich von Bordeaux, Dünen-Gebiete mit Quercus ilex-Wäldern,
und zwar alten, 200jährigen, die ganz sicher außerhalb des
Q u e r c i o n i l i c i s liegen, die ganz eindeutig zum
Q u e r c i o n r o b o r i - p e t r a e a e gehören. Quercus ilex ist dort eine Art, die in das Q u e r c i o n r o -
b o r i - p e t r a e a e hineingeht. Natürlich bleibt sie eine
mediterrane Art. Das ist ein Parallelfall zu Slowenien.

E.BURRICHER:
Es ist eine bekannte Tatsache, die man immer wieder im Mediter-
ran-Gebiet beobachten kann und die meines Wissens auch von einigen Autoren beschrieben worden ist, daß bei der Degradation
z.B. durch Beweidung des submediterranen Waldes der mediterrane

Wald des Q u e r c i o n   i l i c i s nach oben in die Degra-
dations-Stufen einspringt, daß also die Höhen-Grenze des mediterranen Raumes durch den anthropogenen Einfluß nach oben ver-
legt wird. Natürlich steigen nicht alle mediterranen Arten
nach oben, sondern nur verhältnismäßig widerstandsfähige. Dazu
gehört auch Quercus ilex, wahrscheinlich auch Spartium junceum
und verschiedene andere mehr, während empfindlichere Arten wie
z.B. Arbutus unedo das im allgemeinen nicht mitmachen. Auch
Erica arborea geht sehr hoch. Ich habe E.arborea z.B. auf Korsika noch in 1000 m Höhe gesehen. Dabei spielt die anthropogene
Einwirkung bestimmt eine große Rolle. Sehr wahrscheinlich wird
das Q u e r c i o n   i l i c i s durch die Konkurrenz des submediterranen Waldes nach unten gedrückt, und, wenn diese Konkurrenz wegfällt, wenn z.B. der Wald geschlagen wird, kann sich
das Q u e r c i o n   i l i c i s ausbreiten. Es ist wahrschei
lich die natürliche Konkurrenz, die hier durch den menschlichen
Faktor etwas in Unordnung gerät.
Die potentiellen,klimatisch bedingten Höhengrenzen der Q u e r -
c i o n   i l i c i s - Arten liegen also höher als die aktuellen durch die Konkurrenz der Q u e r c e t a l i a   p u b e s -
c e n t i s bedingten. Das schließt nicht aus, daß mediterrane
Arten wie Spartium junceum und Quercus ilex an lokal begünstigten Stellen unter natürlichen Bedingungen im submediterranen
Wald-Bereich wachsen können.

R.CARBIENER:
Ich möchte kurz darauf hinweisen, daß die genannten Arten Quercus ilex, Pistacia terebinthus und Spartium junceum keine eigentlich mediterranen Arten sind, sondern da man die mediterrane Region sehr unterteilt hat, sollte man im weitesten Sinne
von submediterranen Arten sprechen. Im Rhône-Tal unterhalb von
Lyon sind es die drei ersten Arten, die uns als Vorposten der
mediterranen Region begegnen, und so ist es im Grunde genommen
auch in Jugoslawien, wo diese drei Arten die ersten Anklänge an
die eigentliche mediterrane Hartlaub-Region darstellen, weil es
eben diese Arten sind, die am weitesten aus dem mediterranen
Gebiet nach Norden ausstrahlen.

DIE ERMITTLUNG VON VEGETATIONSGRENZEN BEI DER KONSTRUKTION
VON KARTEN KLEINEREN MASSTABS (BAYERNKARTE)

P. S e i b e r t

Im Gang, gegenüber der Buch-Ausstellung habe ich eine "Übersichtskarte der natürlichen Vegetationsgebiete von Bayern 1 :
500 000" aufgehängt. In dieser Karte sind eine ganze Menge Grenzen. Die sind natürlich nicht alle abgegangen, sondern auf andere Weise ermittelt worden. Wie ich das gemacht habe, möchte ich
jetzt in meinem Vortrag erzählen.

Einleitung
Prof.KÜCHLER (1963) hat in seinem Vortrag beim Internationalen
Symposion über Vegetationskartierung in Stolzenau 1959 die ein-
zelnen Schritte erläutert, die bei einer Zusammenstellung von
Vegetationskarten kleinen Maßstabs zu tun sind. Sie lassen sich
kurz folgendermaßen skizzieren:

1. Sammlung von Vegetationskarten der einzelnen Gebietsteile,
   so daß schließlich das ganze Gebiet auf solchen Teilkarten
   studiert werden kann;
2. Suche nach einer geeigneten Klassifikation;
3. Übersetzung der Legenden der Teilgebietskarten in die erwähl-
   te Klassifikation z.T. unter Beiziehung von geologischen und
   Bodenkarten;
4. Herstellung des Manuskripts der neuen Vegetationskarte.

Die Entwicklung der Vegetationskarte von Bayern
Als ich die Möglichkeit erkannte und mich dazu entschloß, eine
Vegetationskarte von Bayern zu entwickeln, war ich mir von vor-
neherein darüber im klaren, daß ein Vorgehen nach Punkt 1 nicht
weit führen würde. Zwar gab es eine Menge von Teilgebietskarten,
aber fast nur in sehr großen Maßstäben 1 : 5000, 1 : 10 000 und
nur für einen Bruchteil der Gesamtfläche. Außerdem stellen sie
meist die reale Vegetation, vor allem Grünland dar, deren Wiedergabe nicht das Ziel unserer Übersichtskarte sein konnte. Karten für größere Teilgebiete oder für ganz Bayern gab es nur in
sehr kleinen Maßstäben 1 : 1 300 000 bei HUECK (1936),1 : 2 700 000
bei RUBNER (1953/1955) und beide in einer sehr allgemeinen Klassifikation. Über diesen Punkt 2 von KÜCHLER hatte ich aber von
Anfang an eine feste Vorstellung. Die Karte sollte eine Karte
der potentiell natürlichen Vegetation sein, und es sollte ihr

die neueste pflanzensoziologisch-systematische Einteilung der
westdeutschen Phanerogamen- und Gefäßkryptogamen-Gesellschaften
von OBERDORFER (1967) zugrunde liegen. Dadurch, daß ich an dieser Einteilung mitwirkte, konnte ich sie von Anfang an, d.h.
seit 1965, bei meinen Arbeiten berücksichtigen und umgekehrt
aus meinen bayerischen Erfahrungen zu ihrer Ergänzung beitra-
gen. Ich wollte eine allgemein zugängliche Einteilung zugrunde
legen, damit der Leser meiner Karte im Zweifelsfalle nachschla-
gen kann, was ich unter den genannten Pflanzengesellschaften
verstehe, auch auf die Gefahr hin, daß ich mich in manchen Gebieten aus einer noch unsicheren Position heraus festlegen mußte. Eine Karte, die so allgemein ist, daß sie schon deswegen
nicht falsch sein kann, wollte ich nicht anfertigen.

Die Einteilung der Vegetationsgebiete (Klassifikation)
In der Karten-Legende sind die physiognomisch, meist auch sozi-
ologisch-systematisch verwandten Pflanzengesellschaften zu For-
mationsgruppen zusammengefaßt worden.
Die namengebenden Pflanzengesellschaften der Vegetationsgebiete
haben in der Regel den Rang einer Assoziation, nur bei den "Alpinen und subalpinen Vegetationsgebieten" und bei den "Vegetationsgebieten der Bruchwälder und Moore" haben sie den Rang von
Verband, Ordnung oder Klasse. Andererseits sind großflächig verbreitete Pflanzengesellschaften noch weiter unterteilt worden
in edaphisch bedingte Subassoziationen, Höhenformen und geographische Rassen.Solche Unterteilungen sind aber nur dann getroffen, wenn sie floristisch und damit synsystematisch begründet
sind.

Die Abgrenzung der Vegetationsgebiete mit Hilfe der Boden-Karte
und der geologischen Karte
Für die Abgrenzung der Vegetationsgebiete wurde zunächst die
"Bodenkundliche Übersichtskarte von Bayern 1 : 500 000" zu Hilfe genommen. Beim Studium dieser Karte und deren Erläuterungen
konnte ich feststellen, daß es möglich war, schon aus meinen bisherigen Erfahrungen bei der vegetationskundlichen Arbeit und
meiner allgemeinen Kenntnis des Landes Bayern für verschiedene
Gebiete die potentiell natürliche Vegetation bestimmter Boden-
Einheiten zu benennen. Bereits beim ersten Versuch gelang ein
ziemlich vollständiger Schlüssel für die Übersetzung der Boden-
Einheiten in Pflanzengesellschaften. Hierbei mußte selbstver-

ständlich die gleiche Boden-Einheit in klimatisch verschiedenen
Gebieten anders übersetzt werden.Die Entscheidung, wo bei der
Übersetzung einer bestimmten Boden-Einheit in 2 verschiedene
Pflanzengesellschaften die Grenze zu ziehen war, entfiel bei
der großräumigen Differenzierung, weil zwischen den gleichen
Böden großklimatisch verschiedener Gebiete meist große Gebiete
mit anderen Boden-Einheiten lagen, die für diese Trennung sorg-
ten. Zum Beispiel tragen podsolige Sandböden im Spessart ein
L u z u l o - F a g e t u m, im Nürnberger Reichswald und in
der Oberpfalz dagegen ein V a c c i n i o - Q u e r c e t u m.
Die Gebiete sind durch große Eichen-Hainbuchenwald-Gebiete voneinander getrennt.
Dieser Übersetzungsschlüssel wurde im Laufe der Zeit durch Auswertung der Literatur und eigenen unveröffentlichten Materials
ergänzt, so daß eine 1. Manuskript-Karte gezeichnet werden konnte. Bei der Entwicklung dieser Karte wurden alle Gebiete erkannt,
in denen eine Übersetzung aus der Boden-Karte nicht möglich oder
durch die Beachtung anderer Kriterien zu ergänzen war. Das galt
vor allem für den Alpen-Raum und den Bayerischen Wald. Auch gab
es größere Gebiete, in denen der Übersetzungsschlüssel unsicher
war, weil in der Boden-Karte nahe verwandte Boden-Typen ausgeschieden waren, und man nicht recht wußte, bei welchen man die
Grenze für die Zuordnung zu dem einen oder anderen Vegetations-
Gebiet ziehen sollte. Das galt vor allem für das Tertiär-Hügel-
land mit seinen vielfach wechselnden Böden. Hier waren Kontroll-
Fahrten nötig, auf deren Methodik ich nachher zu sprechen komme.

Im Frankenwald zeigten solche Kontroll-Fahrten und im Spessart
die Arbeit von LEIPPERT (1962), daß für die Abgrenzung weniger
die Boden-Karte als die geologische Karte geeignet war. Auf dem
oberen Buntsandstein des Spessarts stockt das P o a   c h a i -
x i i - C a r p i n e t u m von LEIPPERT, das ich als Spessart-
Rhön-Rasse des G a l i o - C a r p i n e t u m auffasse, auf
dem mittleren Buntsandstein dagegen das L u z u l o - F a g e-
t u m. In der Boden-Karte sind beide Gebiete als podsolige Sande dargestellt.

Klima und Relief als begrenzender Faktor
Klima- und reliefbedingte Vegetationsgrenzen konnten nicht einfach anhand von Klima- und topographischen Karten ausgeschieden
werden. Vielmehr war es notwendig, durch Ermittlung der Koinzi-

denzen zwischen Vegetation einerseits und Klima und Relief andererseits die für die Vegetation ausschlaggebenden Grenzwerte
dieser Faktoren kennenzulernen. Erst dann konnten diese Grenzwerte aus der Klima- und topographischen Karte für die Abgrenzung der Vegetationseinheiten in der Karte nutzbar gemacht werden. Auch das geschah auf zahlreichen Reisen.
Dem Verlauf bestimmter Niederschlags-Isohyeten wurden die West-
grenze des   L u z u l o - F a g e t u m   im Bayerischen Wald
und die Grenzen der Buchenwald-Gebiete auf der mainfränkischen
Platte angepaßt: sie sind unsicher und nur im allgemeinen richtig, wenn sie auch auf einigen Reisen nachgeprüft wurden. Auf
der mainfränkischen Platte ist die Abgrenzung noch dadurch erschwert, daß auch die Buchenwald-Gebiete infolge mittelwaldar-
tiger Bewirtschaftung als reale Vegetation Eichenwälder tragen.

Durch die Höhenlage ist die Abstufung   V a c c i n i o - Q u e r -
c e t u m - V a c c i n o - A b i e t e t u m,   Hügelland- und
Bergland-Form und   S o l d a n e l l o - P i c e e t u m   im
Bayerischen Wald und die Abstufung Hügelland- und Bergland-Form
des   A s p e r u l o - F a g e t u m   im Allgäu bedingt. Im Bayerischen Wald wurde als ausschlaggebende Höhen-Grenze zwischen
Hügelland- und Bergland-Form des   V a c c i n i o - A b i e -
t e t u m   die 700 m-Linie auf Reisen festgestellt, die untere
Höhen-Grenze des   S o l d a n e l l o - P i c e e t u m   ist
aus TRAUTMANN's (1952) Dissertation mit 1200 m bekannt. Dazwischen schiebt sich in einer thermisch begünstigten Zone der
Zahnwurz-Tannen-Buchenwald. Ich habe diesen hier eigentlich
nicht erwähnen wollen, aber nachdem Herr CARBIENER diese thermisch begünstigte Stufe für die Vogesen so schön erläutert hat,
kann ich das tun, ohne daß es mich Zeit kostet. Im Allgäu konnte die ausschlaggebende Höhen-Grenze aus den Tabellen einer laufenden Dissertation (PETERMANN) abgeleitet werden. Auf Litera-
tur-Studien und Reisen beruhen auch die Höhen-Grenzen zwischen
G a l i o - C a r p i n e t u m   und den   E u - F a g i o n -
Gesellschaften. Das   L u z u l o - F a g e t u m   geht überall
tiefer und ist edaphisch bedingt.
Expositionsbedingt sind   L u z u l o - Q u e r c e t u m   und
die thermophilen Gesellschaften   C l e m a t i d o - Q u e r -
c e t u m, A n e m o n o - und   C y t i s o - P i n e t u m.   Die
letzteren sind bei GAUCKLER (1938) und HOHENESTER (1960) genau

belegt, und es war nur notwendig, die von sekundären Trocken-
Rasen und ̄-Gebüschen und Weinbergen bestockten analogen Standorte in die Vegetationsgebiete einzubeziehen. Alle diese extremen Gesellschaften sind kleinflächig verbreitet, in der Karte
schematisch eingezeichnet und überrepräsentiert, damit man sie
überhaupt darstellen kann.

Bei den Reisen zeigte sich, daß in kontinental getönten Gebieten auf den ärmeren Boden-Einheiten die Form des Oberflächen-
Reliefs dafür den Ausschlag gibt, ob ein L u z u l o - F a g e -
t u m oder ein V a c c i n i o - Q u e r c e t u m zur Ausbildung kommt. Der Buchenwald bevorzugt die hügeligen Gelände-
formen, während die spätfrostgefährdeten ebenen Tal-Lagen vom
Eichenwald besetzt sind. Auch die kontinentalen Rassen des
G a l i o - C a r p i n e t u m auf der Münchener Schotter-Ebe-
ne und im Straubinger Becken sind an ebene Lagen gebunden. In
all diesen Fällen wurden die Vegetationsgrenzen mit Hilfe von
Höhenschichtlinien-Karten festgelegt.

Für den Alpen-Raum wurde die Vegetationskarte im wesentlichen
aus der topographischen Karte 1 : 50 000 abgeleitet, deren Baum-,
Gebüsch-, Rasen- und Fels-Signaturen Hinweise auf die Verbreitung der Formationen Wald, Gebüsch und gehölzfreie Vegetation
geben. Auf den Kalken der alpinen Trias, des Jura und der Krei-
de sind als Waldgesellschaften A p o s e r i d o - F a g e -
t u m und P i c e e t u m s u b a l p i n u m kartiert, als
Gebüsch-Gesellschaft das E r i c o - R h o d o d e n d r e -
t u m h i r s u t i und in der gehölzfreien alpinen Stufe
P o t e n t i l l i o n c a u l e s c e n t i s, T h l a s p i -
o n r o t u n d i f o l i i und E l y n o - S e s l e r i -
e t e a, auf den Mergeln des Lias und Flysch dagegen die Serie:
G a l i o - A b i e t e t u m, B a z z a n i o - P i c e e t u m,
A l n e t u m v i r i d i s, N a r d i o n und R h o d o -
d e n d r o - V a c c i n i o n. Wo die eine oder andere Assoziation der jeweiligen Formation einzutragen war, ergab sich also aus der geologischen Karte.

Methode der Kontroll-Fahrten
Zu Kontroll-Fahrten wurden alle in den Jahren 1965 und 1966 für
andere Zwecke notwendigen Dienstreisen in Bayern benutzt, wobei,
soweit es möglich war, die Reise-Route so ausgewählt wurde, daß
sie für die Fragen der Vegetationskarte günstig lag. Danach noch
unklare Gebiete wurden eigens bereist.

Bei diesen Reisen wurde an allen Stellen gehalten, an denen auf
einem anscheinend mittleren Standort (Klimax-Standort) Wald-
oder Forst-Gesellschaften vorhanden waren. Nach Ansprache des
Bestandes wurde der Gesellschaftsname notiert,und, wo eine synsystematische Zuordnung nicht eindeutig möglich war, wurden
auch die charakteristischen Pflanzenarten aufgeschrieben. Zu
vollständigen Vegetationsaufnahmen fehlte allerdings meistens
die Zeit. Auf diese Weise wurden mehr als 450 Punkte im ganzen
Lande aufgenommen und in die Deutsche Generalkarte 1 : 200 000
eingetragen. Nachdem zu Hause Höhenlage und zugehörige Boden-
einheit der "Bodenkundlichen Übersichtskarte" festgestellt und
dazugeschrieben waren, konnten die Koinzidenzen, die für die
Auffindung mancher oben genannter Grenzen bekannt sein mußten,
gefunden werden.
Nach diesen Vorarbeiten wurde unter nochmaliger kritischer Wür-
digung der vorhandenen Literatur die 2.Manuskript-Karte angefer-
tigt, die für den Druck die Vorlage lieferte.

Zusammenfassung und Schluß
Die Erfahrungen bei der Anfertigung der Vegetationskarte von
Bayern haben gezeigt,daß es bei kleinmaßstäblichen Karten in einem erstaunlich großen Umfang möglich ist, topographische, geologische und bodenkundliche Karten für die Abgrenzung von Vegetationseinheiten zu verwenden. Das setzt natürlich voraus, daß
gute Kartenwerke vorliegen. Ihr Maßstab soll demjenigen ähnlich
sein, der für die Vegetationskarte vorgesehen ist; er soll auf
keinen Fall kleiner sein.
Voraussetzung ist allerdings, daß der Bearbeiter nicht nur einzelne Pflanzengesellschaften und Standortsfaktoren zu paralle-
lisieren versteht, sondern das gesamte Beziehungsgefüge zwischen
Vegetation und Umwelt in seinem Arbeitsgebiet gut überblickt.

Aber auch dann kann eine solche Karte nur ein Behelf sein, um
die Zeit zu überbrücken, die noch vergeht, bis gute, ausschließlich im Gelände abgegrenzte Vegetationskarten vorliegen. Sie ist
daher nur in einem vegetationskundlich unterentwickelten Land
sinnvoll, wie Prof. TÜXEN so schön gesagt hat.
Ich möchte noch einen zweiten Vortrag halten, und zwar möchte
ich noch etwas ergänzen zu meinem Vortrag von vorgestern. Hier
sehen Sie in der Projektion noch einmal die ganze Karte. Doch
hat diese noch etwas anderes als die Karte draußen, nämlich ro-

te und schwarze Abgrenzungen. Diese machen eine vegetationsgeo-
graphische Gliederung deutlich, die aus der Karte induktiv abgeleitet werden kann. Rot umrandet sind die Vegetations-Bezirke
nach der Definition von SCHMITHÜSEN (1959), schwarz umrandet
die Wuchs-Distrikte. Es können 7 Vegetations-Bezirke unterschieden werden:

1. die fränkische Eichenwald-Landschaft
2. die fränkische Buchenwald-Landschaft
3. die oberpfälzisch-obermainische Kiefern- und
   Eichenwald-Landschaft
4. die nordost-bayerische Nadelwald-Landschaft
5. die süd-bayerische Eichen-Hainbuchenwald-Landschaft
6. die süd-bayerische Buchen- und Tannen-Buchenwald-
   Landschaft
7. die subalpin-alpine Vegetations-Landschaft.

Sie sind großklimatisch und durch die groben Höhenstufen bedingt,
während die Wuchs-Distrikte mehr durch edaphische Unterschiede
ihrer dominierenden Gesellschaft und eine feinere Höhenstufen-
Gliederung verursacht sind. Zu diesen Wuchs-Distrikten und Vege-
tations-Bezirken kommt man durch Komplex-Bildungen, durch Bildung von Dominanz-Komplexen im einen und von Mosaik-Komplexen
im anderen Falle. Verkleinert man die Karte auf den Maßstab, sagen wir, 1 : 2 500 000 und legt die Wuchs-Distrikte mit der
Farbe ihrer Leitgesellschaft an, so kommen wir zu diesem Kärtchen, das bereits so klein ist, daß es beinahe als Vorlage für
eine bayerische Briefmarke dienen könnte.

ZUSAMMENFASSUNG
Bei der Entwicklung der "Übersichtskarte der natürlichen Vegetationsgebiete von Bayern 1 : 500 000" konnte nicht auf die Vegetationskarten einzelner Teilgebiete zurückgegriffen werden.
Vielmehr wurden für die Abgrenzung der Vegetationsgebiete andere Kartenwerke zu Hilfe genommen, nämlich die Boden-Karte, geologische Karte, Niederschlags-Karte, topographische Karte für
Höhenlage, Exposition, Form des Oberflächen-Reliefs und im Alpen-Raum für die Abgrenzung der verschiedenen Formationen.
Durch Auswertung der eigenen Erfahrungen, der Literatur und
durch zahlreiche Kontroll-Fahrten im gesamten Arbeitsgebiet konnte ein Schlüssel für die Übersetzung der Karten und die Abgrenzung der Vegetationsgebiete erarbeitet werden.

SUMMARY

The ascertainment of boundaries of plant communities construc-
ting maps of smaller scales (for instance the map of plant communities of Bavaria).

Developping the "Übersichtskarte der natürlichen Vegetationsgebiete von Bayern 1 : 500 000" (= general map of the natural areas of plant communities of Bavaria 1: 500 000) is wasn't possible to rely on maps of plant communities of single parts of areas. Rather there were used and evaluated other special maps and
collections of special maps such as map of soils, map of geolo-
gy, map of precipitation, topographical map for altitude, exposition and surface relief for marking off the areas of plant
communities and in the Alps for marking off the different formations.

By exploiting the own experiences and literature and by nume-
ruos controlling trips all over the area it was possible to eva-
luate a key for translation of the special maps and for marking
off the areas of plant communities.

LITERATUR

GAUCKLER,K.   -1938- Steppenheide und Steppenheidewald der Frän-
     kischen Alb in pflanzensoziologischer, ökologischer und
     geographischer Betrachtung.- Ber.bayer.bot.Ges.23.
     München.
HOHENESTER,A.   -1960- Grasheiden und Föhrenwälder auf Diluvial-
     und Dolomitsanden im nördlichen Bayern.- Ber.bayer.bot.
     Ges.33. München.
HUECK,K.   -1936- Pflanzengeographie Deutschlands.- Berlin-Lich-
     terfelde.
KÜCHLER,A.W.   -1963- Die Zusammenstellung von Vegetationskar-
     ten kleinen Maßstabs.- In: Tüxen,R.(Edit.): Vegetations-
     kartierung. Ber.Intern.Symp.Stolzenau/Weser 1959. Wein-
     heim.
LEIPPERT,H.   -1962- Waldgesellschaften und ihre Böden im Spes-
     sart-Rhön-Vorland.- Diss.Würzburg.
OBERDORFER,E.   -1967- Systematische Übersicht der westdeutschen
     Phanerogamen- und Gefäßkryptogamen-Gesellschaften.-
     Schriftenr.Vegetationskde.2. Bad Godesberg.
RUBNER,K.& REINHOLD,F.   -1953- Das natürliche Waldbild Europas.-
     Hamburg und Berlin.
RUBNER,K.   -1955- Versuch einer waldgeographischen Gliederung
     Bayerns.- Allg.Forstz.10. München.
SCHMITHÜSEN,J.   -1959- Allgemeine Vegetationsgeographie.- Ber-
     lin.
SEIBERT,P.   -1968- Übersichtskarte der natürlichen Vegetations-
     gebiete von Bayern 1: 500 000 mit Erläuterungen.-
     Schriftenr.Vegetationskde.3. Bad Godesberg.
-- --   -1968- Vegetation und Landschaft in Bayern.- Erdkunde
     22. Bonn.

TRAUTMANN,W. -1952- Pflanzensoziologische Untersuchungen der
     Fichtenwälder des Bayerischen Waldes.- Forstwiss.Cbl.71.
     Hamburg und Berlin.
VOGEL,F. -1961- Erläuterungen zur Bodenkundlichen Übersichts-
     karte von Bayern 1 : 500 000.- München.

J.MOORE:

Ich möchte Herrn Dr. SEIBERT zu seiner schönen Arbeit gratulieren. Es wäre schön, wenn jemand bei uns in Irland mit einer solchen Energie eine vorläufige Karte anfertigen könnte.

R.TÜXEN:

Ich habe vor langen Jahren zweimal etwas Ähnliches für Nordwest-
Deutschland gemacht, eine deduktiv abgeleitete Karte. Es gab damals noch kaum eine Boden-Kartierung, es gab manche andere Karten nicht, die Sie benutzen konnten. Ich war also gar nicht imstande, eine so gut durchdachte Methodik anwenden zu können,
wie wir das bei Ihnen gesehen haben. Ich habe dann, nachdem diese Karten gemacht waren und wir viel induktiv kartiert haben,
d.h. also wirklich im Gelände die Karten aufgenommen haben, mir
die Meinung gebildet, ich werde nie wieder eine deduktive Karte
machen. Ich bin durch Ihre Arbeit, die ich aufrichtig bewunde-
re, belehrt worden. Erstens habe ich gelernt, wie man eine gründliche, vielseitige Methodik entwickeln kann, und zweitens, daß
es selbst in Deutschland oder in Mitteleuropa Gebiete gibt, wo
durchaus deduktiv erarbeitete Karten einen hohen Wert haben. Gewiß ist Ihre Karte wahrscheinlich nicht in allen Punkten richtig. Das wird sich herausstellen, wenn man nachher im Gelände
die Kartierung wiederholt. Aber auch diese Karte, die induktiv
im Gelände gewonnen wird, wird nicht in allen Punkten richtig
sein, sondern wird von verschiedenen Bearbeitern, deren Kenntnisse sich nach und nach entwickeln, in verschiedener Weise hergestellt werden. Aber daß wir jetzt überhaupt von Bayern eine
so klare Karte haben, das ist Ihr Verdienst, und dazu wollte
ich Ihnen herzlich gratulieren.

P.SEIBERT:

Ich habe nur einen Wunsch, daß Ihre aufrichtige Bewunderung
nicht zusammenbricht, wenn Sie mit der Karte 'mal in Bayern
spazieren fahren!

ZUR ABGRENZUNG VON EINHEITEN DER HEUTIGEN POTENTIELL
NATÜRLICHEN VEGETATION IN WALDARMEN GEBIETEN NORDWEST-
DEUTSCHLANDS

H. Dierschke

EINLEITUNG

Nordwest-Deutschland gehört heute, nach 40 Jahren kleinräumiger
Untersuchungen und weiträumiger Beobachtungen durch R.TÜXEN und
zahlreiche Mitarbeiter zu den pflanzensoziologisch am besten be-
kannten Gebieten Mitteleuropas. Es ist gleichzeitig im Bereich
nördlich der Mittelgebirge eines der an natürlicher oder doch
noch naturnaher Vegetation ärmsten Gebiete, da die hier einst
außerhalb der Hochmoore fast alle Standorte bedeckenden Wälder
einer zunächst extensiven, in den letzten 150 Jahren zunehmend
intensivierten bäuerlichen Bewirtschaftung weichen mußten.
So wurden gerade in Nordwest-Deutschland immer wieder die Beziehungen zwischen der realen, größtenteils aus Ersatzgesellschaften bestehenden Vegetationsdecke und den standortsgemäßen potentiellen Schluß-Gesellschaften erörtert. Diese Beziehungen sind
heute zumindest im größeren Zusammenhang für das Gebiet geklärt
und verschiedentlich eingehender dargestellt worden (BUCHWALD
1953, KRAUSE 1955, MEISEL-JAHN 1955, PREISING 1954, TRAUTMANN
1966, TÜXEN s.Schriften-Verzeichnis 1956).
Mit der Definition der "potentiell natürlichen Vegetation"(TÜXEN
1956) wurden viele Unklarheiten, die auf dem in verschiedener
Weise gebrauchten Klimax-Begriff beruhten, weitgehend beseitigt.
Die heutige potentiell natürliche Vegetation eines bestimmten
Gebietes ist demnach die Summe aller gedanklich konstruierten
Dauer- und Schluß-Gesellschaften, die den derzeitigen Standortsbedingungen entsprechen.
Die Methoden der Ableitung von Einheiten der heutigen potentiell
natürlichen Vegetation sind für Nordwest-Deutschland vor allem
von TÜXEN (1956, vgl.a. TRAUTMANN 1966) eingehend beschrieben
worden, so daß hierauf nicht weiter eingegangen werden muß. Bei
diesen großräumiger ausgerichteten Betrachtungen wurde jedoch
mehr Wert auf das Erkennen dieser Einheiten im Gelände als auf
ihre Abgrenzung gelegt.
Bei eigenen Kartierungsarbeiten im Bereich der Verdener Geest
(DIERSCHKE 1969) ergaben sich mancherlei Probleme, die mein Au-

genmerk auf die Frage nach möglichst vielseitigen Kriterien für
die Abgrenzung der vorkommenden Einheiten der potentiell natürlichen Vegetation richteten. Einige Hilfsmittel möchte ich hier
am Beispiel des Holtumer Moores und seiner Randgebiete, das den
Ausgangspunkt der Kartierung bildete, kurz aufzeigen.

DAS UNTERSUCHUNGSGEBIET

Das Holtumer Moor liegt in einer weiten Mulde zwischen Rotenburg
und Verden südöstlich von Bremen, die als Absenkung über einem
in der Tiefe lagernden Salzstock gedeutet wird (ILLIES 1952).
Umgeben von Grundmoränen-Platten der Saale-Vereisung bildet sie
ein vermoortes Niederungsgebiet, daß durch den zentral liegenden
Heidberg und einige weitere kleine Rücken vor allem im westlichen Teil untergliedert ist. Das Moor wird durch die Aue nach
Norden zur Wümme entwässert. Während im Süden und Westen an den
flachen Hängen die Grundmoräne bis an die Niederung heranreicht,
treten an etwas steileren Hängen im Norden und Osten die darun-
terliegenden fluvioglazialen Sande aus.
Hoch-, Nieder- und Anmoor, podsolierte Gleye und Gley-Podsole,
trockene Eisenhumus-Podsole und Sand-Parabraunerden sind die
wichtigsten Bodentypen dieses Gebietes. Hochmoor bedeckte früher den gesamten Ostteil, während Niedermoor vorwiegend im Westen verbreitet ist. Das Holtumer Moor ist schon, wie alte Karten zeigen, seit langer Zeit landwirtschaftlich genutzt worden.
Zahlreiche Torfbänke und einzelne Wasserlöcher deuten noch darauf hin, daß der obere Teil des Hochmoores teilweise abgestochen
worden ist, bevor es in Grünland umgewandelt wurde.
Weite Ackerflächen auf der Grundmoräne und Wiesen und Weiden in
den Niederungen bestimmen heute das Bild der bäuerlichen Kultur-
Landschaft, nur auf den trockenen Sandböden unterbrochen durch
Kiefern-Forsten, die größtenteils erst in jüngster Zeit die vor-
mals herrschenden C a l l u n a -Heiden ersetzt haben. In den
nach Westen und Osten angrenzenden alten Staats-Forsten sind dagegen noch naturnahe buchenreiche Waldbestände in größerem Umfange erhalten geblieben.

Artenverteilung und Säuregrad der Gewässer als Indikatoren verschiedener Wuchsbereiche
Gegenüber den sehr einheitlich aufgebauten Geest-Platten, die
bei der Kartierung der heutigen potentiell natürlichen Vegetation keine größeren Schwierigkeiten ergaben, waren die Einhei-

ten in den grundwasserbeeinflußten Gebietsteilen mit ihrem oft
kleinflächigen Wechsel verschiedener Standorte nicht immer
leicht abzugrenzen. Wichtigstes Hilfsmittel für das Erkennen
und Abgrenzen dieser Einheiten ist die heutige Pflanzendecke
mit allen ihren Merkmalen.

a) Verteilung diagnostisch wichtiger Arten
Da im Frühjahr noch keine vollständigen Vegetationsaufnahmen
der vorkommenden Ersatzgesellschaften gemacht werden konnten,
wurde zunächst dem Vorkommen und der Verteilung einzelner Pflanzen-Arten nachgegangen, die für bestimmte Einheiten der potentiell natürlichen Vegetation vermutlich von diagnostischem Wert
sind.
In den Grünland-Gebieten des Holtumer Moores sind Betula pubescens und Alnus glutinosa die häufigsten Bäume, denen sich auf
weniger nassen Böden Quercus robur zugesellt.
Betula pubescens (Abb.1) ist am weitesten verbreitet, läßt aber
Schwerpunkte im östlichen Teil und in einigen Randbezirken im
Westen und Südwesten erkennen. Alnus glutinosa (Abb.2) tritt dagegen gehäuft im Westteil und im Aue-Tal auf. Im Osten kommt
die Erle nur entlang der zügiger fließenden Gräben vor. Die beiden Baum-Arten lassen somit bereits eine Gliederung des Untersuchungsgebietes in einen Wuchsbereich von Betula pubescens und
in einen solchen von Alnus glutinosa erkennen. In diesem kommt
vereinzelt Padus avium (Abb.2) vor, dessen Verbreitung weitere
Hinweise auf die potentiell natürliche Vegetation gibt.
Schon früh fielen bei ersten Begehungen im April die zahlreichen
gelben Blüten von Primula elatior auf, einer Art, die in bestimmten Teilen westlich und nordwestlich des Heidberges in den Wiesen recht häufig ist. (Abb.4) Ähnliche Verbreitung zeigt der
etwas später blühende Ranunculus auricomus (Abb.3). Das Vorkommen beider Arten in den Wiesen ist in diesem Gebiet ein guter
Hinweis auf eine A l n o - P a d i o n - Gesellschaft, hier
vermutlich auf ein erlenreiches P r u n o - F r a x i n e t u m
als heutige potentiell natürliche Vegetation (vgl. DIERSCHKE
1968). Wo die Erle allein vorkommt, ist eher ein C a r i c i
e l o n g a t a e - A l n e t u m möglich, während ein Wald
aus Betula pubescens auf größeren Flächen des Ostteils sowie in
einigen Randbezirken im Westen zu erwarten ist. Auf den Bereich
dieses Birkenwaldes ist auch das Vorkommen von Eriophorum vaginatum und E. angustifolium (Abb.5) beschränkt.

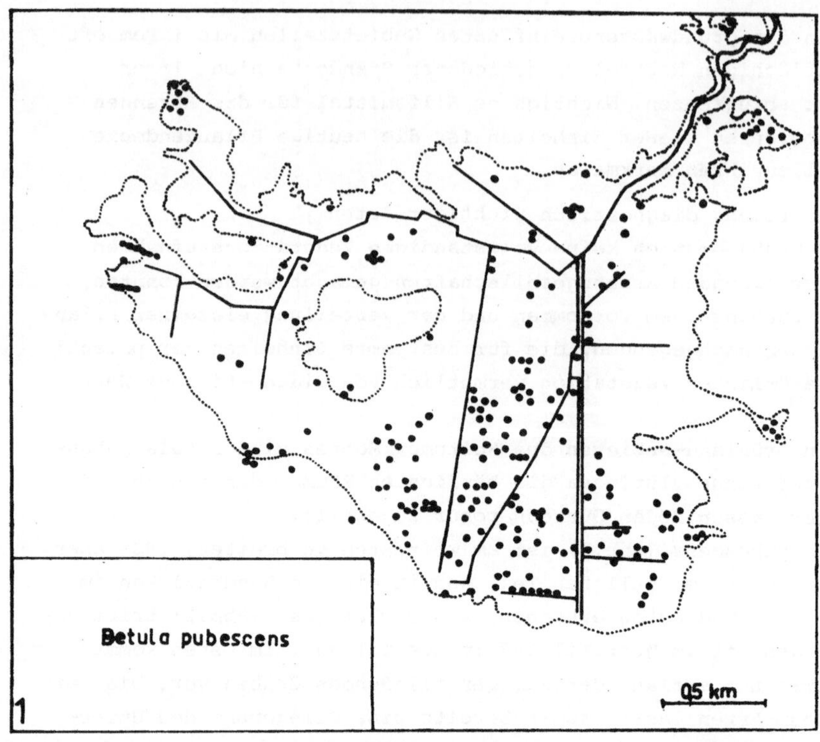
Betula pubescens
Q5 km
1

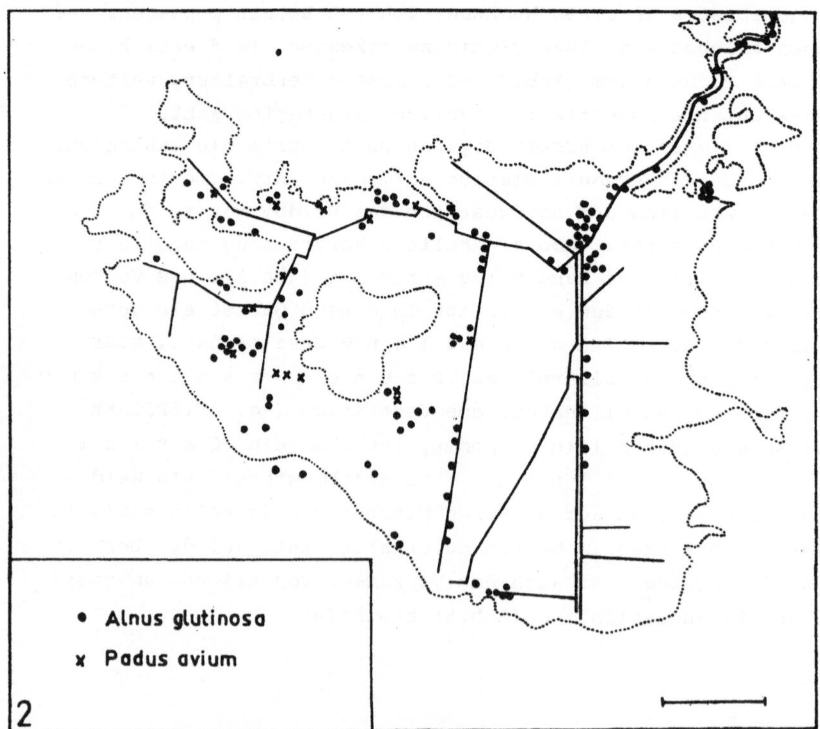
Alnus glutinosa
Padus avium
2

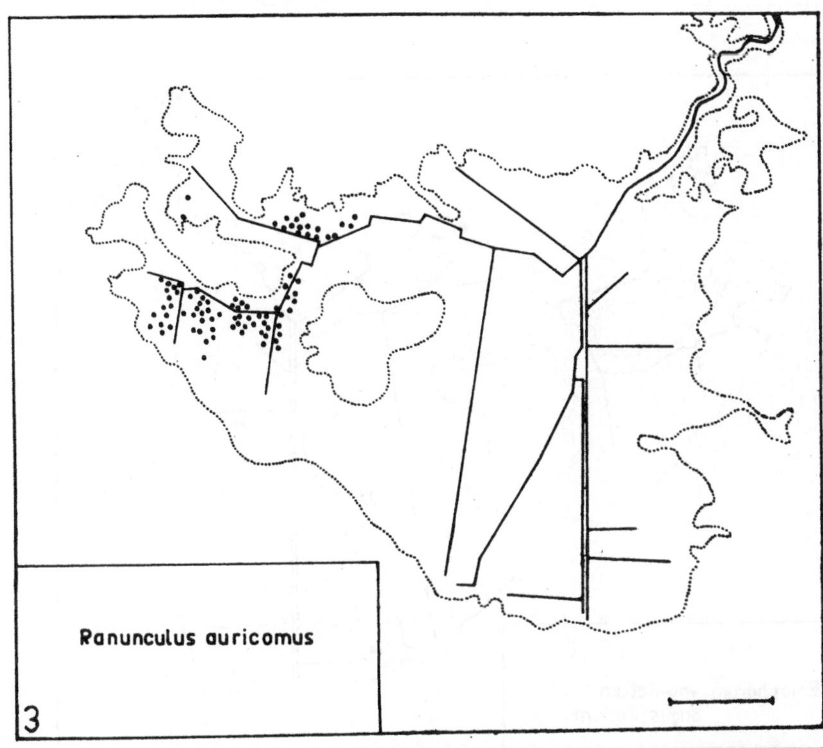

Ranunculus auricomus
3

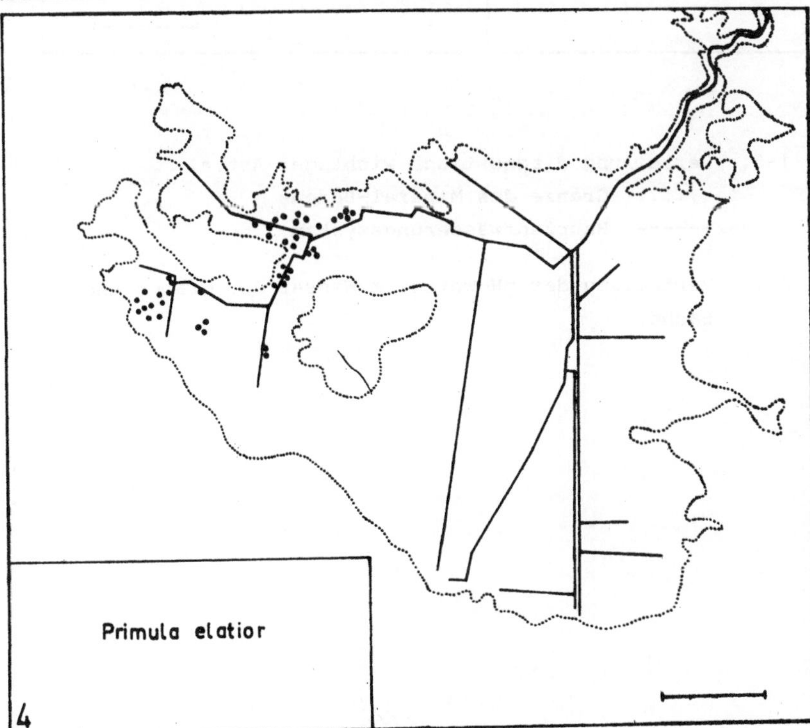

Primula elatior
4

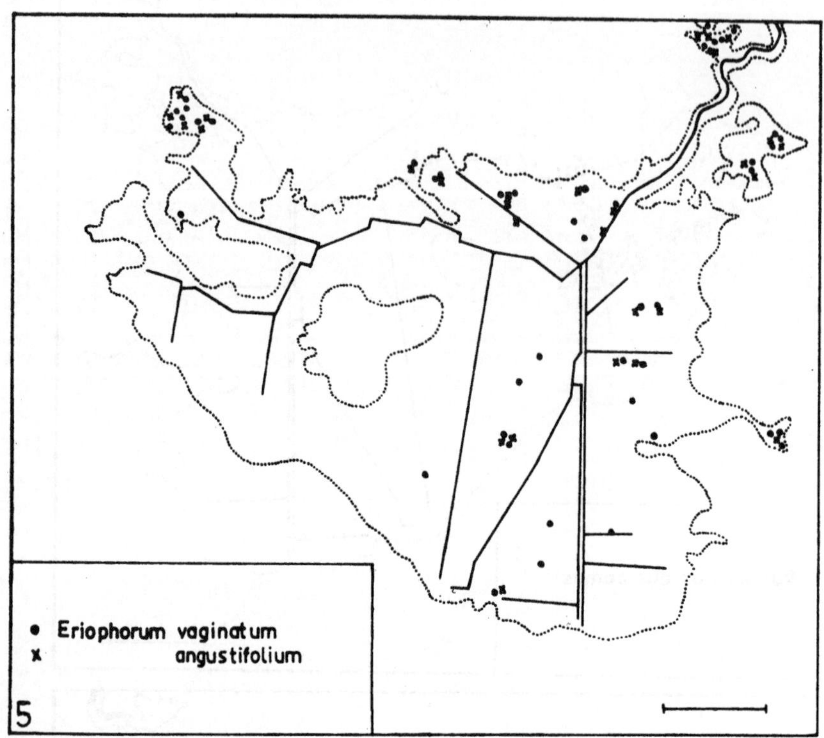

Abb. 1-5. Verteilung diagnostisch wichtiger Arten.
           ...... Grenze des Mineral-Bodens
           ------ Hauptentwässerungssystem

Abb. 6-7. Verteilung der pH-Werte im Wasser der Gräben und
           Bäche.

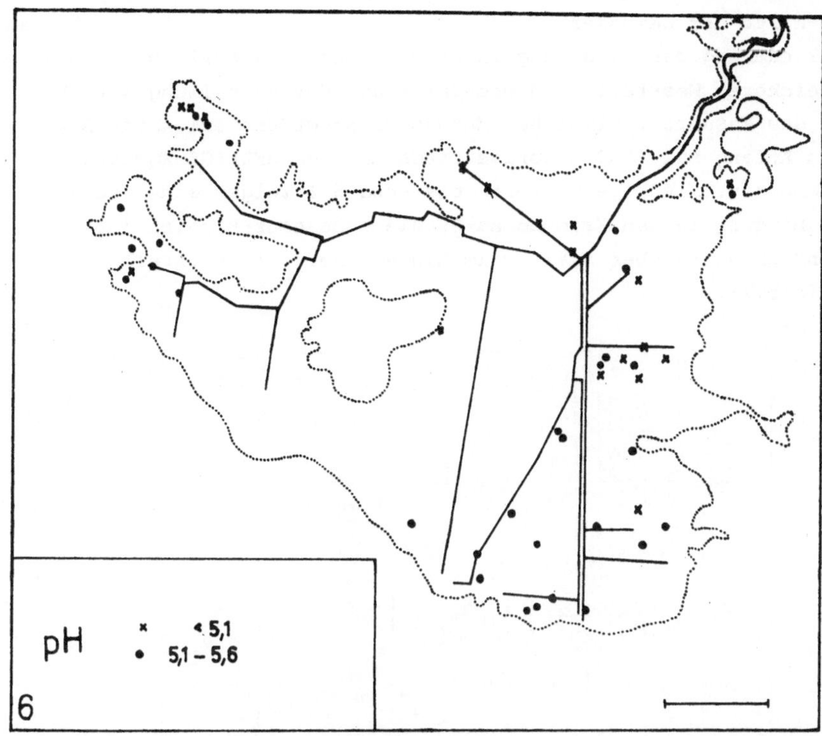
pH
x   < 5,1
•   5,1 – 5,6
6

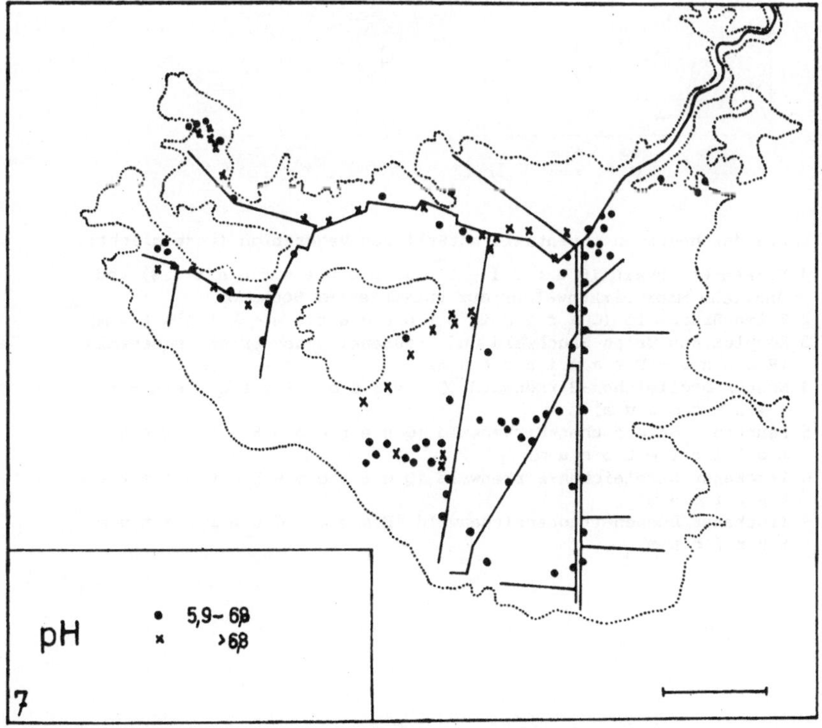
pH
•   5,9 – 6,3
x   > 6,3
7

b) pH-Werte der Gewässer

Die Gliederung der Niederung in einen ärmeren Ostteil und in einen reicheren Westteil wird ebenfalls aus der Verteilung der pH-Werte des Wassers zahlreicher Entwässerungsgräben erkennbar.Sie wurden Anfang Mai 1963 kolorimetrisch mit dem HELLIGE-Komparator gemessen. Niedrige pH-Werte bis etwa 5,6 zeigen eine deutliche Bindung an den Eriophorum-Betula pubescens-Bereich (Abb.6), während pH-Werte über 5,8 nur im Alnus-Bereich festzustellen sind (Abb.7).

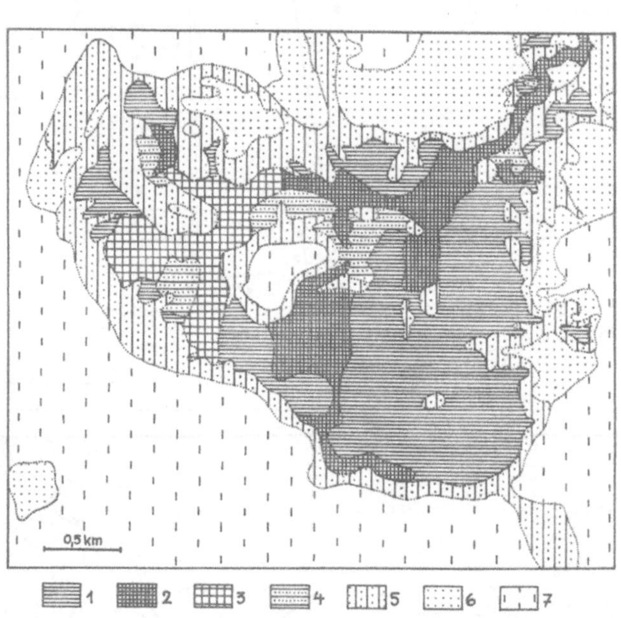

Abb.8. Karte der heutigen potentiell natürlichen Vegetation (vereinfacht).

    1 Birken-Bruchwald (B e t u l e t u m   p u b e s c e n t i s) und
      ähnliche Moor-Birkenwälder auf entwässertem Hochmoor.
    2 Erlen-Bruchwald (C a r i c i   e l o n g a t a e - A l n e t u m)
    3 Komplex aus Erlen-Bruchwald und Traubenkirschen-Erlen-Eschenwald
      (P r u n o - F r a x i n e t u m)
    4 Nasser Steileichen-Birkenwald (Q u e r c o - B e t u l e t u m
      a l n e t o s u m)
    5 Feuchter Stieleichen-Birkenwald (Q u e r c o - B e t u l e t u m
      m o l i n i e t o s u m)
    6 Trockener Stieleichen-Birkenwald (Q u e r c o - B e t u l e t u m
      t y p i c u m)
    7 Trockener Buchen-Traubeneichenwald (F a g o - Q u e r c e t u m
      t y p i c u m)

DIE EINHEITEN DER HEUTIGEN POTENTIELL NATÜRLICHEN VEGETATION

Die Karte der heutigen potentiell natürlichen Vegetation (Abb.8)
gibt in etwas vereinfachter Form die Standortsgliederung des
Untersuchungsgebietes wieder. Auf den Grundmoränen-Platten und
dem oberen Teil des Heidberges ist fast überall F a g o -
Q u e r c e t u m   p e t r a e a e   t y p i c u m   zu erwarten,
wie naturnahe Waldreste zeigen. Die an den Hängen austretenden,
unter Calluna-Heiden podsolierten Sande sind Wuchsgebiete des
Q u e r c o - B e t u l e t u m   t y p i c u m, welches nach unten im Einflußbereich des Grundwassers in das   Q u e r c o -
B e t u l e t u m   m o l i n i e t o s u m   übergeht, das auch
noch auf kleinen Sand-Inseln im Moor vorkommt. Den Übergang zu
erlenreichen Waldgesellschaften der Niederung bildet vermutlich
auf nassen Sand- und Anmoor-Böden das   Q u e r c o - B e t u -
l e t u m   a l n e t o s u m, über dessen genaue Arten-Zusammensetzung noch wenig bekannt ist.
In den Niederungen mit organischen Böden wird die bereits durch
die Verteilung einzelner Pflanzenarten aufgezeigte Gliederung
sichtbar. Der gesamte ehemals von Hochmoor-Vegetation bedeckte
Ostteil ist, wie kleine Gebüsche von Betula pubescens erkennen
lassen, das Wuchsgebiet eines Moor-Birkenwaldes, der gewisse
Ähnlichkeit mit dem B e t u l e t u m   p u b e s c e n t i s
aufweist(vergl.TRAUTMANN 1966), wenn auch auf den stärker aus-
trocknenden, höher gelegenen Torfbänken nässeliebende Arten, insbesondere Sphagnum-Arten fehlen. Birken-Bruchwald ist auch in
einigen nassen, nährstoffarmen Randgebieten der Niederungen im
Kontakt zum Q u e r c o - B e t u l e t u m   m o l i n i e t o-
s u m   zu erwarten.
Standorte des   C a r i c i   e l o n g a t a e - A l n e t u m
sind die weniger nährstoffarmen Anmoor- und Niedermoor-Böden
südlich des Heidberges, die sich, in einem quelligen Randbereich
der Grundmoräne im Süden beginnend, nach Norden fortsetzen. Zumindest teilweise dürfte dieses Gebiet früher, wie kleine Torfbänke zeigen, zum Hochmoor gehört haben. Durch tiefe Abtorfung
haben die Böden jedoch heute Anschluß an das Grundwasser und
damit einen etwas günstigeren Nährstoffhaushalt. Auch das Tal
der Aue und die Niederungen beiderseits ihrer Zuflüsse sind
Wuchsgebiete des Erlen-Bruchwaldes.
Westlich des Heidberges wird der Bereich der Anmoor- und Nieder-
moorböden durch vom Rande her hineingreifende Sand-Rücken stär-

ker gegliedert. Auch zwischen diesen Rücken ist das Kleinrelief
stärker wechselnd und mit ihm verändern sich Torfmächtigkeit
und Wasserhaushalt. Die weniger nassen Stellen, wo Primula elatior und Ranunculus auricomus gehäuft vorkommen, gehören vermutlich zum Wuchsbereich eines erlenreichen P r u n o - F r a x i -
n e t u m, die nassen Teile zum C a r i c i   e l o n g a t a e
- A l n e t u m. Der zu erwartende kleinflächige Wechsel beider
Waldgesellschaften mit ihren gleitenden Übergängen läßt eine genauere Unterscheidung und Abgrenzung nach den heute vorhandenen
Merkmalen nicht zu, so daß nur eine zusammenfassende Darstellung
dieses Komplexes möglich ist.
Als weitere natürliche Waldgesellschaften sind kleinflächig
Q u e r c o - C a r p i n e t u m   s t a c h y e t o s u m   und
F a g o - Q u e r c e t u m   m o l i n i e t o s u m   möglich.
Standorte des Feuchten Eichen-Hainbuchenwaldes können sich als
schmaler Streifen zwischen Q u e r c o - B e t u l e t u m
m o l i n i e t o s u m   und P r u n o - F r a x i n e t u m
einschieben, sind jedoch im Untersuchungsgebiet nicht klar zu
erkennen. Standorte des Feuchten Buchen-Traubeneichenwaldes konnten ebenfalls nicht deutlich ausgeschieden werden, erscheinen
aber vor allem im Süden und Südwesten am Rande der Grundmoräne
denkbar, wo in der Karte das Q u e r c o - B e t u l e t u m
m o l i n i e t o s u m   verzeichnet ist.

HILFSMITTEL FÜR DIE ABGRENZUNG DER EINHEITEN
a) Blüh-Aspekte
Die Abgrenzung der sieben unterschiedenen Einheiten geschah unter Berücksichtigung möglichst aller erkennbaren Merkmale. Zu
diesen gehören im Frühjahr besonders die Blüh-Aspekte bestimmter Pflanzen des Grünlandes, die für die Kartierung der Feucht-
standorte von großer Bedeutung sind. Für die Abgrenzung einzelner Einheiten spielt hier mehr das gehäufte Vorkommen einzelner Arten mit dominanter Farbgebung eine Rolle, als das Blühen verstreut vorkommender Einzelpflanzen.
Einige wichtige Farb-Aspekte sind in Tabelle 1 zusammengestellt.
Einzelne der angegebenen Arten treten auch in hier nicht angegebenen Gesellschaften auf, ohne dort jedoch durch ihre Blüten
stärker farbgebend zu wirken. Besonders wichtig sind die Blüh-
Aspekte im Bereich des B r o m o - S e n e c i o n e t u m
a q u a t i c a e, das die weitest verbreitete Wiesen-Assozia-
tion im Untersuchungsgebiet darstellt. Im Moorbirken-Bereich

Tab.1. Blüh-Aspekte feuchter Standorte und heutige potentiell natürliche Vegetation

	1	2	3	4	5
Anemone nemorosa (Ende 4 - Anfang 5)		●	●		
Primula elatior (Ende 4 - Anfang 5)			●		
Caltha palustris (Anfang 5)		●	●	O	
Ranunculus auricomus (Anfang 5)		O	●		
Cardamine pratensis (Anfang bis Mitte 5)	O	●	O	O	
Padus avium (Anfang bis Mitte 5)			●		
Taraxacum officinale (Anfang bis Mitte 5)					●
Ranunculus acris (ab Mitte 5)	●	●	●	●	●
Eriophorum vaginatum, Fruchtaspekt (ab M.5)	●				
Lychnis flos-cuculi (Ende 5)	●	●	O	O	
Rumex acetosa (Ende 5)	●	●	O	O	
Holcus lanatus (Anfang 6)	●	●	●	●	O
Senecio aquaticus (ab Mitte 7)	O	●	O	O	
Angelica sylvestris (ab Mitte 7)		●	●	O	
Cirsium oleraceum (Ende 7)		●			
Leontodon autumnalis (Mitte 8)					●

1 B e t u l e t u m   p u b e s c e n t i s   und ähnliche Moorbirkenwälder,
2 C a r i c i   e l o n g a t a e - A l n e t u m, 3 P r u n o - F r a x i -
n e t u m, 4 Q u e r c o - B e t u l e t u m   a l n e t o s u m, 5 Q u. -
B. m o l i n i e t o s u m.
● häufig und sehr auffällig   O selten, meist nur Einzelpflanzen

finden sich nur artenarme Wiesen, in denen u.a. Anemone nemorosa, Caltha palustris und Cardamine pratensis fehlen oder doch
nur selten und vereinzelt auftreten. Diese drei Arten bilden
dagegen im Alnus-Bereich oft deutliche Blüh-Aspekte, so daß zu
ihrer Blütezeit eine Abgrenzung der erlenreichen Gesellschaften
gegenüber dem Moor-Birkenwald gut möglich ist.
Primula elatior und Ranunculus auricomus trennen den P r u n o -
F r a x i n e t u m/A l n e t u m-Komplex nährstoffreicher Standorte von denen des reinen Erlen-Bruchwaldes. Auch die weithin
leuchtenden weißen Blüten des Padus avium, der als Strauch hie
und da zu finden ist, sind auf den Bruchwald/Auwald-Komplex beschränkt.
Je weiter das Frühjahr voranschreitet, desto weniger ausgeprägt
sind die farblichen Unterschiede der Feuchtwiesen, Lychnis flos-
cuculi, Rumex acetosa, Ranunculus acris und Holcus lanatus sind
überall reichlich vorhanden. Lediglich die weißen Fruchtstände
von Eriophorum vaginatum und E.angustifolium kommen in nassen
Wiesen oder auf Ödland nur im Betula pubescens-Gebiet in größerer Menge vor.
Der Bereich des Q u e r c o - B e t u l e t u m   a l n e t o -
s u m ist nicht deutlich durch bestimmte Aspekte unterscheid-

bar. Eine Abgrenzungsmöglichkeit gegenüber den höher anschließenden Standorten des Q u e r c o - B e t u l e t u m m o l i -
n i e t o s u m bietet vereinzelt der Blüh-Aspekt von Cardamine
pratensis.

Die vorwiegend vom L o l i o - C y n o s u r e t u m eingenom-
menen Flächen des Q u e r c o - B e t u l e t u m m o l i -
n i e t o s u m und F a g o - Q u e r c e t u m m o l i n i e -
t o s u m lassen sich im Mai oft sehr gut durch den gelben
Blüh-Aspekt und den folgenden weißen Frucht-Aspekt von Taraxum
officinale gegen die Wiesen abgrenzen, in denen der Löwenzahn
nur in geringer Menge wächst. Den Weiden auf entwässertem Hochmoor fehlt dieser Aspekt ebenfalls. Eine weitere Abgrenzungsmöglichkeit bietet im August der Blüh-Aspekt von Leontodon au-
tumnalis.

Im Sommer, nach dem ersten Schnitt, fallen in den Wiesen Senecio
aquaticus und Angelica sylvestris auf. Letztere fehlt im Bereich
des Moor-Birkenwaldes. Im Untersuchungsgebiet nicht vorkommend,
im weiteren Bereich aber vorhanden, weisen die grünlich-gelben
Blütenstände von Cirsium oleraceum meist auf das potentielle
P r u n o - F r a x i n e t u m hin.

Alle diese Aspekte kommen zwar nicht immer so deutlich ausgeprägt, wie hier geschildert, vor, die Kenntnis ihrer Beschrän-
kung auf bestimmte Einheiten der heutigen potentiell natürlichen
Vegetation erleichtert aber besonders im Frühjahr wesentlich die
Kartierung der feuchten Standorte. Weitere auffällige Farben wie
das frische Grün nacheinander sich neu belaubender Bäume und
Sträucher oder auch die verschiedene Herbstfärbung der Blätter
können oft von Nutzen sein.

b) Verteilung der Ersatz-Gesellschaften

Mehr bei kleinräumiger Erfassung oder doch nur bei Ausgehen von
kleinräumigen Untersuchungen ist die unmittelbare Abgrenzung
nach der Verteilung der Ersatz-Gesellschaften möglich. Zeigte es
sich doch im Holtumer Moor, daß die früheren Angaben über die
Zugehörigkeit bestimmter Ersatz-Gesellschaften zu den potentiellen Schluß-Gesellschaften (PREISING 1954, TÜXEN 1956 u.a.) oft
viel zu grob sind, als daß sich aus diesen Angaben genaue Kartie-
rungsgrundlagen ableiten ließen. Die ins Einzelne gehende Vegetationsanalyse, die nur kleinräumig möglich ist, ergab im Holtumer Moor, daß meist nur einzelne Subassoziationen oder Varianten einer bestimmten Schluß-Gesellschaft zugeordnet werden

können. In Tabelle 2 sind die wichtigsten Beziehungen dargestellt.

Tab. 2. Ersatzgesellschaften und heutige potentiell natürliche Vegetation

	1	2	3	8	4	5	6	7
Betuletum pubescentis und Moor-Birkenwald (S)	+							
Molinia-Erica-Hochmoor-Entwässerungsstadien (E1)	+							
Myricetum gale ericetosum (E1)	+							
Ericetum tetralicis sphagnetosum (E1)	+							
Carici canescenti -Agrostietum(E2)	O	O						
Juncus effusus-Feuchtwiese(E2)	O	O						
Bromo-Senecionetum,Subass.v.Carex nigra	O	O			O			
Lolio-Cynosuretum lotetosum,Var.v.Ran.flammula(E3)	O	O			O			
Lolio-Cynosuretum lotetosum,typische Var.(E3)	O		O	O	O	O		
Carici elongatae-Alnetum(S)		+						
Myrico-Salicetum cinereae (E1)		+						
Myricetum gale peucedanetosum(E1)		+						
Cardaminetum amarae(D)		+						
Calamagostis canescens-Feuchtwiese(E1)		+						
Caricetum gracilis(E1)		+						
Filipendulion(E1)	O	O						
Juncus acutiflorus-Feuchtwiese(E2)		+						
Scirpetum silvatici(E2)	O	O						
Junco-Molinietum(E2)	O			O	O	O		
Bromo-Senecionetum,Subass.v.Ran.auricomus(E2)			+					
Arrhenatheretum elatioris(E2)				+				
Bromo-Senecionetum,Subass.v.Trifolium dubium(E2)			O	O				
Lolio-Cynosuretum typicum(E3)			O		O			
Nardo-Galion saxatilis(E2)					O	O		
Ericetum tetralicis typicum, cladonietosum(E1)					+			
Calluno-Genistetum cladonietosum,Var.v.Molinia(E1)					+			
Spergulo-Panicetum cruris galli,Var.v.Mentha(E2)					+			
Lycopsetum arvensis,Var.v.Mentha arvensis(E2)					+			
Aperion spica-venti,Ausb.v.Mentha arvensis(E3)					+			
Fichten-Forsten(E2)						O		O
Kiefern-Forsten(E2)						O	O	O
Lolio-Cynosuretum luzuletosum(E3)						O		O
Calluno-Genistetum cladonietosum,typische Var.(E1)							+	
Spergulo vernalis-Corynephoretum(E1)							+	
Artenarme Chenopodietalia-Gesellschaften(E2)							+	
Aperion spica-venti, reine Ausbildung(E3)							O	O
Fago-Quercetum petraeae typicum(S)								+
Calluno-Genistetum sieglingietosum(E1)								+
Spergulo-Panicetum cruris galli,typische Var.(E2)								+
Lycopsetum arvensis,typische Var.(E2)								+

S Schluß-Gesellschaft, D Dauer-Gesellschaft, E1-E3 Ersatz-Gesellschaften
ersten bis dritten Grades.

+ Charakter-Gesellschaften (25): 3S, 1D, 12E1, 8E2, 1E3
O Übrige Gesellschaften   (15):   1E1, 9E2, 5E3

1 Betuletum pubescentis und ähnliche Moor-Birkenwälder
2 Carici elongatae-Alnetum, 3 Pruno-Fraxi-
netum, 4 Querco-Betuletum alnetosum, 5 Qu.-B.
molinietosum, 6 Qu.-B. typicum, 7 Fago-Querce-
tum petraeae, (8 Querco-Carpinetum stachye-
tosum)

Von den insgesamt 40 ausgeschiedenen Gesellschaften kommen 25
nur in einer Einheit der potentiell natürlichen Vegetation vor.
Von diesen 25 sind drei (Moor-Birkenwald, C a r i c i   e l o n -
g a t a e - A l n e t u m   und   F a g o - Q u e r c e t u m) natürliche Schlußgesellschaften (S) oder diesen doch nahekommende
Bestände. Das fragmentarisch entwickelte C a r d a m i n e t u m
a m a r a e   wächst als Dauer-Gesellschaft (D) kleinflächig auf
quelligen, nassen Standorten im Erlen-Bruchwald. Von den 21 übrigen sind 12 Ersatz-Gesellschaften 1.Grades (E_1) (vergl. TÜXEN
1956), acht sind Ersatz-Gesellschaften 2.Grades (E_2) und eine
3.Grades (E_3). Von den 15 im Bereich mehrerer Einheiten der potentiell natürlichen Vegetation vorkommenden Ersatz-Gesellschaften ist nur eine 1.Grades, neun sind solche 2.Grades und fünf
solche 3.Grades. Wie bereits TÜXEN (1956) gezeigt hat, nimmt
also der diagnostische Wert der Ersatz-Gesellschaften für die
Kartierung der potentiell natürlichen Vegetation von solchen 1.
Grades bis zu denen 3. oder gar 4.Grades ab. Je intensiver ein
Standort genutzt wird, desto weniger lassen sich aus seiner Vegetation heute noch Aussagen über die zu erwartende Schluß-Gesellschaft und ihre Grenzen machen.
Der Wuchsbereich des Moor-Birkenwaldes weist im Untersuchungsgebiet vier ihm eigene Ersatz-Gesellschaften auf, die allerdings
nur kleinflächig zwischen den feuchten Weiden und nassen Wiesen
als Reste des früher vorherrschenden Ödlandes eingestreut sind.
Lichte Wäldchen und Gebüsche von Betula pubescens, kleine artenarme Myrica-Bestände, Molinia-Erica-Stadien und das   E r i c e -
t u m   s p h a g n e t o s u m   wachsen noch auf einzelnen ste-
hengebliebenen Torfbänken oder in alten Torfstich-Löchern.
Die meisten ihm eigenen Ersatz-Gesellschaften hat bei uns das
C a r i c i   e l o n g a t a e - A l n e t u m (7). Neben kleinen naturnahen Wald-Resten fallen Gebüsche von Myrica gale auf,
in denen Calamagrostis canescens, Peucedanum palustre, Agrostis
gigantea und Lycopus europaeus eine artenreicherere Subassoziation kennzeichen (DIERSCHKE 1969). In alten Torfstichen, die
heute in Verbindung mit dem Grundwasser stehen, wachsen dichte
Bestände des M y r i c o - S a l i c e t u m   c i n e r e a e.
An quelligen Standorten kommt im Erlen-Bruchwald kleinflächig
das C a r d a m i n e t u m   a m a r a e   vor. In engem Kontakt
finden sich dichte, nicht genutzte Bestände von Calamagrostis
canescens mit zahlreichen Arten der Feuchtwiesen. Die eigentli-

chen Wiesen, die größtenteils vom B r o m o - S e n e c i o -
n e t u m   a q u a t i c a e, Subass.von C a r e x   n i g r a
gebildet werden, enthalten vereinzelt an nassen Stellen Rest-Be-
stände des C a r i c e t u m   g r a c i l i s, das heute infolge Entwässerung jedoch stark zurückgegangen ist. Auf nicht mehr
bewirtschafteten Wiesen des Aue-Tales, wo vom höheren Terrassen-
Rand Wasser heraussickert, hat nach dem Aufhören der Mahd Juncus
acutiflorus dichte Bestände entwickelt, die nach ihrer gesamten
Arten-Zusammensetzung zu den M o l i n i e t a l i a - Gesellschaften zu rechnen sind. Ebenso wie im Bereich des Moor-Birkenwaldes kommen hier vereinzelt das C a r i c i   c a n e s c e n -
t i - A g r o s t i e t u m   c a n i n a e, eine Juncus effusus
-Feuchtwiese und eine nasse Variante des L o l i o - C y n o -
s u r e t u m   l o t e t o s u m   mit Ranunculus flammula und
Glyceria fluitans vor.
Das P r u n o - F r a x i n e t u m   wird gekennzeichnet durch
die R a n u n c u l u s   a u r i c o m u s - Subass. des B r o -
m o - S e n e c i o n e t u m   a q u a t i c a e, die fast den
gesamten Wuchsbereich dieser Gesellschaft einnimmt. Das Q u e r -
c o - C a r p i n e t u m   s t a c h y e t o s u m   kommt vermutlich kleinflächig am Rande der Niederung vor, wo neben der
T r i f o l i u m   d u b i u m - Subass. des B r o m o - S e -
n e c i o n e t u m   mit Heracleum sphondylium und Crepis bien-
nis auch fragmentarische Bestände des A r r h e n a t h e r e -
t u m   e l a t i o r i s   wachsen.
Das Q u e r c o - B e t u l e t u m   a l n e t o s u m   läßt
sich aus den Ersatz-Gesellschaften kaum erkennen, da es auf Über-
gangs-Standorten zwischen den Bruchwäldern und dem feuchten
Stieleichen-Birkenwald Ersatz-Gesellschaften beider Bereiche
enthält. Das gemeinsame Auftreten von Alnus glutinosa und Quercus robur sowie die Böden sind hier bessere Merkmale.
Auf den flachen, feuchten Sand-Rücken spielen die Ersatz-Gesellschaften nur zur Abgrenzung gegenüber den von Grundwasser nicht
beeinflußten Gebieten eine Rolle, während zum nasseren Bereich
hin die Grenze zwischen feuchten Mineral-Böden und Niedermoor
wichtiger ist. L o l i o - C y n o s u r e t u m   verschiedener
Ausbildung und die M e n t h a   a r v e n s i s - V a r i a n -
t e n   verschiedener Acker-Unkrautgesellschaften sowie kleinflächige Reste des ‚E r i c e t u m   t y p i c u m   und c l a d o -
n i e t o s u m   und der M o l i n i a - V a r i a n t e   des

C a l l u n o - G e n i s t e t u m   c l a d o n i e t u m
kennzeichen den Wuchsbereich des   Q u e r c o - B e t u l e -
t u m   m o l i n i e t o s u m.

Auf den trockenen Grundmoränen-Platten und den am Hang austretenden Sanden, sowie auf den trockenen Dünenrücken im Aue-Tal,
führt, soweit Ackerland vorherrscht, eine Trennung nach der Unkraut-Vegetation nur teilweise zu brauchbaren Ergebnissen. Die
zunehmende Anwendung von Herbiciden macht oft eine gute Unterscheidung der verschiedenen Standorte schwierig. Anstelle des
Q u e r c o - B e t u l e t u m   t y p i c u m   sind nur artenarme Fragmente des   A p e r i o n   s p i c a - v e n t i   und
der C h e n o p o d i e t a l i a   zu finden. Auf Standorten des
F a g o - Q u e r c e t u m   p e t r a e a e   t y p i c u m
wachsen neben etwas artenreicheren Ausbildungen des   A p e r i -
o n   s p i c a - v e n t i   das   S p e r g u l o - P a n i c e -
t u m   c r u r i s - g a l l i   und das   L y c o p s e t u m   a r -
v e n s i s. Im Q u e r c o - B e t u l e t u m - Bereich sind
Kiefern-Forsten und eingestreute Reste der früher großen Heide-
Flächen (C a l l u n o - G e n i s t e t u m   c l a d o n i e -
t o s u m, t y p i s c h e   V a r i a n t e), auf offenem Sand
vereinzelt auch das   S p e r g u l o   v e r n a l i s - C o r y -
n e p h o r e t u m   zu finden.
Neben naturnahen Buchen-Traubeneichen-Wäldern sind dagegen in
Nachbarschaft vor allem Fichten-Forsten verbreitet. Auf dem
Heidberg wächst ein kleiner Rest des etwas anspruchvolleren
C a l l u n o - G e n i s t e t u m   s i e g l i n g i e t o -
sum mit Sieglingia decumbens, Festuca capillata, Scorzonera humilis, Carex pilulifera, Nardus stricta und Hypochoeris radica-
ta als Trennarten.
Die Kenntnis dieser ins Einzelne gehenden Beziehungen zwischen
heutiger realer und potentiell natürlicher Vegetation ist ein
sehr wertvolles Hilfsmittel für die Abgrenzung potentieller Einheiten, besonders wenn, wie im Holtumer Moor, auch noch eine genaue Karte der Ersatz-Gesellschaften erarbeitet wird. Für groß-
flächige Kartierungen sind die Grenzen der Ersatz-Gesellschaften nur in beschränktem Umfange brauchbar (vergl.TRAUTMANN 1966).

c) Bodentyp, Farbe und Struktur der Boden-Oberfläche
Eine wichtige Rolle für die Kartierung der heutigen potentiell
natürlichen Vegetation spielen die Böden. Gute Boden-Karten oder
eigene Untersuchungen können in verschieden starkem Maße mit

eine Grundlage für die Abgrenzung der Einheiten sein. Auf die
Beziehungen zwischen Bodenprofilen und potentiell natürlicher
Vegetation ist bereits in vielen Arbeiten hingewiesen worden
(ELLENBERG 1958, TÜXEN bes.1956,1957 u.a.). Da das Erbohren von
Profilen zu viel Zeit erfordert und nicht überall Aufschlüsse
vorhanden sind, kann auch diese Methode zumindest für die Ab-
grenzung einzelner Einheiten nur selten herangezogen werden. Im
Untersuchungsgebiet bot nur die Grenze zwischen Mineral-Böden
und Anmoor/Niedermoor von Fall zu Fall ein brauchbares Merkmal.

Viel wichtiger als Boden-Profile sind für die Kartierung die
äußerlich im Frühjahr und Herbst sichtbare Färbung und Struktur
der Boden-Oberfläche in Acker-Gebieten. Trockene und feuchte
Mineral-Böden, im Randbereich des Holtumer Moores Standorte des
Trockenen und Feuchten Stieleichen-Birken- bzw.Buchen-Trauben-
eichen-Waldes, lassen sich meist sehr gut nach der Boden-Farbe
unterscheiden und abgrenzen. Die an der Oberfläche gekrümelten
Para-Braunerden der Grundmoräne, also der Wuchsbereich des F a g o -
g o - Q u e r c e t u m   p e t r a e a e, zeigen eine violett-
braune, nach Abtrocknung mehr bräunlichgelbe Färbung. Dagegen
sind die unter Heide podsolierten trockenen Sande mit Einzel-
korn-Struktur heute grau oder gelbgrau und weisen auf Standorte
des Q u e r c o - B e t u l e t u m   hin. Die feuchten Böden
sind meist grauschwarz und setzen sich teilweise sehr scharf
von den trockeneren Bereichen ab. Besonders für Kartierungen
großer Gebiete ermöglicht die Kenntnis der Beziehungen zwischen
Boden-Farbe und Einheiten der heutigen potentiell natürlichen
Vegetation eine wesentliche Hilfe, die allerdings von Ort zu
Ort überprüft und nie allein für die Grenzziehung ausschlaggebend sein sollte.

d) Land- und forstwirtschaftliche Nutzungsweise
Da sich die bäuerliche Nutzung weitgehend im Laufe vieler Jahrhunderte den natürlichen Gegebenheiten angepaßt hat, kann auch
die Verteilung der verschieden bewirtschafteten Flächen oft als
Hilfsmittel herangezogen werden. Im Untersuchungsgebiet besteht
der auffälligste Unterschied zwischen den Ackerflächen der trok-
kenen Gebiete und dem Grünland auf Böden mit Grundwasser-Einfluß.
Hier fallen die Wiesen mit dem Wuchsbereich der Bruch- und Auwälder sowie des Q u e r c o - B e t u l e t u m   a l n e t o-
s u m   zusammen. Weiden sind im Westteil fast nur auf Standor-

ten des Q u e r c o - B e t u l e t u m m o l i n i e t o -
s u m vorhanden. Im Ostteil, auf ehemaligem Hochmoor, ist dieses
Verteilungsmuster jedoch nicht gegeben. Hier finden sich häufiger kleine Ödland-Reste in alten Torfstich-Gebieten, die auf
anderen Feuchstandorten fehlen. Größere Wald- und Forst-Gebiete
sind nur im trockenen Bereich vorhanden, wobei Laubwald hier
stets auf Standorte des F a g o - Q u e r c e t u m p e t -
r a e hinweist.

Die Nutzungsgrenzen gehen vor allem im Übergang von trockenen
zu feuchten Mineral-Böden und von diesen zu organischen Naßbö-
den sehr oft mit Grenzen der potentiell natürlichen Vegetation
überein und sind in diesen Fällen eine gute Hilfe bei der Kartierung.

Noch besser als Topographische Karten lassen Luftbilder die verschiedene Nutzungsweise und andere wichtige Merkmale erkennen
(vergl.KRAUSE 1955, LOHMEYER 1963). Die größeren Ackerflächen
und Wälder mit Laubbäumen sind Standorte des F a g o - Q u e r -
c e t u m p e t r a e a e. Die dunklen Nadel-Forsten lassen
deutlich den Bereich des Q u e r c o - B e t u l e t u m t y -
p i c u m erkennen, der nach unten, wo Acker- und Grünland
wechseln, in das Q u e r c o - B e t u l e t u m m o l i n i e -
t o s u m übergeht. Die weiten Wiesenflächen zeigen den Wuchsbereich der Bruch- und Auwälder an, während die kleinflächig
auftretenden Ödland-Reste auf das ehemalige Hochmoor-Gebiet hinweisen.

SCHLUSS

Neben den hier geschilderten Methoden und Hilfsmitteln zur Abgrenzung von Einheiten der heutigen potentiell natürlichen Vegetation gibt es eine Reihe weiterer Einzelmerkmale, die von
Fall zu Fall zu berücksichtigen sind. Auffällige und diagnostisch
wichtige Arten, verschiedene Farb-Aspekte von Pflanzengesellschaften und Einzelpflanzen, Ersatz-Gesellschaften, Böden und
heutige Nutzungsweise bilden jedoch einen Grundstock, der zur
Ansprache und Abgrenzung dieser Einheiten von wesentlicher Bedeutung ist. Erst die Berücksichtigung aller oder doch möglichst
vieler dieser Merkmale ergibt die Möglichkeit einer befriedigen-
den Kartierung, wobei jeweils bestimmte Merkmale den Gegebenheiten entsprechend im Vordergrund stehen.

Die Tatsache, daß die hier dargestellten Methoden und Hilfsmit-

tel auch außerhalb des engeren Untersuchungsgebietes bei der
Kartierung eines weiteren Bereiches verwendbar waren, zeigt,wie
wichtig die kleinräumige Erforschung der Beziehungen zwischen
den heute in der Kultur-Landschaft erkennbaren Merkmalen und
der potentiell natürlichen Vegetation für die Bearbeitung größerer Gebiete ist.

ZUSAMMENFASSUNG

Am Beispiel des Holtumer Moores und seiner Randgebiete werden
Methoden und Hilfsmittel zur Abgrenzung von Einheiten der heutigen potentiell natürlichen Vegetation in waldarmen Gebieten
Nordwest-Deutschlands erläutert. Die Verbreitung einzelner diagnostisch wichtiger Arten, verschiedene Farb-Aspekte der Wiesen und Weiden, sowie die Grenzen der Ersatz-Gesellschaften,
sind wichtige Merkmale, die aus der Vegetation selbst erkennbar
werden. Von großer Bedeutung sind auch die Bodentypen und vor
allem die wechselnde Boden-Farbe und Struktur der Ackerflächen,
die oft Standortsgrenzen deutlich werden lassen. Schließlich
gibt auch die Verteilung der verschieden genutzten Gebiete (Wiesen, Weiden, Äcker, Wälder, Forsten und Ödland) Anhaltspunkte
für die Grenzen von Einheiten der potentiell natürlichen Vegetation, die bereits aus topographischen Karten und Luftbildern
in Verbindung mit guter Geländekenntnis abgeleitet werden können .

SUMMARY

Using the Holtum moor and its surrounding areas as an example
methods and aids for the delimitation of units of the present
potential natural vegetation are explained. The distribution of
individual important diagnostic species, different color aspects
of the meadows and pastures as also the boundaries of the replacement communities are important characteristics which are re-
cognisable from the vegetation itself. Of considerable importance also are the soil types and especially the differing soil color and structure in the tillage areas as these often indicate
where the habitat boundaries lie. Finally the proportions of the
differently used areas (meadows, pastures, tillage, woods, forests and wasteland) give clues for the delimitation of the units
of potential natural vegetation which can be derived already
from topographical maps and air fotographs in connection with a
good knowledge of the terrain.

LITERATUR

BUCHWALD,K.   -1953-   Erläuterungen zur Naturlandschaftskarte des
      Naturschutzgebietes "Blankes Flat" bei Vesbeck und seiner
      näheren Umgebung.- Mitt.flor.-soz.Arbeitsgem.N.F.4: 125-
      136. Stolzenau/Weser.
DIERSCHKE,H.   -1969-   Die naturräumliche Gliederung der Verde-
      ner Geest. Landschaftsökologische Untersuchungen im nord-
      westdeutschen Altmoränengebiet.- (Diss.Göttingen).
      Forsch.z.dt.Landeskunde 177. Bad Godesberg.
-- --   -1968-   Zur synsystematischen und syndynamischen Stellung
      einiger Calthion-Wiesen mit Ranunculus auricomus L. und
      Primula elatior (L.) Hill im Wümme-Gebiet.- Mitt.flor.-
      soz.Arbeitsgem.N.F.13: 59-70. Todenmann/Rinteln.
-- --   -1969-   Natürliche und naturnahe Vegetation in den Tälern
      der Böhme und Fintau in der Lüneburger-Heide.- Mitt.flor.-
      soz.Arbeitsgem.N.F.14: 377-397. Todenmann/Rinteln.
ELLENBERG,H.   -1958-   Über die Beziehungen zwischen Pflanzenge-
      sellschaft, Standort, Bodenprofil und Bodentyp.- Angew.
      Pflanzensoz.15: 14-18. Stolzenau/Weser.
ILLIES,H.   -1952-   Eisrandlagen und eiszeitliche Entwässerung
      in der Umgebung von Bremen.- Abh.nat.Ver.Bremen 33 (1).
      Bremen.
KRAUSE,W.   -1955-   Pflanzensoziologische Luftbildauswertung.-
      Angew.Pflanzensoz.10. Stolzenau/Weser.
LOHMEYER,W.   -1963-   Erfahrungen bei der Verwendung von Luft-
      bildern.- In: Tüxen,R.(Edit.): Vegetationskartierung.
      Ber.Intern.Symposium 1959 in Stolzenau/Weser: 129-137.
      Weinheim.
MEISEL-JAHN,Sofie   -1955-   Die Kiefern-Forstgesellschaften des
      nordwestdeutschen Flachlandes.- Angew.Pflanzensoz. 11.
      Stolzenau/Weser.
PREISING,E.   -1954-   Übersicht über die wichtigsten Acker- und
      Grünlandgesellschaften NW-Deutschlands unter Berücksich-
      tigung ihrer Abhängigkeit vom Wasser und ihres Wirt-
      schaftswertes.- Angew.Pflanzensoz.8: 19-30. Stolzenau/
      Weser.
-- --   -1956-   Erläuterungen zur Karte der natürlichen Vegeta-
      tion der Umgebung von Göttingen.- Angew.Pflanzensoz. 13:
      43-55. Stolzenau/Weser.
TRAUTMANN,W.   -1966-   Erläuterungen zur Karte der potentiellen
      natürlichen Vegetation der Bundsrepublik Deutschland
      1:200 000. Blatt 85 Minden.- Schriftenr.Vegetationskde.
      1. Bad Godesberg.
TÜXEN,R.   -1956-   Die heutige potentielle natürliche Vegetation
      als Gegenstand der Vegetationskartierung.- Angew.Pflan-
      zensoz.13: 3-42. Stolzenau/Weser.
-- --   -1957-   Die Schrift des Bodens.- Angew.Pflanzensoz. 14.
      Stolzenau/Weser.

R.TÜXEN:

Es gehört für einen Lehrer zu den größten Genugtuungen, glaube
ich, wenn einer seiner Schüler in Grundzügen das, was ihn ge-
lehrt worden ist, bestätigen kann und wenn er es so weit assi-
miliert hat, daß er dann mit eigenen Ideen etwas Neues hinzufügen kann. Dies war mein Eindruck, den ich von den Ausführungen
meines Freundes HARTMUT DIERSCHKE eben bekommen habe, und ich
möchte ihm danken und sehr herzlich gratulieren zu dieser aus-
gezeichneten Arbeit, die sauber analysiert und scharfsinnig aus-
gewertet war. Ich will zwei kleine Ergänzungen machen. Betula
pubescens ist nicht ein so scharfer Zeiger, wie man vielleicht
glauben könnte (es ist nicht gesagt worden). Betula pubescens
hat ihr Optimum natürlich im Birken-Gürtel - B e t u l e t u m
p u b e s c e n t i s einschließlich Moor-Birkenwald. Sie kommt
aber auch, und das ging aus den Karten auch deutlich hervor, im
Q u e r c o - B e t u l e t u m m o l i n i e t o s u m und
im F a g o - Q u e r c e t u m m o l i n i e t o s u m vor.
Und zweitens könnte man vielleicht hinzufügen, ist zwischen dem
B e t u l e t u m p u b e s c e n t i s und dem C a r i c i:
e l o n g a t a e - A l n e t u m noch eine Zwischenstufe ein-
zuschalten, ein birkenreiches A l n e t u m, ein A l n e t u m
b e t u l e t o s u m oder A l n e t u m m o l i n i e t o -
s u m. BUCHWALD hat das kurz nach dem Kriege studiert und auch
veröffentlicht. Dort kann Betula pubescens in gewissen Fazies
fast ebenso dominieren wie im B e t u l e t u m p u b e s -
c e n t i s, so daß diese Einheit zwischen A l n e t u m und
B e t u l e t u m vielleicht doch mindestens gedacht werden
müßte, wenn man ihre Grenzen vielleicht auch nicht feststellen
kann. Und dann noch ein kurzes Wort zum Moor-Birkenwald, der ist
wahrscheinlich 1929 in den Mitteilungen der Provinzialstelle
für Naturdenkmalpflege für das Alt-Warmbüchner-Moor zum ersten-
mal geprägt worden. Das ist in jedem Fall ein alter Begriff.
Synsystematisch gehört der Moor-Birkenwald ganz zweifellos zum
B e t u l e t u m p u b e s c e n t i s.

POSSIBLE MAPPING SCHEMES FOR THE HEATH AND BOG VEGETATION
OF NORTH-WEST IRELAND

A.M. O'S u l l i v a n

INTRODUCTION

The western part of Country Mayo (Lat. 54°N.,Long.$9-10^{\circ}$W) is
bordered on three sides by the Atlantic ocean. It is a sceni-
cally attractive region with towering mountains, long wide val-
leys, vast undulating or flat lowlands and numerous lakes and
rivers. It measures about 50 km. from west to east and about
70 km. from north to south.

There are two main upland areas in West Mayo. One is centered
on Killary harbour - an 18 metre deep fiord - and includes
Mweelrea, the Sheefry hills and the Partry mountains. The other
is to the north of Clew Bay and includes the Nephin Beg mountain
range. By continental standards these mountains are only hills
since they only range in altitude from 300 to 650 metres. Nearly 70 per cent of West Mayo is lowland and less than 150 metres
above sea level. Only 0.5 per cent of the region is above 600
metres.

The mountains are mainly composed of quartzites and gritty sandstones. The lowlands are mainly underlain by schists, gneisses
and conglomeritic sandstones.

The whole region has been heavily glaciated. The lowland coastal
area centered around Clew Bay in the middle of the region is
filled by hundreds of pear-shaped drumlins. These drumlins together with scattered alluvial strips along present rivers compose the bulk of the useful farmland in the region.

The climate is an extreme atlantic one with mild, moist and windy weather for most of each year. Annual rainfall varies from
between 1,250 mm and 2,000 mm in the lowlands to between 2,000
mm and 3,000 mm in the hills and mountains. The effect is that
the whole region is in an 'organic zone' where blanket bog is
the principal natural vegetation. Only in a few local areas,
where steep slopes carry well drained, coarse-textured mineral
soils, is the growth of bog plants prevented. Such areas usually have a heathy grassland vegetation instead or, on rare occasions, oak woodland.

The human population of the region is mainly concentrated
around the coastal fringes and is steadily declining. The area

of enclosed and improved land attached to each farm is usually
between 4 and 8 hectares and may often be less (GILLMOR 1967).
The only crops grown are hay, potatoes and oats.
The major portion of West Mayo is in common land with only a
few of the better hills in private ownership. Vast areas of hill
and lowland are unfenced. They are grazed by small numbers of
cattle and numerous small flocks of Scottish Blackface sheep.

The region is of considerable botanical interest. It has the
largest single area of lowland blanket bog in Ireland. This
blanket bog has as two of its principal and most constant species, Molinia coerulea and Schoenus nigricans.
Intensive grazing and the use of artificial fertilisers is almost unknown. Further, peat harvesting is mainly confined to
the vicinity of inhabited areas. Burning of bog and heath vegetation is not common either since dry spells are so seldom.

State of knowledge on bog and heathland vegetation
The blanket bogs of the west of Ireland have already been described by OSVALD(1949), TANSLEY(1939), BRAUN-BLANQUET & TÜXEN(1952)
and most recently by Moore (1964).   BRAUN-BLANQUET & TÜXEN classified them in an new Association, the P l e u r o z i a p u r-
p u r e a - E r i c a t e t r a l i x - Ass. Br.Bl.&Tx.1952.
Later MOORE (1964) in his "Classification of the bogs and wet
heaths of northern Europe" demonstrated that the western lowland blanket bogs differed substantially both from the raised
bogs of the midlands and from the high-level blanket bogs. He
showed that the following species were only found in the western
lowland blanket bog type:

Schoenus nigricans	Pleurozia purpurea
Molinia coerulea	Campylopus atrovirens
Potentilla erecta	Polygala serpyllifolia
Pedicularis sylvatica	

With the exception of Schoenus nigricans all the above-named species also occur in the lowland blanket bogs of Scotland.
That the western lowland blanket bog on flattish terrain is the
most uniform of all Irish terrestrial vegetation formations was
borne out by the recent work of this author. The ecology of this
formation is easily defined and its distribution mapped.
The plant communities of the upland areas in the west of Ireland
are not so clearcut or as well studied. There is a greater range
of soils, climate and topography in the uplands. This is expres-

sed in the wide variety of vegetation types found in the hills
and mountains. The author is not aware of any publication dealing specifically with the plant communities of the uplands of
the west of Ireland.

BRAUN-BLANQUET & TÜXEN (1952) have introduced and described the
following heath associations which occur mainly in the west of
Ireland:

N a r d o - G a l i o n:
  N a r d o - C a r i c e t u m   b i n e r v i s   Br.-Bl.& Tx.
                                            1950
    J u n c u s   s q u a r r o s u s -
  R h y t i d i a d e l p h u s   l o r e u s - Ass. Br.-Bl.&
                                            Tx.1950

U l i c i o n   n a n a e:
  E r i c e t o - C a r i c e t u m   b i n e r v i s   Br.-Bl.&
                                            Tx.1950

None of these associations are well defined by good character-
species and differential-species. They appear to have a doubtful validity in West Mayo.

The present study

Some staff of the Irish Agricultural Institute are at present
engaged in a hill-land survey of the West Mayo region. The major
task of this survey is the identification and mapping of the
different soil types present. The flat lowland areas are easily mapped since they are covered by a deep 'blanket' of wet peat
dominated by Schoenus nigricans, Molinia coerulea and Narthecium
ossifragum.

In contrast, the soils on sloping ground are very variable. There are peats of a wide variety of depths, degree of humificati-
on and moisture concent. In addition there are Peaty Gleys, Peaty Podzols, Podzols, Podzolised Gleys, Brown Podzolics, Acid
Brown Earths and Lithosols.

The present author was requested by the soil surveyors to see
if a correlation could be found between soil and vegetation type in the survey area. To this end 119 relevés were made in
West Mayo during July, 1967. The location of these relevés is
shown in Fig.1. They are more concentrated in the upland than
in the lowland areas because the greatest botanical variation
occurs in the uplands.

The species lists from the relevés have been summarised in a
constancy table (Table 1). About half the species listed

in this table characteristically occur only in peatland. The
other half typically occur in heaths and heathy grasslands.

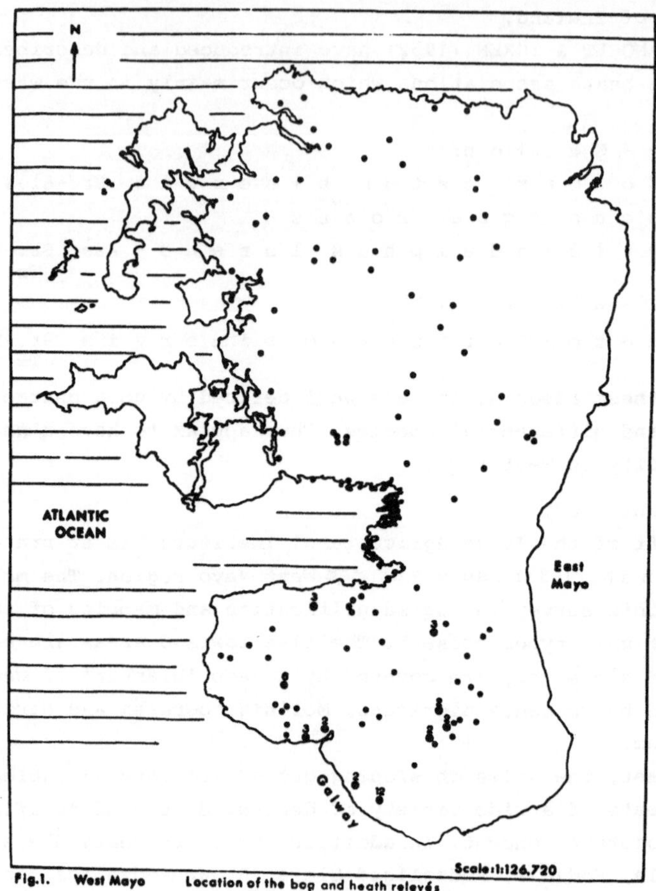

Fig.1.

Cryptogamic plants are of lesser importance in the blanket bogs
than in the raised bogs of the midlands. While Sphagnum, nota-
bly Sphagnum rubellum, S. papillosum and S.cuspidatum are common in the blanket bogs of West Mayo they are not regarded as
the principal peat formers. The phanaerogamic species Molinia
coerulea, Schoenus nigricans, Calluna vulgaris, Trichophorum
caespitosum, Erica tetralix and Eriophorum spp. are the usual
peat builders instead.

In trying various methods of classification of the relevés the

Table 1. Constancy table of the phanerogamic species occuring
         in more than ten per cent of the stands and of the mo-
         re easily identified cryptogamic species

No.of relevés	119
Calluna vulgaris	V
Potentilla erecta	V
Molinia coerulea	IV
Trichophorum caespitosum	IV
Erica tetralix	IV
Carex panicea	III
Erica cinerea	III
Drosera rotundifolia	III
Nardus stricta	III
Narthecium ossifragum	III
Eriophorum angustifolium	III
E.vaginatum	II
Schoenus nigricans	II
Rhynospora alba	II
Myrica gale	II
Juncus squarrosus	II
Succisa pratensis	II
Pedicularis sylvatica	II
Polygala serpyllifolia	II
Festuca vivipara	II
Galium saxatile	II
Agrostis tenuis	II
Sieglingia decumbens	II
Drosera anglica	I
D. intermedia	I
Carex echinata	I
Eleocharis multicaulis	I
Juncus bulbosus	I
Agrostis canina	I
Carex demissa/ C.lepid.	I
C. caryophyllea	I
C. binervis	I
Anthoxanthum odoratum	I
Pteridium aquilinum	I
Blechnum spicant	I
Hypnum cupressiforme	IV
Sphagnum spp.	IV
Cladonia spp.	III
Rhacomitrium lanuginosum	III
Campylopus atrovirens	III
Pleurozia purpurea	III
Rhytidiadelphus squarrosus	II
Hylocomium splendens	II
Thuidium tamaricinum	II

author has refrained from using moss, liverwort and lichen species. The only exception has been the use of Rhacomitrium lanu-
ginosum which is a most distinctive moss, even from a considerable distance. It was felt that a classification scheme of
bogs and heaths based primarily on the higher plants would be

more acceptable and useful to a soil surveyor. A drawback is
that the number of higher plants occuring in the Irish bogs and
heaths is rather limited.In Table 1 it has been shown that only
35 higher plants occur in more than 10 per cent of the relevés
from West Mayo.

CLASSIFICATION BY DOMINANT SPECIES

Classification by this method has the advantage of simplicity
and only requires a knowledge of a limited number of species. The
major interest in hill-land too is often in the grazing value
rather than the ecology of the different vegetation types. The
dominant species are of particular interest in this context.

The relevés were examined and the species which occured with
cover/abundance values of from 3 to 5 (i.e.,25 - 100 per cent
cover) on the BRAUN-BLANQUET scale were tabulated. Only seven
species were found to occur with values greater than 3 in five
or more relevés. Also, only sixty five per cent of the relevés
had a species with a cover/abundance value of 3 or over.
The seven dominant species are listed in Tables 2 and 3 and a
summary of some of the ecological factors of the stands which
they dominated is also given.

CLASSIFICATION BY GROUPS OF CONSTANT OR INDICATOR-SPECIES

An alternative to classification by dominant species is to group
relevés using the commonly occuring and not necessarily dominant
species. The relevés were tabulated in the usual phytosociological manner. Study of the relevés revealed that only twenty eight
species seemed to have a potential classificatory value. This
number included the seven dominant species listed in Table 2.
All the other species either occured in every relevé - e.g.,
Potentilla erecta - or in only a few relevés, e.g., Daboecia
cantabrica.
The order of the relevés and of the species was changed many
times and the final arrangement is shown in Table 4. This table
is also presented schematically as Table 5.
In the schematic table the relative extent of the different vegetation types as judged by the numbers of relevés belonging
to each type is shown diagrammatically.
All of the relevés originally made have been used in crystalli-
sing this two-dimensional scheme of classification. These relevés are considered representative of the range of variation

TABLE 2.   Summary of the ecology of dominant species in the bogs and heath I

	Schoenus	Molinia	Trichophorum	Calluna	Rhacomitrium	Juncus squarrosus	Nardus
No.of stands in which the species dominates:	12	6	7	26	10	5	11
Average altitude (m.)	45	195	209	187	269	180	298
" slope	0.4	9	4	16	8	8	19
Average pH (0-10 cm zone)	4.5	4.5	4.5	4.4	4.4	4.3	4.6
" % Carbon (0-10 cm)	34	32	32	26	24	29	17
Drainage Categories (%)							
Moderately free	–	–	–	27	40	–	18
Impeded	–	33	14	54	50	20	73
Strongly impeded	100	67	86	19	10	80	9
Groundwater Level Categories (%)							
At or over surface	50	20	14	–	–	–	–
5- 25 cm below surface	33	40	29	–	–	33	–
26- 50 cm " "	8	–	14	4	10	–	–
51- 75 cm " "	–	–	14	4	–	–	–
76-100 cm " "	–	–	–	–	–	–	–
100 cm+ " "	9	40	29	92	90	67	100

TABLE 3.   Summary of the ecology of dominant species in the bogs and heath,II

	Average Peat Depth (cm)	Soil Type
Schoenus nigricans	200+	Very deep, wet, flat blanket peat.
Molinia coerulea	100	Deep, rolling, blanket peat.
Trichophorum cespitosum	150	Deep, gently sloping,wet, blanket peat.
Calluna vulgaris	50	Skeletal peats; shallow,eroded or outover peats; Moderately deep, dryish peats on steep slopes; Podzols.
Rhacomitrium lanuginosum	70	Skeletal peats; deep eroding peats on steep slopes or exposed summit plateaus.
Juncus squarrosus	60	Shallow to moderately deep disturbed peats - (due to cutting, erosion, grazing).
Nardus stricta	20	Skeletal peats, peaty podzolised gleys, peaty podzols.

TABLE 4. Survey table of the bog, heath and heathy-grassland communities

Column No:	1	2	3	4	5	6
No. of relevés per column	48	11	18	19	16	7
Molinia coerulea	V	V	IV	IV	-	-
Trichophorum caespitosum	IV	V	IV	III	I	-
Narthecium ossifragum	V	III	IV	I	I	-
Eriophorum angustifolium	V	II	III	-	-	-
Drosera rotundifolia	V	II	II	-	-	-
Schoenus nigricans	IV	-	II	-	-	-
Rhynospora alba	IV	I	I	-	-	-
Eriophorum vaginatum	III	I	I	-	-	-
Myrica gale	III	II	I	-	-	-
Eleocharis multicaulis	II	-	II	I	I	-
Menyanthes trifoliata	I	-	-	-	-	-
Calluna vulgaris	V	V	V	V	IV$^{\circ}$	III$^{\circ}$
Erica tetralix	V	V	IV	III	-	-
Rhacomitrium lanuginosum	IV	IV	V	III	I	-
Erica cinerea	II	IV	IV	V	III	-
Juncus squarrosus	I	IV	V	III	II	-
Carex binervis	-	II	I	II	III	-
Nardus stricta	-	I	V	IV	V	III
Sieglingia decumbens	-	-	II	III	V	IV
Festuca vivipara	-	I	II	II	V	-
Agrostis tenuis	-	-	I	III	V	V
Anthoxanthum odoratum	-	I	-	II	III	V
Festuca rubra	-	-	I	I	I	V
Holcus lanatus	-	-	-	-	I	V
Trifolium repens	-	-	-	-	I	V
Pteridium aquilinum	-	I	-	I[1]	I	II[5]
Vaccinium myrtillus	-	I	-	I[1]	-	-
Deschampsia flexuosa	-	I	-	II[1]	I	-

found in topography, soil type and vegetation cover in the survey area.

Table 5.

Table 5.   Classification of the blanket bogs and of the shrub- and grass-heaths of West Mayo, 1968.

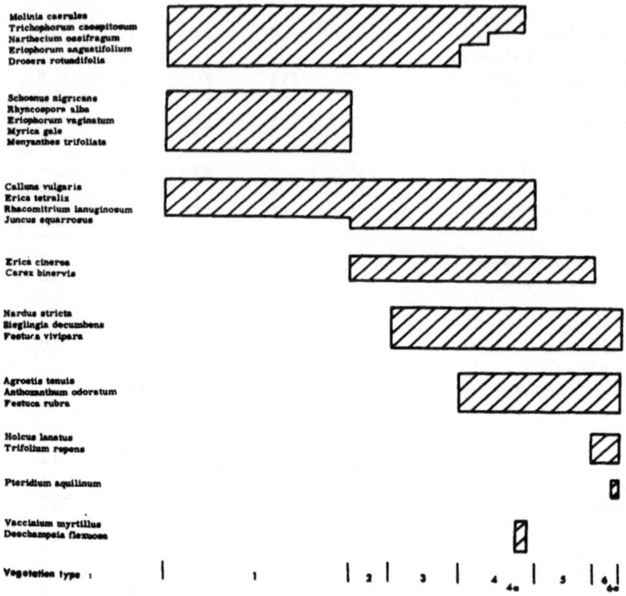

The relevés in vegetation types 1 and 6 represent the two most distinct plant communities. They are defined by groups of species which are mutually exclusive. Between these extremes lie a complex of four other vegetation types .Going from left to right in Tables 4 and 5 they have a decreasing number of bog plants and a corresponding increase in the number and abundance of heath and grassland species. A summary of the more important ecological factors of the stands belonging to each of the six types is given in Table 6. Major differences exist between the six vegetation types in relation to: carbon content of the peat (or A_o horizon), drainage status, groundwater level (July) and peat depth. The soil profile type is also rather different between the six vegetation types:

    Veg.type 1   On very deep,wet,undisturbed blanket peat.
    "   "   2   On moderately deep,firm,rolling blanket peat.
    "   "   3   On shallow,wet peat - often eroded,disturbed
                    or humified in the surface.

Veg.type 4   Mainly on Skeletal Peats but also on Peaty Podzols, Podzols and Peaty Podzolised Gleys.

"   "   5   Mainly Peaty Podzols but also Peaty Podzolised Gleys and very shallow Peat overlying Podzols.

"   "   6   On Brown Podzolics and Acid Brown Earths.

The influence of the mineral sub-soil in types 4 and 5 is reflected by the appearance of Sieglingia decumbens, Agrostis tenuis and Anthoxanthum odoratum. Vegetation type 6 is confined to mineral soils and these are also the most fertile soils in the region.

Table 6.   Summary of the ecology of the bog, heath and heathy grassland communities (c.f.Table 4).

	1	2	3	4	5	6
Column No:	1	2	3	4	5	6
No.of relevés per column:	48	11	18	19	16	7
Average Altitude (m.)	88	133	218	225	202	131
" Slope ($^{\circ}$)	2	8	10	19	21	14
Average pH (O=10cm zone)	4.5	4.4	4.5	4.4	4.7	5.1
" % Carbon (O=10cm)	33	31	26	19	13	8
Drainage Categories (%)						
Free	-	-	-	-	6	86
Moderately free	-	27	11	21	25	14
Impeded	19	27	61	63	63	-
Strongly impeded	81	46	28	16	6	-
Groundwater Level Categories (%)						
At or over surface	40	9	-	-	-	-
5-25 cm below surface	27	-	6	-	6	-
26-50 cm " "	8	18	11	-	-	-
51-75 cm " "	4	-	-	11	-	-
76-100 cm " "	2	-	-	-	6	-
100 cm+ " "	19	73	83	89	88	100
Average peat depth (cm)	200+	78	45	23	14	O
" " " (in.)	80+	30	18	9	5.5	O
Peat Depth categories	Deep	Mod.Deep.	Shal.	Skl.	Skl.	O
Range in depth category (in.)	40-80+	2o-40	8-30	4-16	O-8	O

DISCUSSION

The phytosociological position of vegetation type I is clear. It belongs to the Pleurozia purpurea-Erica tetralix-Association of the Oxycocco-Sphagnetea. The phytosociological position of vegetation types 2 to 6 is however far from clear. The present author is doubtful if the shallow bog, heath and heathy grassland vegetation represented by types 2 to 6 lends itself to the hierarchical phytosociological classification of the Zürich-Montpellier type. There seems not to be enough species in the sward. The small number that do occur are often only moderately respon-

sive to the environment. Only when there is a major change in
the habitat conditions does a real discontinuity in the composition of the vegetation occur. Mostly variation appears to be
continuous. The presence of Nardus stricta and Juncus squarro-
sus in many stands may in fact often have a biotic rather than
an ecological explaination (KING 1962; WELCH 1966).

A classification based on dominant species has been shown earlier in the paper to take account only of about half the number
of stands examined. It would seem to be of limited value in the
survey area.

A scheme based on a strict Zürich-Montpellier approach also appears impractical since the vegetation is too poor in species.
Moss, liverwort and lichen species could of course be used for
such a classification. The resulting mapping scheme would however be unattractive to anyone lacking a knowledge of bryophytes
and lichens.

The scheme of classification presented in Tables 4 and 5 has
been derived using all the relevés originally collected. It
therefore is a true reflection of the natural situation in the
field. The Tables have been divided into six units or vegetation types at points where groups of species with similar ecological preferences either fade out or appear for the first time.
These six vegetation types have been shown in Table 6 to differ
in their ecology. They have not as yet been tested in the field.
The author has also refrained from naming all but vegetation type I.

1. He would prefer for the moment to regard the types as provisional and to use them as a basis for discussion on hill-land
vegetation at this Symposium.

The need as this author sees it is to find a National classification of hill-land vegetation based on the dominant and also
the indicator-species of the sward. This might fulfil the two-
fold rôle of providing information on the grazing value and the
ecology of the different vegetation types.

ZUSAMMENFASSUNG

Für diese Studie wurden die Hochmoore, die Heiden und die Magerrasen des westlichen Teiles von Mayo, Irland, einbezogen.
Die Artenlisten wurden auf Dominanz, Stetigkeit und Zeigerwert hin analysiert.
Die Vegetationsklassifizierung auf Grund der vorherrschenden Arten erlaubt eine beschränkte Anwendung, weil große Flächen keine eigentlich dominierenden Arten besitzen.
Eine phytosoziologische Klassifizierung der Gesellschaften ist schwierig, da die Variationen sich eher durch Veränderungen der Abundanz einer kleinen Zahl von höheren Arten als durch absolute Veränderungen in der Arten-Zusammensetzung zeigen.
Die vorgeschlagene Klassifizierung für West-Mayo ist zur Zeit nur provisorisch. Sie ist auf die steten und Zeiger-Arten begründet. Sechs Vegetationstypen sind aus den gesammelten Aufnahmen abgeleitet. Sie unterscheiden sich auch in ihrer Ökologie.

SUMMARY

The bog, heath and heathy grassland vegetation of the western half of county mayo, Ireland has been studied. The relevés collected have been analysed for their dominant, their constant and their indicator-species.
Classification of the vegetation on the basis of dominant species has a limited use because large areas have no single dominant species.
A phytosociological classification is difficult because variation mainly takes the form of changes in the abundance of the small number of phanaerogamic species rather than as absolute changes in the species composition of the sward.
The classification proposed for West Mayo is a provisional one. It is based on the constant-species and the indicator-species. Six vegetation types have been derived and these are shown to differ in their ecology.

LITERATUR

BRAUN-BLANQUET,J.& TÜXEN,R.   -1952-   Die Pflanzengesellschaften
        Irlands.- Veröff.geobot.Inst.Rübel,Zürich 25: 224-415.
        Bern.
GILLMOR,D.A.   -1967-   The Agricultural regions of the Republic
        of Ireland.- Irish Geogr.5 (4): 245-261. Dublin.
KING,J.   -1962-   The Festuca-Agrostis grassland complex in
        south-east Scotland.- J.Ecol.50: 321-355. Oxford.
MOORE,J.J.   -1968-   A classification of the bogs and wet heaths
        of northern Europe.- In: Tüxen,R.(Edit.): Pflanzensozio-
        logische Systematik. Ber.Intern.Sympos.1964 Stolzenau/
        Weser. Den Haag.
OSVALD, H.   -1949-   Notes on the vegetation of British and
        Irish mosses.- Acta Phytogeogr.Suecica 26. Uppsala.
TANSLEY,A.G.   -1939-   The British Islands and their vegetation.-
        Cambridge.
WELCH,D.   -1966-   Biological Flora of the British Isles: Juncus
        squarrosus L.-   J.Ecol.54 (2): 535-548. Oxford.

ZUR ABGRENZUNG VON ÜBERGANGSMOOR-KOMPLEXEN

G. K a u l e

Die komplexe Vegetation der Übergangs- und Hochmoore kann man
von verschiedenen Seiten aus betrachten. Die gesellschaftssyste-
matische Gliederung ordnet die echte Hochmoor-Vegetation verschiedenen Klassen zu, die Schlenken-Gesellschaften werden als
"Niedermoorfenster" von den eigentlichen Hochmoor-Gesellschaf-
ten abgetrennt. Die dynamische Betrachtungsweise versucht das
Moor-Wachstum und den Wechsel von Bulten und Schlenken zu berücksichtigen und zu deuten. PAUL & LUTZ (1941) stellen Lysimachia thyrsiflora, Carex rostrata und Equisetum fluviatile zur
Initialphase der S c h e u c h z e r i a   p a l u s t r i s -
Gesellschaft. Das Einwandern von Hochmoorbult-Arten bedeutet
dann die Degradationsphase dieser Gesellschaft. Die von PAUL &
LUTZ (1941) angegebenen Arten des Initialstadiums können im echten Hochmoor nicht existieren, sie kommen nur in den damals von
ihnen untersuchten Übergangsmooren vor. (Zur Nomenklatur der
allgemeinen Moorbegriffe siehe ALETSEE,1967,154-155). Mit der
Einführung der Mineralbodenwasserzeigergrenze durch THUNMARK
(1940) und der Ausarbeitung des Begriffes durch DU RIETZ (1954)
rückt ein weiterer Gesichtspunkt der Moorforschung in den Vordergrund des Interesses. Die Moorpflanzen werden in Mineralbodenwasser unabhängige (Ombrominerobionten) und Mineralbodenwasserzeiger (Eum18inerobionten) eingeteilt, also in Pflanzen, die
ausschließlich von Regenwasser leben können, und solche, die
eine, wenn auch noch so geringe Mineralboden-Wassermenge benö-
tigen. Dabei ist es zunächst gleichgültig, ob es sich um Arten
der Bulte oder Schlenken handelt.
Das vielzitierte Schema des Hochmoor-Wachstums durch den regelmäßigen Wechsel von Bulten und Schlenken wird nach neueren Untersuchungen in Frage gestellt. In Torf-Profilen ist häufig anhand der fossilen Moose ein Wechsel der Torfmoos-Vereine festzustellen. TANSLEY (1965) beschreibt nach Untersuchungen von
Oswald etwa 7 Wechsel von Bult- und Schlenken-Gesellschaften seit
der Eiszeit. AVERDIECK (1957) kommt bei Untersuchungen in Nord-
Deutschland zu ähnlichen Ergebnissen. EUROLA (1962,S,195) nimmt
wie PAASIO (1933) an, daß bei Bult-Schlenken-Komplexen die klimatischen Faktoren ausschlaggebend für die Struktur sind; "Die
Kerminis der Hochmoore im südwestlichen Finnland (Schärenfinn-

land) sind als labile, die Kermini-Hochmoore von Küstenfinnland
(AARIO 1932, PAASIO 1933) und der Hochmoore von Binnenfinnland
als stabile Gebilde zu betrachten". CASPARIE (1969) kommt bei
Untersuchungen im Moor-Gebiet von Emmen (Niederlande) zu dem
Schluß, daß dort von zyklischen Sukzessionen beim Moor-Wachs-
tum keine Rede sein kann.
Betrachtet man den Bult-Schlenken-Komplex zumindest für längere
Zeit als Dauerzustand, als ein Nebeneinander von verschiedenen
Gesellschaften, die voneinander abhängig sind, so ist es als
weiterer Gesichtspunkt auch interessant festzustellen, welche
Bult- und Schlenken-Gesellschaften miteinander vorkommen, und
welche sich unter Umständen gegenseitig ausschließen. Das ein-
gangs erwähnte Beispiel der Initialphase der Schlenken-Gesell-
schaft zeigt, daß allein durch Berücksichtigung der Mineralbodenwasserzeigergrenze die Schlenken-Gesellschaften im Hoch- und
Übergangsmoor getrennt betrachtet werden müssen. JENSEN (1961)
führt als erster konsequent bei einem Übergangs- und Hochmoor
im Harz eine Gliederung durch, bei der die Komplexe als ganzes
zu nährstoff-ökologischen Einheiten zusammengefaßt werden, die
er im Unterschied zu den vom Moorwasserstand abhängigen Komplexen (Stillstandskomplex, Wachstumskomplex, Erosionskomplex etc.)
als Vegetations-Stufenkomplexe bezeichnet. Die Untersuchungen
von JENSEN (1961) fanden in einem Hangmoor statt; das Mineral-
Bodenwasser rieselt von einigen Quellen im Moor hangabwärts,
der Mbw.-Einfluß nimmt mit zunehmender Entfernung von der Quelle ab. Bei den von mir untersuchten Mooren im Chiemsee-Gebiet
(KAULE 1969) handelt es sich um Verlandungsmoore, die Grundwas-
ser-Bewegungen sind ohne exakte Messungen nicht feststellbar.
Die Übergangsmoore im Harz (Urgestein) beginnen bei Braunseggen-
Sümpfen nahestehenden Vegetationskomplexen, die Komplexe im
Inn-Chiemsee-Gebiet dagegen bei der Durchdringung von Niedermoor-Gesellschaften wie Davallseggen-Sümpfen (C a r i c i o n
d a v a l l i a n a e Klika 1934), Großseggenrieden (C a r i -
c i o n   r o s t r a t a e (Géhu 1961) Bal.-Tul.1963) und
Schneidebinsen-Sümpfen (C l a d i e t u m   m a r i s c i Zobrist 1935) mit hochmoorartigen Bulten. Die Schlenken der nährstoffreichsten Stufen-Komplexe, insbesondere deren Moosschicht,
stehen diesen Niedermoor-Gesellschaften noch nahe.
Komplizierter als im echten Hochmoor ist die Zonierung in manchen Übergangsmooren; hier schiebt sich zwischen die Bulte und

Schlenken eine Bultfuß-Gesellschaft, die zwar nicht streng durch
eigene Arten gekennzeichnet wird, aber im Komplex schon oberflächlich durch eine höhere Artmächtigkeit der Phanerogamen auf-
fällt und meist einen eigenen Torfmoos-Verein besitzt. Typisch
ist die Durchdringung von Arten der Bulte und Schlenken.
Die hier geschilderten Untersuchungen wurden, wie schon erwähnt,
in Mooren im Moränen-Gebiet des ehemaligen Inn-Chiemsee-Glet-
schers im oberbayerischen Alpen-Vorland durchgeführt. Die Größe der Probeflächen lag zwischen 1/4 und 1 m^2, ca. 500 Aufnahmen wurden verarbeitet. Die Pflanzenlisten der Schlenken- Bult-
fuß- und Bult-Gesellschaften wurden unabhängig von einander
differenziert, und dann anhand der Nummern festgestellt, welche
Gruppen zusammengehören. Für den gesamten Komplex sind oft Arten bedeutend,deren soziologischer Wert nur gering ist. Bei den
Bulten und Schlenken kann man leicht homogene Probeflächen aus-
wählen; die oft nur wenige cm breite Bultfuß-Gesellschaft ist
dagegen nur schwer gegen Bult und Schlenke abzugrenzen. Die Probeflächen müssen oft zusammengesetzt werden. In jedem Fall wurde das Flächen-Verhältnis der einzelnen Gesellschaften notiert.
Genau so wichtig wie die Pflanzen-Tabelle sind Struktur und
Morphologie der einzelnen Moor-Komplexe, die mit einer Reihe
von Kleinkartierungen erfaßt wurden. Zur Arten-Zusammensetzung
ist allgemein zu bemerken, daß mit Ausnahme von Eriophorum vaginatum und Sphagnum cuspidatum alle wichtigen Hochmoor-Arten,
also die Ombrominerobionten, mit gleicher Stetigkeit auch in
den Übergangsmoor-Komplexen vorkommen. Ausschlaggebend für die
Differenzierung ist in erster Linie das Neuhinzukommen von Arten, je weiter sich die Komplexe von der Mineralbodenwasserzeigergrenze in Richtung auf das Niedermoor entfernen.
Im folgenden Fall soll versucht werden, die von JENSEN (1961)
im Harz entwickelte Vegetationsgliederung durch Stufen-Komplexe
auch in den Mooren des Inn-Chiemsee-Vorlandes anzuwenden.
Tabelle 1 zeigt die Verbreitungsschwerpunkte der Arten in den
Übergangsmoor-Stufenkomplexen des Inn-Chiemsee-Vorlandes. Die
Hochmoor-Stufenkomplexe sind in dieser Tabelle nicht mit erfaßt,
da sie nur noch in Relikten vorkommen; so konnte kein echter
Hochmoor-Wachstumskomplex mehr gefunden werden. In den untersuch-
ten Hochmoor-Stillstandskomplexen fehlen die Schlenken-Arten
Carex limosa und Scheuchzeria palustris, wahrscheinlich sind
diese Arten in diesem Gebiet Mineralbodenwasserzeiger (vergl.

auch ALETSEE 1966,S.255). In den Bulten sind Cladonia-Arten,
Sphagnum acutifolium und Calluna vulgaris häufig (vergleiche
Komplex-Kartierung 4). Die Originaltabellen der Schlenken-,
Bultfuß- und Bult-Gesellschaften, aus denen die Übersichtsta-
belle 1 entwickelt wurde, enthält die Arbeit von KAULE (1969).

Wir unterscheiden neben dem ombrotrophen Hochmoor 3 große Gruppen von mineralbodenwasserbeeinflußten Stufen-Komplexen. Die
nährstoffreichsten Komplexe (B r a u n m o o s - S t u f e n -
k o m p l e x e) zeichnen sich durch Arten des C a r i c i o n
d a v a l l i a n a e in den Schlenken aus: Scorpidium scorpioides, Drepanocladus revolvens, Campylium stellatum u.a. Die
Bultfuß-Gesellschaft wird von einem Sphagnum contortum- oder S.sub.-
secundum-Verein gebildet.In den Bulten können die Hochmoor-Sphag-
nen durch Sp.centrale, Sp.palustre oder Sp.plumulosum ersetzt
sein. Von den ombrotraphenten Arten fehlen Eriophorum vaginatum und Sp.cuspidatum.
Dann folgt eine Gruppe von Komplexen, in denen die Arten des
C a r i c i o n   d a v a l l i a n a e fehlen, die Hochmoor-
Arten aber noch nicht vollständig vertreten sind. Eriophorum
vaginatum und Sphagnum cuspidatum tauchen nur ganz vereinzelt
auf. Nach den hier wichtigsten Mineralbodenwasserzeigern nennen
wir diese Komplexe U t r i c u l a r i a (i n t e r m e d i a)
- S p h a g n u m   s u b s e c u n d u m - S t u f e n k o m -
p l e x e. Schließlich gibt es eine Reihe von Moor-Komplexen,
in denen alle ombrotraphenten Arten vertreten sind; nur die
mehr oder weniger häufigen minerotraphenten Arten kennzeichnen
sie als Übergangs-Moore.In Anlehnung an DU RIETZ (1956) fassen
wir sie unter dem Begriff P s e u d o h o c h m o o r - S t u -
f e n k o m p l e x e zusammen.
1. Pflanzengesellschaften
1. 1 Schlenkengesellschaften
Der Begriff Schlenke ist moor-morphologisch definiert, es ist
der nasseste, fast ständig überflutete Bereich eines Moor-Kom-
plexes. Soziologisch kann es sich um sehr verschiedene Gesellschaften handeln. Zusammengefaßt werden sie nur durch einige
wenige Arten: Rhynchospora alba, Carex limosa, Scheuchzeria palustris, Drosera anglica, Drosera intermedia. Die Moos-Schicht
ist sehr unterschiedlich. Schlenken mit Scorpidium scorpioides,
Drepanocladus revolvens und Campylium stellatum gehören in den
C a r i c i o n   d a v a l l i a n a e - Verband, die nährstoff-

reichsten Ausbildungen kennzeichnen Schoenus ferrugineus, Parnassia palustris, Carex panicea und Carex lepidocarpa.
Den Übergang zu den Schlenken des R h y n c h o s p o r i o n a l b a e - Verbandes bilden moosfreie Schlenken mit Utricularia intermedia und Mineralbodenwasserzeigern wie Carex elata, Trichophorum alpinum u.a..
Demgegenüber stehen die Torfmoos-Schlenken des R h y n c h o s p o r i o n a l b a e - Verbandes. Hier treten Sphagnum cuspidatum, Sp.angustifolium fo. fallax und seltener Sp.dusenii sowie Sp.papillosum auf. Sphagnum papillosum gilt als Art der Hochmoor-Bulte (vgl. z.B. SCHWICKERATH 1940, OBERDORFER 1957, JES TÜXEN 1969). Im Untersuchungsgebiet überwiegen jedoch in den S p h a g n u m p a p i l l o s u m - Vereinen die Arten der Schlenken gegenüber den Bultarten, so daß das Moos hier nur eine Fazies in den Schlenken bildet. Ebenso können auch Sp.magellanicum et rubellum als Fazies im Schlenken-Horizont auftreten (rote Schlenken bei GAMS 1932). Die wichtigsten Mineralbodenwasserzeiger der Torfmoos-Schlenken sind Eriophorum angustifolium, Drosera intermedia, Carex rostrata, Carex lasiocarpa und Rhynchospora fusca. Im echten Hochmoor kommen in den Stillstands-Komplexen Schlenken-Fragmente vor, die nur durch Sp.cuspidatum und Rhynchospora alba gekennzeichnet sind. Scheuchzeria palustris und Carex limosa fehlen.

1.2 Bultfuß-Gesellschaften

Im allgemeinen zeichnen sich die Bultfuß-Gesellschaften durch wesentlich höhere Artenzahlen als die anderen Gesellschaften eines Komplexes aus. Bult-Arten und Schlenken-Arten sind etwa gleich stark vertreten. Eindeutig ihren Verbreitungsschwerpunkt haben hier die Torfmoose der Subsecunda-Gruppe, wobei Sp.contortum gegenüber Sp.subsecundum die ionenreicheren Standorte bevorzugt und meist mit C a r i c i o n d a v a l l i a n a e - Trennarten wie Schoenus ferrugineus und dem S c o r p i d i u m - Verein vergesellschaftet ist. In den U t r i c u l a r i a - S p h a g n u m s u b s e c u n d u m - S t u f e n - K o m - p l e x e n klingt die S p h a g n u m s u b s e c u n d u m - B u l t f u ß - G e s e l l s c h a f t aus. Die steilen Bultränder werden von einer Lebermoos-Kleingesellschaft eingenommen (siehe Tabelle 1).

1.3 Bult-Gesellschaften

In den Bulten der untersuchten Moore kommen eine ganze Reihe
von Torfmoosen vor. Nach der jeweils dominanten Art können die
Bulte in verschiedene Gesellschaften gegliedert werden. Diese
Bult-Gesellschaften haben dann homologe Ausbildungen. Der Ver-
breitungs-Schwerpunkt der einzelnen Arten liegt jedoch oft in
unterschiedlichen ökologischen Bereichen.
Wir beginnen mit den minerotraphenten Arten unter den Bult-Bild-
nern: Sp.plumulosum, Sp.centrale und Sp.palustre können in
nährstoffreichen Komplexen die ombrotraphenten Torfmoose erset-
zen. In diesen Bulten fehlt Eriophorum vaginatum, die anderen
Kennarten der Hochmoor-Bulte sind vertreten, in der Artenzahl
überwiegen jedoch Mineralbodenwasserzeiger.
Weitaus die häufigsten bultbildenden Torfmoose im Untersuchungsgebiet sind Sp.magellanicum, rubellum und angustifolium. Dabei
zeigt Sp.angustifolium eine stärkere Bindung an die Bergkiefer.
Im ombrotrophen Hochmoor kommt die Art fast ausschließlich im
Schutz der Latschen vor. Diese Bult-Gesellschaften gliedern
sich in zwei große Untereinheiten. Die Ausbildung von Rhynchospora alba bevorzugt gegenüber der von Eriophorum vaginatum
stärker vom Mineralbodenwasser beeinflußte Standorte. Die weitere Untergliederung der Bult-Gesellschaften nach den Mineralbodenwasserzeigern ist aus Tabelle 1 zu ersehen.

2. Die Stufen-Komplexe

2.1 Braunmoos-Stufenkomplexe

Die Trennarten der Braunmoos-Stufenkomplexe sind Scorpidium
scorpioides, Drepanocladus revolvens, Campylium stellatum,
Sp.centrale, Sp.palustre, Sp.plumulosum und Sp.contortum.
Die weitere Untergliederung der Braunmoos-Stufenkomplexe ist:

        1. Schoenus ferrugineus-Stufenkomplex: Die Hochmoor-Torf-
moose fehlen, Arten des C a r i c i o n   d a v a l l i a n a e
-Verbandes sind zahlreich vertreten, im Bultfuß herrscht Sp.
concortum vor.

        2. Carex panicea-Stufenkomplex: Die kennzeichnenden Ar-
ten sind: Carex panicea, Carex lepidocarpe und Parnassia palustris.

        3. Typischer Braunmoos-Stufenkomplex: Die extremen Kalk-
zeiger fehlen. Als Bult-Gesellschaft überwiegt die S p h a g -
n u m   m a g e l l a n i c u m   et   r u b e l l u m - G e -
s e l l s c h a f t, A u s b i l d u n g   v o n   R h y n -
c h o s p o r a   a l b a. In der Bultfuß-Gesellschaft tritt

Sp.subsecundum auf.

        4. Drepanocladus exannulatus-Stufenkomplex: Hier han-
delt es sich um einen relativ seltenen Sonderfall, der oben
genannte Braunmoos-Verein in den Schlenken wird durch Drepanocladus exannulatus ersetzt.
Karte 1 und Tabelle 3 zeigen als Beispiel einen Carex panicea-
Stufen-Komplex.

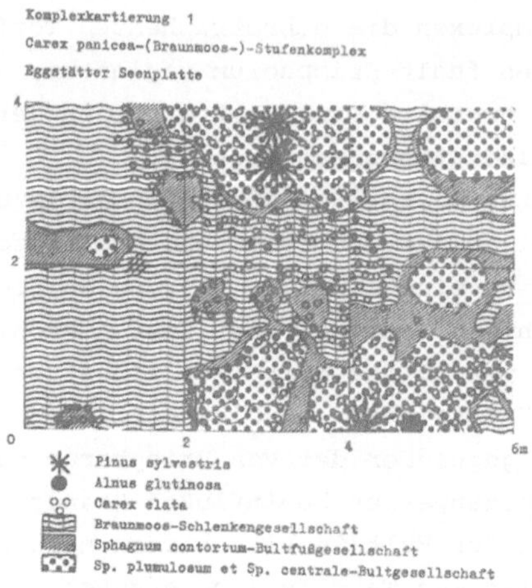

2.2 Utricularia intermedia-Sphagnum subsecundum-Stufenkomplexe

Wichtigstes Kennzeichen ist das Fehlen der ombrotraphenten Arten Eriophorum vaginatum und Sp.cuspidatum sowie der Mineralbodenwasserzeiger des C a r i c i o n d a v a l l i a n a e -
Verbandes. Neben den namengebenden Trennarten dieser Komplexe
können alle Mineralbodenwasserzeiger der Pseudohochmoor-Stufen-
komplexe vorkommen (siehe 2.3). Die Utricularia-Sphagnum sub-
secundum-Stufenkomplexe des Untersuchungsgebietes unterscheiden
sich von den anderen Übergangsmooren durch ihren morphologischen
Aufbau. Es handelt sich um eine fast ebene Fläche, in die das
Schlenken-Netz wie ein System von Entwässerungsrinnen einge-
schnitten ist. Die Schlenken sind fast kryptogamen-frei, auch
die Phanerogamen sind nur durch Utricularia intermedia und einige Cyperaceen vertreten. Der Bult-Fuß fällt durch den dichten
Bewuchs von Phanerogamen auf, dominant ist Rhynchospora alba,
die Moos-Schicht wird von Sp.subsecundum gebildet. In manchen
Komplexen kann diese Bultfuß-Gesellschaft auch durch einen Le-

Tabelle 3

Carex panicea-(Braunmoos-)Stufenkomplex

	1	2	3	4	5	6	7	8
Laufende Nummer	1	2	3	4	5	6	7	8
Artenzahl	17	15	15	23	18	15	15	13
Ombrotraphente Bultarten								
Vaccinium oxycoccus	1	3	2	+	1	.	+	.
Andromeda polifolia	2	+	1	.	+	.	.	.
Aulacomnium palustre	2	+	1	.	.	.	.	.
Drosera rotundifolia	+	.	+	+	.	.	.	.
Polytrichum strictum	+	.	.	.	.	.	.	.
Minerotraphente Arten VS. in Bultgesellschaften								
Sphagnum centrale	1	5	5	+	.	.	.	.
Sphagnum plumulosum	5	.	.	.	.	.	.	.
Trichophorum alpinum	1	+	2	2	2	r	.	.
Molinia coerulea	2	1	1	.	+	.	.	r
Calliergon stramineum	+	1	2	+	+	.	.	.
Peucedanum palustre	+	+	.	.	.	.	.	.
Viola palustris	.	1	+	.	.	.	.	.
Potentilla erecta	.	.	+	.	.	.	.	.
Pinus sylvestris K.	+	.	.	.	.	.	.	.
Minerotraphente Arten VS. in Bultfußgesellschaften								
Sphagnum contortum	.	1	.	4	4	1	+	.
Sphagnum obtusum	.	.	.	1	.	.	.	.
Ombrotraphente Schlenkenarten								
Rhynchospora alba	+	+	+	+	.	+	.	.
Drosera anglica	.	.	.	1	.	.	+	.
Minerotraphente Arten VS. in Schlenkengesellschaften								
Carex limosa (omb?)	.	.	+	1	+	2	1	1
Scorpidium scorpioides	.	.	.	+	1	2	1	2
Campylium stellatum	.	.	.	1	1	+	.	1
Equisetum fluviatile	.	.	.	+	.	+	+	+
Riccardia pinguis	.	.	.	+	+	+	.	+
Scheuchzeria palustris (omb?)	.	+	.	+	.	.	+	+
Utricularia intermedia	.	.	.	.	+	1	1	2
Drosera intermedia	.	.	.	+	+	+	.	.
Utricularia minor	.	.	.	.	.	+	.	.
Weitere minerotraphente Arten								
Carex elata	+	2	+	+	1	+	+	2
Carex lasiocarpa	.	.	1	+	+	+	2	1
Carex panicea	.	+	.	.	+	+	+	.
Rhynchospora fusca	.	+	.	+	.	.	+	1
Menyanthes trifoliata	+	.	+	+	.	.	+	.
Eriophorum angustifolium	.	.	+	.	.	.	+	.
Parnassia palustris	.	.	.	+	.	.	.	+
Carex lepidocarpa	.	.	.	+	.	.	.	+
Dactylorchis fuchsii	.	.	.	+	.	.	.	+
Carex rostrata	.	.	.	+	.	.	.	.
Lysimachia vulgaris	.	.	.	.	+	.	.	.
Comarum palustre	+	.	.	.	.	.	.	.
Alnus glutinosa	+	.	.	.	.	.	.	.
Phragmites communis	.	.	.	.	+	.	.	.
Pinus sylvestris	.	1	.	.	.	.	.	.

Lage des Stufenkomplexes: Eggstätter Seenplatte zwischen Schloßsee und Blassee.

Größe der Probeflächen: 1/4-1/2 m², Aufnahmejahr: 1968

Gesamtartenzahl: 42, davon ombrotraphent 7, minerotraphent 33, ökologisches Verhalten noch ungeklärt 2.

a Sphagnum centrale et plumulosum-Bultgesellschaft

b Sphagnum contortum-Bultfußgesellschaft

c Scorpidium scorpioides-(Braunmoos-)Schlenkengesellschaft

bermoos-Verein vertreten werden (Tabelle 1,B,2,F und 5,C 5).
Häufig durchdringen sich jedoch beide Bultfuß-Gesellschaften
in den Komplexen. Die Bultflächen werden in erster Linie durch
Rhynchospora alba und Eriophorum angustifolium locker bewach-
sen. In den Komplexen kommen einige Bergkiefern (Pinus mugo
ssp.pumilo) und Waldkiefern (Pinus sylvestris) vor. Karte 2
und Tabelle 4 stellen einen Utricularia-Sphagnum subsecundum-
Stufenkomplex dar.

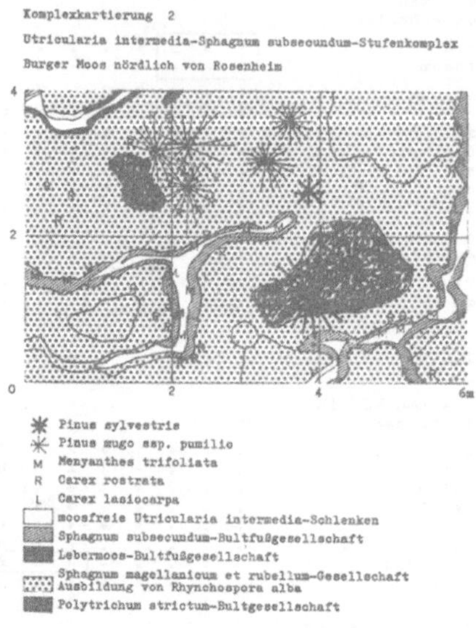

2.3 Pseudohochmoor-Stufenkomplexe

In den Pseudohochmoor-Stufenkomplexen sind alle Hochmoor-Arten
vertreten. Die Mineralbodenwasserzeiger dieser Komplexe sind:
Eriophorum angustifolium, Drosera intermedia, Carex rostrata,
Menyanthes trifoliata, Carex lasiocarpa und Rhynchospora fusca.
Entweder kommen mehrere Arten dieser Gruppe gemeinsam vor, oder
es tritt nur eine dieser Arten vereinzelt auf. Bisher ist es
jedoch nicht möglich, eine ökologische Wertigkeit der Arten
untereinander aufzustellen. Dem Hochmoor am nächsten stehen
jedenfalls Komplexe, in denen nur Carex limosa, Scheuchzeria
palustris und Carex pauciflora vorkommen. Bei diesen Arten ist
noch ungeklärt, ob es sich um eine ombrotraphente oder minerotraphente Art handelt. Im Sinne von JENSEN (1961) kann man die-

Tabelle 4

Utricularia intermedia-Sphagnum subsecundum-Stufenkomplex

	a		b		c	d		e	
Laufende Nummer	1	2	3	4	5	6	7	8	9
Artenzahl	9	10	8	9	15	13	12	6	7
Strauchschicht									
Pinus mugo ssp. pumilio	5	.	.	.	.	.	.	.	.
Pinus sylvestris	.	4	.	.	.	.	.	.	.
Ombrotraphente Bultarten									
Vaccinium oxycoccus	1	3	2	1	+	+	+	.	.
Andromeda polifolia	+	+	2	+	1	+	+	.	.
Sphagnum magellanicum	4	3	5	5	4	1	+	.	.
Drosera rotundifolia	+	1	1	+	+	.	.	.	.
Aulacomnium palustre	1	+	+	.	+	.	.	.	.
Sphagnum angustifolium	.	3	.	2	.	.	.	.	.
Polytrichum strictum	.	4	.	.	.	.	.	.	.
Sphagnum rubellum	2	.	.	.	.	.	.	.	.
Drosera obovata	.	.	.	.	.	+	.	.	.
Calluna vulgaris	.	.	.	1	.	.	.	.	.
Ombrotraphente Schlenkenarten									
Rhynchospora alba	.	+	1	2	4	3	3	.	+
Drosera anglica	.	.	.	.	.	.	.	.	+
Lebermoose (ombrotraphent)									
Cephalozia macrostachya	.	.	.	.	2	.	+	.	.
Mylia anomala	.	.	.	.	2	.	.	.	.
Cephalozia fluitans	.	.	.	.	1	.	.	.	.
Telaranea setacea	.	.	.	.	+	.	.	.	.
Calypogeia sphagnicola	.	.	.	.	+	.	.	.	.
Diff.-A. Bultfuß (minerotraphent)									
Sphagnum subsecundum	.	.	.	.	.	5	5	.	.
Minerotraphente Schlenkenarten									
Drosera intermedia	.	.	.	.	+	+	1	1	.
Carex limosa (omb?)	.	.	.	.	.	.	+	1	+
Utricularia intermedia	.	.	.	.	.	.	.	1	4
Scheuchzeria palustris	.	.	.	.	.	.	.	+	.
Übrige minerotraphente Arten									
Eriophorum angustifolium	1	.	2	2	2	1	1	2	1
Carex rostrata	.	+	1	.	+	+	+	.	2
Menyanthes trifoliata	.	.	.	.	.	2	1	1	1
Carex lasiocarpa	+	.	.	.	+	+	.	.	.
Calliergon stramineum	.	.	.	+	.	1	+	.	.
Trichophorum alpinum	.	.	.	.	.	1	.	.	.

Lage des Stufenkomplexes: Burger Moos nördlich von Rosenheim
Größe der Probeflächen: 1/4-1/2 m^2 ;Aufnahmejahr: 1968

Gesamtartenzahl: 30, davon ombrotraphent 18, minerotraphent 10,
                 ökologisches Verhalten noch ungeklärt 2.

a 1 Sphagnum magellanicum et rubellum-Gesellschaft,
    Fazies von Pinus mugo ssp. pumilio

a 2 Polytrichum strictum-Gesellschaft Fazies von Pinus sylvestris

b   Sphagnum magellanicum et rubellum-Gesellschaft
    Ausbildung von Rhynchospora alba

c   Lebermoos-Bultfußgesellschaft

d   Sphagnum subsecundum-Bultfußgesellschaft

e   Utricularia intermedia-Schlenkengesellschaft

se Komplexe als Übergangs-Hochmoor-Stufenkomplexe bezeichnen.
Karte 3 und Tabelle 5 zeigen einen derartigen Komplex.

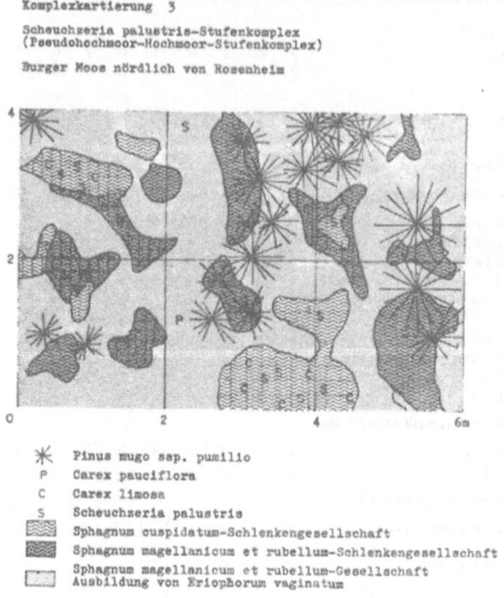

Pinus mugo ssp. pumilio
P   Carex pauciflora
C   Carex limosa
S   Scheuchzeria palustris
Sphagnum cuspidatum-Schlenkengesellschaft
Sphagnum magellanicum et rubellum-Schlenkengesellschaft
Sphagnum magellanicum et rubellum-Gesellschaft
Ausbildung von Eriophorum vaginatum

2.4 Hochmoor-Stufenkomplexe

Die Vegetation der echten ombrotraphenten Hochmoore im Unter-
suchungsgebiet ist nur noch in Resten vorhanden. Die großen
Hochmoore sind fast überall stark degradiert. Auf den trockenen
Flächen breiten sich entweder die Hochmoor-Flechtenheide (SCHU-
MACHER 1937) oder trockene Latschen-Filze aus. Letztere wurden
häufig gerodet; dann konnte sich bei nur am Rand entwässerten
Mooren auch ein Hochmoor-Stillstands-Komplex entwickeln. Sicherlich war jedoch auch in den natürlichen Hochmooren die
Latsche vertreten (PAUL und RUOFF 1927 und 1932). Betrachten
wir die dem Hochmoor am nächsten stehenden Übergangsmoore, so
bekommen wir vielleicht eine Vorstellung, wie die Vegetation
der wachsenden Hochmoore ausgesehen haben mag. Danach müßte
es sich um locker mit Bergkiefer bestockte Flächen handeln, in
denen die S p h a g n u m   m a g e l l a n i c u m   e t   r u -
b e l l u m - G e s e l l s c h a f t weitaus den größten Flä-
chenanteil bedeckte und die Schlenken nur einen sehr geringen
Anteil hatten.
Das noch relativ am besten erhaltene Hochmoor im Inn-Chiemsee-
Vorland ist das Hintermoos bei Endorf. Der zentrale Teil be-

steht aus einem ausgedehnten Latschen-Filz, etwa 1/3 der Fläche wurde gerodet (Unterlagen über den Zeitpunkt liegen nicht

Tabelle 5

Scheuchzeria palustris-(Pseudohochmoor-)Stufenkomplex

	a		b		c	d	
Laufende Nummer	1	2	3	4	5	6	7
Artenzahl	10	9	8	10	8	6	4
Strauchschicht							
Pinus mugo ssp. pumilio	4	.	.	.	.	.	.
Ombrotraphente Bultarten							
Eriophorum vaginatum	3	2	3	3	.	.	.
Vaccinium oxycoccus	2	1	2	2	1	.	+
Andromeda polifolia	1	1	2	+	1	.	.
Sphagnum magellanicum	3	4	2	4	2	.	.
Drosera rotundifolia	+	+	+	1	.	.	.
Sphagnum angustifolium	4	.	2	1	.	1	.
Sphagnum rubellum	.	.	2	2	5	.	.
Calluna vulgaris	+	3	.	1	.	.	.
Cephalozia fluitans	.	.	+	.	.	1	.
Sphagnum acutifolium	2	3	.	.	.	.	.
Aulacomnium palustre	.	1	.	.	.	.	.
Carex pauciflora (min?)	+	.	.	.	.	.	.
Polytrichumg strictum	.	1	.	.	.	.	.
Minerotraphente Schlenkenarten?							
Scheuchzeria palustris	.	.	.	+	2	2	1
Carex limosa	.	.	.	+	1	1	1
Ombrotraphente Schlenkenarten							
Rhynchospora alba	.	.	.	.	1	1	+
Drosera anglica	.	.	.	.	1	.	+
Sphagnum cuspidatum	.	.	.	.	.	5	5

Lage des Stufenkomplexes: Burger Moos nördlich von Rosenheim
Größe der Probeflächen: 1/2 m²   Aufnahmejahr: 1968

Gesamtartenzahl 19 davon 3 vermutlich minerotraphent

a Sphagnum magellanicum et rubellum-Gesellschaft, Fazies von
  Pinus mugo ssp. pumilio

b Sphagnum magellanicum et rubellum-Gesellschaft Ausb. von
  Eriophorum vaginatum

c Sphagnum magellanicum et rubellum-Gesellschaft Ausb. von
  Scheuchzeria palustris (rote Schlenken)

d Sphagnum cuspidatum-Scheuchzeria palustris-Schlenkengesellschaft

vor.) Auf dieser Rodungsfläche wurde ein Ausschnitt des dort
stockenden Stillstand-Komplexes kartiert (Karte 4, Tabelle 6).

Komplexkartierung 4

Hochmoor-Stillstands-Komplex

Hintermoos bei Endorf

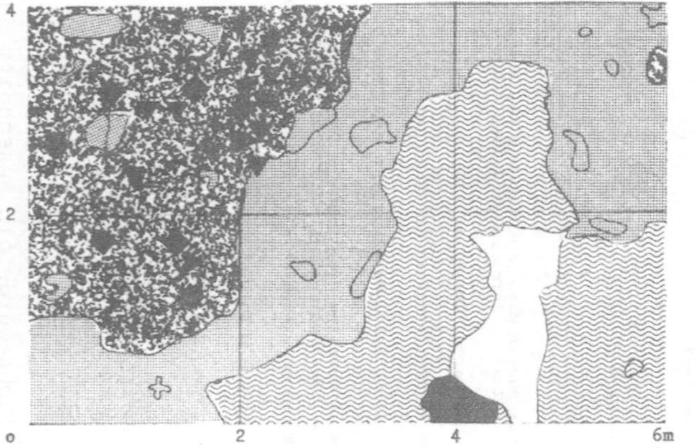

offener Torf

Sphagnum cuspidatum-Rhynchospora alba-Gesellschaft

Lebermoose

Sphagnum magellanicum et rubellum-Gesellschaft
Ausbildung von Eriophorum vaginatum

Dicranum bergeri-Bultgesellschaft

diverse Cladonia-Arten

Tabelle 6

Stillstands-Komplex (Hochmoor, alle Arten ombrotraphent)

	a		b		c	
Laufende Nummer	1	2	3	4	5	6
Artenzahl	6	7	6	7	7	3

Austrocknungszeiger

	a		b		c	
Dicranum bergeri	5	3	.	.	.	.
Cladonia squamosa	+	2	.	.	.	.
Cladonia arbuscula	+	1	.	.	.	.
Cladonia floerceana	+	.	.	.	.	.
Cladonia rangiferina	.	+	.	.	.	.
Entodon schreberi	+	.	.	.	.	.
Calluna vulgaris	3	3	2	.	.	.

Arten der Hochmoorbulte

	a		b		c	
Eriophorum vaginatum	.	1	2	1	3	2
Andromeda polifolia	.	.	+	2	+	.
Vaccinium oxycoccus	.	.	+	1	+	.
Sphagnum magellanicum	.	.	4	5	.	.
Sphagnum rubellum	.	.	2	2	.	.
Drosera rotundifolia	.	.	.	+	.	.

Arten der Hochmoorschlenken

	a		b		c	
Rhynchospora alba	.	+	.	1	1	1
Sphagnum cuspidatum	.	.	.	.	3	5

Lebermoose

	a		b		c	
Cephalozia connivens	.	.	.	.	1	.
Mylia anomala	.	.	.	.	1	.

Lage des Stillstandskomplexes: Hintermoos westlich von Endorf
(Landkreis Rosenheim)

Größe der Probeflächen: 1/2 m^2 **Aufnahmejahr 1969**

Gesamtartenzahl: 17

a **Dicranum bergeri-Bultgesellschaft**

b **Sphagnum magellanicum et rubellum-Bultgesellschaft**

c **Sphagnum cuspidatum-Rhynchospora alba-Schlenkengesellschaft**

Tabelle 2
Zusammenstellung der Artenzahlen in den Stufenkomplexen

	1	2	3	4	5
Braunmoos-Stufenkomplex	42	7	33	2	0,2
Utricularia-...-Stufenkomplex	29	18	10	2	1,5
Pseudohochmoor-Stufenkomplex	19	16	-	3	5,3
Hochmoor-Stillstandskomplex	17	17	-	-	-

1 Gesamtartenzahl
2 ombrotraphente Arten
3 minerotraphente Arten
4 Arten mit ungeklärtem ökologischen Verhalten
5 Verhältnis der ombrotraphenten Arten zu den minerotraphenten
  dabei wurden die Arten mit ungeklärten ökologischem Verhalten wie Mineralbodenwasserzeiger behandelt 2:(3+4).

ZUSAMMENFASSUNG

Die Vegetations-Komplexe der Übergangs- und Hochmoore im Inn-
Chiemseegebiet werden untersucht.Zur Unterscheidung von den
vom Moor-Wasserstand abhängigen Komplexen (Wachstumskomplex,
Stillstandskomplex etc.) werden die nährstoffbedingten Komplexe nach JENSEN (1961) als Vegetations-Stufenkomplexe bezeichnet. Eine solche Stufenkomplex-Reihe wird aufgestellt. Dabei
werden neben dem ombrotrophen Hochmoor 3 Gruppen von Mineralbodenwasser beeinflußten Stufen-Komplexen unterschieden. Tabelle 1 zeigt eine Übersicht über die Verbreitung der Arten in diesen Komplexen.
Die nährstoffreichsten Komplexe, die B r a u n m o o s -
S t u f e n k o m p l e x e, zeichnen sich durch Arten des
C a r i c i o n   d a v a l l i a n a e aus. Zwischen die Bulte und Schlenken schiebt sich hier, ebenso wie bei der nächsten Komplex-Gruppe, ein Verein aus Sphagnen der Subsecunda-
Gruppe. Die Bulte selbst können von minerotraphenten Torfmoosen
wie Sp.centrale und Sp.plumulosum gebildet werden.
Dann folgen in der Reihe die U t r i c u l a r i a i n t e r-
m e d i a - S p h a g n u m   s u b s e c u n d u m S t u f e n-
k o m p l e x e, sie werden in erster Linie negativ charakterisiert. Es fehlen die Arten des C a r i c i o n   d a v a l l i-
a n a e und die Hochmoor-Arten Sphagnum cuspidatum und Erio-

phorum vaginatum, Sphagnum subsecundum oder eine Gruppe von Lebermoosen bilden die Bultfuß-Gesellschaft.

In der dritten Gruppe sind alle ombrotraphenten Arten vertreten, sie unterscheiden sich vom Hochmoor nur durch einige Mineralbodenwasserzeiger wie Eriophorum angustifolium oder Carex lasiocarpa. In Anlehnung an DU RIETZ (1956) werden sie als P s e u - d o h o c h m o o r - S t u f e n k o m p l e x bezeichnet.

LITERATUR (1969 abgeschlossen)

ALETSEE,L. -1967- Begriffliche und floristische Grundlagen zu einer pflanzengeographischen Analyse der europäischen Regenwassermoorstandorte.- Beitr.Biol.Pflanzen 43: 170-283. Berlin.

AVERDIECK,F.-R. -1957- Zur Geschichte der Moore und Wälder Holsteins.- Nova Acta Leopoldina 19 (130). Leipzig.

CASPARIE,W.A. -1969- Bult- und Schlenkenbildung in Hochmoortorf.(Zur Frage des Moorwachstums-Mechanismus).- Vegetatio 19: 146-180. The Hague.

DU RIETZ,G.E. -1954- Die Mineralbodenwasserzeigergrenze als Grundlage einer natürlichen Zweigliederung der Nord- und Mitteleuropäischen Moore.- Vegetatio 5-6: 571-585. Den Haag.

EUROLA,S. -1962- Über die regionale Einteilung der südfinnischen Moore.- Ann.Bot.Soc."Vanamo" 33 (2): 1-243. Helsinki.

GAMS,H. -1932- Beiträge zur Kenntnis der Alpenmoore.- Abh. Naturwiss.Ver.Bremen 28. Bremen.

JENSEN,U. -1961- Die Vegetation des Sonneberger Moores im Oberharz und ihre ökologischen Bedingungen.- Naturschutz u.Landschaftspflege in Niedersachsen 1. Hannover.

KAULE,G. -1969- Vegetationskundliche und landschaftsökologische Untersuchungen zwischen Inn und Chiemsee.- Diss. Techn.Hochsch.München.

OBERDORFER,E. -1957- Süddeutsche Pflanzengesellschaften.-Jena.

PAASIO,I. -1933- Über die Vegetation der Moore Finnlands.- Acta for.fenn.39: 1-210. Helsinki.

PAUL,H.& LUTZ,J. -1941- Zur soziologisch-ökologischen Charakterisierung von Zwischenmooren.- Ber.bayer.bot.Ges.25: 1-28. München.

-- -- & RUOFF,S. -1927-1932- Pollenstatistische und stratigraphische Mooruntersuchungen in Bayern. Teil I: Moore im außeralpinen Gebiet des diluvialen Salzach-, Chiemsee- und Inngletschers.- Ber.bayer.bot.Ges.19: 1-84. Teil II: Moore in den Gebieten der Isar-, Allgäu- und Rheinlandgletscher.- Ibid.20: 1-264. München.

SCHUMACHER,A. -1937- Floristisch-soziologische Beobachtungen in Hochmooren des südlichen Schwarzwaldes.-Beitr.naturk. Forsch.Südw.Dtschl.2: 221-283. Karlsruhe.

SCHWICKERATH,M. -1944- Das hohe Venn und seine Randgebiete; Vegetation, Boden und Landschaft.- Pflanzensoziologie 6. Jena.

TANSLEY,A.G. -1965- The British Islands and their Vegetation.- Cambridge.

THUNMARK,S. -1940- Orientierungen über die Exkursionen des IX. internationalen Limnologenkongresses im Anebodagebiet.-

          Verh.Int.Ver.Limnologie 9: 59-68. Stuttgart.
TÜXEN,J. -1969- Gedanken über ein System der Oxycocco-Sphag-
     netea Br.-Bl.& Tx.43.- Vegetatio 19: 181-192. The Hague.

P.SEIBERT:

Haben Sie auch schon in etwas kleineren Maßstäben Karten ange-
fertigt, die ein größeres Gebiet umfassen, und wenn ja, wie haben Sie das gemacht?

G.KAULE:

Ich habe bisher noch keine größeren Komplexe kartiert, habe das
aber natürlich vor. Ich werde es diesen Sommer machen und zwar
analog der Arbeit von KLAUS MÜLLER in den norddeutschen Mooren,
also über einen größeren Komplex. Dann werde ich nicht Gesellschaften kartieren, sondern die Kryptogamen und die Phanerogamen einzeln, also die Arten einzeln eintragen.

J.J.BARKMAN:

In den Niederlanden kommen Carex limosa und Scheuchzeria palustris nie zusammen vor. Auch Drosera anglica findet sich nie
in diesen Gesellschaften. In Süd-Schweden aber bilden sie eine
feste Kombination. Im kontinentalen, baltischen Gebiet mangelt
es dieser Gesellschaft an R h y n c h o s p o r e t u m - Elementen. Interessant ist nun, daß im subatlantischen und subalpinen Gebiet, welches Herr Dr.KAULE untersuchte, dieses
S c h e u c h z e r i e t u m p a l u s t r i s sich völlig
mit dem R h y n c h o s p o r e t u m a l b a e (Rhynchospora alba, Drosera intermedia, Rhynchospora fusca die in den
holländischen Mooren fehlt) vermischt. Eine derartige Assozia-
tions-Fusion habe ich auch bei Epiphyten-Gesellschaften beobachtet und beschrieben. Der von Herrn JAKUCS erwähnte Fall einer Verschmelzung von T r i f o l i o - G e r a n i e t e a mit
dem Q u e r c e t a l i a p u b e s c e n t i s in Ungarn
ist eine ähnliche Erscheinung.
Es ist weiter festzuhalten, daß Rhynchospora alba in den Niederlanden nie in Niedermooren anzutreffen ist. Die Bultfuß-Gesellschaft ist ein typisches Beispiel eines limes divergens,
also eine unscharfe Übergangszone zwischen zwei Gesellschaften.
Derartige Grenzbereiche sind meistens artenreicher als die angrenzenden Gesellschaften und dies findet sich auch hier: in

Tab.2[1]hat Spalte A 18 Arten, in 1 und 3 nur 13 bzw.14. Spalte
F enthält 36, Tabelle 1 und 2 nur 23 bzw. 25 Arten. Ein limes
divergens ist zeitlich sehr stabil, was hier auch stimmt(nach
neueren Untersuchungen existiert in den meisten Gebieten der
Bulten-Schlenken-Zyklus überhaupt nicht!). Die Stabilität, wie
sie hier zu finden, ist darin begründet, daß die Bulten oligo-
troph, die Schlenken dagegen meso- bis eutroph sind.
[1]Vergl.Anm.auf p.355.

A.APINIS:

Über das Bult- und Schlenken-Problem ist eine reiche Literatur
aus Skandinavien vorhanden, wo dieses Problem sehr eingehend
von norwegischen, finnischen, schwedischen und estnischen Moor-
forschern behandelt worden ist. Vielleicht könnten diese Erfahrungen behilflich sein, um hier in einem pflanzengeographischen
Randgebiet, diese Probleme endgültig zu lösen. Denn was die soziologischen Einheiten betrifft, sind diese Bult- und Schlen-
ken-Probleme noch nicht gelöst. Ökologisch ist es ja sehr klar.
Ich habe z.B. in Estland in Flachmooren auf dem silurischen
Kalkstein oder in dem Gebiet des devonischen Dolomits diese
alkalischen Sphagnen-Schlenken beobachtet. Dort wachsen Sphagnum magellanicum und Sphagnum-Arten auf den Bulten, können aber
in den Schlenken gar nicht vorkommen, weil die Azidität dort
manchmal im Sommer pH 9 oder 10 beträgt, oder wo die totale
Aziditäts-Grenze erreicht wird, durch die Assimilation der
Schlenken-Moose, besonders Scorpidium scorpidioides oder Depla-
nocladus revolvens und anderen Arten, die eine Ausscheidung von
Kalzium-Karbonat an der Oberfläche erzeugen. So ist dort eine
sehr scharfe ökologische Grenze zwischen Schlenken und Bulten.
Auf diesen Bulten kann Sphagnum recurvum oder sogar Sphagnum
fuscum wachsen, die sehr empfindlich gegen eine neutrale oder
schwach saure Reaktion sind. Die geographische Lage dieser
Standorte in Bayern und in Skandinavien und im ostbaltischen
Gebiet wird auch durch manche Zwergsträucher wie Pinus mugo
und andere gekennzeichnet. Ich hoffe, daß diese schöne Arbeit
von Herrn KAULE auf einem anderen Gebiet erweitert und abgeschlossen werden kann.

H.DIERSCHKE:

In Nordwest-Deutschland kommen Scheuchzeria palustris und Carex
limosa getrennt vor, es gibt aber auch Wuchsorte, wo beide Ar-

ten zusammen auftreten.

In einem kleinen nährstoffarmen Schlatt im Emsland fand ich im
vergangenen Jahr optimal entwickelte Horste von Carex elata im
offenen Wasser. Dazwischen schwimmen im Wasser Utricularia minor, Sphagnum obesum, Sp.cuspidatum fo.plumosum u.a. M a g n o -
c a r i c i o n - oder P h r a g m i t a l i a - Arten fehlen
ganz. Gibt es in Süd-Deutschland neben den beschriebenen Gesellschaften ein richtiges C a r i c t u m   e l a t a e und mit
welcher Wuchskraft kommt diese Art in den Moor-Gesellschaften
vor?

F.RUNGE:

Ich möchte Herrn KAULE fragen, was er unter Niedermoor versteht.
Ich habe früher gelernt, unter Niedermoor, Flachmoor, versteht
man Phragmites-Sümpfe. Sie wissen, welche Gesellschaften ich
meine. Es sind die nährstoffreichen Flachmoore, im Gegensatz zu
den Hochmooren. Wenn ich aber nun hier lese "Sphagnum magella-
nicum-Niedermoor" dann ist mir das nicht klar, denn Sp.magella-
nicum gilt für mich eigentlich als Hochmoor-Art, während andere
Pflanzen,die hier aufgeführt sind,wie Carex rostrata, Menyanthes
trifoliata, Comarum palustre, von mir zu den Zwischenmoor-Arten
gerechnet werden. Verstehen Sie unter Niedermoor vielleicht
Zwischenmoor? Dann bin ich einverstanden.

G.KAULE:

Ich nehme Niedermoor in diesem Zusammenhang in den Definitionen
von ALETSEE (1967).

V.WESTHOFF:

Ich hatte das Glück, mit unserem Freund THEO MÜLLER vergleich-
bare Gebiete in Baden-Württemberg zu studieren, vor allem im
Wurzacher Ried.Wir haben solche auch in Finnland und in Schweden sehen können: also Gradienten zwischen eutrophem Niedermoor
und oligotrophen Bulten, die sich darin bilden, nicht nur Gradienten von unten nach oben, sondern auch horizontal vom eutrophen zum oligotrophen Bereich. Ich habe da genau dieselben
Verhältnisse gefunden, wie sie hier dargestellt worden sind,
und ich möchte betonen, daß,wenn man diese Schlenken und Bulten
und Bult-Füße ganz kleinflächig aufnimmt, man immer zu solchen
Ergebnissen kommen wird, obwohl sie vielleicht etwas merkwürdig

aussehen, wenn man das selbst nicht gemacht hat. Wenn wir z.B.
in Tabelle 2[1]in Spalte E, wo sich also die Ombrominerobionten
halten, Vaccinium oxycoccus mit Präsenz IV und Andromeda polifolia mit II finden, dann könnte man sich fragen, ob das wirklich so ist. Das ist aber wirklich so! Herr KAULE hat sich auch
darüber gewundert und gefragt, ob es nicht das sei, was nach
NORDHAGEN Vizinismus genannt wird, also daß die Arten sich nur
aus der Umgebung in diese Mikroaufnahmen einmischen, wie man
das von V.oxycoccus erwarten könnte, das ja mit Ausläufern
kriecht. Das ist aber nicht so: es kriecht zwar aus der Umgebung, es wurzelt aber auch in diese Mikrogesellschaft, und Andromeda polifolia, die ja nicht kriecht, benimmt sich auch so.
Das ist auch bis an den Polarkreis in Finnland ebenso festzustellen.
Eine Frage hätte ich noch zu der Tabelle 2[1]zu den Spalten A-C.
Es sind da nicht die Hochmoor-Lebermoose angegeben worden, My-
lia anomala, Odontoschisma sphagni, Cephalozia connivens, Caly-
pogeia sphagnicola. Es wäre zu erwarten, daß es die da doch
auch gibt. Es wäre interessant zu sehen, ob sie auch so weit
in den T o f i e l d i e t a l i a - Bereich vordringen, wie
es die Phanerogamen anscheinend tun. Ich möchte auch gern wissen, ob Tofieldia sich in dieser Gesellschaft findet. Das tut
sie in Baden-Württemberg. Sie wäre dann etwa in der Spalte 2 E
zu erwarten.
Über Rhynchospora alba möchte ich noch etwas bemerken in Tabelle 3[1]. Es ist ganz auffällig, daß Rh.alba in der Spalte a Präsenz I zeigt, und dann in allen anderen Spalten, den Bulten
also, mit hoher Präsenz vorkommt. Das tut sie in echten Hoch-
moor-Bulten nicht. Solche sind es hier auch nicht. Es sind Bulten, die aus dem T o f i e l d i e t a l i a - Bereich hervor-
ragen, und es sieht so aus, als ob ein gewisser Nährstoff-Ein-
fluß bis in die Bulten hinauf auftritt, also eine gewisse Mine-
rotrophie hier die Ombrotrophie der Bulten derart beeinflußt,
daß die Rhynchospora alba eben noch konstant vorkommen kann.
Ich möchte gern wissen, ob Herr KAULE derselben Meinung ist.
1) Vergl.Anm.auf p.355.

R.TÜXEN:
Die Bilder sowohl wie die Tabellen drängen mir einen Vergleich
mit japanischen Mooren auf. In Japan sprechen die Japaner von
ihren "Hochmooren". Wenn man dort hinkommt, findet man keine,

In Japan gibt es genau diese Moore, wie wir sie eben gesehen haben, mit den gleichen Komplexen, also Flach- oder Niederungs-
moore mit Sphagnum magellanicum,Sph.papillosum und Sph.fuscum-
Bulten. Aber diese nur einzeln verstreut; keine Hochmoore, wie
wir sie etwa vorher bei O'SULLIVAN aus Irland gesehen haben.
Die Gesellschafts-Komplexe in den japanischen Mooren sind sehr
ähnlich denen, wie wir sie eben aus Bayern kennengelernt haben.
Sie haben nur etwa statt Molinia Moliniopsis japonica (jeder
von uns würde die Art sofort als Molinia ansprechen, so ähnlich
sehen sie beide aus), oder sie haben statt Carex elata, die
hier genannt wurde, Carex middendorfii. Ein paar Arten vikari-
ieren, aber jeder von uns, der hier Moore aufnehmen kann, kann,
wenn er 2 oder 3 Arten sammelt, in Japan genau die gleichen Aufnahmen machen, so ähnlich sind sie.
Dann zum Problem Bult-Schlenke. Ich habe durch OVERBECK zum ersten Mal von der von ihm ausgesprochenen Kritik an der schwedischen Auffassung der Wechsel-Beziehung dort erfahren, und vor 2
oder 3 Jahren hat ELLENBERG im Harz auf einer Naturschutz-Ex-
kursion auf diese Problematik hingewiesen. Beide haben mich
nicht ganz überzeugen können, weil wir doch sehr häufig sehen,
wie von den Bulten die Sphagnen in die Schlenken hineinwachsen,
einfach herüberkriechen, herüberquellen. Man kann ganze Polster
von den dickblättrigen Sphagnen, also Sph.papillosum etwa oder
Sph.magellanicum abheben, die auf der Schlenke liegen, die hinein gewachsen sind. Ich glaube, seitdem ich die Ausführung von
Herrn KAULE gehört habe, daß man beide Auffassungen miteinander
vereinigen kann. Die schönen kleinen Karten, die er gezeigt hat,
die also das Klein-Mosaik der Schlenken, der Bult-Füße, der
Bult-Mitten und der Bult-Spitzen zeigten, kann man sich ganz
gut konstant vorstellen, wenn man annimmt, daß sie nicht ganz
genau in der gleichen Lage in die Höhe wachsen, sondern daß sie
sich beim Moor-Wachstum ein wenig seitlich verschieben, so daß
der Komplex Bult - Schlenke dauernd bestehen bleibt, aber sich
bald mehr nach dort, bald mehr nach hier hin neigt. Ich glaube,
daß hier eine Lösung für die verschiedenen Auffassungen liegen
dürfte.
Ich fand die Arbeit und die Betrachtungsweise und die Zusammenfassung von Herrn KAULE unseren systematischen Vorstellungen
recht nahe. Ich finde,daß die Schlenken im großen zum S c h e u c h-
z e r i e t u m s.l. hinneigen. (Nicht alle Spalten der Tabelle

sind gleichwertig.Man müßte die Einzelaufnahmen haben.) Ein
Teil   ähnelt deutlich dem R h y n c h o s p o r e t u m.
WESTHOFF hat sich gewundert, daß Rhynchospora in Flachmooren
vorkommt. Ich habe mein erstes R h y n c h o s p o r e t u m
in der Schweiz mit WALO KOCH im Jahre 1926 in dieser Situation
kennengelernt und habe mich gewundert, als ich wieder nach Hause kam, daß Rhynchospora ohne diese anspruchsvollen Begleiter
hier bei uns wachsen kann. Also neigt Tabelle 2[1])im großen zum
R h y n c h o s p o r e t u m (Spalte E ist vielleicht zum
C a r i c i o n d a v a l l i a n a e zu stellen). Tabelle 3[1])
ist ein ausgesprochenes "S p h a g n e t u m m a g e l l a n i-
c i" r h y n c h o s p o r e t o s u m, mit Ausnahme der Spalte
a (und Spalte F,wo eher an ein C a r i c i o n c a n e s c e n-
t i s - f u s c a e zu denken wäre.) Das S p h a g n e t u m
m a g e l l a n i c i r h y n c h o s p o r e t o s u m kommt
sehr wohl bei uns in Hochmooren vor. Die Esterweger Dose hatte
diese Gesellschaft. Die Bulten dort waren auf großen Flächen
ganz weiß von R h y n c h o s p o r a a l b a, auf dem Hochmoor von etwa 12 m Mächtigkeit. Die Tabelle 4[1])ist wiederum ein
reines S c h e u c h z e r i e t u m, so scheint es mir, in
einer ausgesprochen armen und sehr extremen Ausbildung.
Ich finde diese Arbeit deswegen so bemerkenswert, weil sie viele Verbindungen mit anderen Betrachtungsweisen erlaubt und anregend ist für eben Probleme,die von anderer Seite her betrachtet
worden sind.
[1])Vergl.Anm.auf p.355.
S.SEGAL:
In den Niederlanden kommen in gleichen Mikro-Zonierungen auch
bestimmte Übergangsgesellschaften zwischen Scorpidium-Schlenken
und Sphagnum-Bulten, oftmals mit Pellia-,Fissidens-,Bryum- und
Mnium-Arten und auch mit Linum catharticum, Sagina nodosa, Li-
paris loeselii und Epilobium palustre vor. Gleiche Gesellschaften finden sich auch in Mittel- und Ost-Europa. Die Bultfuß-Gesellschaften sind oftmals nicht stabil, und in den Niederlanden
verschieben sich in Dauerquadraten die Grenzen.Liegen keine Untersuchungen über längere Zeitabschnitte vor, so sollte man mit
dem Begriff Stabilität vorsichtig umgehen. Ich glaube nicht,
daß eine hohe Artenzahl ein Indiz für Stabilität ist. Es gibt
wahrscheinlich eine Relation zwischen Ausprägung des Horizon-
talgefüges und Stabilität oder der Sukzessions-Geschwindigkeit.

Verhältnismäßig scharfe Grenzen zeigen sich z.B. öfters bei
"M a g n o c a r i c i o n"-Bulten. Drepanocladus-Arten sind
in diesem Zusammenhang wahrscheinlich wertvolle Milieu-Zeiger.

In Bayern treffen wir des öfteren Drepanocladus intermedius an,
ein Moos, das sich auch äußerlich gut von Drepanocladus revolvens unterscheiden läßt. Möglicherweise handelt es sich hier
um Ökotypen.

G.KAULE:

Carex elata ist an dem Standort, wo ich sie aufgenommen habe,
äußerst schlecht ausgebildet. Sie ist nicht einmal richtig bult-
förmig. Es sind einzelne Carex elata-Halme. Es gibt in der Gegend natürlich viel bessere, viel schönere Großseggen-Rieder,
auch ein typisches Caricetum elatae. Es gibt aber auch diese
einzelnen im Wasser stehenden Bulten,die Sie erwähnt haben.
Dann können verschiedene Sphagnen, Utricularia usw. dazwischen
schwimmen. Der schönste Standort, den ich in dieser Form kenne,
ist allerdings wieder untypisch, er ist durch Möwen eutrophiert
oder nitrifiziert. Durch die große Möwen-Kolonie kommen noch
weitere andere Arten dazu, wie Epilobium, Eupatorium cannabium
usw.
Lebermoose kommen im Untersuchungsgebiet außerhalb von Mikro-
Standorten äußerst selten in den Bulten vor.
Ich habe 2 Aufnahmen mit Tofieldia, die ich in diesen zusammen-
fassenden Tabellen weggelassen habe, weil die Stetigkeit nicht
groß genug war.
Über das Verhalten von Rhynchospora stimme ich mit Herr WEST-
HOFF überein, daß es vermutlich das höhere Nährstoff-Angebot in
den von eutrophem und mineralreichem Wasser umgebenen Bulten ist,
das ihr Auftreten erlaubt.
Zum Problem Bult-Schlenke: Es liegen einige Bohrungen vor, aus
England z.B. von OSVALD, in TANSLEY: The Vegetation of the British Islands veröffentlicht. Ich glaube nicht, daß diese Bohrungen irgendwie falsch sind, oder daß die Sphagnen an der Stelle, wo OSVALD die Bohrungen durchgeführt hat, nicht bestimmt
sind. Wenn also die Untersuchungen stimmen, hat tatsächlich ein
Wechsel der Moos-Vereine stattgefunden. Genauso nach den Untersuchungen über die Frage der Rekurrenz-Flächen,die zu ähnlichen
Wechseln in den Moos-Vereinen seit der Eiszeit kommt. Es geht
natürlich sehr langsam. OSVALD beschreibt in seinem Buch seit

der Eiszeit 7 Wechsel zwischen Bulten und Schlenken. Wahrscheinlich,ich stimme mit Herrn TÜXEN überein, liegt die Wahrheit irgendwo in der Mitte.
Dann zur Frage des Drepanocladus intermedius und D.revolvens.
Ich halte das persönlich noch für eine Glaubensfrage. So lange
die Gattung Drepanocladus nicht taxonomisch bearbeitet ist, un-
terscheide ich D.revolvens und D.intermedius nicht. Ich kann
nicht beurteilen, ob es echte Arten sind, aber ich glaube nicht
daran.

Anm.   Die Tabellen- und Spaltenbezeichnungen in den Diskussi-
        onsbemerkungen beziehen sich auf die während des Referats
        verteilten Stetigkeits-Tabellen, die hier durch neugeord-
        nete Original-Tabellen ersetzt worden sind.

BEMERKUNGEN ZUM PROBLEM DER ASSOZIATIONSBEGRENZUNG

D e n i s a   B l a ž k o v á

Es ist nicht schwierig, in eine abstrakte Einheit, z. B. eine
Assoziation, eine optimal entwickelte Phytozönose einzureihen,
mit einer Reihe von Kennarten, mit einer gut ausgebildeten Arten-Verbindung, oder mit Trenn-Arten, mit einer typischen Struktur usw. Solche Phytozönosen kommen jedoch, insbesondere in Mitteleuropa, leider verhältnismäßig selten vor. Ihr Rückgang ist
auf die Störung des regelmäßigen ökologischen Haushaltes der
Standorte in beträchtlichem Maße zurückzuführen. Als solche Störungen sind entweder drastische einmalige Eingriffe, wie z.B.
tiefe Entwässerung, oder die Unterbrechung von regelmäßigen in-
standhaltenden Maßnahmen anzusehen. Als Reaktion erscheinen
dann gestörte, schwankende Bestände, die in eine der Einheiten
des Vegetationssystems nur schwierig, oder höchstens als Phasen
mit einer Randstellung im Rahmen der Einheiten, eingereiht werden können. Andererseits muß man in Betracht ziehen, daß die
phytozönologische Klassifizierung eine typologische Klassifizierung darstellt, d.h. daß bei einer jeden konkreten Phytozönose
durch Vergleich festgestellt wird, inwieweit sie einem abstrakten Typus entspricht, der den vollständigen Komplex von Merkmalen der Einheit besitzt. Die gut entwickelten, ungestörten Phytozönosen nähern sich mehr oder weniger dem Typus oder entfernen sich von demselben, und das Problem besteht darin, festzustellen, bis zu welcher Stufe der Abweichung vom Typus (und bei
welcher Art der Abweichung) von derselben Assoziation noch die
Rede sein darf. Dieses Problem ist besonders brennend bei der
Vegetationskartierung.
Da für jeden Begriff seine Definition die Grenze zieht, sollten
wir die Antwort darin suchen. Keine Definition der Assoziation
liefert jedoch eine scharfe Grenze; denn in den Definitionen
spricht man von der Notwendigkeit einer bestimmten Übereinstimmung oder Ähnlichkeit der Eigenschaften von Phytozönosen - mögen
es floristische, ökologische oder physiognomische Eigenschaften
sein. Bestimmte Verschiedenheiten der Zönosen bestehen natürlich
immer, es gibt so gut wie keine zwei vollkommen identische Phytozönosen. Der Begriff der Übereinstimmung oder Verschiedenheit
ist demnach sehr vage. Andererseits ist auch eine bestimmte Kon-

tinuität der Vegetation unbestreitbar. Nicht selten kommt der
Fall eines kontinuierlichen Gradienten vor, dessen beide Enden
typische Assoziationen darstellen, aber ein großer Teil der Bestände zwischen beiden Enden steht dazwischen.
Wodurch sollte aber die Grenze zwischen Einheiten festgestellt
werden? Welche Merkmale aus ihrer Gesamtheit sollten überein-
stimmend sein und welche dürfen schwanken oder kontinuierlich
ineinander übergehen? Die Antwort ist davon abhängig, zu welchem Zwecke das gebildete System angewandt werden soll. Die
Klassifizierung bildet eine Art von "Gedanken-Automat" zur Er-
werbung von Informationen über den klassifizierten Gegenstand.
Das heißt, daß man durch richtige Einreihung von Phytozönosen
in das Klassifikationssystem automatisch eine Reihe von Informationen über ihre Eigenschaften gewinnt. Über welche Eigenschaften - das hängt davon ab, welche Merkmale man für die Klassifizierung anwenden wird, und demnach wird das Bestreben sein, die
für die nötige Information wichtigsten Merkmale zu benutzen.
Für verschiedene Bereiche menschlicher Tätigkeit werden die wichtigsten Eigenschaften desselben Begriffs verschieden sein, z.B.
bestimmte Aspekte der Vegetation interessieren den Maler, andere den Landwirt und wieder andere den Phytozönologen. Jeder wird
also für sein Bedürfnis eine andere Einteilung fordern. So kann
man sich schwer vorstellen, daß es möglich wäre, ein einziges,
allseitig anwendbares System zu schaffen. In dem riesigen Komplex der Klassifizierungs-Möglichkeiten läßt sich jedoch diejenige deutlich unterscheiden, in der die für das Bestehen der
Phytozönose wesentlichen Merkmale angewandt sind. Solche Eigenschaften sollte eben die phytozönologische Klassifizierung der
Vegetation besitzen.
Vergleichen wir nun diese theoretische Voraussetzung mit der
Weise, wie diese Einheiten begrenzt werden. BRAUN-BLANQUET
(1964) schreibt: "Jede Einzelsiedlung muß, um als Beispiel des
Assoziationstypus gelten zu können, die wesentlichen Assoziati-
onsmerkmale aufweisen. Vor allem muß die normale charakteristische Artenverbindung vorhanden sein, das heißt, eine Mindest-
zahl von Kenn- und Trennarten und wichtigeren Begleitern. Ferner
darf die Gesamtartenzahl nicht unter ein für jede Assoziation
spezifisches Minimum sinken." Diese Merkmale der Assoziation
(d.h. Kenn- und Trennarten usw.) werden aus der synthetischen
Bearbeitung der Aufnahmen in den Tabellen abgeleitet. Es wird

jedoch nicht in allen Fällen in genügendem Maße unterschieden,
welches Merkmal wesentlich für die Existenz der Phytozönose
als Mitglied einer Assoziation - und welches Merkmal wichtig
für die Identifizierung der Gesellschaft als diakritisches Merkmal ist. So z.B. können die Assoziations-Kennarten für das Erkennen einer Assoziation von großer Bedeutung sein - als dia-
kritische Merkmale, nach denen man die Assoziation leicht iden-
tifiziert. Doch das bedeutet nicht, daß sie auch in ontologi-
scher Hinsicht, d.h. für das Bestehen der Pflanzengesellschaft
selbst ebenso bedeutend sein müssen. Diese Tatsache spiegelt
sich übrigens auch darin wider, daß in die Assoziatons-Tabellen,
die eine bestimmte Einheit abgrenzen sollen - also mit typischen
Aufnahmen - auch solche Aufnahmen eingereiht werden, in welchen
die Kennarten fast oder völlig fehlen. Die Anwendung solcher
Aufnahmen zur Feststellung des Typus deutet schon an, daß auch
andere Merkmale respektiert werden, und daß auch die Teilnahme
anderer Arten für wesentlich gehalten wird. Das wird auch von
BRAUN-BLANQUET festgestellt, wie es sich aus dem zitierten Auf-
satz ergibt. Aus diesem Grunde wird auch ein so großer Nachdruck
auf die Gruppe von konstanten Arten gelegt, was sich in der charakteristischen Arten-Verbindung ergibt. In den Grenzfällen, wo
die Einreihung nicht einmal nach der charakteristischen Arten-
Verbindung eindeutig ist, geht man in den meisten Fällen von
der gesamten Arten-Zusammensetzung aus. Die Phytozönose wird
dann in diejenige Einheit eingereiht, aus der (oder der ihr ü-
bergeordneten Einheit) eine größere Artenzahl anwesend ist.
Durch verschiedene mathematische Methoden läßt sich dieses Verfahren regeln und verfeinern - z.B. mit Rücksicht auf den Dek-
kungsgrad einzelner Arten - doch das Grundprinzip bleibt unver-
ändert. Ein solches Verfahren scheint sehr objektiv zu sein, da
man zum Endresultat durch Berechnung angelangt ist. Auch die
Einreihung der Arten ist nicht subjektiv - wenn sie auch in verschiedenen Gebieten schwankt - sondern ergibt sich aus der objektiven Bearbeitung der Aufnahmen in den Tabellen. In solchen
Fällen halten wir jedoch alle Arten für gleichwertig, nehmen
wir keine Rücksicht auf das Verhalten der Arten in Zönosen,auf
deren Anteil in Bau und Struktur der Bestände, auf ihre Reaktion gegenüber verschieden mächtiger Veränderungen der Umwelt sowie auf das resultierende sogenannte "Puffer-Vermögen" der Gesellschaft. Eben in diesen Eigenschaften kommt es jedoch deut-

lich zu Tage, welche Art - oder, allgemein gesagt, welches Merkmal - für die Gesellschaft mehr oder weniger wesentlich ist.Es
ist also von großer Wichtigkeit, die entscheidenden Merkmale
für die Feststellung der Grenze unter den Einheiten eben für
die Fälle derjenigen Phytozönosen objektiv festzustellen, die
nicht typisch entwickelt sind und die Grenzfälle im Rahmen der
gesamten Amplitude der Gesellschaft darstellen. Vom methodischen
Standpunkt aus wird freilich ein solches Verfahren schwierig
sein: der verschieden starke Nachdruck, den man auf einzelne
Arten legt, wird einen Eindruck von subjektiver Wertung hervor-
rufen. Geht man jedoch von einer wirklich objektiven Erkenntnis
aus, dann kann man das Vegetationssystem hierdurch präzisieren
und auch seine bessere Ausnützung ermöglichen. Daraus geht also
hervor, daß die Grenzen unter den Assoziationen vor allem durch
die Diskontinuität der für ihr Bestehen an sich wesentlichen
Merkmale, bzw. durch die Diskontinuität des Gradienten dieses
Merkmales, gegeben sind.
Ich möchte dieses Problem noch an einigen konkreten Beispielen
erläutern. Betrachten wir zwei verschiedene Kennarten, Lathyrus
versicolor - Lathyrus pannonicus (Kram.) Garcke subsp. versicolor (Gmel.)Janch. und Carex gracilis. Carex gracilis ist eine
Kennart der Assoziation C a r i c e t u m   g r a c i l i s
Greabner et Hueck 1931, Lathyrus versicolor der Q u e r c u s
p u b e s c e n s - L a t h y r u s   v e r s i c o l o r - Assoziation Klika 1933. Beim Einreihen der Phytozönosen in diese
Assoziationen wertet man jedoch die Anwesenheit oder Abwesen-
heit, die Quantität und Vitalität der erwähnten Kennarten nicht
auf dieselbe Weise. Die Bestände der Flaumeichen-Wälder ohne
Lathyrus versicolor können noch sehr typische Vertreter der
Assoziation von Q u e r c u s   p u b e s c e n s - L a t h y-
r u s   v e r s i c o l o r sein. In C a r i c e t u m g r a-
c i l i s ist jedoch nicht nur die Anwesenheit von Carex gracilis, sondern auch ihre dominierende bestand-bildende Stellung
obligat. Wenn Carex gracilis fehlt oder nur in geringer Menge
vorkommt, wird die Phytozönose in die Assoziation C a r i c e-
t u m   g r a c i l i s nur dann eingereiht, wenn die Funktion
von Carex gracilis als aufbauende Art (Edifikator), der die ü-
brige Arten-Zusammensetzung mitbestimmt, durch andere, zönotisch
nahestehende Seggen-Arten (am häufigsten durch Carex disticha,
Carex acutiformis oder Carex vesicaria) ersetzt oder ergänzt

ist. Andererseits wird nicht jeder Bestand mit hohem Deckungsgrad der physiognomisch auffälligen Carex gracilis der Assoziation C a r i c e t u m   g r a c i l i s angehören. Die Schlank-Segge dringt vor allem in verschiedene Assoziationen der Ordnung M a g n o c a r i c e t a l i a ein. So beschrieb schon W.KOCH (1926) eine Variante von Carex gracilis im C a r i c e t u m e l a t a e. In anderen Fällen ist Carex gracilis eine Kodominante im C a r i c e t u m   v u l p i n a e usw. Man begegnet jedoch auch solchen Phytozönosen, die an der Grenze der Einheiten aus verschiedenen Klassen variieren, im gegebenen Falle insbesondere mit dem Verband C a r i c i o n   f u s c a e. In solchen Grenzfällen nehmen wir zur Entscheidung, ob es sich noch um die Assoziation C a r i c e t u m   g r a c i l i s oder um die Variante von Carex gracilis aus irgendwelcher anderen Assoziation handelt, dann ein anderes wesentliches Merkmal in Betracht, vor allem die charakteristische Arten-Verbindung.Es mag jedoch geschehen (z.B. bei der Unterscheidung von Assoziationen aus der Ordnung M a g n o c a r i c e t a l i a), daß auch dieses Merkmal das Problem der Grenze eindeutig nicht löst. Die Arten-Zusammensetzung einzelner Großseggen-Gesellschaften besitzt eine so weite Amplitude, daß sie sich in Grenzfällen überdecken und ihre Unterschiede mehr als eine statistisch ermittelbare Tendenz gültig sind. Im konkreten Falle besitzt z.B. das C a r i c e t u m   e l a t a e - im Vergleich mit dem C a r i c e t u m   g r a c i l i s - einen mächtigeren Anteil der Klassen-Kennarten, einen geringeren Anteil von Begleitern aus der Klasse M o l i n i o - A r r h e n a t h e t a l i a und sogar auch die Kennarten des von BALÁTOVÁ (1963) als C a r i c i o n r o s t r a t a e benannten Verbandes.Man begegnet jedoch auch Phytozönosen mit dominierender Carex gracilis, die hinsichtlich des Typus der Assoziation C a r i c e t u m   g r a c i l i s dieselben Abweichungen zeigen - sogar samt den Kennarten des Verbandes C a r i c i o n   r o s t r a t a e, die größtenteils die Trennarten der Subassoziation C a r i c e t u m   g r a c i l i s   c o m a r e t o s u m bilden. Ich bin der Ansicht, daß in solchen Fällen - also z.B. beim Bestimmen, ob es sich um einen Grenzfall der Ass.C a r i c e t u m   g r a c i l i s oder C a r i c e t u m   e l a t a e, Variante mit Carex gracilis handelt - die Kenntnis der Struktur, Ökologie und Dynamik der Gesellschaft, sowie die Rolle einzelner Arten beim Aufbau der

Zönosen eine wichtige Hilfe für die genauere Entscheidung vermittelt. In dem angeführten konkreten Falle wird es also ratsam sein, in Erwägung zu ziehen, ob der Bestand eine Mosaik-Struktur von Bulten und Schlenken besitzt - wie es für C a r i c e- t u m e l a t a e üblich ist - ob die Oberfläche mehr oder weniger homogen ist, ob die Verteilung der Pflanzen dem sublitoralen oder eulitoralen Haushalt angepaßt ist, usw.Oft handelt es sich freilich um Fälle mit unregelmäßigem, unausgewogenem Haushalt, der als Ursache der Unausgeprägtheit einer solchen Phytozönose anzusehen ist - oder um eine erst unlängst begonnene Entwicklungs-Tendenz von einer Gesellschaft zur anderen. In solchen Fällen wird es nötig sein, den Wert einzelner Merkmale sorgfältig zu schätzen und zwar auch mit Rücksicht auf eine weitere Dimension - die Zeit. Die Entscheidung, welches Merkmal vorübergehend ist, und welches eine große Dauer besitzt und sich auch unter den Bedingungen eines schwankenden Haushaltes der Zönose zäh durchsetzt - solche Entscheidung erfordert gründliche Kenntnisse der inneren Gesetzmäßigkeiten des Lebens einzelner Phytozönosen. Ich vermute, daß solche nur scheinbar subjektive Wertungen oft imstande sind, die durch mechanische Berechnung der Ähnlichkeits-Koeffizienten erworbenen Wertungen zu verbessern oder zu präzisieren, die (wie sie heutzutage allgemein angewandt werden) die Hierarchisierung der Merkmale, ihren verschiedenen Wert und insbesondere eben die Bedeutung von Arten beim Variieren der Phytozönosen nicht berücksichtigen. Hiermit will ich nicht behaupten, daß die mathematischen Methoden verworfen werden sollten. Ich beabsichtige nur, auf die Tatsache aufmerksam zu machen, daß es uns zugute kommt, wenn wir bei der Klassifizierung - und insbesondere beim Ziehen der Assoziations-Grenzen - auch die Erkenntnisse aus anderen Gebieten der Vegetationskunde, vor allem aus der Ökologie, Dynamik usw. bewußt benutzen.

ZUSAMMENFASSUNG

Dieser Beitrag beschäftigt sich mit dem Problem der Kriterien zur Assoziations-Begrenzung, insbesondere bei solchen Phytozönosen, die an der Grenze einiger Assoziationen stehen und die doch auf irgend eine Weise klassizifiziert werden müssen (z.B. für die Kartierung.

Da es sich um die phytozönologische Klassizifizierung handelt,
sollten die Grenzen unter den Einheiten durch die Diskontinuität insbesondere der für die Existenz der Assoziation wesentlichen Merkmale, bzw. durch die Diskontinuität des Gradienten des
quantitativen Wertes dieser Merkmale, gekennzeichnet werden.
Die gebrauchten Entscheidungs-Kriterien stellen jedoch manchmal
Merkmale dar, die zwar von gnoseologischer Bedeutung sind, doch
die gleichzeitig vom ontologischen Gesichtspunkte nicht wesentlich sein brauchen. Das gilt z.B. für manche Kennarten. Die Bedeutung der Anwesenheit einer Art, der Art ihrer Vertretung oder
anderer Assoziations-Merkmale tritt vor allem in der Struktur
und der Dynamik der Phytozönosen hervor, wie es sich aus vielen
synökologischen und syndynamischen Studien ergibt. Eben die auf
diesem Fachgebiet erworbenen Kenntnisse sind oft behilflich beim
Unterscheiden, welche Assoziations-Merkmale mehr oder weniger
wesentlich sind für das Wesen der Assoziation an sich, und
wo demgemäß eine Grenze unter den Einheiten zu ziehen ist.

SUMMARY
The paper deals with the range of criteria used in the delimitation of the association. This problem occurs particularly in
the situation where phytocoenoses at the boundary of several associations need to be delimited and classified for the purpose
of vegetation mapping. In phytocoenological classification, the
boundary of the unit should be marked by discontinuities in the
same quantitative aspects of vegetation used to define the association. The criteria used are often scientifically important
features which, however are not necessarily also ontologically
essential features. This aspect applies to character species.
The relevance of species presence and the importance of other
features of the association are seen particularly in the structure and dynamics of the phytocoenose. The true decision on which
of the features of the association are greater or lesser importance for its existence and, therefore, for its delimitation is
enabled by investigations into these areas of study. The problem
is illustrated with example of the association C a r i c e -
t u m   g r a c i l i s Graebner et Hueck 1931.

LITERATUR

BALÁTOVÁ-TULÁČKOVÁ,Emilie   -1963- Zur Systematik der europäischen
     Phragmitetea.- Preslia 35: 118-122. Praha.
BRAUN-BLANQUET,J. -1964- Pflanzensoziologie.3.Aufl.- Wien-
     New-York.
KLIKA,J. -1933- Studien über die xerotherme Vegetation Mittel-
     europas. II. Xerotherme Gesellschaften in Böhmen.- Beih.
     bot.Zbl.50,Abt.2: 7o7-773. Dresden.
KOCH,W. -1926- Die Vegetationseinheiten der Linthebene.- Jb.
     St.Gall.naturwiss.Ges.61 (2): 1-46. St.Gallen.

H.DIERSCHKE:

Ich habe mich sehr gefreut, daß dieses Problem hier angeschnitten wurde und zwar aus dem Grund, weil man wirklich bei den
Großseggen-Gesellschaften, auch bei einigen anderen, oft im Zweifel sein kann, ob es sich um eine eigenen Assoziation, oder um
Fazies-Bildungen verschiedener Ausmaße handelt, wenn eine Art
dominiert. Ein anderes Beispiel gab schon, was verschiedentlich
anklang, das S c i r p o - P h r a g m i t e t u m, wo die Lage wohl noch nicht ganz klar ist. Ein anderes sind die Großseggen-Gesellschaften, bei denen dies ebenfalls der Fall ist. Ferner gibt es die verschiedenen Juncus-Bestände von J.effusus, J.
filiformis und auch J.acutiflorus, die verschiedentlich als Fazies auftreten, wo also eine Art dominiert und sicherlich für
den Bestandes-Aufbau entscheidend wirksam wird. Aber die Frage
ist nun, handelt es sich in diesem Fall um Assoziationen, oder
sind es nur Fazies innerhalb einer anderen Gesellschaft? In den
Wiesen ist das besonders problematisch, auch dort können die
Großseggen bestandsbildend auftreten, aber immer mit Wiesen-Ar-
ten zusammen. Und dort dürften eben andere Merkmale, wie hier
hervorgehoben wurde, oft eine wesentliche Rolle spielen, um zu
entscheiden, wie man diese Bestände soziologisch einordnet.

P.JAKUCS:

Die Abgrenzung einiger Einheiten ist sicher eins der schwierig-
sten Probleme der Phytosoziologie. Wir in Ungarn arbeiten immer
so, wie wir das jetzt gehört haben. Nicht nur die Kennarten,son-
dern die Ökologie und alle anderen Untersuchungen werden berücksichtigt. Ich weiß,daß es in einem Fall,wo die Abgrenzung nicht
klar ist, schwer zu machen ist. Dort kann man zusätzlich mit diesen Tatsachen arbeiten. Aber die pflanzensoziologische Systematik muß auf Kennarten beruhen.

R.TÜXEN:

Die Überlegungen von Fräulein BLAŽKOVÁ beleuchten klar den Wert der Erfahrung, der höher einzuschätzen ist als ein objektiv genanntes Schema.

ZUR SYNCHOROLOGIE DER DÜNENVEGETATION AN DER CHILENISCHEN KÜSTE
(Kurzfassung)

A. K o h l e r

An der Küste Chiles wurden Pflanzengesellschaften der offenen
Dünen zwischen 27° und 42° S analysiert, ihre wichtigsten geographischen Typen ermittelt und deren Verbreitung dargestellt.
Das Untersuchungsgebiet erstreckt sich über verschiedene Klima-
zonen und natürliche Vegetationsgebiete. Von N nach S nehmen
die ariden Monate von 12 auf 0 ab; dementsprechend folgen die
Vegetationsgebiete des Wüstenrandes (mit ephemeren und Zwerg-
strauchformationen), der Dornstrauch-Sukkulenten-Formation, des
Hartlaubgebietes und der südchilenischen Lorbeerwälder aufein-
ander.
Als Grundlage für die Abgrenzung der Dünengesellschaften sind
fünf Gruppen von Pflanzenarten mit jeweils ähnlichem Areal von
Bedeutung:
1. Arten mit ausgesprochen nordchilenischer Verbreitung mit
   Schwerpunkt im vollariden Küstenbereich (Beispiel: Nolana
   divaricata).
2. Arten mit relativ weiter klimatischer und geographischer Amplitude, deren Verbreitung zwischen (voll) aridem und voll-
   humiden Bereich liegt (Beispiel: Ambrosia chamissonis).
3. Arten mit Hauptverbreitung im ariden Bereich (Beispiel:
   Cristaria glaucophylla).
4. Arten, die aus dem humiden S in den semiariden Bereich einstrahlen, und im Gebiet von dessen Nordgrenze ausklingen
   (Beispiel: Euphorbia portulacoides).
5. Arten, die ebenfalls aus dem S nach N einstrahlen, die aber
   nicht oder nur in Vorposten die Grenze zwischen den semihu-
   den und dem semiariden Gebiet überschreiten (Beispiel:
   Panicum urvilleanum).
Als wichtigste geographische Typen wurden folgende vier Pflanzengesellschaften ermittelt (vgl.Tabelle):
A: N o l a n a   d i v a r i c a t a - T e t r a g o n i a   m a-
   r i t i m a -Gesellschaft
B: Gesellschaften der Ambrosia-Sekundärdünen (Assoziationsgrup-
   pe):
   1: C r i s t a r i o - A m b r o s i e t u m

2: P o o - A m b r o s i e t u m
3: C a r i c i - A m b r o s i e t u m

Die N o l a n a d i v a r i c a t a - T e t r a g o n i a
m a r i t i m a - Gesellschaft ist verbreitet im vollariden Küstenbereich am Südrand der Atacama-Wüste (ca.27-29 1/2^{O}S). Im
Grenzbereich vollarid/arid, am Rand des Zwergstrauchgürtels,
setzen nach S Dünengesellschaften ein, die durch Ambrosia chamissonis beherrscht werden. Ihren nördlichsten Flügel stellt das
C r i s t a r i o - A m b r o s i e t u m dar, das bis ca.
32 1/2^{O}S nach S ins mittelchilenische Hartlaubgebiet hineinreicht.
Ihm schließt sich südlich das P o o - A m b r o s i e t u m
an, die typische Sekundärdünengesellschaft des mittelchilenischen Hartlaubgebietes. Im Gegensatz zur Nordgrenze ist ihre
Südgrenze undeutlich; sie geht zwischen 34° und 36° S allmählich
in das C a r i c i - A m b r o s i e t u m Südchiles über. Dies
entspricht auch den Durchdringungen des mittel- und südchilenischen Vegetationskreises im Bereich der Küste.

LITERATUR

KOHLER,A. -1970- Geobotanische Untersuchungen an Küstendünen
     · Chiles zwischen 27 und 42 Grad südl.Breite.-
     Bot.Jb.9o (1/2): 55-200. Stuttgart.

SUMMARY

The plant communities of the secondary dunes of the Chilean
coast between 27° and 42° S. latitude have been investigated.
From the arid north up to the humid south the following communities were recorded and delimited:

A: N o l a n a d i v a r i c a t a - T e t r a g o n i a
        m a r i t i m a c o m m u n i t y in the perarid coa-
        stal region.

B 1:   C r i s t a r i o - A m b r o s i e t u m in the arid
        coastal region

  2:    P o o - A m b r o s i e t u m in the semi-arid coastal
        region.

  3:    C a r i c i - A m b r o s i e t u m ranging from the
        semihumid to the perhumid coastal regions.

Gliederung der Sekundärdünengesellschaften der chilenischen Küste zwischen
27 und 42 Grad S (vereinfacht)

	A	B 1	B 2	B 3
Arten der Nolana divaricata-Tetragonia maritima-Gesellschaft:				
Nolana divaricata	V	I	.	.
Tetragonia maritima	V	.	.	.
Alona carnosa	III	.	.	.
Ephedra breana	II	.	.	.
Skytanthus acutus	II	.	.	.
Suaeda fruticosa	III	.	.	.
Spergularia arbuscula	III	.	.	.
Frankenia glabrata	II	.	.	.
und andere				
Arten der Ambrosia chamissonis-Dünengesellschaften:				
Ambrosia chamissonis	.	V	V	V
Carpobrotus chilensis	.	V	V	V
Distichlis spicata	.	III	III	III
Kennarten des Cristario-Ambrosietum:				
Cristaria glaucophylla	.	IV	.	.
Senecio munnozii et aristianus et spec.	.	IV	.	.
Solanum heterantherum	.	III	.	.
Gemeinsame Arten von B 2 und 3:				
Poa affin lanuginosa	.	.	IV	IV
Polygonum sanguinaria	.	I	III	III
Hypochoeris toltensis et radicata	.	.	III	II
Rumex maricola	.	.	I	III
Euphorbia portulacoides	.	.	II	II
Calystegia soldanella	.	.	I	II
Alstroemeria ligtu var.	.	.	II	I
und andere				
Kenn- und Trennarten des Carici-Ambrosietum:				
Carex pumila var.urvillaei	.	.	.	V
Panicum urvilleanum	.	.	.	II
Fragaria chiloensis	.	.	.	I
Juncus lesueurii	.	.	.	I
Scirpus olneyi	.	.	.	I
Weitere Arten:				
Baccharis concava	.	II	IV	.
Margyricarpus setosus	.	I	II	II
Scirpus nodosus	.	I	III	(I)
Chamissonia dentata	.	.	I	I
Astragalus valparadisiensis et amatus	.	.	III	.
Sisyrhinchium striatum	.	.	II	.

	A		B	
		1	2	3
Oenothera stricta	.	.	I	.
Phacelia magellanica	.	.	I	.
Leuceria seneciodes	.	.	I	.
und andere				

A: Nolana divaricata-Tetragonia
          maritima-Gesellschaft

B 1:   Cristario-Ambrosietum
  2:      Poo-Ambrosietum
  3:      Carici-Ambrosietum

Die Untereinheiten der Gesellschaften sind nicht berücksichtigt.

V.WESTHOFF:
Eine Frage zur Namenbildung der Gesellschaften. Warum ist die
Assoziation B 2 als P o o - A m b r o s i e t u m bezeichnet
worden, obwohl in den Assoziationen B 2 und B 3 Poa gleich
stark vertreten ist, und nicht nach den guten Differenzialarten
Astragalus valparadisiensis und A. amatus? Wäre die Bezeichnung
"A s t r a g a l o - A m b r o s i e t u m" nicht angebrachter?

H.SUKOPP:
Hat der Neophyt Ambrosia chamissonis einheimische, chilenische
Arten verdrängt, oder hat sie der Vegetation neue Standorte er-
schlossen?

H.DOING:
In West-Frankreich setzen sich submediterrane Vegetationstypen
an der Küste von der Bretagne in immer schmaler werdenden Streifen weiter nach Norden fort als weiter ins Inland. In gleicher
Weise setzen sich in Australien tropische bzw. subtropische Vegetationstypen an der Küste viel weiter nach Süden fort als im
Inland. Man kann dies sehr leicht durch den direkten Einfluß
des Meeres erklären. In Australien z.B. ist das Küsten-Gebiet,
sogar in großen Teilen von Tasmanien, vollkommen frostfrei. Nur
wenige Kilometer landeinwärts gibt es Frost. Also ein sehr schmaler Streifen, der pflanzengeographisch-vegetationskundlich gesehen noch zu einem wärmeren Klima gehört. Vielleicht hat Herr
KOHLER in Chile ähnliche Beobachtungen machen können.

A.KOHLER:
Zunächst zu der Frage von Herrn Prof.WESTHOFF, warum ich die mit-
telchilenische Gesellschaft nach dieser Poa benannt habe. Ich habe mir darüber sehr viel Gedanken gemacht und die Nomenklatur-
Frage noch nicht abgeschlossen. Das ist also erst ein Vorschlag.
Ich habe sie deshalb nach Poa benannt, weil Poa in den meisten
Phasen an der gesamten Küstenstrecke in der Gesellschaft vertreten ist. Astragalus valparadisiensis ist ungefähr auf einer Küstenstrecke von 50 km Länge, wo die Gesellschaft allerdings ihr
Optimum hat, endemisch. Er ist ein Endemit, der im Vergleich zu
der gesamten mittelchilenischen Dünen-Vegetation räumlich aber
doch nur eine geringe Rolle spielt. Und deshalb habe ich diese
Gesellschaft vorerst nach dieser vorherrschen Poa benannt.

Auf die Frage von Herrn Dr. SUKOPP möchte ich sagen:es ist eine
sehr schwierige Frage, ob Ambrosia die alten Dünen-Gesellschaften zum Teil verdrängt hat oder nicht, oder ob sie sich eingefügt hat. Wir können das nur mit Sicherheit aus einem Gebiet sagen, nämlich aus dem Gebiet der Maule-Mündung, wo die einzige
Dünen-Arbeit von REICHE vorliegt. Ambrosia ist ja gegen Sturm-
fluten und gegen Salz-Einfluß relativ resistent. Bei einer Sturm-
flut werden zwar die Blätter abgeschlagen und verwelken, aber
die Sprosse haben einen Kork-Mantel, treiben wieder aus und ver-
mögen die Düne schnell wieder zu besiedeln. Dort, wo der Strand
sehr flach ist, formt Ambrosia unter Umständen sogar eine Zone,
die salzempfindliche Arten wie Calystegia soldanella und Euphorbia portulacoides u.a geschaffen haben könnte. Sie ist mit Sicherheit in die bestehenden Gesellschaften eingedrungen. Im Norden, also im Bereich des C r i s t a r i o - A m b r o s i e -
t u m, glaube ich, daß Ambrosia sich dort eingefügt hat in das
schon bestehende Dünen-Gefüge der dortigen Vegetation mit Cristaria glaucophylla und Solanum heterantherum, Chamaephyten,die
einen verhältnismäßig hohen Bauwert in der Dünen-Vegetation besitzen. Man kann diese Frage nicht pauschal beantworten. In dem
einen Fall wird Ambrosia eine neue Zone geschaffen haben, da
sie viel salz-resistenter ist als die einheimischen Dünen-Arten.
Im anderen Fall ist sie soweit eingebürgert, daß sie das ehemalige soziologische Gefüge verändert hat.
Zu der Homologisierung der Küstendünen-Vegetation und der zonalen Vegetation möchte ich Herrn DOING antworten. Ich habe hier
nur die offenen Dünen betrachtet. Schon der Verbreitungs-Cha-
rakter des nächsten Stadiums der Zwergstrauch-Gesellschaften
ist völlig anders. Und auch die Strand-Gesellschaft - das N o -
l a n e t u m p a r a d o x a e - umfaßt das Areal fast aller
drei Gesellschaften der Ambrosia-Dünen, also des C r i s t a -
r i o - A m b r o s i e t u m, des P o o - A m b r o s i e -
t u m und des C a r i c i - A m b r o s i e t u m. Im übrigen
sind auch meine Gesellschaften hier, z.B. das C r i s t a r i o
- A m b r o s i e t u m nicht allein auf die Gesellschaften des
Kleinen Nordens, also die Dornstrauch-Sukkulenten-Gesellschaf-
ten der Klasse G u t i e r r e z i o - T r i c h o c e r e -
t e a beschränkt,sondern diese typische Gesellschaft des Kleinen Nordens dringt noch weit nach S in das mittelchilenische Hart-
laub-Gebiet ein .Die Homologisierung der Areale einzelner Gesell-

schaften ist sehr schwierig.

F.G.SCHROEDER:

Zu der Frage von Herrn DOING: Es scheint mir zwischen Chile einerseits und den beiden anderen Gebieten andererseits ein wesentlicher Unterschied zu bestehen. Auch in Chile ist der wesentliche Faktor die Aridität, die sich auf die Klimax-Vegetation und die Dünen-Gesellschaften in gleicher Weise auswirkt, während die Wärme an der Küste relativ gleich ist. In Frankreich und Australien hingegen ändert sich hauptsächlich die Wärme. Nur im engsten Küsten-Bereich wirkt sich der ausgleichende Einfluß des Meeres aus.

R.TÜXEN:

Noch eine "Dünen-Bemerkung". Ich will sagen, daß auf der 2000 km langen Küsten-Ausdehnung von Japan im wesentlichen zwei Dünen-Gesellschaften vorkommen, zwei Ausbildungen der Carex kobomugi-Assoziation: eine nördliche und eine südliche. Die nördliche ist durch Elymus mollis differenziert, ganz ähnlich wie bei uns in Europa durch Elymus arenarius. Und im Süden treten dafür eine Reihe von südlichen Arten auf, die im Norden fehlen. (Näheres in OHBA,T., MIYAWAKI,A.& TÜXEN,R. -1972- Pflanzengesellschaften der japanischen Küsten-Dünen.- Vegetatio 26 (1-3): 3-141. The Hague).Auch hier dürfte es die Wärme sein, die ausschlaggebend ist für diese Differenzierung der Dünen von Südwest- gegen Nord-Japan.

BEOBACHTUNGEN ÜBER ABGRENZUNG UND GRENZEN DER
TOMENTHYPNUM-TUNDRA SÜDOST-SPITZBERGENS

G. P h i l i p p i

Im folgenden soll über Beobachtungen berichtet werden, die der
Verf. zusammen mit Dr. HOFMANN (Schweinfurt) auf der von Prof.
BÜDEL geleiteten Deutschen Spitzbergen-Expedition 1967 machen
konnte. Das Beobachtungsgebiet erstreckte sich längs des Free-
man-Sundes zwischen der Barents- und der Edge-Insel in Südost-
Spitzbergen und liegt ungefähr 78° 15' nördlicher Breite und
21°-22° Länge. Klimatisch ist das oft von Eis umschlossene Südost-Spitzbergen gegenüber West-Spitzbergen deutlich benachtei-
ligt, was in der Zusammensetzung der Pflanzenwelt deutlich zum
Ausdruck kommt. So fehlen auf Südost-Spitzbergen die in West-
Spitzbergen vorkommenden Betula nana oder Cassiope tetragona.
Andere Arten wie Stellaria crassipes oder Ranunculus hyperbore-
us wurden im Untersuchungsgebiet nur steril angetroffen.
Die auf Südost-Spitzbergen gerade in küstennahen Gebieten verbreitete und physiognomisch besonders auffallende Pflanzengesellschaft ist eine Moos-Tundra. Sie bestimmt den goldbraunen Farb-
ton weiter Gebiete. Tomenthypnum nitens (= Camptothecium n.)ist
die dominierende Art dieser Gesellschaft, des T o m e n t h y p-
n e t u m n i t e n t i s Hadač 1946). Daneben finden sich
Dicranum angustum, Aulacomnium palustre, A.turgidum, Philonotis
tomentella, Polytrichum alpinum und Hylocomium splendens, die
als lokale Kennarten des T o m e n t h y p n e t u m gelten
können. Phanerogamen spielen in der Gesellschaft meist keine
Rolle; von ihnen sind Ranunculus sulphureus, Salix polaris, Alopecurus alpinus und Saxifraga-Arten stet anzutreffen, können jedoch nicht zur Differenzierung der Tomenthypnum-Tundra herangezogen werden, da sie auch in anderen Pflanzengesellschaften Südost-Spitzbergens vorkommen.- Der Boden des T o m e n t h y p -
n e t u m ist basenreich, jedoch kalkarm (meist Basalt und Mer-
gel-Verwitterungsböden). Die Moos-Rasen bleiben während der Sommermonate gut durchfeuchtet, jedoch nicht überschwemmt und können gut betreten werden.
a. Zur Abgrenzung der Gesellschaft
An nassen, durchsickerten Stellen wird die Tomenthypnum-Tundra
von einer Gesellschaft mit Bryum cryophilum und Calliergon gigan-

teum abgelöst, an nassen, stagnierenden Stellen von einer solchen mit Paludella squarrosa, Meesea triquetra und Calliergon
sarmentosum. Die Grenzen dieser beiden Gesellschaften gegenüber
der Tomenthypnum-Tundra sind sehr scharf ausgebildet. Tomenthypnum wurde kaum zusammen mit Meesea triquetra angetroffen. Diese
bereits von der Arten-Kombination her deutlichen Grenzen wurden
oft durch das Klein-Relief noch verstärkt: die M e e s e a - Gesellschaft siedelt oft über den Eis-Keilen, die im Gelände als
langgestreckte, untereinander netzartig verknüpfte, flache Wan-
nen hervortreten, während die Tomenthypnum-Tundra die Felder
zwischen den Eis-Keilen einnimmt.
An trockenen Stellen schließt an die Tomenthypnum-Tundra eine
von Luzula confusa, L.arctica und Salix polaris beherrschte
Pflanzengesellschaft an, die prov. als S a l i x - L u z u l a
-Gesellschaft bezeichnet wird. Moose spielen am Aufbau der Gesellschaft keine Rolle. Jedoch kommen immer wieder Tomenthypnum
so wie die anderen für das T o m e n t h y p n e t u m genannten Moose vor. So erscheinen viele Bestände der S a l i x -
L u z u l a - Gesellschaft nur quantitativ, jedoch nicht qualitativ vom T o m e n t h y p n e t u m geschieden. Die Abgrenzung der Tomenthypnum-Tundra von der S a l i x - L u z u l a -
Gesellschaft war meist recht schwierig und an vielen Stellen unmöglich. Bei einer Kartierung 1 : 10 000 konnten nur Komplexe
mit der jeweils vorherrschenden der beiden Gesellschaften kartiert werden.
Die starke Durchdringung des T o m e n t h y p n e t u m und
der S a l i x - L u z u l a -Gesellschaft ist sicher konkurrenz-
bedingt: die Moose des T o m e n t h y p n e t u m können sich
in der S a l i x - L u z u l a - Gesellschaft gut halten, da
diese nur lückig den Boden bedeckt und Konkurrenten der Moose
fehlen. Das Fehlen der Moose des T o m e n t h y p n e t u m
in der B r y u m c r y o p h i l u m- und in der M e e s e a
t r i q u e t r a - Gesellschaft ist wohl darauf zurückzuführen,
daß sie an den nassen Standorten in den geschlossenen Moos-Rasen dieser beiden Gesellschaften nicht konkurrenzfähig sind.

Eine ähnlich unscharfe Grenze zwischen der S a l i x - L u z u-
l a - Gesellschaft und dem T o m e n t h y p n e t u m konnte
in Südost-Spitzbergen auch zwischen der S a l i x - L u z u l a
-Gesellschaft und den an trockeneren Stellen anschließenden
Schuttfluren mit Papaver dahlianum festgestellt werden: die

Kennarten des P a p a v e r e t u m d a h l i a n i Hadač,wie
Papaver dahlianum, Poa abbreviata und Festuca brachyphylla konnten immer wieder in der S a l i x - L u z u l a - Gesellschaft
angetroffen werden. Lückige Vegetations-Deckung und Fehlen von
Konkurrenten ermöglichen diesen Arten das Vorkommen in der S a-
l i x - L u z u l a - Gesellschaft. Trotz dieser häufigen Durchdringung waren beide Gesellschaften in ihren typischen Ausbildungen sehr gut voneinander geschieden.
In Mittel-Europa finden sich ähnlich unscharfe Grenzen zwischen
C a r i c i o n d a v a l l i a n a e- und C a r i c i o n
c a n e s c e n t i - f u s c a e - Gesellschaften einerseits,
(die ökologisch etwa den T o m e n t h y p n e t u m entsprechen),und den Gesellschaften des M o l i n i o n- und des V i-
o l i o n c a n i n a e-Verbandes andererseits: die für die
S c h e u c h z e r i o - C a r i c e t e a kennzeichnenden
Carex-Arten wie C.davalliana, C.pulicaris und C.fusca können
sich in den niederwüchsigen und meist etwas lückigen Gesellschaften des M o l i n i o n- und des V i o l i o n c a n i n a e-
Verbandes gut halten, offensichtlich,da Konkurrenten fehlen.

b.   Die Grenzen der Moos-Tundra
Das T o m e n t h y p n e t u m zeigt in Meeresnähe vollkommen geschlossene Moos-Rasen. Bereits in den Höhen zwischen 50
und 100 m ü.M. neigen diese zur Bulten-Bildung, besonders in
Dicranum angustum-reichen Beständen. Diese Bulten-Bildung ist
wohl auf Kryoturbation zurückzuführen, also auf eine durch den
Wechsel von Frieren und Auftauen bedingte Boden-Bewegung. In höheren Lagen - meist über 100 m Höhe - reißen die Moos-Decken
auf und machen Feinerde-Kernen Platz. Diese haben meist Durchmesser zwischen 0,5 und 1 m; die Moos-Decke bleibt dazwischen
nur noch als Netzwerk unterschiedlicher Stärke übrig. Diese Art
von Moos-Tundra - die Flecken-Tundra - ist die häufigste Ausbildung in Südost-Spitzbergen. Die Grenzen zwischen den Feinerde-
Auffrierungen und den Moos-Rasen des T o m e n t h y p n e t u m
bleiben überall sehr scharf und offensichtlich auch über längere Zeiten hin konstant. Zwar kann man vereinzelt die Wiederbe-
siedlung eines Feinerde-Kernes beobachten: hier stellten sich
dann Draba-Arten, Alopecurus alpinus, Anflüge von Stereocaulon
oder Einzelpflanzen von Aulacomnium turgidum ein. Wir konnten
jedoch nirgends beobachten, daß sich auf diesen Feinerde-Kernen
wieder ein T o m e n t h y p n e t u m ausbildete.

Mit zunehmender Meereshöhe vergrößern sich die Feinerde-Kerne,
während die Moos-Decke einen immer geringer werdenden Anteil
einnimmt. Die Obergrenze geschlossener Moos-Tundren wurde in
Südost-Spitzbergen bei ungefähr 300 m festgestellt. Fragmentarische, kleinflächige Moos-Tundren konnten jedoch noch in Höhen
bis 440 m (höchster besuchter Punkt des Untersuchungsgebietes)
festgestellt werden. Auch die meisten Kennarten des T o m e n t -
h y p n e t u m reichten bis in diese Höhen.- Die Obergrenze
geschlossener T o m e n t h y p n e t e n dürfte nicht allein
klimatisch bedingt sein, sondern auf die Empfindlichkeit von
Tomenthypnum nitens und anderer Kennarten des T o m e n t -
h y p n e t u m gegenüber Kryoturbations-Erscheinungen zurückzuführen sein. Infolge dieser Boden-Bewegungen kann Tomenthypnum keine Rasen aufbauen.
In oberen Lagen kommt auch Rhacomitrium canescens im T o m e n t-
h y p n e t u m vor. Das Nebeneinander von Rhacomitrium canescens und Tomenthypnum nitens enspricht - auf mitteleuropäische
Verhältnisse übertragen - einem Nebeneinander von C a r i c i -
o n d a v a l l i a n a e - und F e s t u c o - S e d e t a -
l i a - Gesellschaften, die in Mittel-Europa jedoch nicht im
Kontakt angetroffen wurden.

ZUSAMMENFASSUNG

Die Tomenthypnum-Tundra Südost-Spitzbergens läßt sich scharf gegenüber den Sicker-Fluren mit Bryum cryophilum und den Gesellschaften mit Paludella squarrosa und Meesea triquetra, dagegen
nur unscharf gegenüber den an trockenen Stellen anschließenden
Gesellschaften mit Luzula div.spec. und Salix polaris abgrenzen.
Auch zeigt das T o m e n t h y p n e t u m eine recht unscharfe Obergrenze, die in Südost-Spitzbergen bei ungefähr 300 m
liegt und weniger direkt klimatisch als durch Kryoturbations-Erscheinungen bedingt ist.

SUMMARY

T o m e n t h y p n e t u m -Tundra is sharply outlined against
communities with Bryum cryophilum in percolated areas and other
communities with Paludella squarrosa and Meesea triquetra,where-
as it is very difficult to fix the limits of T o m e n t h y p-
n e t u m in dry places with its S a l i x - L u z u l a -
communities. So the upper limit of T o m e n t h y p n e t u m

are not to be defined. In Southeastern Spitsbergen there may be
a border within the region of 300 m. This border is more dependent on Kryoturbation than on climatic facts.

I.ZONNEVELD:

Es gibt seit einigen Jahren in Holland eine Theorie (von MAARLE-
VELD) über die verhältnismäßig scharfe Grenze zwischen den LÖß-
und den Decksand-Ablagerungen. In Belgien gehen die Sand-Abla-
gerungen im Norden ziemlich schroff über den LÖß. Es ist schwer,
das rein klimatisch zu erklären, darum ist die Theorie entwik-
kelt worden, daß die Vegetation auf die Sedimentation Einfluß
gehabt hätte, und zwar folgendermaßen: Die etwas besseren Böden
trugen eine etwas höhere Vegetation, die den LÖß fixieren konnte. Auf dem ärmeren Untergrund konnten nur Lichenen oder ähnlich
anspruchslose Arten wachsen. Hier konnte der LÖß sich wohl ab-
lagern, wurde aber bald wieder weggeblasen. Nun ist meine Frage
an Herrn PHILIPPI: gibt es auf Spitzbergen auch diese aeolischen
Ablagerungen, und gibt es irgendwo einen ähnlichen Zusammenhang?
Es gibt ein ganz typisches Grenz-Problem das via die Vegetation
geologisch wird.

G.PHILIPPI:

Der Boden zeigt unter den Moos-Rasen keine Unterschiede gegenüber dem der Feinerde-Auffrierungen.

R.TÜXEN:

Ich habe mich viele Jahre hindurch mit fossilen oder rezenten
Boden-Profilen und ihren Beziehungen zur Vegetation im Altmorä-
nen-Gebiet von Nordwest-Deutschland beschäftigt. Sie wissen,
unser Q u e r c o - B e t u l e t u m oder unser F a g o -
Q u e r c e t u m, unsere C a l l u n a-Heiden,die feuchten
und trockenen, die E r i c a - Heiden, die F e s t u c a
o v i n a - Rasen und andere haben ihre sehr charakteristischen
Boden-Profile. Wir fanden dabei natürlich viele geologische Erscheinungen, u.a. Eis-Keile, die pseudo-morphosiert sind, indem
die ehemaligen Eis-Füllungen durch nachfallenden Sand von oben
ausgefüllt wurden, also mit allochthonem Material deutlich von
der autochthonen Schichtung sich abheben. Da ich versucht habe,
aus den Profilen und ihren Einzelheiten auf die Vegetation zu
schließen, die dazu gehört, oder dazu gehörte, wie z.B. bei

Alleröd-Profilen, wo wir P i n e t u m - Profile sehr klar erkennen können, möchte ich fragen, ob über den Eiskeil-Füllungen,
die wir bei uns allerdings sehr selten in größerer Breite finden, bei ihrer Bildung eine andere Vegetation gewesen sein kann
als daneben auf den Flächen, die nicht mit massivem Eis erfüllt
waren, wo allerdings der Boden auch gefroren war.

M.WRABER:
Ist es möglich, daß sich auf diesen Flächen auf derselben Strek-
ke diese Streifen und Flecken allmählich zusammenschließen und
eine größere Bedeckung erreichen, oder nicht? Sind es also Pionier-Stadien, die sich ausbreiten können, oder sind das Degra-
dations-Stadien?

G.PHILIPPI:
Über den Eis-Keilen selbst siedelt eine etwas feuchtigkeitslie-
bende Vegetation. Wenn wir ein feuchtes T o m e n t h y p n e -
t u m haben, haben wir über den Eis-Keilen eine M e e s e a -
Gesellschaft. Haben wir eine S a l i x - L u z u l a - Gesellschaft, dann wächst über den Eis-Keilen die nächst feuchtere
Gesellschaft, nämlich das T o m e n t h y p n e t u m. Allerdings kann man keine Beziehungen zwischen Eis-Keil und der Vegetation im allgemeinen herstellen, Es gibt Eis-Keile unter nassen T o m e n t h y p n e t e n, unter feuchten T o m e n t-
h y p n e t e n, und es gibt auch Eis-Keile in vegetationsfrei-
en Gebieten.
Diese Frage kann ich schwer beantworten. Wir waren zwei Monate
dort, und haben uns das auch immer wieder überlegt. Es scheint
so zu sein, daß dieses Verhältnis zwischen Feinerde-Kern und
zwischen Moos-Rasen sehr konstant ist. Wir haben nicht gesehen,
daß eben diese Feinerde-Kerne etwa seitlich vom T o m e n t -
h y p n e t u m wieder erobert werden, und wir haben den Eindruck, daß die Grenzen zwischen Feinerde-Kern und T o m e n t-
h y p n e t e n sehr konstant sind. Da wäre vielleicht noch zu
erwähnen, daß auch angenommen wurde, daß diese Feinerde-Kerne
eben sich unter dem heutigen Klima gar nicht mehr bilden, sondern daß es fossile Erscheinungen sind. Aber das ist nur eine
Hypothese, die geäußert wurde. Hier fehlen noch genaue Untersuchungen.

THE COLONIZATION OF PLANTS ON SURTSEY AND ITS VALUE IN
ECOLOGICAL STUDIES OF AN ISLAND-FLORA

St. F r i d r i k s s o n

Floras of various islands differ widely in origin, richness and
composition. Distant islands which may be hundreds of kilometers
from other land-masses are often rich in endemics, if they are
of ancient geological origin. Islands of more recent origin are
often poor in species and have but few endemics. Those which
have undergone recent nudation or have recently emerged from
under the ice or out of the sea are thus often ecologically
poorly advanced.
Phytogeographically such islands are however interesting subjects
and,due to their floristic simplicity,they are well suitable for basic
ecological investigation. The following is a brief account of a
few aspects concerning the flora of Iceland, including that of
the Westman Islands and the most recent member, the newborn volcanic island of Surtsey. There one may compare the condition of
smaller islands with those of the large ones; young floras with
old; primitive communities with the complex and more advanced
in succession.

ICELAND

In the postglacial flora of Iceland there are approximately 430
species of vascular plants. This assembly of species and communities should be and is in accordance with the climatic conditions of the country, which is sub-arctic to arctic but at the
same time oceanic with relatively mild winters, moist and windy.

It has been commonly accepted by most Icelandic botanists that
more than half the species in the indigenous flora have survided at
least the last glaciation. This element of the flora is considered to have reached the island by a landbridge connecting Iceland with the continent of Europe during the last interglacial
period, and then survived the last glaciation as small refugia
on nunataks and other icefree regions while the greater part of
the country was covered by glaciers. The remainder of the species in the present Icelandic flora must accordingly have reached Iceland in post-glacial times.
This theory is partially supported by the fact that a number of
species seem to have a centric distribution corresponding to re-

gions which, according to some geologists, might possibly have
remained icefree during the last glaciation.
An earlier opinion was that Iceland had been completely covered
with ice and that all species in the present flora would have
to have dispersed post-glacially over the ocean.
Botanically it will be difficult to prove either theory. But if
it is assumed that half the number of species in the present flora
could have dispersed to the island over the ocean, one might
therefore just as well argue that a greater part of or even all
the species could possibly have come in that way, across the
ocean.
For one thing the Icelandic flora is young in the country. No
true endemic plant species are encountered and the centric distribution does not necessarily have to be an indication of glacial survival. The species may have a centric distribution due
to a late sporadic immigration or because of certain environmental conditions at the centric locations, favouring certain elements of the flora. Thus, in the north central mountain ridge
alpine-arctic plants predominate. In the south central region
the temperate element is found and so on. These centers have de-
finate environmental boundaries.
Since the time of the Settlement of Iceland 1000 years ago when
the vegetation was in perfect balance with other natural forces,
there has been great deforestation accompanied by a decrease in
the total vegetation, followed by extensive erosion, for which
man with his domestic animals and activities, is for the most
part, responsible.
However parallel with a deplection of the vegetation resources
there is also a continuous invasion of plants in the denuded
areas and a colonization of plants on the vast areas of coastal
sands and gravel soils washed out by the glacial rivers. Vegetation invades the gravel and rocky nunataks or moraines exposed
by the retreating glaciers. New lava flows are gradually being
invaded by mosses, lichens and higher plants, and even new islands emerge from the ocean to be gradually colonized by flora
and fauna. The Icelandic vegetation may thus be considered quite
dynamic or unstable in nature. And there are numerous opportu-
nities for recording the pioneer invaders on the different substrata and for studying the successions and their boundaries. It
is possible, to compare vegetation on lava flows of different

ages, glacial river beds and the extreme temperature limits of
the freezing glaciers and almost boiling waters of the hot springs
or even volcanic islands, and to follow the succession step by
step to climax communities.
From the more economical point of view there is an increasing
necessity for the ecological studies of the Icelandic rangeland
which is partly becoming overstocked. An extensive plan for mapping pastureland is being carried out and an attempt is being
made to classify plant communities, trace their boundaries and
observe their limitations. An organized survey for investigating
growth rate of grasses in various districts and at different altitudes has also been made for a number of years. Seeding trials
are being carried out in the eroded areas. These are preliminary tests, a prerequsite to a more extensive reseeding program.
These seeding trials have made it possible to demonstrate that
in Iceland the upper limits of economic establishment of grassland extend approximately to an elevation of 600-700 m in the
southern part and in the central highlands of the country. Above
this limit there is no continuous cover of higher plants, and
these are virtually the upper boundaries of the grassland zone,
above which are barrens and fell-field types of vegetation. The
black basaltic gravel and sand substrata common in the open habitats are rather unfertile and low available phosphate, so that
both nitrogen and phosphate fertilizers have to be applied with
the grass seed if the seeding is to be successful. A natural colonization of these substrata is an extremely slow process. Areas have been protected for dozens of years without any marked
vegetal improvements.
The extensive draining program of the moorland which has been
carried out during the last decades has lowered the waterlevel
of large areas in the lowland, causing a gradual transformation
of the marshland vegetation from being dominated by sedges and
rushes into dry grassland communities. These changes, which are
of economic importance, have been studied and some environmental factors experimented with.
There is no significant grain production in Iceland and no fruit
production (except in greenhouses), and land is ploughed up mainly for use as grassfields. Green fodder crops my be raised on
the land for a couple of years, but eventually the land will be
turned into permanent hayfields or pastures. The general practi-

ce is to use imported seed such as that of Phleum pratensis and
Alopecurus pratensis for the sowing, but gradually the native
grasses invade and finally dominate the permanent grassfield.
At this stage Festuca rubra, Poa pratensis and Agrostis sp.will
be the dominating species in the more fertile field, whereas
Deschampsia caespitosa may occupy the fields of poorer drainage.

When a field is abandoned the grass community may be invaded by
forbs, rushes or woody species such as heath and willows, and
may, if protected, reach the climax of a low birch forest. The
upper limits of such a naturel forest are in the south approximately at the elevation of 450 m.
Man has drastically affected the original biota of Iceland. His
major influence on the vegetation was primarily the deforestation and other destructive measures or modifications by his
agricultural activities, which in extreme cases have led to
erosion, but man has also stimulated growth by cultivation and
accidental or deliberate plant introductions.
There were no herbivorous mammals in Iceland prior to the Settlement, and the vegetation may be considered to have been in
perfect balance. However, with the introduction of domestic animals there was a drastic change in the biotic harmony and an
upset of the equilibrium which the vegetation had reached with
the environment. The vulnerable borderlines of the vegetation
were soon spoiled and devastated by the influence of this new
biotic introduction, man and his animals. The total vegetation
cover decreased and the forest retreated before the advancing
grassland. Culture plants and weeds were introduced, some of
which spread rapidly and invaded the native communities. Ever
since there has been a sporadic introduction of species. Since
the turn of this century, for example, 190 new accidental introductions have been recorded. In addition, deliberate plant
introduction is made by the various gardeners, the Agricultural
Research Institute and the Forestry Department.
Many of the introduced varieties have become well established
and may be considered permanent members of the present flora,
whereas others are dependent on rather artifical man-made habitats and have only become temporary immigrants.
It is obvious that a greater number of species are able to sur-
vive in the country than those that were present in the original
native Icelandic flora.

The poverty of species is to some extent caused by the recent
geological origin of Iceland or recent emersion from the glacial dome, as well as by the edaphic and climatic conditions, the
latter being rather selective due to the sub-arctic location of
the country. But the oceanic barrier and the distance from the
source of available species must also have affected greatly the
quantity of species in the native flora. The effect of the oceanic barrier in general can to some extent be studied on islands
which have previously been devoid of life, such as the islands
of Krakatoa and Bikini, and now recently on the volcanic island,
Surtsey, off the coast of Iceland.

Surtsey is a new member of a group of islands named Vestmanney-
jar (Westman Islands) off the southern coast of Iceland. All
the islands are of volcanic origin with palagonite tuff or basaltic lava. Their soil is loessy, often rich in organic matter
due to extensive bird dropping.

In comparison with the mainland their climate is relatively warm
and moist. The precipitation is the seventh highest and the mean
temperature the third highest in Iceland. The vegetation is correspondingly more European than Arctic in nature, ranging from
a few species of vascular plants on the smaller skerries to 30
species of vascular plants on the larger islands and to 150 species on Heimaey, the biggest member. On these islands the environment is quite selective, and topographic conditions limited
and comparatively stable. Their vegetation has also become sta-
bilized in four major types of climax communities in their re-
spective habitats. These have to some extent formed zones exten-
ding from the splash zone at the lower level and they differ in
composition due to distance from the ocean and effects of birds
which colonized the island. Surtsey is the southernmost and the
most recent member of these islands.

The eruption which gradually built up the island started on 14th
November, 1963. In the spring of 1964 the first biological observations were made and detailed studies have continued ever
since at fairly regular intervals. In order to encourage and
organize research a Surtsey Research Society was established.
This is supported by the Islandic government and has also received foreign grants. The research of the terrestrial biology
was financially supported by the U.S.Atomic Energy Commission,
Biology Branch, under contract number (30-1)-3549.

Extensive geological observations were carried out during the
first phases of the eruption, and geologists have ever since
followed closely all changes in the volcanic activities which
continued until June 4th, 1967. By then the island had attained
a size of 2.5 sq.km, half of which was covered with lava, the
remainder being mostly volcanic ash and cinder. The island rose
to a height of 180 m above sea level, but has since eroded a
great deal and undergone marked topographical changes.
When discussing the subject of colonization of plants on the
island of Surtsey we are up against a number of interacting
factors, such as 1) the location of the territory, 2) the source of available species, 3) the means of dispersal, and 4) growth
conditions on the island.
As Surtsey is an island in the North Atlantic Ocean, it already
occupies a special position not encountered elsewhere. This primarily limits a great number of possible colonizers of the island to arctic and subarctic plants. Secondly, all the invading
plants have to be transported over an ocean barrier, which again
obviously excludes a vast number of species that might otherwise have a chance to colonize the island, for instance those
species that depend exclusively on dispersal by land animals.
The colonization of Surtsey is in that respect not comparable
to isolated areas on land or areas with similar substrata on
the mainland, such as new lava flows of barren sand stretches
to which plants can be carried attached to the walking animals,
or to which they might be blown by rolling over a land surface,
or simply to which they could spread by vegetative growth. The
colonizing plants on Surtsey have to overcome other obstacles
and a more selective barrier.
The amount of plant material dispersing to Surtsey should be
roughly in proportion to the distance of source of available
plants. This might, however, be biased by some special and local conditions, such as strong air and ocean currents and the
selective long distance dispersal by migratory birds. Surtsey
is one of a group of islands, the nearest being the skerry of
Geirfuglasker, at a distance of 5,5 km. It is reasonable to
consider this and the other outer islands to be the most likely
habitats from which plants could colonize Surtsey. In our study
we have thus conducted a thorough examination of the vegetation
of these individual islands. And as we have found their vegeta-

tion to vary in number of species, the survey enables us to determine the minimal distance a species has to be transported in
order to reach Surtsey. This again reflects the spreading poten-
tiality of various species.
A dispersal from the mainland of Iceland is also extremely probable, where the vegetation of the southern part has an obvious
advantage over the more arctic element of the interior and the
northern districts. Finally there is the possibility of a long
distance dispersal. This would most likely be from other European countries, although America and other sources should not
be excluded.
Although the study of the colonization of Surtsey is of great
interest per se, it could thus also furnish information on the
long distance dispersal of plants in the North Atlantic basin
and give valuable answers to many a riddle in the great confusion and dilemma of the argument between those who believe that
all life was completely eradicated during the last Glaciation,
or the tabula rasa theory, and those, who oppose the theory and
believe that life in Scandinavia and Iceland survived the last
Glaciation on ice-free centers or nunataks.
So far the species of vascular plants that have colonized Surtsey have all been coastal plants, which have dispersed by ocean
as drifting seeds. These are Cakile edentula, Honckenya peploides, Elymus arenarius and Mertensia maritima.
The discovery of these individuals on Surtsey shows that living
seed of at least coastal plants can disperse over such oceanic
distances as those between Surtsey and the nearest colonies,
which are at least 20 km away. Viability tests of seed after
immersion in sea water also indicate that seed of various other
Icelandic species may possibly disperse over still greater distances.
It was considered probable that dispersal by ocean would be the
most likely way of plant invasion on Surtsey. However, airborne
seeds such as Senecio vulgaris and various seeds carried by
birds, have also been recorded in the island, thus proving that
other means of dispersal are possible.
The second phase of the problem is the growth condition on an
island for the newly arrived plant part. i.e. the limitation ex-
tended by the edaphic and climatic factors. These are being
thoroughly studied on Surtsey and compared with conditions on

other islands in the Westman Island group, as well as with selected areas on the mainland of Iceland. The necessary prerequisite for an effective colonization for pioneer plant invaders
of a nude area is the presence of favourable substrata. Characteristic for the substrata of Surtsey is its volcanic origin
and its extremely low water retention capacity. The substrata
is mainly of three types: The lava, the tephra and the secondary beach substrata.

Lava

Under normal circumstances the formation of soil on Icelandic
lava is an extremely slow process.

On the east and the south side of Surtsey there is lava of varying age, the older part being covered by a thin layer of volcanic ash and cinder which has filled hollows and crevices.
This cinder substrata may accelerate the formation of soil on
the Surtsey lava, but these dry and bare substrata without any
organic material are extremely hostile habitats, and it will
presumably take some time for biota to accomplish there a successful establishment. The most likely pioneers of the lava will
presumably be mosses and lichens forming the primary lithosere.

TEPHRA

The central part of Surtsey is largely formed of loose tephra
which will later harden into tuff. This substrata was formed
during the first phase of the eruption and was continously mixed
with ocean water. When tested for soluble mineral content, it
still had a high salt content (2850 mg/l of Cl in the leachate).
This substrata is still practically sterile.
Sea lime grass may possible be the first to establish itself in
the dry sand forming the primary psammosere. However, only time
will show in this respect.

THE BEACH

On the beach of Surtsey, however, conditions favourable for
plant growth can develop rapidly. Organic matter is washed ashore, e.g. seaweed and driftwood, as well as the remains of various marine organisms. This organic matter mixes with the washed
out volcanic cinders, and gradually decomposes. Large numbers
of seabirds frequent the shore, supplementing the organic content of the sand with their excreta. This initial step in the
formation of soil on the beach, however, is not stable. The

heavy winter seas cause considerable disturbances to such an extent that organic material may be churned up.The most soluble nutrients are washed away, while other organic matter is buried by the sand and shingle. However, though a great deal of organic matter is thus lost, the sea compensates with a fresh supply of drift material. This cycle is repeated annually, so there is no question of a really stable formation of soil on the shore. However, some organic matter may be carried up the beach beyond the highest tide mark and thus supply the basis for a more consistent soil formation with an inprovement in water retention capacity. Such conditions now exist on the northern shore of Surtsey, where the pioneer invaders of vascular plants are growing.

This island is being thoroughly mapped, and we have marked a grit over the total area. On the different substrates we have permanent quadrates of one hectare each, which again are divided up and will be watched closer . So far we have registered and plotted every single vascular plant, which has started to grow on Surtsey.

The first to arrive was Cakile edentula. It was in 1965,in the second summer of the existence of the island,that over 20 individuals were found growing on the coast in the nitrogen rich organic substrate. The following year Elymus arenarius was added. Then in the summer 1967 Honckenia peploides showed up and finally Mertensia maritima. These species were making up a well represented coastal community. The Cakile plants flowered in 1967 and set hundreds of seeds, and will thus be adding native inhabitants to this new community of pioneers.

It will take a long time for these different habitats to be completely colonized, but it is an intriguing task for ecologists to follow these events, and to compare them with the climax communities of the older Westman Islands and the communities of various successional orders on the mainland of Iceland.

SUMMARY

The volcanic island Surtsey which emerged November 14th 1963 off the coast of Iceland has offered ecologists a unique opportunity for studying how biota can disperse over long distances and colonize an island, which was previously completely void of life. The first steps in the colonization have been carefully

recorded and will be compared with the vegetation of their neigh-
bouring islands in the Westmanneyjar group as well as isolated
areas on the mainland of Iceland.
The substrate of the island is lava and tephra extremely porous
in nature. Already four coastal species of vascular plants have
dispersed to the island and started growth on the beach, but it
will take a long time until the vegetation reaches the same
climaxes as are present on the older volcanic islands in the
group.
Iceland has immensely greater variation in topography. However,
there are vast areas of barrens which in many respects are comparable with Surtsey, such as moraines which have recently been
freed of the retreating glaciers or recently exposed nunataks,
lava flows of various ages, glacial river beds and eroded sands
and gravel substrates. The study of these areas is ecologically
intersting and of economic value for supplying information for
range management and cultivation.
See also SURTSEY BUTTERWORTHS 1974.

Während des Empfangs-Abends, den wie üblich der Kreis Grafschaft
Schaumburg, die Stadt Rinteln und die Glashütte Stövesandt für
die Symposion-Teilnehmer und geladene Gäste gaben, zeigte und
erläuterte Dr. STURLA FRIDRIKSSON, Reykjavik einen Farbfilm von
der Entstehung der Vulkan-Insel Surtsey.

V.WESTHOFF:
Thank you very much for your intersting lecture. Can you tell
us in some words what you have done to avoid that any infection
from human or animal sources brought into the island?

St.FRIDRIKSSON:
This is a very delicate question. It is very difficult to observe the island without bringing in the human element. We have
to come there ourselves, and as we step in, we are intruders into the biotic element of the island. But we have taken some measures to prevent any tourism although this was very difficult.
These volcanic eruptions were quite a spectular event and it
was difficult to hold traffic out. The tourist bureaus even wan-
ted to sell the admittance to the island, to take tourists over
to Surtsey, but we could keep them away. We may have been a

little careless in the first year, but the small islands, spread
ash over Surtsey and helped us in sterilizing the island again. So
actually it was completely sterile from the second eruption, in
1965, when these first Cakile-plants were wiped out and it was
actually covered by a thick layer of ash again. But we have prevented the tourists and our Society has the legal right to al-
low certain scientists to come to the island and to refuse
others. It is protected, and we are careful ourselves when we
go to the island. There are strict rules in not having our meals
all over the island. We have a little hut there on the island,
and we were very precautious when we built the hut by spraying
the timber that was used to construct the hut, so we did not
bring any insect to the island. And we try to clean our clothes,
especially our shoes - behind the heels, the common place where
bacteria and seeds could be stuck-, the sleeves of the trousers,
clean them out, etc. And then a number of us fall into the ocean when they go ashore - the landing is quite rough - so nature
cleans us by dipping us in. We have tried to keep the island
clean, but do not forget that the island is in the ocean and
there are fishing boats around us, sometimes. I have walked on
the shore and found an occasional cucumber or orange. But I
suppose that we have to admit that man is to stay there and
that he is also one of the biotic factors.

V.WESTHOFF:
Thank you very much. We see that you are very careful to prevent
any unexpected bringing in of for foreign material. I think this
is a very good example of what we should do in such an investigation, to avoid some of the mistakes which have been made in
the research of Krakatoa, so long time ago.

G.H.SCHWABE:
Ich denke, wir sind uns alle einig darüber, daß Surtsey eine
einmalige Kostbarkeit für uns gerade vom ökologischen Gesichtspunkt her ist. Ich glaube nun, daß man diese Chance auch experimentell ausnutzen sollte und zwar aus folgendem Grunde: Das
juvenile Substrat, aus dem Surtsey aufgebaut ist, ist sowohl
aus chemischen wie aus physikalischen Gründen in mehrerer Hinsicht ausgesprochen lebensfeindlich. Infolgedessen wird viel
von den Keimen, die dort ankommen, nur deswegen sich nicht ent-

wickeln können, weil das Substrat nicht taugt. Infolgedessen
scheint mir erwägenswert, daß man auf Surtsey Substrate anbringt,
die andere Vergleichbedingungen schaffen. Ich denke z.B. an
fraktioniert - dampfsterilisierte Bodenflächen, die man z.B. von
Heimaey,also von den Westmänner-Inseln, herüberbringt und geschlossen einsetzt. Damit ergibt sich die Möglichkeit des Ver-
gleiches.

Eine zweite Sache, die mir erwägenswert erscheint, sehe ich in
der Anlage von Erd-Kegeln. Ich habe selbst einmal mit derartigen Installationen gearbeitet. Die Boden-Entwicklung hängt ja
in hohem Grade von der Exposition ab. Es wird von Interesse
sein, in einem zur Zeit weitgehend ungeschichteten und unzonier-
ten Material kontrollierbare Expositions-Bedingungen zu schaffen. Wir haben das so gemacht, daß wir Erd-Kegel von 1,5 m Höhe
aufgeschüttet haben, eine Stange angebracht und dann Kegel aufgeschüttet, die nach verschiedenen Seiten exponiert, dann in
Sektoren geerntet, bzw. kontrolliert werden. Wir müssen ja damit rechnen, daß sowohl Regen-, Wind- und Sonnen-Exposition
einen mehr oder weniger intensiven Einfluß auf den Vorgang sowohl der Zersetzung wie in der Mikro-Besiedlung haben werden.
Ich denke bei diesen beiden zur Diskussion gestellten Vorschlägen natürlich in erster Linie algologisch bzw. mikrobiologisch,
um die Voraussetzungen für die Etablierung einer höheren Vegetation verfolgen zu können.

St.FRIDRIKSSON:

We have sometimes discussed whether or not to do any experiments
on Surtsey, but we have decided that the idea will not be taken
up to bring anything over to the island that could affect its
highly unique nature. We are trying to harm it as little as
possible and we are not doing any experiments on the island.We
might bring in soil and then some might be interested in seeing
whether a certain species wouldn't grow, whether tomatoes would'
t grow, whether we couldn't raise this or that, the entomolo-
gists might be interested in doing experiments with banana flies,
and the ecologists in bringing in other animals. So we have decided to do the opposite and to take the substrate from Surtsey
over to the mainland and tried to grow different plants in the
experimental stations on the mainland.

St.FRIDRIKSSON:

We have collected soil samples ever since the biologists got to
the island. We have made a transect from the shore-line up to
the top and over from the North towards the South, and regular
microbial counts are being taken from these soil samples. There
might be organisms that are difficult to grow on regular agar
substrates, as some sulphur and iron bacteria. We have stored
these samples and some of them might be investigated at a later
date. Actually our German collegue Professor SCHWARTZ is taking
care of that part of the study; he has actually tried to grow
some of these bacteria.

E.VAN DER MAAREL:

I can appreciate the plans of Dr.SCHWABE, and I have thought
for a while whether we could organize such investigations to
study these developments of micro-communities on some of the
many artifical islands in the North Sea of the oil companies.
But please, let us keep this unique opportunity of Surtsey free
from any interference by people with other interests. I should
like to ask Dr.FRIDRIKSSON if he is sure that the island will
be safe in this respect. Should it be of some help when we, as
the International Association of Plant Geography and Ecology,
should address to the Icelandic Government that the island Surtsey should be kept free from any interference except science.

St.FRIDRIKSSON:

It is true that for a while it was rather difficult for us to
keep the island from being visited by large groups of people,
but now, after the volcanic activity has ceased the pressure is
not as great as it was before. Moreover there are great difficulties in getting to the island, we have to go by rubber boat
and not everybody likes to take the risk of going there by this
way. But the island is situated in that part of the ocean which
has very good fishing grounds and some day somebody might think
whether the island was a good place to establish there a canning fac-
tory. However, there is no water on the island, so this might cause
some difficulties in all that. But I would appreciate greatly,
and would be grateful to have an assembly like yours to support
us in keeping the island secure and recognised as a centre of
biological interest. I welcome that highly.

G.H.SCHWABE:

Dann bleibt nur noch das Erdkegel-Experiment. Denn da bringe ich
ja nichts anderes auf die Insel, sondern schaffe eben nur ver-
gleichbare Bedingungen. Ich kann dieselben Kegel auf Island
selbst an irgendeiner geeigneten Stelle aufstellen und dann beobachten, wie unter verschiedenen Expositionen die Boden-Entwicklung vor sich geht. Ich bin davon überzeugt, daß eine Reihe
von solchen Kegeln wesentliche Aufschlüsse bringen können über
die erste Boden-Bildung.

V.WESTHOFF:

Ich glaube, Herr Dr. SCHWABE hat ganz recht, wenn er sagt, die
Insel ist eine einzigartige Kostbarkeit und damit sollen wir
sehr vorsichtig sein. Wissenschaftlich hat Dr.SCHWABE völlig
recht, man kann hier ein Experiment ausführen, aber politisch
sieht die Sache anders aus. Die Frage ist eigentlich eine poli-
tische Frage. Wenn man auf so einer Insel wie Surtsey einen
Versuch anstellt, wie sehr er auch begründet sei, wie vorsichtig er gemacht sei, und wie wenig wissenschaftlich dagegen ein-
zuwenden sei, dann kann später jeder sich darauf berufen und
sagen: es hat einer doch schon etwas gemacht, nun will ich To-
maten pflanzen oder Bananen-Fliegen einschleppen und ich will
'mal sehen, was geschieht, wenn ich einen Käfer oder sonst ein
"tolles" Tier einführe. Und wenn die Sache aus der Hand läuft,
gibt das immer Probleme, denn die Welt wird ja nicht von Wis-
senschaftlern beherrscht, sondern, sagen wir von - Laien. Dann
ist es ja immer so, daß man nicht weiß, was geschieht. Und man
ist stärker, wenn man sagen kann, es darf überhaupt nichts geschehen, und es ist auch nichts geschehen. Daher glaube ich,
daß die Isländer recht haben, wenn sie das nicht gestatten.

R.TÜXEN:

Die meisten Küsten-Gesellschaften bestehen aus einer oder ganz
wenigen Arten: S a l i c o r n i e t u m, die S u a e d a- oder
S p a r t i n a - Gesellschaften und andere.Nur dort, wo organi-
sches Material hinkommt, in den therophytischen C a k i l e -
t e a und in den ausdauernden Spülsaum-Gesellschaften, die wir
ja erst seit kurzem richtig verstanden haben, ist die Artenzahl
größer. Ich denke an die H o n c k e n y a - E l y m u s - Gesellschaften mit ihren Begleit-Pflanzen,im Norden Mertensia

z.B. und manchen anderen, die weltweit um das boreale Küsten-
Gebiet der ganzen Holarktis herum reichen. Es ist mir von höch-
stem Interesse gewesen, als ich die ersten Publikationen über
Surtsey bekommen habe von Herrn FRIDRIKSSON und anderen, zu sehen, daß diese relativ artenreichen Küsten-Saumgesellschaften
auch diejenigen sind, die sich zuerst einstellen, nicht etwa
Puccinellia, nicht etwa Salicornia oder Spartina. Dafür ist es
auch zu nördlich dort. Die nitrophile Cakile war die erste, und
dann kam Elymus und dann Honckenya. Also die nitrophilen Initial-
und die nitrophilen Dauergesellschaften sind die artenreichsten
und diejenigen, die sich am schnellsten einstellen können. Das
ist für mich soziologisch ein starker Eindruck von der Besiedlung dieser wunderbaren Insel.

J.J.MOORE:
Would you please tell us if any of the work has been published,
and where you intend publishing?

St.FRIDRIKSSON:
The Surtsey Research Society publishes regular annual reports.
They are called Surtsey Research Progress Reports. No.I,II,III
and IV have been issued as mimeographed papers. In addition a
few papers on the pioneer plants have been published by me in
an Iceland Journal "Natturufræ dingurinn"Reykjavik. The algolo-
gical work has been reported on by S.JOHNSSON in a French botanical journal. The entomological research is reported on by K.
LINDROTH in "Naturen" 1967, and you may find a paper by me in
"The New Scientist" April issue 1968.

A.APINIS:
I would like to ask Dr.FRIDRIKSSON especially about the primary
succession. We had this fine occasion on Krakatoa, but actually
at that time the soil biology was at a different approach. The
geographical and floristic approach was very nicely done, but
they completely missed this soil developmental stage at very
early dates. The first mycrologist who visited Krakatoa was
BOEDIJN in 1933/1934, that is 50 years after the eruption. Is
there any specific soil primary succession research planned or
intended to be done in this fine island?

DIE NORD- UND SÜD-GRENZE DER BUCHENWÄLDER IN DER OSTASIATISCHEN INSELKETTE

S u z u k i - T o k i o

Vergleichende Untersuchungen über die Buchenwälder in der Nord-
Hemisphäre sind pflanzensoziologisch sehr wichtig, weil sie den
Normal-Vegetationsgürtel auf allen Kontinenten bilden. Der Nor-
mal-Gürtel soll als Maßstab, um die Zonation jeden Kontinents
miteinander zu vergleichen, nutzbar sein. Dafür muß er allen
Kontinenten gemeinsam sein, und es ist zu wünschen, daß er um
die Mitte der horizontalen sowie auch der vertikalen Zonation
liegt.
Die borealen Nadel-Wälder der V a c c i n i o - P i c e e t e a
sind auf dem eurasischen und dem nordamerikanischen Kontinent
gleich, aber sie liegen leider an der Nord- und Höhengrenze der
Wald-Vegetation. Mit ihrem Maßstab kann man nicht die südlich-
sten und die untersten Gürtel genau vergleichen, denn die Ein-
beziehung dazwischen liegender Gürtel würden die entsprechenden Gesellschaften schwer verfolgen lassen. Die Buchenwälder
sind dafür geeignet, denn sie liegen in der temperierten Region
eines jeden Kontinents, und es ist leichter, die entsprechenden
Beziehungen zu beiden Seiten aus ihnen heraus zu verfolgen.
Die Buchenwälder bilden in der Nord-Hemisphäre eine floristisch
einheitliche Gesellschaftsgruppe durch das gesellige Vorkommen
von zwei Gattungen, Fagus und Acer (16). In Europa aber finden
sich viele Buchen-Bestände ohne Ahorn, obwohl Acer pseudo-pla-
tanus und A.platanoides oft mit der Buche wachsen. Das ist meines Erachtens die Ausnahme. Die ost- und westwärts streichenden
Gebirge: Alpen, Pyrenäen und Karpaten haben vielleicht diese
beiden geselligen Baum-Gattungen während der Eiszeit in den
Floren-Verschiebungen verloren gehen lassen. In Ost-Asien und
Nord-Amerika liegen aber die Haupt-Gebirgsketten in Nord-Süd-
Richtung, sie haben somit nicht die Floren-Verschiebung verhin-
dert, wie es in Europa geschehen ist.
Die Buchenwälder in Ost-Asien sind durch ihren Zwergbambus-Un-
terwuchs ausgezeichnet. Sasa und Sasamorpha sind endemische
Gattungen auf den ostasiatischen Inseln. Jedoch sind sie nicht
nur in den Buchenwäldern zuhause. Sie bilden oft eigene Strauch-
Bestände und wachsen auch unter Nadelbäumen, wie Abies.
In Japan und Formosa haben wir drei Fagus-Arten: Fagus crenata

Bl., F.sieboldii und F.hayatae Palib. Die erste ist die wichtigste Art und dominiert in den meisten Buchenwäldern Japans. Sie
reicht vom Südteil von Hokkaidô südwärts bis Südkyûsýû. Der Insel Yaku aber fehlt nicht nur diese Art, sondern selbst die Gattung. Fagus sieboldii ist etwas wärmeliebender als die erste
Art, und sie fehlt auf Hokkaidô und auch in den Gebieten an der
Seite des Japanischen Meeres. In dem unteren Teile der montanen
Stufe an der pazifischen Seite gedeiht F.sieboldii oft sehr üppig. Sie wächst oft in reinen Beständen, aber in den meisten
Fällen gemischt mit F.crenata und anderen Bäumen, wie Tsuga
sieboldii und Abies firma (19,22).
Fagus hayatae kommt nur in Nord-Formosa vor, obwohl sie auch in
Japan als Fossil verbreitet ist (5).
Der Verfasser untersuchte die F a g u s c r e n a t a - Wälder in Hokkaidô (11,18), und die F a g u s h a y a t a e -
Wälder in Formosa (13). Die obere und die untere Grenze der
F a g u s c r e n a t a - Wälder wird hier kurz behandelt.

Nordgrenze
Die Nordgrenze der Fagus crenata-Bestände liegt an der Kuromatunai-Talzone, die die Osima-Halbinsel vom Hauptteile von Hokkaidô abtrennt. M.TATEWAKI (17) schreibt über die nördlichsten
Bestände, unter denen der Utasai-Schutzwald am typischsten ist.
Aus der floristischen Zusammensetzung ist zu entnehmen, daß
dieser Bestand ohne Zweifel mit dem S a s o - F a g e t u m
c r e n a t a e identisch ist.
Fagus crenata besetzt 90% der Baumschicht, Quercus crispula,
Kalopanax septemlobum und Acer japonicum wachsen dazwischen.
Sasa kurilensis und S.paniculata dominieren in der Strauchschicht, auch Laub-Sträucher wie Viburnum furcatum, Daphniphyllum humile und Skimmia japonica var. repens sind auch vorhanden.
Im Vergleich mit den typischen Beständen vom Hiyama-Gebiet, wo
die Buchen-Bestände von Hokkaidô sich sehr üppig entwickeln und
vom Verfasser pflanzensoziologisch untersucht worden sind, kann
man keine Abschwächung der charakteristischen Arten-Kombination
finden. Nur Betula ermani, ein Begleiter der Nadelwälder, ist
zuweilen anzutreffen. Nördlich von der Kuromatunai-Talzone verschwinden die Buchenwälder plötzlich.
Die Osima-Halbinsel Japans kann geotektonisch als Fortsetzung
der Hauptinsel Japans angesehen werden. Man kann somit die erd

geschichtliche Ursache für die Nord-Grenze der Buchenwälder unseres Landes nicht verneinen.

Die Grenze ist jedoch auch klimatisch bedingt. R.KATO hat die
KIRAS-Indices verändert (KIRAS originaler (4) Nullpunkt für die
vegetativen Erscheinungen ist $5^{o}C$, nach KATO jedoch $0^{o}C$. Die
Isothermen vom Wärme-Index bei 110° stimmt sehr gut mit der
Südgrenze von Abies sachalinensis und der Nord-Grenze von Fagus
crenata überein. Tatsächlich dürfte der Wettbewerb zwischen Buche und Tanne den Verlauf der Grenz-Linie bestimmen.

Die Winter- und Frühlings-Kälte bedingen auch die Buchen-Nord-
Grenze, denn die Isotherme des Kälte-Index -20° stimmt mit der
Grenz-Linie überein. Der Spätfrost wirkt besonders schädlich
auf die Buche. Forst-Botaniker verweisen auf die Tatsache, daß
einige sommergrüne Baum-Arten, wie Quercus crispula, Q.dentata,
Tilia japonica, Cercidiphyllum japonicum und Ulmus laciniata,
die Grenz-Linie nach Norden überschreiten. Es ist pflanzensoziologisch sehr erwähnenswert, daß die subarktischen Nadel-Wälder,
so von japanischen Botanikern angesprochen, viele Kenn-Arten der
Buchenwälder enthalten.

Nach meinen eigenen Aufnahmen im Hauptteil von Hokkaidô sind typische Nadelwald-Kennarten in Hokkaidô ziemlich selten. Es sind
die Dominanten Abbies sachalinensis, Picea jezoensis und P.gleh-
ni, wenn man die Tabelle der Nadelwälder von Hokkaidô mit denen
der subalpinen Nadelwälder von Mittel-Japan vergleicht.

1. Sasa kurilensis (Sasa cerna umfassend), Sasamorpha purpura-
scens, Sasa paniculata und Sasa nipponica sind die wichtigsten
Arten der Strauchschicht in den F a g u s   c r e n a t a - Wäldern; sie kommen aber auch in den Nadelwäldern in Hokkaidô, sowie in der subalpinen Region Mittel-Japans vor. Die Zwergbambus-
Arten sind somit als Differenzial-Arten gegen andere Buchenwälder der nördlichen Hemisphäre aufzufassen, obgleich sie auch in
Nadelwälder übergreifen.

2. Viburnum furcatum findet sich in beiden Buchen-Gesellschaf-
ten sowohl auf der pazifischen als auch auf der Seite des Japanischen Meeres, aber auch oft in den Nadelwäldern von Hokkaidô.

3. Immergrüne Laub-Sträucher von kriechendem Habitus, wie Daphniphyllum humile und   Skimmia japonica var. repens sind charakteristisch in den japanischen Buchenwäldern an der Seite vom
Japanischen Meer, aber sie wachsen auch teilweise in den Nadelwäldern von Hokkaidô (2).

4. Die eigentümliche Ahorn-Art Acer japonicum in den Buchenwäldern des Japanischen Meer-Gebietes verhält sich wie die oben
genannten immergrünen Sträucher.

Diese Tatsachen zeigen die Verwandtschaft der Nadelholzwälder von
Hokkaidô mit den Buchenwäldern des Japanischen Meer-Gebietes
und sind darum pflanzensoziologisch beachtenswert, wenn man die
Buchenwälder von Nordost-Japan in einer höheren soziologischen
Einheit vereinigen wollte. Es ist jedoch klar, daß die Klasse
V a c c i n i o - P i c e e t e a j a p o n i c a mit den
panborealen Kryptogamen-Arten Pleurozium schreberi, Hylocomium
splendens, Rhytidiadelphus triquetrus u.a. und den Pinaceen und
Ericaceen von der Klasse S a s o - F a g e t e a, den ostasiatischen Buchenwäldern, unterschieden werden muß.

Südgrenze

Hier sollen zwei Probleme behandelt werden: Die Süd-Grenze der
Fagus crenata-Bestände in Kyûsyû, Südwestjapan; und das Vorkommen von Fagus hayatae-Beständen in Formosa.

Die Nord-Grenze bildet das S a s o - F a g e t u m c r e n a -
t a e, der der Japanischen Meer-Seite eigentümliche Buchenwald. Aber
nur das S a s a m o r p h o - F a g e t u m c r e n a t a e,
der pazifische Buchenwald, kommt in Kyûsyû vor. In den Gebirgen
entlang der Grenzlinie gegen diesen Bereich liegt der Gürtel
des S a s o - F a g e t u m mit Sasa nipponica-Unterwuchs (9,
20).

Diese Bestände finden sich in Hikosan, Nordkyûsyû, aber der
Hauptteil des Buchenwaldes in Kyûsyû, sogar in Hikosan, hat
auch Sasamorpha-Unterwuchs (1). Das bedeutet, daß das Klima der
Insel Kyûsyû kaum vom Japanischen Meer beeinflußt wird, so daß
man hier schneearme Winter genießt.

Die Süd-Grenze vom S a s a m o r p h o - F a g e t u m liegt
im Gebiet der Kirisima-Vulkan-Gruppe (6), obwohl einige Buchen-
Bestände noch weiter südlich vorkommen (7,4). Der Berg Takakuma
und der Sibi-Berg sind als Süd-Grenze des japanischen Buchenwaldes bekannt. Anstatt der Buche, die schnee-ertragend ist, dominiert die Hemlock-Tanne, Tsuga sieboldii, in der pazifischen Region Japans, wo das Klima schneearm ist. Im Katamuki- und Sobo-
Gebirge in Mittel-Kyûsyû kann man noch einen montanen Buchenwald-
Gürtel über dem submontanen Tsuga-Waldgürtel finden. Südwärts
bis an den Itihusa-Berg, bildet der Buchenwald einen geschlossenen Gürtel. Im Süden erlaubt der schneeärmere Winter den immer-

grünen Laubbäumen, höher hinaufzusteigen. Cyclobalanopsis acuta ist die immergrüne Eiche, die am höchsten aufsteigt. Schon
im Nakama-Tonobata-Berg, Nord-Kyûsyû, kann man die Buchen-Bestände, die viel immergrüne Bäume enthalten, pflanzensoziologisch nicht mehr zur Assoziation S a s a m o r p h o - F a -
g e t u m zählen, obwohl hier gute Ausbildungen des S a s a -
m o r p h o - F a g e t u m auch noch vorkommen. Südlich in
Kirisima bildet das S a s a m o r p h o - F a g e t u m keinen Gürtel. Es bedeckt nur die Nordwest-Seite des alten Vulkan
Oonami (6).
Auf der Insel Yaku , mit dem höchsten Gipfel von Kyûsyû, befin-
det sich weder ein Buchenwald noch eine Buche. Der Tsuga-Wald,
der mit Trochodendron aralioides und oft mit Cryptomeria japonica gemischt ist, bildet hier die Waldgrenze. Darüber liegt
der Sasa owatarii-Gürtel (21). Diese Art ist auf der Insel endemisch und hat im Habitus keine Ähnlichkeit mit irgendeiner
dominanten Zwergbambus-Art der F a g u s c r e n a t a Wälder.
Die Verbreitung von Fagus hayatae ist auf Nord-Formosa beschränkt.
Diese Buche fehlt in der Daiton-Vulkangruppe, wo ihr Höhe und
klimatische Bedingungen zusagen würden. Ihr Areal umfaßt das
Sôten-Gebirge und das nördlichste Ende der zentralen Gebirgs-
kette.
Fagus hayatae ist mit keiner Acer-Art vergesellschaftet, hat
aber Arundinaria (od.Pleioblastus) niitakayamensis-Unterwuchs
(12). Die submontane Stufe von Nordost-Formosa ist mit immergrünen Laubwäldern bedeckt. Bis 600 - 700 m Höhe erstrecken
sich subtropische Regenwälder mit Shiion stipitatae. Darauf folgen die Nadelwälder des C y c l o b a l a n o p s i o n p a u-
c i d e n t a t a e. Um 1000 m wird plötzlich der Farn-Unter-
wuchs durch Zwerg-Bambus ersetzt. Es gibt auch Laubwälder mit
dem Zwergbambus-Unterwuchs in der Zwischenzone, aber auf über
1000 m, wo der Nordost-Monsun von Ende Oktober bis Ende Februar
starken Einfluß ausübt, kommen F a g u s h a y a t a e - Wälder vor. An den Lee-Seiten, wo Nebel fehlt, finden sich Tsuga
sinensis-Wälder. Wo der Wind nicht zu stark ist, wird der Unterwuchs von Plagiogyria formosana anstelle von Pleioblastus niitakayamensis beherrscht.
Pleioblastus niitakayamensis fehlt dagegen im F a g u s h a-
y a t a e - Wald. Dieser Zwerg-Bambus ist weit in den Nadel-Wäl-

dern Formosas aspektbildend, so z.B. oft unter Abies kawakamii
in der subalpinen Stufe (9). Er verhält sich pflanzensoziologisch nicht neutral, sondern ist eine ökologische Zeigerpflanze,
die auf windexponierte Standorte weist.
Zum Schluß wollen wir kurz die alpine Waldgrenze, die mit Fagus
crenata erreicht wird, erwähnen. Im Gassan-Gebirge fehlt der
subalpine Nadelwald mit Abies mariesii als zonale Gesellschaft.
In dem sehr schneereichen Gebirge oberhalb des Buchenwaldes entwickelt sich das C a r i c i - S a s e t u m k u r i l e n -
s i s, ein Zwergbambus-Gestrüpp (14,15). Wegen ihrer hochgebirgs·
ähnlichen Struktur hat T.SHIDEI (8) die Stufe oberhalb des Buchenwaldes pseudoalpin genannt, aber man kann sie als eine subalpine Stufe ohne Nadelholz ansehen. Die Bestände, die nahe an
der Waldgrenze liegen, umfassen einige Arten, die gewöhnlich in
den subalpinen Nadelwäldern vorkommen:

Sorbus commixta	Olopanax japonicus
Menziesia pentandra	Lycopodium serratum var.
Acer tschonoskii	thunbergii
Ilex sugerokii	Ephippianthus schmidtii
Vaccinium smallii	Clintonis udensis u.a.

Der Zwergbambus-Unterwuchs steigt also etwas höher als der Buchenwald und die Buchen-Bestände ohne Sasa-Strauchschicht sogar allgemein in den typischen Buchenwäldern an der Seite des
Japanischen Meeres. Aber sie sind stets mit den immergrünen Laub·
gehölzen in der Strauchschicht ausgestattet.
Die Standorte sind auch schneereich, vor starkem Wind aber oro-
graphisch geschützt.

ZUSAMMENFASSUNG
Im Vergleich mit den Buchenwäldern anderer Kontinente sind die
der ostasiatischen Inselkette als Vergesellschaftung von sommer-
grünen Buchen- und Ahorn-Beständen aufzufassen. Ihre Eigenart
liegt vornehmlich im Vorhandensein eines Zwergbambus-Unterwuch-
ses. Daneben sind die immergrünen Laubgehölze und Sträucher in
der zweiten Baumschicht und in der Strauchschicht bezeichnend.

Ihre Nord-Grenze liegt an der Kuromatunai-Talzone, Hokkaidô.
Sie ist hauptsächlich erdgeschichtlich bedingt, aber auch das
Klima ist bedeutsam. Die Südgrenze des F a g i o n c r e n a -
t a e finden wir in der Kirisima-Vulkangruppe. Die schneearmen
und wärmeren Winter sind hier für die immergrünen Nadelbäume,

wie Tsuga sieboldii und die immergrünen Laubbäume wie Cyclobala-
nosis acuta günstiger.
Der Insel Yaku fehlen Buchenbestände. Das Klima ist dort für
die Buchenwälder wohl nicht geeignet, aber das Fehlen von Fagus
und Sasamorpha ist vielleicht auch irgendeiner erdgeschichtli-
chen Ursache zuzuschreiben.
Fagus hayatae-Bestände wachsen in Formosa auf Gipfeln, die dem
Nordost-Wintermonsun über 1000 m ausgeliefert sind.Hier fehlen
Acer-Arten. Der Pleioblastus niitakayamensis-Unterwuchs ist auch
in den subalpinen Nadelwäldern Formosas vorhanden.
Es gibt keine eigene Art, die in allen Buchenwäldern der ostasiatischen Inseln gemeinsam vorkommt. Die Zwergbambus-Arten reichen über die Buchengrenze nach Nord und Süd hinaus. Die Verge-
sellschaftung von Fagus und Bambusoideen ist jedoch in diesem
Bereich ein Charakteristikum für ostasiatische Buchenwälder im
Vergleich zu denen anderer Kontinente.

SUMMARY
The northern limit of the beech forests in eastasiatic Archipelago runs along the valley zone of Kuromatunai, Hokkaidô. The
southern limit of the F a g i o n   c r e n a t a e stands is
found in the Kirisima Volcanoes, though some stands exist far-
ther south. Here the climate is warmer end less snowy, so that
Fagus crenata retreats by the competition with evergreen conifers
like Tsuga sieboldii and broad-leafed evergreens like Cyclobalanopsis acuta.
In spite of the Sasa owatarii-belt, Yaku Island has neither
beech forest nor individual. The lack of the beech forest in
Yaku Island, where there the highest peak in Kyûsyû, is not only the result of the unsuitable climate, but also brought by
the historical factor.
The occurrences of the Fagus hayatae-forests in Formosa are restricted to peaks facing to the northeastern monsoon in winter.
This beech is combined with Pleioblastus niitakayamensis. This
dwarf bamboo is however not confined to the Fagus hayatae-fo-
rest and grows commonly under the subalpine coniferous forest
in Formosa, and here we can find no Acer species growing with
Fagus.
No species is common to all the beech forest of the East Asia-
tic Archipelago, and the dwarf bamboo species are also not con-

fined to the beechforest, but the most remarkable character or
the beech forest of this region is the commensale occurrence of
East Asia with those of other continents in the Northern Hemi-
sphere.

LITERATUR

HOSOKAWA,T. -1956- An introduction of On the structure of the
          beechforests,Mt.Hiko of S.W.Japan.- Jap.Ecol.5 (3).
KATO,R. -1952- An attempt to the rational elucidation of so-
          cial life-hydrature and Abies-Picea-Climax.- Japan J.
          For.Soc.Jap.32 (1):16. Tokyo.
MAEDA,M. -1965- Flora und Vegetation des Berges Sibi.-
          Kanran 1: 18-39. Kagosima.
MIKI,S. -1953- Metasequoia-Fossil and Living.- Kyoto.
ODA,T.& SUMATA,H. -1966- Pflanzengesellschaften des Kirisima-
          gebirges, Südwestjapan und ihre Kartierung.- Jap.J.Ecol.
          16(4): 149-157. Sendai.
SAKO,S. -1960- On the beechforst (Fagus crenata Bl.) in the
          district of its southern limit of distribution.- Bull.
          Fac.Agr.Kagoshima Univ.9: 128-135. Kagoshima.
SHIDEI,T. -1952- .On the forest zone of Ou District..- J.Toho-
          ku Branch For.Soc.Jap.2 (2): 2-8. Sendai.
SUZUKI,S. -1961- Ecology of the Bambusaceous genera Sasa and
          Sasamorpha in the Kanto and Tohoku Districts of Japan,
          with special reference to their geographical distributi-
          on.- Ecol.Rev.15 (3): 131-147. Sendai.
SUZUKI,T. -1939- A vegetational sketch of the eastern slopes
          of Mt.Tyûôsenzan, one of the peaks in the Central Range
          of Taiwan.- Bull.Biog.Soc.Jap.8 (13): 77-195. Tokyo.
-- -- -1954- The forest vegetation of the North-Formosan
          mountains.- J.Jap.Ecol.4 (1): 7-13. Sendai.
-- -- -1959 (1962)- Schneetälchen-Gesellschaften des Gassan-Ge-
          birges.- In: Tüxen,R.(Edit.): Vegetationskartierung. Ber.
          Intern.Sympos.1959 Stolzenau/Weser. Weinheim/B. p.219-230.
-- -- -1966- The highest vegetation units of Japan and their
          areas.- Pedologist 10 (2): 1-7.
TATEWAKI,M. -19 - Northern limit of Fagus crenata.- Ecol.1
          (1): Rev.11 (1-2): 46-51. Sendai.
-- -- INOKUMA,T., SISIDO,M.& SUZUKI,T. -1951- Sciadophythis-
          Bestände vom Berg Koya.- Plant.Ecol.1 (1): 31-35. Sendai.
TOHYAMA,M. -1965- Synecological study of the Fagus japonica
          forest on Mt.Ohmuro, Pref.Yamanashi.- Jap.J.Ecol.15 (4):
          139-142. Sendai.
USUI,H. -1961- Phytosociological revisions of the dominant spe-
          cies of Sasa-type undergrowth.- Sp.Bull.Coll.Agr.Utsono-
          miya Univers.11: 1-35. Utsonomiya.
-- -- & KASIWA,H. -1962- Über die floristische Zusammenset-
          zung der Naturwälder in der Insel Yaku und Standort von
          Cryptomeria japonica.- 72.Ref.Jap.For.Soc.126-129.
YOSIOKA,K. -1952- Ex.Bot.27. Sendai.

ÜBER AREALE (GRENZEN) EINIGER BUCHENWALD-GESELLSCHAFTEN
JAPANS

Y. S a s a k i

Der Vortrag unseres am 8.August 1972 verstorbenen jungen Freundes YOSHIYUKI SASAKI erschien in erweiterter Form unter dem Titel:
            Versuch zur systematischen und geographischen
        Gliederung der Japanischen Buchenwald-Gesellschaften
in Vegetatio 20 (1-4). The Hague.

BOUNDARIES ON VEGETATION MAPS

A. W. K ü c h l e r

INTRODUCTION

The vegetation map as a scientific instrument forces the mapper
to a degree of exactness that can hardly be achieved by words
alone. The determination of the boundaries is the only way by
which location and extent of a phytocenose can be presented accurately; it directly affects the correlation of phytocenoses
with specific qualities of their biotopes.

The possibilities of drawing boundaries on vegetation maps vary
within wide limits. At one extreme, a boundary may be sharp as
in the case of some substitute communities, e.g. between a vine-
yard and a wheatfield. In less extreme cases, the determination of the location of a boundary becomes problematical. This
can go so far that at the opposite extreme, vegetation boundaries cease to exist, and logically can not be shown on maps. To
map vegetation boundaries thus becomes a problem and the ultimate cause of this problem lies in the very nature of vegetation.

VEGETATION

Vegetation has been defined as "the mosaic of phytocenoses in
the landscape."(KÜCHLER 1967a). Phytocenology, the science of
vegetation (German: Vegetationskunde), is therefore concerned
with the nature of individual phytocenoses including their boundaries. However, in order to appreciate the basic features and
problems of boundaries, it is first necessary to recall some
characteristics of phytocenoses.

All plant communities consist of growth forms, and all plant
communities consist of taxa. This dual approach to the study of
vegetation implies the complexities that led to the evolution
of diverging points of view but the insistence on growth form
composition opens the way to a richer choice of vegetation boundaries. All such boundaries have in common that they refer exclusively to vegetation and that they are not based on ecological observations, although these may be implied. In dealing
with boundaries on vegetation maps, it is justifiable and indeed
preferable to adhere strictly to phytocenotic boundaries. The
ecological meaning of the boundaries need, however, not be
suppressed and may be added as supplementary information in an

elaboration of the map legend.

PHYTOCENOSES AND CONTINUA

The above definition of vegetation implies the existence of phytocenoses but this has been questioned. In the United States, CURTIS (1959), WHITTAKER (1962) and others have maintained that phytocenoses might be replaced by continua.

Climatic conditions change imperceptibly and the manner of expressing them rests on statistical averages. Where the character of vegetation depends exclusively on climate, i.e. where all other environmental factors remain uniform, a perfect vegetational continuum may be expected, with a complete lack of boundaries. However, if a phytocenose is an integrated expression of the entire biogeocenose, then such a continuum is not likely to occur so universally because changes in the physical and chemical qualities of the substratum will break the continuity of the change. Topographic features may have similar effects. Certainly, the prairie in the heart of North America lends itself well to the establishment of continua if only species are considered. But if the biotopes are considered, too, and if only comparable biotopes are selected, i.e. sites with similar topography and substratum, the prairie shows a remarkable lack of continuity in the floristic composition of its local divisions.

Taxa do not behave uniformly throughout their area. Thus, Aster oblongifolius indicates more xeric conditions in the prairies of Ohio and more mesic conditions in Kansas. In a limited area, character species may be restricted to a given well defined association. The larger the region, the more blurred are the phytocenoses and the character species spread into different associations. Floristic units like associations may therefore be difficult to define in larger areas and hence their boundaries present serious problems. A regional limitation seems a prerequisite for meaningful floristic units. WHITTAKER (1962, p.91) expresses the same opinion when he says: "Associations as definable combinations of species are local phenomena." Phytocenoses, like taxa, have a definite range of tolerance, it is not astonishing that TÜXEN and his collaborators could so precisely define the phytocenoses of a relatively small area like northwestern Germany (approximately 1/2 the size of Louisana). FRIEDEL (1967) independently confirmed that phytocenoses have their

ranges of tolerance. He observed in the Austrian Alps that even
though the environment may change gradually, phytocenoses tend
to remain relatively uniform throughout their areas but change
rather rapidly when a certain threshold value of their sites is
passed. In other words, the environmental gradient of a given region may be gradual and continuous in contrast to vegetational
changes in the same area which tend to alternate between very
gentle and much steeper gradients. The various phases of a hy-
drarch succession frequently offer fine examples of this tenden-
cy. Ecoclines and cenoclines (coenoclines) sensu WHITTAKER
(1967) are therefore not necessarily parallel. GOODALL (1963)
cites a number of researches who make similar observations. Undoubtedly, however, most phytocenoses have ill defined borders.

From the point of view of the vegetation mapper it seems desirable to distinguish between a transition and a continuum. A
transition may be said to be the species population observable
along a gradient between two unlike phytocenoses. Thus a beech-
maple forest may grade into an elm-ash forest(KÜCHLER,1964). The
gradient between these types may be steep or gentle but it seems
obvious that contrasting biotopes necessarily lead to contrasting species combinations. In a transition, therefore, a type
ends, albeit gradually.
On the other hand, a continuum may be said to arise from chan-
ging conditions within a vegetation type. Thus the bluestem
prairie of Iowa can be expected to differ from the bluestem
prairie of Oklahoma simply because of a difference of eight degrees latitude. However, a mapper is hard-put in finding a
boundary within this area as the gradient between north and
south is very gentle, and the bluestem prairie continues from
one end to the other. Some arbitrary boundary can always be
established, based for instance on the frequency of certain
taxa. But the significance of such a boundary is debatable and
the taxa in question must not include the dominants because
their wide range of tolerance permits them to maintain their
dominant position througtout the area.
This floristic change within the boundaries of the bluestem
prairie continuum differs therefore from the transition between the beech-maple forest and the elm-ash-forest. As a result,
a continuum as a population cline within a given vegetation type may be differentiated from a transition between two unlike

types.
Used in this sense, between transitions and continua is reflected in boundary problems. The contrast between the two ends of
a transition is such as to permit a relatively clear and simple
definition of boundaries, and they will be relatively acceptable, no matter how arbitrary they are. In a continuum, however,
there are no boundaries. Any boundary introduced into a continuum has little meaning and it is usually best to omit it. However, if a vegetation map of a small segment of the bluestem
prairie is prepared at a large scale, local differentiation of
structure and floristic composition result from unlike substra-
tal conditions, and vegetational boundaries can be justified
more readily.
Clearly, discrete phytocenoses and continua are not mutually
exclusive. GOODALL (1963), DAUBENMIRE (1966) and others have
demonstrated this principle satisfactorily and subsequent pu-
blications like WHITTAKER's (1967) do not alter it.

CRITERIA
An accurate description of phytocenoses requires appropriate.
The number of such criteria is so large that many different
approaches to vegetation are possible and justifiable. ELLENBERG
(1956,p.38) has tabulated the essential criteria, revealing that
vegetation maps may differ widely depending on what criteria are
chosen. KÜCHLER (1956), DANSEREAU (1960) and BUELL & DANSEREAU
(1966) have illustrated this point by mapping the vegetation
of a given area according to different criteria, thus obtaining
a series of vegetation maps with unlike boundaries. A comparison of the various maps reveals that it is not so much a difference of boundaries as a difference of combinations. Thus floristic map (Map I) in DANSEREAU's series has quite the largest
number of vegetation units. All subsequent maps reinterpret
these units on the basis of different criteria. This leads to
different combinations. The boundaries do not shift their location but not all boundaries of Map I are shown, and different
boundaries are selected on different maps. There are therefore
no new boundaries on any of the subsequent maps even though
that may be a reader's first impression. However, many boundaries on the vegetation maps of the Swiss comparative studies
(MOOR & SCHWARZ 1957; ELLENBERG 1967) disagree as a result of
differences in the interpretation of the field observations.

It is easy to select criteria according to which the location
of a boundary can be established unequivocally. For instance,
in the southeastern United States, the Forest Service classified forests as pine forests when 50% or more the trees are pi-
nes. But an arbitrary decision is subject to challenge. For example, one of the generally recognized vegetation types of the
United States is the oak-hickory forest, but it is difficult
to draw its boundaries because there is disagreement on what
should be inclused. The boundaries of this forest type have been
shown on many maps but their location has differed widely.

The interpretation of observations made in the field depends
therefore on the purpose for which a vegetation map is prepared.
For instance, a floristically oriented map designed to reveal
the qualities of biotopes is not likely to agree everywhere
with a vegetation map of the same area emphasizing structural
features that are important in military considerations. KUHN-
HOLTZ-LORDAT (1949) suggested that vegetation maps might be
drawn from the point of view of sheep, and in a state like Ore-
gon, such a map would differ basically from one prepared for
lumbermen. Such divergent interpretations of the observed plant
communities affect the location and character of vegetation
boundaries, and a boundary which is sharp and clear on one map
may be blurred or absent on another.
Boundaries based on arbitrarily selected criteria are practically inevitable in cases of gentle gradients between phytocenoses.
But the mapper must choose the criteria for the boundaries of
ecotones and their subdivisions so as to be in harmony with the
purpose of the vegetation map if the latter is to be uniformly
meaningful throughout its area.

BOUNDARIES ON VEGETATION MAPS OF VERY LARGE SCALE
In order to observe the nature of vegetation boundaries in detail, it is the best to begin with the areas with the steepest
gradients between contrasting phytocenoses.They are recorded
on vegetation maps of very large scales because only on these
is it possible to present the vegetation without significant
generalization.
Sharp boundaries occur in the cultural landscape as mentioned
in the introduction. They also occur in the natural landscape.
For instance, the plant communities on recent lava flows differ

from those on older lava with an abrupt change along the borders
of the recent lava flow; Hawai offers many good examples of this
kind. Similarly, contacts between adjacent out-crops of lime-
stones and igneous rock may well be revealed by vegetational
discontinuities. In the mountains west of Chiengmai, Thailand,
the tropophyllous forests of the lower altitudes change almost
abruptly into the evergreen forests above (KÜCHLER & SAWYER,
1967); many other examples can be cited from all continents.

A phytocenose may be assumed to be sufficiently extensive that
it can be mapped at least at very large scales. Thus, MOLINIER
and his collaborators (1951) mapped the Forêt de la Sainte Baume
in southeastern France at a scale of 1 : 2 000 accurately and
without significant generalization. A scale of 1 : 5 000 is also large enough to show the location of all phytocenoses correctly although the use of symbols may become desirable in order to
avoid cluttering the map. This is especially true if the vegetation is herbaceous. Clearly, vegetation maps can be prepared
which are not generalized.

The extent of a phytocenose does therefore not present a problem,
but the nature of its boundaries is not so clear. Even the same
author may hold more than one view of it. Thus TÜXEN (1954,p.
477) explained that phytocenoses can be established so precisely that their boundaries can be drawn accurately basing this
on the presence or absence of a single species. Therefore, where a given species ceases to occur there is the boundary of the
phytocenose. Only on year later, TÜXEN (1955,p.1960) discused
types of vegetation and accepted what he considered a significant by Kretschmer:"Das Klare und Präzise an einem Typus ist
immer sein Kern, nicht sein Rand. Es gehört zum Wesen eines
Typus, daß er keine scharfen Grenzen haben darf." Undoubtedly,
most phytocenologists will agree more readily with the more recent statement.

BOUNDARIES ON GENERALIZED VEGETATION MAPS

Most vegetation maps are more or less generalized. This directly affects the problems of boundaries as well as of gradients
and transitions. Basically, a transition is secondary to the
types which is links and all is well if the area of an ecotone
is so large that it can be shown at a reduced scale. If it is
too small, it is suppressed. When a transition is suppressed

it is concealed and, in fact, replaced by the boundary between
two plant communities.

In many instances, types of vegetation can be treated like the
transitions: when the area of a type is too small to be shown
on the map, the type is suppressed. Another approach is the
combination of two or more vegetation units into larger ones,
but here the drawing of boundaries may become problematical and
even controversial. For instance, the classification of BRAUN-
BLANQUET assigns all communities to a particular rank in a hierarchy. The logical approach to generalization is to combine
units of low rank into units of the next higher rank. The number
of units is thereby reduced, their areas are enlarged and hence
better adjusted to a smaller scale. However, this procedure may
be inapplicable because it may be the high ranks of the hierarchy that must be suppressed! It is one of the inherent weaknes-
ses of a hierarchical floristic approach to the classification
of vegetation that it may create very difficult problems for
mapping at medium and smaller scales.

Generalization is usually simpler if the vegetational units are
physiognomic-structural rather than floristic. It is equivalent
to a reduction of structural detail and such a reduction can be
adjusted to any scale.

However, if the vegetation units are ecological types, the generalization must be along ecological lines. Essentially, this
amounts to a generalization of the environmental features. Such
a generalization will end in the broad formations of the climatic climax.

The merger several units into a larger one should be accompli-
shed according to a logical procedure. Phytocenoses can be combined only when they are related somehow. Unrelated ones are
suppressed if too small and insignificant. If they should be
indicated on the map, they can be represented by over-printed
symbols with appropriate comments in the legend. This difference
between being suppressed and being merged is this: suppression
due to small scale signifies that the area is too small to be
shown and hence remains ignored on the small scale map. Merger
implies the establishement of a new vegetational unit which is
heterogeneous and the heterogeneity of which should somehow
find expression in its name, directly or by implication. The
map reader usually assumes some degree of vegetational unifor-

mity within a boundary. The boundary must therefore be drawn so
as not to mislead the reader.
The boundaries relatively "ungeneralized" vegetation maps enclose therefore units that may differ essentially from those on
generalized maps. The chief difference is that on such "ungeneralized" maps the boundaries enclose exactly described and outlined phytocenoses that can be verified in the field. On the
other hand, generalization results in a more blurred condition,
and what the map shows may or may not exist in a given location.

It is not possible to say at what scale generalization begins
because this varies with the individual phytocenoses. Thus, the
location of a forest boundary can be shown accurately at 1 :
100 000 especially in a much disturbed landscape where it stands
out clearly. At this scale, it is unlikely that grassland communities can be shown without generalization, and of course,
this may apply even to a number of forest types. Certainly, if
the vegetation map of Schlitz (SEIBERT 1954) were to be reduced
to 1 : 100 000 many forests would be generalized almost beyond
recognition. By contrast, small scales imply such a degree of
generalization that boundaries cannot be expected to imply the
same precision as the boundaries on large scale maps. However,
the decline in precision and in scale are not necessarily parallel, and the more abrupt the change in site qualities and in
vegetation, the longer can accurate boundaries be maintained on
maps of shrinking scales. For instance, if the Mississippi
alluvium is studied in detail, the various terraces and other
local topographic features are reflected in the vegetation; such
details desappear when the map scale is reduced. But the flood-
plain forest as a major unit of vegetation continues to have
justifiably sharp boundaries even on maps of less than 1:3,000,000
(KÜCHLER 1964). The accuracy of the boundaries on vegetation
maps of different scales is therefore at least in part a matter
of degree in contrast. The greater the contrast between unrelated vegetation types the smaller the scale at which sharp boundaries may be drawn.
VAN LEEUWEN (1966) considered boundary problems and distinguished between "limes convergens" or "ecotone" and "limes díver-
gens" or "ecocline". The limes convergens (ecotone) is said to
be characterized by boundaries which are frequently sharp and
straight; the surrounding vegetation pattern is coarse and

highly unstable; the plant communities are poor in species. On
the other hand, the limes divergens (ecocline) implies faint
lines of demarkation, a finely granulated pattern of stable floristically rich phytocenoses.

These suggestions are interesting because they seem to apply in
densely populated areas of northwestern Europe and elsewhere.
They do give the impression of being over-generalized because
there are too many exceptions to the rule. In less urbanized regions, and for instance along sharp geological borders, the vegetation may be stable and floristically rich, as in northwestern
California. On the other hand, the features of the limes divergens also apply to highly dynamic conditions as for example in
the midgrass prairies of Nebraska, Kansas and Oklahoma.CLEMENTS
introduced the term ecocline in 1936 and to change its meaning
after it has been accepted for so long seems debatable.

DYNANISM OF VEGETATION

As soon as a boundary is drawn on a map, it fixes the location
of the phytocenose which it encloses. This fixication is an
illusion in at least two cases. In the first instance, in primary succession, a series of phytocenoses "moves" along a gradient, for instance from hydric to mesic in the case of a hy-
drosere, producing shifting boundaries. Invasions resulting
from a change in the environment imply similarly unstable boundaries.

Succession and invasion result in a form of dynamic vegetation
boundary which shifts its position in a given direction an ecological gradient. The other major form of dynamic boundary is
the one which results from fluctuations of one or more environmental factors. An example of this kind occurs where the boundary between humid and arid climates passes across extensive
plains covered with a grassland vegetation. This is characteristic of the interior of the United States, of central Argentina,
southern Russia, northern China, and large parts of Australia
and Africa. In theory, the boundaries fluctuations result from
fluctuations in rainfall. In practice, however, the underlying
causes are highly complex and it is therefore extremely difficult or impossible to predict rate and direction of the boundary shifts.

The vegetation mapper's simplest method is to draw boundaries
that separate the dynamic types from the surrounding stable ve-

getation, and omit all boundaries within the dynamic section of
his map. However, this results in new features that must be
dealt with. First, there may be a number of boundaries within
the dynamic region that, while omitted, should nevertheless be
brought to the attention of the reader in some fashion. Second,
the boundaries surrounding the dynamic region are partly temporary, yet they do not shift. At times, they are true vegetation
boundaries as well as regional boundaries, bounding the dynamic
region. At other times, such a boundary is only regional. For
instance, during a dry cycle, a boundary may separate the stable mesic grassland from the dynamic more xeric grassland. It is
thus a vegetation boundary and also a regional boundary (of the
dynamic region). But during a rainy cycle, the mesic vegetation
crosses this boundary into the more arid region. The vegetational boundary moves whereas the regional boundary remains stati-
onary. The nature and the effects of this type of dynamism in
the North American prairie have been discussed in some detail
elsewhere (KÜCHLER,1967b); suffice it here to say that these
dynamic boundaries have so far received inadequate attention,
largely due to their complex nature and the difficulties in
examining them. A date is an item of basic information and must
be indicated if the map includes any dynamic boundaries within
its area,or else one of the basic values of the vegetation map
can not be exploited.

BOUNDARIES LINES ON VEGETATION MAPS
The actual drawing of boundary lines on vegetation maps leades
to some further considerations. On most vegetation maps, the various vegetation units are separated by lines. Such boundaries
offer a double advantage in that they 1) indicate where on vegetation type ends and another one begins, and 2) increase the
contrast between colors and patterns of the units of vegetation.
The second point is especially important where similar colors
or patterns are contiguous. HUECK (1959), on his vegetation map
of Venezuela used three different kinds of boundaries to indicate they were established, approximate or doubtful. The drawing
of different kinds of boundary lines on vegetation maps does
not usually present a problem and mappers will know how to adapt
the lines to their particular circumstances.
But as many phytocenoses have no distinct boundaries, it may
seem more logical and correct not to show boundaries on vegeta-

tion, maps. Boundaries have indeed been omitted in many instances, for example on the vegetation map of France (GAUSSEN 1945)
a number of Swiss maps and many others. The custom of drawing
boundaries around vegetation units has become so well established that it is generally accepted as a matter of course. But
in fact, a strong case can be made for the omission of boundaries on vegetation maps just because in the field the actual
boundaries are often blurred and must be arbitrary. On the other
hand, from a strictly technical point of view, the presence of
boundaries is desirable because it aids the printer considerably in assuring perfect registration of the different colors.

The degree of desirability of boundary lines on vegetation maps
can therefore vary greatly. A number of authors have separated
major vegetation units from one another by lines whereas minor
units are shown by printing patterns over the colors, and these
minor units are left without borders. The patterns can thus be
made to end "gradually" or they may overlap more or less, per-
mitting them a varying density and at the same time a strict
correlation with the classification of vegetation. Symbols can
be placed individually and in varying concentrations, reflecting
the actual distribution most realistically. Some examples of
this type have become well known, like the Vegetationskundliche
Karte des oberen Wutachgebietes (LANG and OBERDORFER,1956-1958)
or the map of Madagascar by HUMBERT and COURS-DARNE (1964). The
combination of few boundary lines and unbounded patterns is not
only convincing and effective but has the added advantage of
being applicable at all scales.

CONCLUSION
The above observations lead to the conclusion that vegetation
boundaries imply problems which the vegetation mapper must appreciate in order to make his map as meaningful as possible. He
has the choice of using boundary lines around every unit of vegetation or around the major ones only, or none at all. This
choice cannot be standardized and rests essentially on the mapper's experience and judgment.

SUMMARY
Phytocenoses exist as definable entities; these include their
exolutionary phases as well as transitions between them. Phytocenoses and continua are not mutually exclusive. Boundaries are

established according to the criteria used in describing the
phytocenoses. They are best avoided within the area of a continuum. Vegetation boundaries indicate the extent of vegetation
units and increase the contrast between similar colors or patterns.

Boundaries on vegetation maps of very large scale correspond to
field checks whereas on more generalized maps they surround heterogeneous vegetation units of logically combinable components.
Generalization of physiognomic vegetation units is a matter of
the amount of structural detail to be shown whereas generalization of ecological maps rests on the generalization of environmental features.

Significant characteristics of dynamic boundaries can be explained in the map legend. Vegetation maps with dynamic boundaries
must have a date. Boundaries need not be shown on a vegetation
map but it is often best to use them in outlining the major vegetation types and to omit them around subdivisions shown by
overprinted patterns.

LITERATURE

BUELL,P.F.& DANSEREAU,P. -1966- Analysis and mapping of the
        Roosevelt Road area.- In: Studies on the vegation of
        Puerto Rico 1: 46-287. Mayagüez,P.R.
CURTIS,J.T. -1959- The vegetation of Wisconsin.- Madison Univ.
        of Wis.Press.
CLEMENTS,F.E. -1936- Nature and structure of the climax.- J.
        Ecol.24: 252-284. London.
DANSEREAU,P. -1960- Essai de représentation cartographique
        des éléments structuraux de la végétation.- In:Gaussen,H.
        (Edit.): -1961- Méthodes de la cartographie de la végé-
        tation. C.N.R.S: pp.233-255. Paris.
DAUBENMIRE,R. -1966- Vegetation: identification of typal com-
        munities.- Science 151: 291-298. Lancaster,Pa.
ELLENBERG,H. -1956- Aufgaben und Methoden der Vegetationskunde.-
        In: Walter,H.: Einführung in die Phytologie 4 (1).
        Stuttgart.
-- --   -1967- Vegetation und bodenkundliche Methoden der forst-
        lichen Standortskartierung.- Veröff.geobot.Inst.Rübel
        Zürich 39. Zürich.
FRIEDEL,H. -1967- Der Verlauf der alpinen Waldgrenze.- Mitt.
        forstl.Bundesversuchsanstelt 75. Wien.
GAUSSEN,H. -1945- Tapis végétal. 1:1,000,000. Editions géogra-
        phiques, Atlas de France pp.30-33. Paris.
GOODALL,D.W. -1963- The continuum and the individualistic asso-
        ciation.- Vegetatio 11 (5-6): 297-316. Den Haag.
HUECK,K. -1959- Vegetación de Venezuela. 1:2,000,000.- Inst.
        Forestal Lat.-Amer.de Investigación y Capacitación.Mérida.

HUMBERT,H.& COURS DARNE,G. -1965- Carte internationale du ta-
        pis végétal: République Malgache. 1:1.000.000.-C.N.R.S.
        Paris.
KÜCHLER,A.W. -1956- Classification and purpose in vegetation
        maps.-Geogr.Rev.46: 155-167. New York.
-- -- -1964- Potential natural vegetation of the conterminous
        United States. 1:3,168,000.⊣ Amer.Geogr.Soz.,Special
        Publ.36. New York.
-- -- -1967a- Vegetation Mapping.- Ronald Press. New York.
-- -- -1967b- Some geographic aspects of the Kansas prairie.-
        Kansas Academy of Science, Transactions 70: 388-401.
        Lawrence,Kan.
-- -- & SAWYER,J.O. -1967- A study of the vegetation near
        Chiengmai, Thailand.- Kansas Academy of Science, Trans-
        actions 70: 281-348. Lawrence,Kan.
KUNHOLTZ-LORDAT,G. -1949- La cartographie parcellaire.- Insti-
        tut National de la Recherche Agronomique. Paris.
LANG,G.& OBERDORFER,E. -1960- Vegetationskundliche Karte des
        oberen Wutachgebietes 1:25,000.- Karlsruhe,
LEEUWEN,C.G.van -1966- A relation theoretical approach to
        pattern and process in vegetation.- Wentia 15: 25-46.
        Amsterdam.
MOLINIER,René, MOLINIER,Roger & PIALOT,H. -1951- Cartes phy-
        togéographiques à diverses échelles de la forêt de la
        Ste.Baume (Var.).- Ass.Fr.Avanc.Sci.,Proceedings 70th
        Congress fasc.4. Tunis.
MOOR,M.& SCHWARZ,U. -1957- Die kartographische Darstellung der
        Vegetation des Creux-du-Vent Gebietes.- Beitr.geobot.
        Landesaufn.Schweiz 37. Bern.
SEIBERT,P. -1954- Vegetationskarte des Graf Görtzischen Forst-
        bezirks Schlitz 1:15,000.- Angew.Pflanzensoz.9. Stolze-
        nau/Weser.
TÜXEN,R. -1954- Über die räumliche, durch Relief und Gestein
        bedingte Ordnung der natürlichen Waldgesellschaften am
        Rande des Harzes.- Vegetatio 5-6: 454-478. Den Haag.
-- -- -1955- Das System der nordwestdeutschen Pflanzengesell-
        schaften.- Mitt.flor.-soz.Arbeitsgem.N.F.5: 155-176.
        Stolzenau/Weser.
WHITTAKER,R.H. -1962- Classification of natural communities.-
        Bot.Rev.28: 1-239. Lancaster,Pa.
-- -- -1967- Gradient analysis of vegetation.- Biol.Rev.42:
        207-264. Cambridge.

        Wir danken Herrn Dr. DAVID SHIMWELL herzlich für die freund-
liche Durchsicht einiger summaries.

SCHLUSSWORT

M.WRABER:

Wir sind am Ende des 12. Symposium angekommen. Wenn wir uns
darüber Rechenschaft geben, dann glaube ich die Überzeugung
aller Anwesenden auszusprechen, daß auch dieses Symposium, wie
auch alle vorhergehenden, mit einem vollen Erfolg zu Ende ging.
Man merkt schon seit vielen Jahren, daß sich bei diesen Symposien ein Freundschaftskreis zusammenfindet, der auf hohem aka-
demischen Niveau die verschiedenen Streitfragen der Vegetationskunde gründlich und ernsthaft diskutiert. Wir können alle be-
friedigt nach Hause gehen und an das nächste, das 13.Symposion,
denken. Wir hoffen, daß wir uns alle hier wiederfinden und auch
neue Gäste zu uns stoßen werden.
Ich kann aber das Schlußwort nicht aussprechen, ohne in Ihrer
aller Namen, meine verehrten Damen und Herren, unseren herzlich-
sten Dank und unsere volle Anerkennung unserem Sekretär, Herrn
Prof. TÜXEN, und seinen Mitarbeitern zu sagen für die große Auf-
opferung und für die anstrengende Arbeit, die Herr Professor
TÜXEN und seine Mitarbeiter in diesen Tagen und schon während
der langen Vorbereitungen geleistet haben. Vielen herzlichen
Dank, Herr Professor TÜXEN, und nun möchte ich Sie auffordern,
das Schlußwort zu sprechen.

R.TÜXEN:

Meine lieben Freunde!

Ich bin meinen Mitarbeitern sehr vielen Dank schuldig, denn ohne
die aufopfernde Tätigkeit von Frau MILBRADT, und der Herren HANS
BÖTTCHER, HARTMUT DIERSCHKE, KLAUS DIERSSEN, KARL-HEINZ HÜLBUSCH
und PETER JANIESCH wäre unser Symposion nicht gelungen. Ihnen
allen meinen herzlichen Dank! Ich darf in meinen persönlichen
Dank, obwohl sie nicht hier ist,auch meine Frau einschließen,
die glücklicherweise wieder in der Lage ist, den Anstrengungen
des Symposium widerstehen zu können.
Wir haben wieder hier vier, und einige von uns auch sechs Tage
fruchtbarer Arbeit mit reichen Ergebnissen, die weit meine Er-
wartungen übertroffen haben, hinter uns gebracht unter dem er-
neuten Erlebnis dieser einzigartigen, über alle trennenden Grenzen hinweg reichenden internationalen Freundschaft. Und ich hoffe,daß diese uns noch recht lange erhalten bleibt. Wir haben
hier eine geistige Insel der Freundschaft gebildet, die mit all

dem Schrecklichen,was rings um uns geschieht und geschehen ist, gar nichts zu tun hat,und ich meine, wir haben etwas Aufbauendes geleistet in diesen Tagen. Ich hoffe, daß wir uns im nächsten Jahre so fröhlich, so aktiv wiedersehen, wie wir es in diesem Jahr getan haben.Noch einmal allen herzlichen Dank und auf Wiedersehen im nächsten Jahre.

Schutz der Insel Surtsey (Island).

Das 12.Symposion der Internationalen Vereinigung für Vegetationskunde beschloß einstimmig die folgende Resolution, die Herrn
Dr.STURLA FRIDRIKSSON, Reykjavík, zur Weiterleitung an die Is-
ländische Regierung übergeben wurde.

The Annual Symposium
of the International Society of Plant Geography and Ecology
held in Rinteln, Germany, from April 8th to 11th 1968 recognizes the extreme significance of Surtsey as a valuable natural
centre where various biologigal phenomen can be studied:
it values the stimulation that Surtsey offers in furthering general interest in basic ecological and plant geographical research.
The Society appreciates the steps taken by the Islandic autho-
rities in protecting this unique island for scientific purposes
and emphasises the need for securing a continuation of its conservation.

St.FRIDRIKSSON:
May I express my kindest gratitude to this Symposium for having
passed this and giving our biological research on Surtsey a mo-
ral support and to support the protection of the island by the
Icelandic Government. Thank you very much.

Additional material from *Tatsachen und Probleme der Grenzen in der Vegetation,*
ISBN 978-94-011-7596-8 (978-94-011-7596-8_OSFO3),
is available at http://extras.springer.com